Neutrino Physics

Series in High Energy Physics, Cosmology, and Gravitation

Series Editors: **Brian Foster,** *Oxford University, UK*
Edward W Kolb, *Fermi National Accelerator Laboratory, USA*

This series of books covers all aspects of theoretical and experimental high energy physics, cosmology and gravitation and the interface between them. In recent years the fields of particle physics and astrophysics have become increasingly interdependent and the aim of this series is to provide a library of books to meet the needs of students and researchers in these fields.
Other recent books in the series:

Neutrino Physics, Third Edition
Kai Zuber

The Standard Model and Beyond, Second Edition
Paul Langacker

An Introduction to Beam Physics
Martin Berz, Kyoko Makino, and Weishi Wan

Neutrino Physics, Second Edition
K Zuber

Group Theory for the Standard Model of Particle Physics and Beyond
Ken J Barnes

The Standard Model and Beyond
Paul Langacker

Particle and Astroparticle Physics
Utpal Sakar

Joint Evolution of Black Holes and Galaxies
M Colpi, V Gorini, F Haardt, and U Moschella (Eds)

Gravitation: From the Hubble Length to the Planck Length
I Ciufolini, E Coccia, V Gorini, R Peron, and N Vittorio (Eds)

The Galactic Black Hole: Lectures on General Relativity and Astrophysics
H Falcke, and F Hehl (Eds)

The Mathematical Theory of Cosmic Strings: Cosmic Strings in the Wire Approximation
M R Anderson

Geometry and Physics of Branes
U Bruzzo, V Gorini, and U Moschella (Eds)

Modern Cosmology
S Bonometto, V Gorini, and U Moschella (Eds)

Gravitation and Gauge Symmetries
M Blagojevic

Gravitational Waves
I Ciufolini, V Gorini, U Moschella, and P Fré (Eds)

Neutrino Physics
Third Edition

Kai Zuber

CRC Press
Taylor & Francis Group
Boca Raton London New York

CRC Press is an imprint of the
Taylor & Francis Group, an **informa** business

Third edition published 2020
by CRC Press
6000 Broken Sound Parkway NW, Suite 300, Boca Raton, FL 33487-2742

and by CRC Press
2 Park Square, Milton Park, Abingdon, Oxon, OX14 4RN

First issued in paperback 2021

First edition published by IOP 2003
Second Edition published by CRC Press 2011

CRC Press is an imprint of Taylor & Francis Group, LLC

Library of Congress Cataloging-in-Publication Data

Names: Zuber, K., author.
Title: Neutrino physics / Kai Zuber.
Description: Third edition. | Boca Raton : CRC Press, 2020. | Series: Series in high energy physics, cosmology & gravitation | Includes bibliographical references and index.
Identifiers: LCCN 2019059117 | ISBN 9781138718890 (hardback) | ISBN 9781315195612 (ebook)
Subjects: LCSH: Neutrinos.
Classification: LCC QC793.5.N42 Z83 2020 | DDC 539.7/215--dc23
LC record available at https://lccn.loc.gov/2019059117

ISBN 13: 978-1-03-224220-0 (pbk)
ISBN 13: 978-1-138-71889-0 (hbk)

DOI: 10.1201/9781315195612

Contents

Preface to the first edition

The last decade has seen a revolution in neutrino physics. The establishment of a non-vanishing neutrino mass in neutrino oscillation experiments is one of the major new achievements. In this context the problem of missing solar neutrinos could be solved. In addition, limits on the absolute neutrino mass could be improved by more than an order of magnitude by beta decay and double beta decay experiments. Massive neutrinos have a wide impact on particle physics, astrophysics and cosmology. Their properties might guide us to theories beyond the standard model of particle physics in the form of grand unified theories (GUTs). The precise determination of the mixing matrix like the one in the quark sector lies ahead of us with new machines, opening the exciting possibility to search for CP-violation in the lepton sector. Improved absolute mass measurements are on their way. Astrophysical neutrino sources like the Sun and supernovae still offer a unique tool to investigate neutrino properties. A completely new window in high astrophysics using neutrino telescopes has just opened and very exciting results can be expected soon. Major new important observations in cosmology sharpen our view of the universe and its evolution, where neutrinos take their part as well.

The aim of this book is to give an outline of the essential ideas and basic lines of developments. It tries to cover the full range of neutrino physics, being as comprehensive and self-contained as possible. In contrast to some recent, excellent books containing a collection of articles by experts, this book tries to address a larger circle of readers. This monograph developed out of lectures given at the University of Dortmund, and is therefore well suited as an introduction for students and a valuable source of information for people working in the field. The book contains extensive references for additional reading. In order to be as up-to-date as possible many preprints have been included, which can be easily accessed electronically via preprint servers on the World Wide Web.

It is a pleasure to thank my students M. Althaus, H. Kiel, M. Mass and D. Münstermann for critical reading of the manuscript and suggestions for improvement. I am indebted to my colleagues S. M. Bilenky, C. P. Burgess, L. diLella, K. Eitel, T. K. Gaisser, F. Halzen, D. H. Perkins, L. Okun, G. G. Raffelt, W. Rodejohann, J. Silk, P. J. F. Soler, C. Weinheimer and P. Vogel for valuable comments and discussions.

Many thanks to Mrs. S. Helbich for the excellent translation of the manuscript and to J. Revill, S. Plenty and J. Navas of the Institute of Physics Publishing for their faithful and efficient collaboration in getting the manuscript published. Last, but not least, I want to thank my wife for her patience and support.

Oxford, **K Zuber** August 2003

Preface to the second edition

Only six years have passed but neutrino physics kept its pace. New major discoveries make a second edition timely and necessary. The solar neutrino problem has basically been solved, the first real-time measurement of solar neutrinos below 1 MeV, the first discovery of geoneutrinos, the start of new long baseline accelerator experiments and many more developments had to be included. To avoid a complete rewriting of the book the intention was to keep the original text as much as possible and implement the new results in a smooth way.

Dresden, **K Zuber** September 2011

Preface to the third edition

Another nine years have passed since the second edition and again major new discoveries suggested a third edition. Some highlights, among others, are that mixing angles of the mixing matrix have been observed, coherent neutrino-scattering has been discovered, the fundamental pp-neutrinos from the Sun have been seen in real-time and the field of ultra-high energy astrophysics using neutrino telescopes experienced an enormous boost. To avoid a complete rewriting of the book, the intention was to keep the original text as much as possible and implement the new results in a smooth way. Furthermore, a few older topics have been left in because they might help in understanding.

First of all, I want to thank Prof. em. E. Sheldon whose comments were very valuable for the third edition. The friendly hospitality of my stay at the MTA Atomki Debrecen (Hungary) helped a lot for writing various parts of the book. An incredible great thanks for their work, which cannot be appreciated highly enough, is going to my students: S. Turkat and H. Wilsenach. Without them this book would not be on this level.

Last but not least I have to thank my dear wife, S. Helbich, for her patience, as she has barely seen me during the last months.

Dresden, **K Zuber** January 2020

Notation

Covering the scales from particle physics to cosmology, various units are used. A system quite often considered is that of natural units ($c = \hbar = k_B = 1$) which is used throughout this book. Other units are used if they aid understanding. The table overleaf gives practical conversion factors into natural units.

In addition, here are some useful relations:

$$\hbar c = 197.33 \text{ MeV fm},$$
$$1 \text{ erg} = 10^7 \text{ J},$$
$$1 \, M_\odot = 1.988 \times 10^{30} \text{ kg},$$
$$1 \text{ pc} = 3.262 \text{ light years} = 3.0857 \times 10^{16} \text{ m}.$$

Among the infinite amount of Web pages from which to obtain useful information, the following URLs should be mentioned:

- https://arXiv.org (*arXiv* preprint server)
- http://adsabs.harvard.edu (Search for astrophysical papers)
- http://inspires.net (Search for High Energy Physics literature)
- http://pdg.lbl.gov (Review of Particle Properties)
- http://www.hep.anl.gov/ndk/hypertext/ (The Neutrino Oscillation Industry)

Conversion factors for natural units.

	s^{-1}	cm^{-1}	K	eV	amu	erg	g
s^{-1}	1	0.334×10^{-10}	0.764×10^{-11}	0.658×10^{-15}	0.707×10^{-24}	1.055×10^{-27}	1.173×10^{-48}
cm^{-1}	2.998×10^{10}	1	0.229	1.973×10^{-5}	2.118×10^{-14}	3.161×10^{-17}	0.352×10^{-37}
K	1.310×10^{11}	4.369	1	0.862×10^{-4}	0.962×10^{-13}	1.381×10^{-16}	1.537×10^{-37}
eV	1.519×10^{15}	0.507×10^{5}	1.160×10^{4}	1	1.074×10^{-9}	1.602×10^{-12}	1.783×10^{-33}
amu	1.415×10^{24}	0.472×10^{14}	1.081×10^{13}	0.931×10^{9}	1	1.492×10^{-3}	1.661×10^{-24}
erg	0.948×10^{27}	0.316×10^{17}	0.724×10^{16}	0.624×10^{12}	0.670×10^{3}	1	1.113×10^{-21}
g	0.852×10^{48}	2.843×10^{37}	0.651×10^{37}	0.561×10^{33}	0.602×10^{24}	0.899×10^{21}	1

Chapter 1

Important historical experiments

DOI: 10.1201/9781315195612-1

With the discovery of the electron in 1897 by J. J. Thomson, a new era of physics - today called elementary particle physics - started. By deconstructing the atom as the fundamental building block of matter the question arose as to what could be inside the atom. Probing smaller and smaller length scales is equivalent to going to higher and higher energies which can be done using high-energy accelerators. With such machines a complete "zoo" of new particles was discovered, which finally led to the currently accepted standard model (SM) of particle physics (see Chapter 3). Here, the building blocks of matter consist of six quarks and six leptons shown in Table 1.1, all of them being spin-$\frac{1}{2}$ fermions. They interact with each other through four fundamental forces: gravitation, electromagnetism and the strong and weak interactions.

In quantum field theory, these forces are described by the exchange of the bosons shown in Table 1.2. Among the fermions there is one species - neutrinos - where our knowledge is still limited. Being leptons (they do not participate in strong interactions) and having zero charge (hence no electromagnetic interactions) they interact only via weak interactions (unless they have a non-vanishing mass, in which case gravitational interaction might be possible). This makes experimental investigations extremely difficult. Hence, neutrinos are the obvious tool with which to explore weak processes and therefore the history of neutrino physics and that of weak interactions are strongly connected.

The following chapters will depict some of the historic milestones. For more detailed discussions on the history, see [Pau61, Sie68].

1.1 The birth of the neutrino

Ever since its discovery, the neutrino's behaviour has been out of the ordinary. In contrast to the common way of discovering new particles, i.e. in experiments, the neutrino was first postulated theoretically. The history of the neutrino began with the investigation of β-decay (see Chapter 6).

After the observation of discrete lines in the α- and γ-decay of atomic nuclei, it came as a surprise when J. Chadwick discovered in 1914 a continuous energy

Table 1.1. (*a*) Properties of the quarks: I, isospin; S, strangeness; C, charm; Q, charge; B, baryon number; B^*, bottom; T, top. (*b*) Properties of leptons: L_i, flavour-related lepton number, $L = \sum_{i=e,\mu,\tau} L_i$.

Flavour	Spin	B	I	I_3	S	C	B^*	T	$Q[e]$
u	1/2	1/3	1/2	1/2	0	0	0	0	2/3
d	1/2	1/3	1/2	−1/2	0	0	0	0	−1/3
c	1/2	1/3	0	0	0	1	0	0	2/3
s	1/2	1/3	0	0	−1	0	0	0	−1/3
b	1/2	1/3	0	0	0	0	−1	0	−1/3
t	1/2	1/3	0	0	0	0	0	1	2/3

Lepton	$Q[e]$	L_e	L_μ	L_τ	L
e^-	−1	1	0	0	1
ν_e	0	1	0	0	1
μ^-	−1	0	1	0	1
ν_μ	0	0	1	0	1
τ^-	−1	0	0	1	1
ν_τ	0	0	0	1	1

Table 1.2. Phenomenology of the four fundamental forces and a hypothetical grand unification theory (GUT) interaction based on SU(5). Natural units $\hbar = c = 1$ are used.

Interaction	Strength	Range R	Exchange particle	Example
Gravitation	$G_N \simeq 5.9 \times 10^{-39}$	∞	Graviton?	Mass attraction
Weak	$G_F \simeq 1.02 \times 10^{-5} m_p^{-2}$	$\approx m_W^{-1}$ $\simeq 10^{-3}$ fm	$W^\pm, Z^0$	μ-decay
Electro-magnetic	$\alpha \simeq 1/137$	∞	γ	Force between electric charges
Strong (nuclear)	$g_\pi^2/4\pi \approx 14$	$\approx m_\pi^{-1}$ ≈ 1.5 fm	Gluons	Nuclear forces
Strong (color)	$\alpha_s \simeq 1$	confinement	Gluons	Forces between the quarks
GUT SU(5)	$M_X^{-2} \approx 10^{-30} m_p^{-2}$ $M_X \approx 10^{16}$ GeV	$\approx M_X^{-1}$ $\approx 10^{-16}$ fm	X, Y	proton-decay

spectrum of electrons emitted in β-decay [Cha14]. The interpretation followed two lines: one assumed primary electrons with a continuous energy distribution (followed mainly by C. D. Ellis) and the other assumed secondary processes, which broaden an initially discrete electron energy (followed mainly by L. Meitner). To resolve this question, a calorimetric measurement has been performed which should result in either the average electron energy (if C. D. Ellis was right), or the maximal energy (if L. Meitner was correct). This approach can be understood in the following way: β-decay is described by the three-body decay

$$M(A, Z) \rightarrow D(A, Z + 1) + e^- + \bar{\nu}_e, \tag{1.1}$$

where $M(A, Z)$ describes the mother nucleus and $D(A, Z + 1)$ its daughter and A being the atomic mass and Z the atomic number. The actual decay is that of a neutron into a proton, electron and antineutrino. For a decay at rest of $M(A, Z)$, the electron energy should be between $E_{min} = m_e$ and using energy conservation

$$E_{max} = m_M - m_D - T_D - E_\nu \approx m_M - m_D. \tag{1.2}$$

In the last step of (1.2) the small kinetic recoil energy T_D of the daughter nucleus was neglected and $E_\nu = 0$ (also: $m_\nu = 0$) was assumed. Hence, if there are only electrons in the final-state, the calorimetric measurement should always result in $E_{max} = m_M - m_D$.

The experiment was done using the β-decay of the isotope RaE (today known as ^{210}Bi) with a nuclear transition Q-value of 1161 keV. The measurement performed in 1927 resulted in a value of $344 \pm 10\%$ keV) [Ell27] clearly supporting the first explanation. L. Meitner, still not convinced, repeated the experiment in 1930 ending up with 337 keV $\pm$ 6% confirming the primary origin of the continuous electron spectrum [Mei30]. To explain this observation, only two solutions seemed to be possible: either the energy conservation law is only valid statistically in such a process (preferred by N. Bohr) or an additional undetectable new particle (later called the neutrino by E. Fermi) is emitted, carrying away the additional energy and spin (preferred by W. Pauli). There was a second reason for Pauli's proposal of a further particle, namely angular momentum conservation. It was observed in β-decay that if the mother atom carries integer/fractional spin then the daughter also does, which cannot be explained by the emission of only just one spin-$\frac{1}{2}$ electron. In a famous letter dated 4 December 1930, W. Pauli proposed his solution to the problem; a new spin-$\frac{1}{2}$ particle (which we nowadays call the neutrino) is produced together with the electron but escapes detection. In this way the continuous spectrum can be understood: both electron and neutrino share the transition energy in a way that the sum of both always corresponds to the full transition energy, called Q-value. Shortly afterwards the neutron was discovered in 1932 [Cha32]. At this point the understanding of atomic nuclei of β-decay changed rapidly and led E. Fermi to develop his successful theory of β-decay in 1934 [Fer34]. The first experiments to support the notion of the neutrino came after the second world war. An experiment called "El Monstro" was considered at Los Alamos [Los97] using a nuclear bomb explosion to prove the existence of antineutrinos, which was replaced by using the first nuclear reactors as antineutrino source. The next sections will discuss some historical conducted experiments in more detail.

1.2   Nuclear recoil experiment by Rodeback and Allen

In 1952 the first experimental evidence for neutrinos was found in the electron capture (EC) of the K-shell e_K^- of ^{37}Ar:

$$^{37}\text{Ar} + e_K^- \rightarrow {}^{37}\text{Cl} + \nu_e + Q \tag{1.3}$$

with a Q-value of 816 keV. As the process has only two particles in the final-state the recoil energy of the nucleus is fixed. Using energy and momentum conservation, the recoil energy T_{Cl} is given by

$$T_{\text{Cl}} = \frac{E_\nu^2}{2m_{\text{Cl}}} \approx \frac{Q^2}{2m_{\text{Cl}}} = 9.67 \, \text{eV} \tag{1.4}$$

because the rest mass of the ^{37}Cl atom is much larger than $Q \approx E_\nu$. This recoil energy corresponds to a velocity for the ^{37}Cl nucleus of 0.71 cm μs^{-1}. Therefore, the recoil velocity could be determined by a delayed coincidence measurement. It is started by the Auger electrons emitted after electron capture and stopped by detecting the recoiling nucleus. By using a variable time delay line, a signal should be observed if the delay time coincides with the time of flight of the recoil ions. With a flight length of 6 cm, a time delay of 8.5 μs was expected. Indeed, the expected recoil signal could be observed at about 7 μs. After several necessary experimental corrections (e.g., thermal motion caused a 7% effect in the velocity distribution), both numbers were in good agreement [Rod52].

Soon afterwards the measurement was repeated with an improved spectrometer and a recoil energy of $T_{\text{Cl}} = (9.63 \pm 0.03)$ eV was measured [Sne55] in good agreement with (1.4). Even though the measured recoil indicates a two-body decay, the neutrino wasn't observed directly.

1.3   Discovery of the neutrino by Cowan and Reines

The discovery finally took place at nuclear reactors, which were the strongest terrestrial neutrino sources available. The basic detection reaction was

$$\bar{\nu}_e + p \rightarrow e^+ + n. \tag{1.5}$$

The detection principle was a coincident measurement of the 511 keV photons associated with positron annihilation and a neutron capture reaction a few μs later. Cowan and Reines used a water tank with dissolved CdCl$_2$ surrounded by two liquid scintillators (Figure 1.1). The liquid scintillators detect the photons from positron annihilation as well as the ones from the ^{113}Cd(n, γ) ^{114}Cd reaction after neutron capture. The detector is shown in Figure 1.2. The experiment was performed in different configurations and at different reactors and finally resulted in the discovery of the neutrino.

In 1953, at the Hanford reactor (USA) using about 300 l of a liquid scintillator and rather poor shielding against background, a vague signal was observed. The

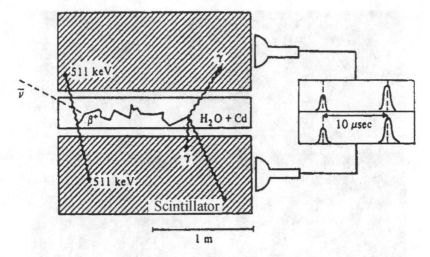

Figure 1.1. Schematic illustration of the experimental set-up for neutrino detection used by Cowan and Reines. A $CdCl_2$ loaded water tank is surrounded by liquid scintillators. They are used for a coincidence measurement of the 511 keV annihilation photons and the γ-rays emitted by the neutron capture on Cd. The water acts as neutron moderator (from [Pau61]). Reproduced with permission of SNCSC.

experiment was repeated in 1956 at the Savannah River reactor (USA) with 4200 l of scintillator, finally proving the existence of neutrinos. For more historical information on this experiment see [Los97]. The obtained energy averaged cross-section for reaction (1.5) was [Rei53, Rei56]

$$\bar{\sigma} = (11 \pm 3) \times 10^{-44} \text{ cm}^2 \qquad (1.6)$$

which, when fully revised, agreed with the V-A theory.

1.4 Difference between ν_e and $\bar{\nu}_e$ and solar neutrino detection

The aim of the experiment was to find out whether neutrinos and antineutrinos are identical particles. If so, the reactions

$$\nu_e + n \rightarrow e^- + p \qquad (1.7)$$
$$\bar{\nu}_e + n \rightarrow e^- + p \qquad (1.8)$$

should occur with the same cross-section. In the corresponding experiment in 1955, Ray Davis Jr. was looking for

$$\bar{\nu}_e + {}^{37}\text{Cl} \rightarrow e^- + {}^{37}\text{Ar} \qquad (1.9)$$

by using the Brookhaven reactor (USA). He was using 4000 l of liquid CCl_4. After some exposure time, produced argon atoms were extracted by flooding helium

(a)

(b)

Figure 1.2. *(a)* The experimental group of Clyde Cowan (left) and Fred Reines (right) of 'Unternehmen Poltergeist' (Project 'Poltergeist') to search for neutrinos. *(b)* The detector called 'Herr Auge' (Mr Eye). With kind permission of Los Alamos Science.

through the liquid and then freezing out the argon atoms in a cooled charcoal trap. Proportional counters were used to detect the EC of ^{37}Ar. By not observing the process (1.8) he could set an upper limit of

$$\bar{\sigma}(\bar{\nu}_e + {}^{37}\text{Cl} \rightarrow e^- + {}^{37}\text{Ar}) \leq 2 \times 10^{-42} \text{ cm}^2 \text{ per atom} \quad (1.10)$$

where the theoretical prediction was $\bar{\sigma} \approx 2.6 \times 10^{-45}$ cm^2 per atom [Dav55].

This showed that ν_e does indeed cause the reaction (1.7). This detection principle was used years later in a larger scale version in the successful detection of solar neutrinos. (1.7). This pioneering effort marks the birth of neutrino astrophysics and will be discussed in detail in Chapter 10.

Later it was found at CERN that the same applies to muon neutrinos by using ν_μ beams. Only μ^-s in ν_μ beam interactions were ever detected in the final state, but never a μ^+ [Bie64].

1.5 Discovery of parity violation in weak interactions

Parity is defined as a symmetry transformation by an inversion at the origin resulting in $\vec{x} \rightarrow -\vec{x}$. It was assumed that parity is conserved in all interactions. At the beginning of the 1950s, however, people were puzzled by observations in kaon decays (the so-called 'τ–θ' puzzle). Lee and Yang [Lee56], when investigating this problem, found that parity conservation had never been tested for weak interactions and this could provide a solution to this problem.

Parity conservation implies that any process and its mirror process occurs with the same probability. Therefore, to establish parity violation, an observable quantity which is different for both processes must be found. This is exactly what pseudoscalars do. Pseudoscalars are defined in such a way that they change sign under parity transformations (see Chapter 3). This can be realised by a product of a polar and an axial vector, e.g., $\vec{p}_e \cdot \vec{I}_{\text{nuc}}, \vec{p}_e \cdot \vec{s}_e$, with $\vec{I}_{\text{nuc}}$ as the polarisation vector of the nucleus and $\vec{p}_e$ and $\vec{s}_e$ as momentum and spin of the electron. Any expectation value for a pseudoscalar different from zero would show parity violation. Another example of a pseudoscalar is provided by possible angular distributions in beta-decays like

$$\Delta\theta = \lambda(\theta) - \lambda(180° - \theta) \quad (1.11)$$

where λ is the probability for an electron to be emitted under an angle θ with respect to the spin direction of the nucleus. Under parity transformation the emission angle changes according to $\theta \rightarrow 180° - \theta$ which leads to $\Delta\theta \rightarrow -\Delta\theta$. In the classical experiment of Wu *et al.* polarized ^{60}Co atoms were used [Wu57]. To get a significant polarization, the ^{60}Co was deposited as a thin layer on a paramagnetic salt and kept at a temperature of 0.01 K.

The polarization was measured via the angular correlation of the two emitted γ-rays from ^{60}Ni decays using two NaI detectors. The decay of ^{60}Co into an excited state (denoted by *) of ^{60}Ni characterised by a nuclear spin I and a parity π, is given by

$$^{60}\text{Co}(5^+) \rightarrow {}^{60}\text{Ni}^*(4^+) + e^- + \bar{\nu}_e, \quad (1.12)$$

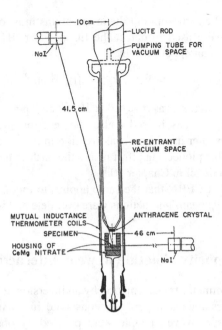

Figure 1.3. Schematic diagram showing the demagnetisation cryostat used in the measurement of the angular distribution of the electrons from the β-decay of ^{60}Co nuclei (from [Wu57]). © 1957 by the American Physical Society.

with the de-excitation sequence $I^{\pi} = 4^{+} \rightarrow 2^{+} \rightarrow 0^{+}$ to the ground state, emitting an 1173 keV and 1332 keV gamma ray, respectively. The electrons emitted under an angle θ were detected by an anthracene detector producing scintillation light. The mirror configuration $(180° - \theta)$ was created by reversing the applied magnetic field. A schematic view of the experiment is shown in Figure 1.3, the obtained data in Figure 1.4. It shows that electrons are preferably emitted in the opposite spin direction to that of the mother nucleus. The observation can be described by an angular distribution

$$W(\cos\theta) \propto 1 + \alpha \cdot \cos\theta \tag{1.13}$$

with a measured $\alpha \approx -0.4$. This was clear evidence that $\Delta\theta \neq 0$ and β-decay does indeed violate parity. The reason is that α is given by $\alpha = -P_{\text{Co}}\frac{\langle v_e \rangle}{c}$ where P_{Co} is the polarisation of the ^{60}Co nuclei and $\langle v_e \rangle$ the electron velocity averaged over the electron spectrum. With the given parameters of $P_{\text{Co}} \simeq 0.6$ and $\langle v_e \rangle/c \approx 0.6$, a value of $\alpha \approx -0.4$ results, showing that parity is not only violated but is maximally violated in weak interactions. The same result has been observed slightly later in pion decay at rest [Gar57]. The positive pion decays via

$$\pi^{+} \rightarrow \mu^{+} + \nu_{\mu}. \tag{1.14}$$

Considering the fact that the pion carries spin-0 and decays at rest, this implies that the spins of the muon and neutrino are opposed to each other (Figure 1.5). Defining

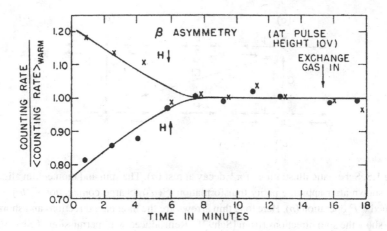

Figure 1.4. Observed β-decay counting rate as a function of time normalized to a warmed-up state. A typical run with a reasonable polarization of ^{60}Co lasted only about 8 minutes. In this time period, a clear difference in the counting rate for the two magnetic field configurations emerges, showing the effect of parity violation (from [Wu57]). © 1957 by the American Physical Society.

the helicity $\mathcal{H}$ as spin projection on the momentum

$$\mathcal{H} = \frac{\vec{\sigma} \cdot \vec{p}}{|\vec{\sigma}||\vec{p}|} \tag{1.15}$$

this results in $\mathcal{H}(\mu^+) = \mathcal{H}(\nu_\mu) = -1$. Applying parity transformation, $\mathcal{H}(\mu^+)$ and $\mathcal{H}(\nu_\mu)$ both become $+1$. Parity invariance would imply that both helicities should have the same probability and no longitudinal polarization of the muon should be observed. Hence, parity violation would already be established if there were some polarization. By measuring only $\mathcal{H}(\mu^+) = +1$ it turned out that parity in this process is maximally violated. These observations finally led to the V-A theory of weak interaction (see Chapter 6).

1.6   Direct measurement of the helicity of the neutrino

The principal idea of the Goldhaber experiment in 1957 was that the neutrino helicity could be determined under special circumstances by a measurement of the polarization of photons in electron capture (EC) reactions. In the classical experiment by Goldhaber *et al.* and a Swedish group, the electron capture of ^{152}Eu was used [Gol58, Mar59]. The decay is given by the sequence

$$^{152m}\text{Eu} + e^- \rightarrow \nu_e + {}^{152}\text{Sm}^*; \quad {}^{152}\text{Sm}^* \rightarrow {}^{152}\text{Sm} + \gamma. \tag{1.16}$$

The experimental set-up is shown in Figure 1.7. Due to momentum conservation, the decay at rest of ^{152m}Eu results in $\vec{p}_{^{152}\text{Sm}^*} = -\vec{p}_\nu$. The emission of photons

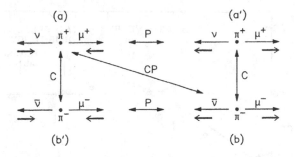

Figure 1.5. Schematic illustration of π^+ decay at rest (a). The spin and momentum alignment is also shown after applying parity transformation $P(a')$, charge conjugation $C(b')$ and the combined CP operation (b). Here long thin arrows are the momenta directions and short thick arrows show the spin directions (from [Sch97]). Reproduced with permission of SNCSC.

K-Capture γ– Emission   Resonance-Scattering

ν_e Eu   Sm*   Sm γ Sm   Sm* Sm γ

$$-\vec{P}_\nu \;=\; \vec{P}_{Sm^*} \;=\; \vec{P}_\gamma \;=\; \vec{P}_{Sm^*} \;=\; \vec{P}_\gamma$$

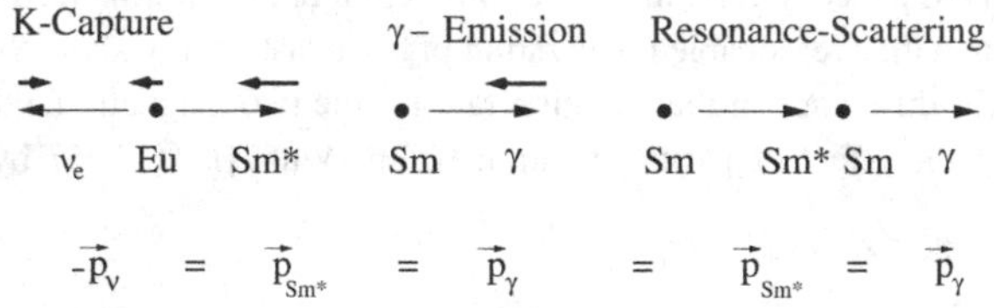

Figure 1.6. Schematic illustration of the neutrino helicity in the Goldhaber experiment. Long thin arrows are the momenta and short thick arrows are the spin directions in the three processes. K-capture means the electron is captured from the K-shell (from [Sch97]). Reproduced with permission of SNCSC.

(961 keV) in forward direction will stop the Sm nucleus, implying $\vec{p}_\gamma = -\vec{p}_\nu$ (Figure 1.6). Such photons also carry the small recoil energy of the ^{152}Sm* essential for resonant absorption (to account for the Doppler effect) which is used for detection. The resonant absorption is done in a ring of Sm_2O_3 and the re-emitted photons are detected under large angles by a well-shielded NaI scintillation detector. The momenta of these photons are still antiparallel to the neutrino momentum. Concerning the spin $\vec{J}$, the initial state is characterized by the spin of the electron $J_z = \pm 1/2$ (defining the emission direction of the photon as the z-axis, using the fact that $J(^{152m}\text{Eu}) = 0$ and that the K-shell electron has angular momentum $l = 0$) the final-state can be described by two combinations $J_z = J_z(\nu) + J_z(\gamma) = (+1/2, -1)$ or $(-1/2, +1)$. Only these result in $J_z = \pm \frac{1}{2}$. However, this implies that the spins of the neutrino and photon are opposed to each other. Combining this with the momentum arrangement implies that the helicity of the neutrino and photon are the same: $\mathcal{H}(\nu) = \mathcal{H}(\gamma)$. Therefore, the measurement of $\mathcal{H}(\nu)$ is equivalent to a measurement of $\mathcal{H}(\gamma)$. The helicity of the photon is nothing else than its circular polarisation. To measure this quantity, it should be noted that the Compton-scattering cross-section depends on the polarisation of the material. Hence a magnetized iron block has been used between source and absorber, resulting

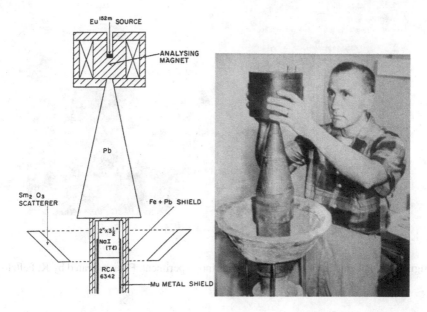

Figure 1.7. Left: Experimental set-up of the Goldhaber experiment to observe the longitudinal polarization of neutrinos in EC reactions. For details see text (from [Gol58]). © 1958 by the American Physical Society. Right: Installation of the Goldhaber experiment. With kind permission of L. Grodzins.

in about 7-8% polarisation. By reversing the polarisation, a different counting rate should be observed. After several measurements a polarization of $67 \pm 10\%$ was observed in agreement with the expected 84% [Gol58]. Applying several experimental corrections, the final outcome of the experiment was that neutrinos do indeed have a helicity of $\mathcal{H}(\nu) = -1$.

1.7 Experimental proof that ν_μ is different from ν_e

In 1959, Pontecorvo investigated whether the neutrino emitted together with an electron in β-decay is the same as the one emitted in pion decay [Pon60]. The idea was that if ν_μ and ν_e are identical particles, then the reactions

$$\nu_\mu + n \rightarrow \mu^- + p \tag{1.17}$$
$$\bar{\nu}_\mu + p \rightarrow \mu^+ + n \tag{1.18}$$

and

$$\nu_\mu + n \rightarrow e^- + p \tag{1.19}$$
$$\bar{\nu}_\mu + p \rightarrow e^+ + n \tag{1.20}$$

should result in the same rate, because the latter could be done by ν_e and $\bar{\nu}_e$; otherwise the last two should not be observed at all. At the same time, the use of

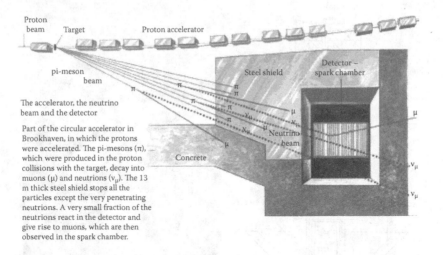

Proton
beam   Target   Proton accelerator

pi-meson
beam

Detector –
Steel shield   spark chamber

The accelerator, the neutrino
beam and the detector

Part of the circular accelerator in
Brookhaven, in which the protons
were accelerated. The pi-mesons (π),
which were produced in the proton
collisions with the target, decay into
muons (μ) and neutrions (ν_μ). The 13
m thick steel shield stops all the
particles except the very penetrating
neutrions. A very small fraction of the
neutrions react in the detector and
give rise to muons, which are then
observed in the spark chamber.

Concrete

Neutrino
beam

ν_μ

ν_μ

Figure 1.8. Schematic view of the AGS neutrino experiment. Figure created by K. Feltzin.

high-energy accelerators as neutrino sources was discussed by Schwarz [Sch60]. Thus, the experiment was done at the Brookhaven AGS using a 15 GeV proton beam irradiating a beryllium target (Figure 1.8) [Dan62]. The created secondary pions and kaons produced an almost pure ν_μ beam. Ten modules of spark chambers with a mass of 1 t each were installed behind a shielding of 13.5 m iron to absorb all the hadrons and most of the primary muons. Muons and electrons could be discriminated in the detector by their tracking properties, meaning muons produce straight lines, while electrons form an electromagnetic shower. In total, 29 muon-like and six electron-like events were observed, clearly showing that $\nu_\mu \neq \nu_e$. Some electron events were expected from ν_e beam contaminations due to K-decays (e.g., $K^+ \rightarrow \pi^0 + e^+\nu_e$). The experiment was repeated shortly afterwards at CERN with higher statistics and confirmed the Brookhaven result [Bie64].

1.8   Discovery of weak neutral currents

The development of the electroweak theory in the 1960s by Glashow, Weinberg and Salam, which will be discussed in more detail in Chapter 3, predicted the existence of new gauge bosons called W and Z. Associated with the proposed existence of the Z-boson, weak neutral currents (NC) should exist in nature. They were discovered in a bubble chamber experiment (Gargamelle) using the proton synchrotron (PS) $\nu_\mu/\bar{\nu}_\mu$ beam at CERN [Has73, Has73a, Has74]. The bubble chamber was filled with high-density fluid freon (CF_3Br, $\rho = 1.5$ g cm^{-3}) and it had a volume of 14 m^3, with a fiducial volume of 6.2 m^3 and a 2 T magnetic field. The search relied on pure hadronic events without a charged lepton (NC events) in the final-state which is

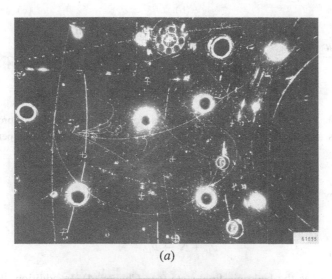

(a)

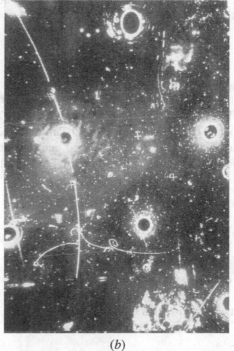

(b)

Figure 1.9. (a) A hadronic NC event with charged hadrons in the final-state as observed by the Gargamelle bubble chamber with a neutrino beam coming from the left. (b) A leptonic NC event $\bar{\nu}_\mu e \to \bar{\nu}_\mu e$ as obtained by Gargamelle. The $\bar{\nu}_\mu$-beam is coming from the right side. With kind permission of CERN.

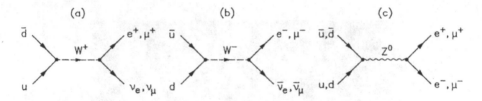

Figure 1.10. Lowest-order Feynman diagrams for $W^{\pm}$- and Z-boson production in $p\bar{p}$-collisions and their leptonic decays (from [Sch97]). Reproduced with permission of SNCSC.

described by

$$\nu_{\mu} + N \rightarrow \nu_{\mu} + X \tag{1.21}$$
$$\bar{\nu}_{\mu} + N \rightarrow \bar{\nu}_{\mu} + X \tag{1.22}$$

where X denotes the hadronic final-state (see Chapter 4). In addition, the charged current (CC) interactions

$$\nu_{\mu} + N \rightarrow \mu^{-} + X \tag{1.23}$$
$$\bar{\nu}_{\mu} + N \rightarrow \mu^{+} + X \tag{1.24}$$

were detected. In total, 102 NC and 428 CC events were observed in the ν_{μ} beam run and 64 NC and 148 CC events in the $\bar{\nu}_{\mu}$ one (Figure 1.9). The total number of pictures taken was of the order 83 000 in the ν_{μ} beam and 207 000 in the $\bar{\nu}_{\mu}$ run. After background subtraction, due to the produced neutrons and K_L^0 which could mimic NC events, the NC/CC ratio turned out to be (see also Chapter 4)

$$R_{\nu} = \frac{\sigma(\text{NC})}{\sigma(\text{CC})} = 0.21 \pm 0.03 \tag{1.25}$$

$$R_{\bar{\nu}} = \frac{\sigma(\text{NC})}{\sigma(\text{CC})} = 0.45 \pm 0.09. \tag{1.26}$$

Purely leptonic NC events resulting from $\bar{\nu}_{\mu} + e \rightarrow \bar{\nu}_{\mu} + e$ were also discovered [Has73a] (Figure 1.9). Soon afterwards, these observations were confirmed by several other experiments [Cno78, Fai78, Hei80].

1.9   Discovery of the weak $W^{\pm}$ and Z^0 gauge bosons

The weak gauge bosons predicted by the Glashow–Weinberg–Salam (GWS) model were finally discovered at CERN in 1983 by the two experiments UA1 and UA2 [Arn83, Bag83, Ban83]. They used the SPS as a $p\bar{p}$-collider with a centre-of-mass energy of $\sqrt{s} = 540$ GeV. The production processes for weak charged and neutral currents is given at the quark level by (Figure 1.10)

$$\bar{d} + u \rightarrow W^+ \rightarrow e^+ + \nu_e \quad (\mu^+ + \nu_\mu, \tau^+ + \nu_\tau)$$
$$\bar{u} + d \rightarrow W^- \rightarrow e^- + \bar{\nu}_e \quad (\mu^- + \bar{\nu}_\mu, \tau^- + \bar{\nu}_\tau)$$
$$\bar{d} + d \rightarrow Z^0 \rightarrow e^+ + e^- \quad (\mu^+ + \mu^-, \tau^+ + \tau^-)$$
$$\bar{u} + u \rightarrow Z^0 \rightarrow e^+ + e^- \quad (\mu^+ + \mu^-, \tau^+ + \tau^-). \tag{1.27}$$

These were difficult experiments because the cross-sections for W and Z production at that energy are rather small. They are including the branching ratio (BR)

$$\sigma(p\bar{p} \rightarrow W^\pm X) \times \mathrm{BR}(W \rightarrow l\nu) \approx 1 \text{ nb} = 10^{-33} \text{ cm}^2 \tag{1.28}$$
$$\sigma(p\bar{p} \rightarrow Z^0 X) \times \mathrm{BR}(Z^0 \rightarrow l^+l^-) \approx 0.1 \text{ nb} = 10^{-34} \text{ cm}^2 \tag{1.29}$$

while the total cross-section $\sigma(p\bar{p})$ is 40 mb, i.e., about 8 orders of magnitude larger![1] The signature for W detection was an isolated lepton ℓ with high transverse momentum p_T balanced by a large missing transverse momentum $\not{p}_T$ due to the escaping neutrino and for Z^0 detection two high p_T leptons with an invariant mass around the Z^0 mass. With regard to the latter, the Z^0 mass could be then determined to be (neglecting the lepton mass)

$$m_Z^2 = 2E^+E^- \left(1 - \cos(\theta)\right) \tag{1.30}$$

with $\cos(\theta)$ being the angle between the two leptons $\ell^\pm$ of energy E^+ and E^-. Both experiments came up with a total of about 25 W or Z events which were later increased. With the start of the e^+e^--collider LEP at CERN in 1989 and the SLC at SLAC the number of produced Z^0's is now several millions and its properties are well understood. Furthermore, the $W^\pm$ properties have been investigated at LEP, the Tevatron at Fermilab and are currently studied at the Large Hadron Collider (LHC). Both gauge bosons are discussed in more detail in Chapter 3.

1.10   Observation of neutrinos from SN 1987A

The observation of neutrinos from a supernova type-II explosion by large underground neutrino detectors was one of the great observations in last century's astrophysics (Figure 1.11). About 25 neutrino events were observed within a time interval of 20 s. This was the first neutrino detection originating from an astrophysical source besides the Sun. The supernova SN1987A occurred in the Large Magellanic Cloud at a distance of about 50 kpc. This event will be discussed in greater detail in Chapter 11.

1.11   Number of neutrino flavours from the width of the Z^0

The number N_ν of light ($m_\nu < m_Z/2$) neutrinos was determined at LEP by measuring the total decay width Γ_Z of the Z^0 resonance. Calling the hadronic decay

---

[1] 1 barn $= 10^{-24}$ cm^2.

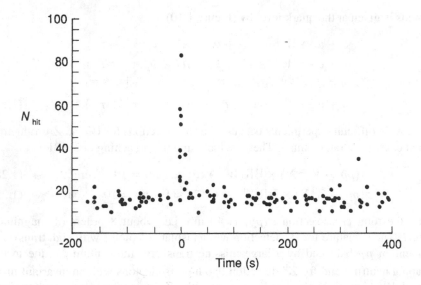

Figure 1.11. Number of struck photomultipliers N_{hit} as a function of time in the Kamiokande II detector on 23 February 1987. The zero on the time axis marks 7:35 UT. The increase in count rate is clearly visible and attributed to SN 1987A (from [Sut92]). © Cambridge University Press.

width Γ_{had} (consisting of $Z^0 \to q\bar{q}$ which materialise in hadronic jets) and assuming lepton universality (implying that there is a common partial width Γ_l for the decay into charged lepton pairs $\ell^+\ell^-$), the invisible width Γ_{inv} is given by

$$\Gamma_{\text{inv}} = \Gamma_Z - \Gamma_{\text{had}} - 3\Gamma_l. \tag{1.31}$$

As the invisible width corresponds to

$$\Gamma_{\text{inv}} = N_\nu \cdot \Gamma_\nu \tag{1.32}$$

the number of neutrino flavours N_ν can be determined. The partial widths of decays in fermions $Z \to f\bar{f}$ are also given in electroweak theory (see Chapter 3) by

$$\Gamma_f = \frac{G_F m_Z^3}{6\sqrt{2}\pi} c_f [(g_V)^2 + (g_A)^2] = \Gamma_0 c_f [(g_V)^2 + (g_A)^2] \tag{1.33}$$

with

$$\Gamma_0 = \frac{G_F m_Z^3}{6\sqrt{2}\pi} = 0.332 \text{ GeV} \tag{1.34}$$

and the mass of the Z^0 as $m_Z = 91.1876 \pm 0.0021$ GeV [PDG16]. In this equation c_f corresponds to a colour factor ($c_f = 1$ for leptons, $c_f = 3$ for quarks) and g_V and g_A are the vector and axial vector coupling constants, respectively. They are closely related to the Weinberg angle $\sin^2 \theta_W$ and the third component of weak isospin I_3 (see Chapter 3) via

$$g_V = I_3 - 2Q \sin^2 \theta_W \tag{1.35}$$

$$g_A = I_3 \tag{1.36}$$

with Q being the charge of the particle. Therefore, the different branching ratios are

$$\Gamma(Z^0 \to u\bar{u}, c\bar{c}) = (\tfrac{3}{2} - 4\sin^2 \theta_W + \tfrac{16}{3}\sin^4 \theta_W)\Gamma_0 = 0.286 \text{ GeV}$$

$$\Gamma(Z^0 \to d\bar{d}, s\bar{s}, b\bar{b}) = (\tfrac{3}{2} - 2\sin^2 \theta_W + \tfrac{4}{3}\sin^4 \theta_W)\Gamma_0 = 0.369 \text{ GeV}$$

$$\Gamma(Z^0 \to e^+e^-, \mu^+\mu^-, \tau^+\tau^-) = (\tfrac{1}{2} - 2\sin^2 \theta_W + 4\sin^4 \theta_W)\Gamma_0 = 0.084 \text{ GeV}$$

$$\Gamma(Z^0 \to \nu\bar{\nu}) = \tfrac{1}{2}\Gamma_0 = 0.166 \text{ GeV}. \tag{1.37}$$

Summing all decay channels into quarks results in a total hadronic width $\Gamma_{\text{had}} = 1.678$ GeV. The different decay widths are determined from the reaction $e^+e^- \to f\bar{f}$ for $f \neq e$ whose cross-section as a function of the centre-of-mass energy $\sqrt{s}$ is measured ($\sqrt{s} \approx m_Z$) and is dominated by the Z^0 pole. The cross-section at the resonance is described in the Born approximation by a Breit-Wigner formula not taking radiative corrections into account:

$$\sigma(s) = \sigma^0 \frac{s\Gamma_Z^2}{(s - m_Z^2)^2 + s^2\Gamma_Z^2/m_Z^2} \quad \text{with } \sigma^0 = \frac{12\pi}{m_Z^2} \frac{\Gamma_e \Gamma_f}{\Gamma_Z^2} \tag{1.38}$$

with σ^0 being the maximum of the resonance. Γ_Z can be determined from the width and $\Gamma_e \Gamma_f$ from the maximum of the observed resonance (Figure 1.12).

Experimentally, the Z^0 resonance is fitted with four different parameters which have small correlations with each other:

$$m_Z, \Gamma_Z, \sigma^0_{\text{had}} = \frac{12\pi}{m_Z^2} \frac{\Gamma_e \Gamma_{\text{had}}}{\Gamma_Z^2} \quad \text{and} \quad R_l = \frac{\Gamma_{\text{had}}}{\Gamma_l} \tag{1.39}$$

σ^0_{had} is determined from the maximum of the resonance in $e^+e^- \to$ hadrons. Assuming again lepton-universality, which is justified by the equality of the measured leptonic decay widths, the number of neutrino flavours can be determined as

$$N_\nu = \frac{\Gamma_{\text{inv}}}{\Gamma_l} \left(\frac{\Gamma_l}{\Gamma_\nu}\right)_{\text{SM}} = \left[\sqrt{\frac{12\pi R_l}{m_Z^2 \sigma^0_{\text{had}}}} - R_l - 3\right] \left(\frac{\Gamma_l}{\Gamma_\nu}\right)_{\text{SM}}. \tag{1.40}$$

This form is chosen because in this way radiative corrections are already included in the Standard Model (SM) prediction. Using the most recent fit to the data of the four LEP experiments, a number of

$$N_\nu = 2.984 \pm 0.008 \tag{1.41}$$

can be deduced [PDG16], in very good agreement with the theoretical expectation of three.

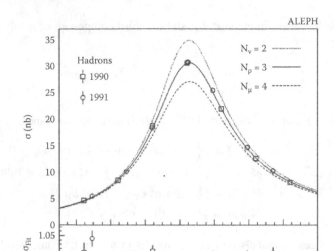

Figure 1.12. Cross-section as a function of $\sqrt{s}$ for the reaction $e^+ e^- \to$ hadrons as obtained by the ALEPH detector at LEP. The different curves show the Standard Model predictions for two, three and four light neutrino flavours (from [Ale19]). With kind permission of the ALEPH collaboration.

1.12   Further milestones in the last 20 years

Of course, more exciting results have appeared over the last decades. They will only be mentioned briefly here as a more detailed presentation is the major content of this book. A key result was to prove that neutrinos have a non-vanishing mass. This is linked to the observation of a muon-neutrino deficit of upward going atmospheric neutrinos measured by Super-Kamiokande and by solving the solar neutrino problem by measuring the charged-current and neutral current reaction on the deuteron by the Sudbury Neutrino Observatory (SNO). Both measurements are accompanied with various neutrino oscillation experiments. In this way a mixing matrix for leptons has been established similar to the CKM-matrix but currently with less accuracy. Additional new features are the discovery of $\bar{\nu}_e$ from the Earth due to radioactive decays (geoneutrinos). Furthermore, ultrahigh energy particles from the Universe at extremely high energies was seen by the IceCube and the Auger experiment opening the door for high energy particle astrophysics. Last but not least, the low energy process of coherent neutrino-scattering on nuclei was observed by the COHERENT experiment. All these new measurements widened the field of neutrino research enormously. Before going into further details, some theoretical formalism will be discussed.

Chapter 2

Properties of neutrinos

DOI: 10.1201/9781315195612-2

In quantum field theory spin-$\frac{1}{2}$ particles are described by four-component wavefunctions $\psi(x)$ (spinors) which obey the Dirac equation. The four independent components of $\psi(x)$ correspond to particles and antiparticles with the two possible spin projections $J_Z = \pm 1/2$ equivalent to the two helicities $\mathcal{H} = \pm 1$. Neutrinos as fundamental leptons are spin-$\frac{1}{2}$ particles like other fermions; however, it is an experimental fact that only left-handed neutrinos ($\mathcal{H} = -1$) and right-handed antineutrinos ($\mathcal{H} = +1$) are observed as mentioned in Section 1.6. Therefore, a two-component spinor description should, in principle, be sufficient (Weyl spinors). In a four-component theory they are obtained by projecting out of a general spinor $\psi(x)$ the components with $\mathcal{H} = +1$ for particles and $\mathcal{H} = -1$ for antiparticles with the help of the operators $P_{L,R} = \frac{1}{2}(1 \mp \gamma_5)$. This two-component theory of the neutrino will be discussed in detail later. The discussion will be quite general; for a more extensive discussion see [Bjo64, Bil87, Kay89, Kim93, Sch97, Fuk03a, Giu07].

2.1 Helicity and chirality

The Dirac equation is the relativistic wave equation for spin-$\frac{1}{2}$ particles and is given by

$$\left(i\gamma_\mu \frac{\partial}{\partial x_\mu} - m \right) \psi = 0. \tag{2.1}$$

Here ψ denotes a four-component spinor and the 4×4 γ-matrices are given in the form[1]

$$\gamma_0 = \begin{pmatrix} \mathbb{1} & 0 \\ 0 & -\mathbb{1} \end{pmatrix} \qquad \gamma_i = \begin{pmatrix} 0 & \tilde{\sigma}_i \\ -\tilde{\sigma}_i & 0 \end{pmatrix} \tag{2.2}$$

where $\tilde{\sigma}_i$ correspond to the 2×2 Pauli matrices. Detailed introductions and treatments can be found in [Bjo64]. An additional matrix γ_5 is defined by

$$\gamma_5 = i\gamma_0\gamma_1\gamma_2\gamma_3 = \begin{pmatrix} 0 & \mathbb{1} \\ \mathbb{1} & 0 \end{pmatrix} \tag{2.3}$$

---

[1] Other conventions of the γ-matrices are also commonly used in the literature, which leads to slightly different forms for the following expressions.

19

and the following anticommutator relations hold:

$$\{\gamma_\alpha, \gamma_\beta\} = 2g_{\alpha\beta} \tag{2.4}$$

$$\{\gamma_\alpha, \gamma_5\} = 0 \tag{2.5}$$

with $g_{\alpha\beta}$ as the Minkowski metric diag$(+1, -1, -1, -1)$. Multiplying the Dirac equation from the left with γ_0 and using $\gamma_i = \gamma_0\gamma_5\sigma_i$ results in

$$\left(i\gamma_0^2 \frac{\partial}{\partial x_0} - i\gamma_0^2\gamma_5\sigma_i \frac{\partial}{\partial x_i} - m\gamma_0 \right)\psi = 0 \qquad i = 1,\ldots,3. \tag{2.6}$$

Another multiplication of (2.6) from the left with γ_5 and using $\gamma_5\sigma_i = \sigma_i\gamma_5$ (which follows from (2.5)) and $\sigma_i = $ diag $(\tilde{\sigma}_i, \tilde{\sigma}_i)$ leads to $(\gamma_0^2 = 1, \gamma_5^2 = 1)$

$$\left(i\frac{\partial}{\partial x_0}\gamma_5 - i\sigma_i \frac{\partial}{\partial x_i} + m\gamma_0\gamma_5 \right)\psi = 0. \tag{2.7}$$

Subtraction and addition of the last two equations result in the following system of coupled equations:

$$\left(i\frac{\partial}{\partial x_0}(1 + \gamma_5) - i\sigma_i \frac{\partial}{\partial x_i}(1 + \gamma_5) - m\gamma_0(1 - \gamma_5) \right)\psi = 0 \tag{2.8}$$

$$\left(i\frac{\partial}{\partial x_0}(1 - \gamma_5) + i\sigma_i \frac{\partial}{\partial x_i}(1 - \gamma_5) - m\gamma_0(1 + \gamma_5) \right)\psi = 0. \tag{2.9}$$

Now let us introduce left- and right-handed components by defining two projection operators P_L and P_R given by

$$P_L = \tfrac{1}{2}(1 - \gamma_5) \qquad \text{and} \qquad P_R = \tfrac{1}{2}(1 + \gamma_5) \tag{2.10}$$

As they are projectors, the following relations are valid:

$$P_L P_R = P_R P_L = 0 \qquad P_L + P_R = 1 \qquad P_L^2 = P_L \qquad P_R^2 = P_R. \tag{2.11}$$

With the definition

$$\psi_L = P_L\psi \qquad \text{and} \qquad \psi_R = P_R\psi \tag{2.12}$$

it is obviously valid that

$$P_L\psi_R = P_R\psi_L = 0. \tag{2.13}$$

Then the following eigenequation emerges:

$$\gamma_5\psi_{L,R} = \pm\psi_{L,R}. \tag{2.14}$$

The eigenvalues ± 1 to γ_5 are called chirality and $\psi_{L,R}$ are called chiral projections of ψ. Any spinor ψ can be rewritten in chiral projections as

$$\psi = (P_L + P_R)\psi = P_L\psi + P_R\psi = \psi_L + \psi_R. \tag{2.15}$$

The equations 2.8 and 2.9 can now be expressed in these projections as

$$\left(i\frac{\partial}{\partial x_0} - i\sigma_i\frac{\partial}{\partial x_i}\right)\psi_R = m\gamma_0\psi_L \qquad (2.16)$$

$$\left(i\frac{\partial}{\partial x_0} + i\sigma_i\frac{\partial}{\partial x_i}\right)\psi_L = m\gamma_0\psi_R. \qquad (2.17)$$

Both equations decouple in the case of a vanishing mass $m = 0$ and can then be depicted as

$$i\frac{\partial}{\partial x_0}\psi_R = i\sigma_i\frac{\partial}{\partial x_i}\psi_R \qquad (2.18)$$

$$i\frac{\partial}{\partial x_0}\psi_L = -i\sigma_i\frac{\partial}{\partial x_i}\psi_L. \qquad (2.19)$$

But these are identical to the Schrödinger equation ($x_0 = t$, $\hbar = 1$)

$$i\frac{\partial}{\partial t}\psi_{L,R} = \mp i\sigma_i\frac{\partial}{\partial x_i}\psi_{L,R} \qquad (2.20)$$

or in momentum space ($i\frac{\partial}{\partial t} = E$, $-i\frac{\partial}{\partial x_i} = p_i$)

$$E\psi_{L,R} = \pm\sigma_i p_i\psi_{L,R}. \qquad (2.21)$$

The latter implies that $\psi_{L,R}$ are also eigenfunctions to the helicity operator $\mathcal{H}$ given by (see Chapter 1)

$$\mathcal{H} = \frac{\boldsymbol{\sigma}\cdot\boldsymbol{p}}{|\boldsymbol{\sigma}||\boldsymbol{p}|} \qquad (2.22)$$

ψ_L is an eigenspinor with helicity eigenvalues $\mathcal{H} = +1$ for particles and $\mathcal{H} = -1$ for antiparticles. Correspondingly, ψ_R is the eigenspinor to the helicity eigenvalues $\mathcal{H} = -1$ for particles and $\mathcal{H} = +1$ for antiparticles. Therefore, in the case of massless particles, chirality and helicity are identical.[2] For $m > 0$ the decoupling of (2.16) and (2.17) is no longer possible. This means that the chirality eigenspinors ψ_L and ψ_R no longer describe particles with fixed helicity and helicity is no longer a good conserved quantum number.

The two-component theory now states that the neutrino spinor ψ_ν in weak interactions always reads as

$$\psi_\nu = \tfrac{1}{2}(1 - \gamma_5)\psi = \psi_L \qquad (2.23)$$

meaning that the interacting neutrino is always left-handed and the antineutrino always right-handed. For $m = 0$, this further implies that ν always has $\mathcal{H} = -1$ and $\bar{\nu}$ always $\mathcal{H} = +1$. The proof that indeed the Dirac spinors ψ_L and ψ_R can be written as the sum of two independent 2-component Weyl spinors can be found in [Sch97].

---

[2] May be of opposite sign depending on the representation used for the γ-matrices.

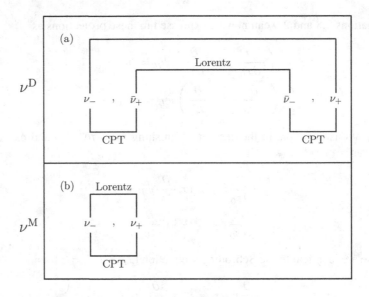

Figure 2.1. Schematic drawing of the difference between massive Dirac and Majorana neutrinos. (*a*) The Dirac case: ν_L is converted via CPT into a $\bar{\nu}_R$ and via a Lorentz boost into a ν_R. An application of CPT on the latter results in $\bar{\nu}_L$ which is different from the one obtained by applying CPT on ν_L. The result is four different states. (*b*) The Majorana case: Both operations CPT and a Lorentz boost result in the same state ν_R, there is no difference between particle and antiparticle. Only two states emerge (from [Boe92]). © Cambridge University Press.

2.2 Charge conjugation

While for all fundamental fermions of the Standard Model (see Chapter 3) a clear discrimination between particle and antiparticle can be made by their electric charge, for neutrinos this property is not so obvious. If particle and antiparticle are not identical, we call such a fermion a Dirac particle which has four independent components. If particle and antiparticle are identical, they are called Majorana particles (Figure 2.1). The latter requires that all additive quantum numbers (charge, strangeness, baryon number, lepton number, etc.) have to vanish. Consequently, lepton number is violated if neutrinos are Majorana particles.

The following derivations are taken from [Bil87, Sch97]. The operator connecting particle $f(\boldsymbol{x}, t)$ and antiparticle $\bar{f}(\boldsymbol{x}, t)$ is charge conjugation C:

$$C|f(\boldsymbol{x}, t)\rangle = \eta_c |\bar{f}(\boldsymbol{x}, t)\rangle. \tag{2.24}$$

If $\psi(x)$ is a spinor field of a free neutrino, then the corresponding charge conjugated field ψ^c is defined by

$$\psi \xrightarrow{C} \psi^c \equiv C\psi C^{-1} = \eta_c C\bar{\psi}^T \tag{2.25}$$

with η_c as a phase factor with $|\eta_c| = 1$. The 4×4 unitary charge conjugation matrix

C obeys the following general transformations:

$$C^{-1}\gamma_\mu C = -\gamma_\mu^T \qquad C^{-1}\gamma_5 C = \gamma_5^T \qquad C^\dagger = C^{-1} = C^T = -C. \qquad (2.26)$$

A possible representation is given as $C = i\gamma_0\gamma_2$. Using the projection operators $P_{L,R}$, it follows that

$$P_{L,R}\psi = \psi_{L,R} \overset{C}{\to} P_{L,R}\psi^c = (\psi^c)_{L,R} = (\psi_{R,L})^c. \qquad (2.27)$$

It is straightforward to show that if ψ is an eigenstate of chirality; ψ^c is an eigenstate too but it has an eigenvalue of opposite sign. Furthermore, from (2.27) it follows that the charge conjugation C transforms a right(left)-handed particle into a right(left)-handed antiparticle, leaving the helicity (chirality) untouched. Only the additional application of a parity transformation changes the helicity as well. However, the operation of (2.25) converts a right(left)-handed particle into a left(right)-handed antiparticle. Here helicity and chirality are converted as well.

To include the fact that $\psi_{L,R}$ and $\psi_{L,R}^c$ have opposite helicity, one avoids calling $\psi_{L,R}^c$ the charge conjugate of $\psi_{L,R}$. Instead it is more frequently called the CP (or CPT) conjugate with respect to $\psi_{L,R}$ [Lan88]. In the following sections we refer to ψ^c as the CP or CPT conjugate of the spinor ψ, assuming CP or CPT conservation correspondingly.

2.3 Parity transformation

A parity operation P is defined as

$$\psi(\boldsymbol{x},t) \overset{P}{\to} P\psi(\boldsymbol{x},t)P^{-1} = \eta_P \gamma_0\psi(-\boldsymbol{x},t). \qquad (2.28)$$

The phase factor η_P with $|\eta_P| = 1$ corresponds for real $\eta_P = \pm 1$ to the intrinsic parity. Using (2.25) for the charge conjugated field, it follows that

$$\psi^c = \eta_C C\bar{\psi}^T \overset{P}{\to} \eta_C \eta_P^* C\gamma_0^T \bar{\psi}^T = -\eta_P^* \gamma_0\psi^c. \qquad (2.29)$$

This implies that a fermion and its corresponding antifermion have opposite intrinsic parity, i.e., for a Majorana particle $\psi^c = \pm\psi$ holds which results in $\eta_P = -\eta_P^*$.

Therefore, an interesting point with respect to the intrinsic parity occurs for Majorana neutrinos. A Majorana field can be written as

$$\psi_M = \frac{1}{\sqrt{2}}(\psi + \eta_C\psi^c) \qquad \text{with } \eta_C = \lambda_C e^{2i\phi}, \ \lambda_C = \pm 1 \qquad (2.30)$$

where λ_C is sometimes called creation phase [Kay89]. By applying a phase transformation

$$\psi_M \to \psi_M e^{-i\phi} = \frac{1}{\sqrt{2}}(\psi e^{-i\phi} + \lambda_C\psi^c e^{i\phi}) = \frac{1}{\sqrt{2}}(\psi + \lambda_C\psi^c) \equiv \psi_M \qquad (2.31)$$

it can be achieved that the field ψ_M is an eigenstate with respect to charge conjugation C

$$\psi_M^c = \frac{1}{\sqrt{2}}(\psi^c + \lambda_C \psi) = \lambda_C \psi_M \tag{2.32}$$

with eigenvalues $\lambda_C = \pm 1$. This means the Majorana particle is identical to its antiparticle; i.e., ψ_M and ψ_M^c cannot be distinguished. With respect to CP, one obtains

$$\psi_M(\boldsymbol{x},t) \overset{C}{\to} \psi_M^c = \lambda_C \psi_M \overset{P}{\to} \frac{\lambda_C}{\sqrt{2}}(\eta_P \gamma_0 \psi - \lambda_C \eta_P^* \gamma_0 \psi^c)$$

$$= \lambda_C \eta_P \gamma_0 \psi_M = \pm i \gamma_0 \psi_M(-\boldsymbol{x},t) \tag{2.33}$$

because $\eta_P^* = -\eta_P$. This means that the intrinsic parity of a Majorana particle is imaginary, $\eta_P = \pm i$ if $\lambda_C = \pm 1$. Finally, from (2.31) it follows that

$$(\gamma_5 \psi_M)^c = \eta_C C (\overline{\gamma_5 \psi_M})^T = -\eta_C C \gamma_5^T \bar{\psi}_M^T = -\gamma_5 \psi_M^c = -\lambda_C \gamma_5 \psi_M \tag{2.34}$$

because $\gamma_5 \bar{\psi}_M = (\gamma_5 \psi_M)^\dagger \gamma_0 = \psi_M^\dagger \gamma_5 \gamma_0 = -\bar{\psi}_M \gamma_5$. Using this together with (2.27) one concludes that an eigenstate to C cannot be at the same time an eigenstate to chirality. Therefore, a Majorana neutrino has no fixed chirality. However, as ψ and ψ^c obey the Dirac equation, ψ_M will also do so as well.

For a discussion of T transformation and C, CP and CPT properties, see [Kay89, Kim93].

2.4 Dirac and Majorana mass terms

Consider the case of free fields without interactions and start with the Dirac mass. The Dirac equation can then be deduced with the help of the Euler–Lagrange equation from a Lagrangian [Bjo64]:

$$\mathcal{L} = \bar{\psi}\left(i\gamma_\mu \frac{\partial}{\partial x_\mu} - m_D\right)\psi \tag{2.35}$$

where the first term corresponds to the kinetic energy and the second is the mass term. The Dirac mass term is, therefore,

$$\mathcal{L} = m_D \bar{\psi}\psi \tag{2.36}$$

where the combination $\bar{\psi}\psi$ has to be Lorentz invariant and Hermitian. Requiring $\mathcal{L}$ to be Hermitian as well, m_D must be real ($m_D^* = m_D$). Using the following relations valid for two arbitrary spinors ψ and ϕ (which follow from (2.10) and (2.11))

$$\bar{\psi}_L \phi_L = \bar{\psi} P_R P_L \phi = 0 \qquad \bar{\psi}_R \phi_R = 0 \tag{2.37}$$

it follows that

$$\bar{\psi}\phi = (\bar{\psi}_L + \bar{\psi}_R)(\phi_L + \phi_R) = \bar{\psi}_L \phi_R + \bar{\psi}_R \phi_L. \tag{2.38}$$

In this way the Dirac mass term can be written in its chiral components (Weyl spinors) as

$$\mathcal{L} = m_D(\bar{\psi}_L\psi_R + \bar{\psi}_R\psi_L) \qquad \text{with } \bar{\psi}_R\psi_L = (\bar{\psi}_L\psi_R)^\dagger. \qquad (2.39)$$

Applying this to neutrinos, it requires both a left- and a right-handed Dirac neutrino to produce such a mass term. In the Standard Model of particle physics only left-handed neutrinos exist; this is the reason why neutrinos remain massless as will be discussed in Chapter 3.

In a more general treatment including ψ^c one might ask which other combinations of spinors behaving like Lorentz scalars can be produced. Three more are possible: $\bar{\psi}^c\psi^c$, $\bar{\psi}\psi^c$ and $\bar{\psi}^c\psi$. The term $\bar{\psi}^c\psi^c$ is also hermitian and equivalent to $\bar{\psi}\psi$; $\bar{\psi}\psi^c$ and $\bar{\psi}^c\psi$ are hermitian conjugates, which can be shown for arbitrary spinors

$$(\bar{\psi}\phi)^\dagger = (\psi^\dagger\gamma_0\phi)^\dagger = \phi^\dagger\gamma_0\psi = \bar{\phi}\psi. \qquad (2.40)$$

This allows an additional hermitian mass term, called the Majorana mass term and is given by

$$\mathcal{L} = \frac{1}{2}(m_M\bar{\psi}\psi^c + m_M^*\bar{\psi}^c\psi) = \frac{1}{2}m_M\bar{\psi}\psi^c + h.c.^3 \qquad (2.41)$$

m_M is called the Majorana mass. Now using again the chiral projections with the notation

$$\psi_{L,R}^c = (\psi^c)_{R,L} = (\psi_{R,L})^c \qquad (2.42)$$

one gets two hermitian mass terms:

$$\mathcal{L}^L = \frac{1}{2}m_L(\bar{\psi}_L\psi_R^c + \bar{\psi}_R^c\psi_L) = \frac{1}{2}m_L\bar{\psi}_L\psi_R^c + h.c. \qquad (2.43)$$

$$\mathcal{L}^R = \frac{1}{2}m_R(\bar{\psi}_L^c\psi_R + \bar{\psi}_R\psi_L^c) = \frac{1}{2}m_R\bar{\psi}_L^c\psi_R + h.c. \qquad (2.44)$$

with $m_{L,R}$ as real Majorana masses because of (2.40). Let us define two Majorana fields (see (2.30) with $\lambda_C = 1$)

$$\phi_1 = \psi_L + \psi_R^c \qquad \phi_2 = \psi_R + \psi_L^c \qquad (2.45)$$

to rewrite (2.43) as

$$\mathcal{L}^L = \frac{1}{2}m_L\bar{\phi}_1\phi_1 \qquad \mathcal{L}^R = \frac{1}{2}m_R\bar{\phi}_2\phi_2. \qquad (2.46)$$

While $\psi_{L,R}$ are interaction eigenstates, $\phi_{1,2}$ are mass eigenstates to $m_{L,R}$.

The most general mass term (the Dirac–Majorana mass term) is a combination of (2.39) and (2.43) (Figure 2.2):

$$2\mathcal{L} = m_D(\bar{\psi}_L\psi_R + \bar{\psi}_L^c\psi_R^c) + m_L\bar{\psi}_L\psi_R^c + m_R\bar{\psi}_L^c\psi_R + h.c.$$

$$= (\bar{\psi}_L, \bar{\psi}_L^c) \begin{pmatrix} m_L & m_D \\ m_D & m_R \end{pmatrix} \begin{pmatrix} \psi_R^c \\ \psi_R \end{pmatrix} + h.c. \qquad (2.47)$$

$$= \bar{\Psi}_L M\Psi_R^c + \bar{\Psi}_R^c M\Psi_L$$

---

[3] $h.c.$ throughout the book signifies Hermitian conjugate.

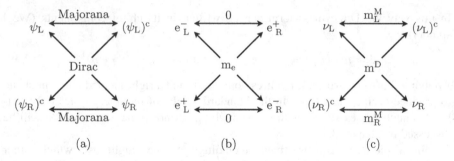

Figure 2.2. Coupling schemes for fermion fields via Dirac and Majorana masses: (*a*) general scheme for left- and right-handed fields and the charge conjugate fields; (*b*) the case for electrons (because of its electric charge, only Dirac-mass terms are possible) and (*c*) coupling scheme for neutrinos. They are the only fundamental fermions that allow all possible couplings (from [Mut88]). Reproduced with permission of SNCSC.

where, in the last step, the following was used:

$$M = \begin{pmatrix} m_L & m_D \\ m_D & m_R \end{pmatrix} \qquad \Psi_L = \begin{pmatrix} \psi_L \\ \psi_L^c \end{pmatrix} = \begin{pmatrix} \psi_L \\ (\psi_R)^c \end{pmatrix} \qquad (2.48)$$

implying

$$(\Psi_L)^c = \begin{pmatrix} (\psi_L)^c \\ \psi_R \end{pmatrix} = \begin{pmatrix} \psi_R^c \\ \psi_R \end{pmatrix} = \Psi_R^c.$$

In the case of CP conservation the elements of the mass matrix M are real. Coming back to neutrinos, in the known neutrino interactions only ψ_L and ψ_R^c are present (active neutrinos) and not the fields ψ_R and ψ_L^c (often called sterile neutrinos, they are not participating in weak interaction); it is quite common to distinguish between both types in the notation: $\psi_L = \nu_L$, $\psi_R^c = \nu_R^c$, $\psi_R = N_R$, $\psi_L^c = N_L^c$. With this notation, (2.47) becomes

$$2\mathcal{L} = m_D(\bar{\nu}_L N_R + \bar{N}_L^c \nu_R^c) + m_L \bar{\nu}_L \nu_R^c + m_R \bar{N}_L^c N_R + h.c.$$

$$= (\bar{\nu}_L, \bar{N}_L^c) \begin{pmatrix} m_L & m_D \\ m_D & m_R \end{pmatrix} \begin{pmatrix} \nu_R^c \\ N_R \end{pmatrix} + h.c. \qquad (2.49)$$

The mass eigenstates are obtained by diagonalizing M and are given as

$$\psi_{1L} = \cos\theta\psi_L - \sin\theta\psi_L^c \qquad \psi_{1R}^c = \cos\theta\psi_R^c - \sin\theta\psi_R \qquad (2.50)$$

$$\psi_{2L} = \sin\theta\psi_L + \cos\theta\psi_L^c \qquad \psi_{2R}^c = \sin\theta\psi_R^c + \cos\theta\psi_R \qquad (2.51)$$

while the mixing angle θ is given by

$$\tan 2\theta = \frac{2m_D}{m_R - m_L}. \qquad (2.52)$$

The corresponding mass eigenvalues are

$$\tilde{m}_{1,2} = \frac{1}{2}\left[(m_L + m_R) \pm \sqrt{(m_L - m_R)^2 + 4m_D^2}\right].\tag{2.53}$$

To get positive masses,[4] we use [Lan88, Gro90]

$$\tilde{m}_k = \epsilon_k m_k \quad \text{with } m_k = |\tilde{m}_k| \text{ and } \epsilon_k = \pm1 \ (k = 1, 2).\tag{2.54}$$

To get a similar expression as (2.45), two independent Majorana fields with masses m_1 and m_2 (with $m_k \geq 0$) are introduced via $\phi_k = \psi_{kL} + \epsilon_k\psi_{kR}^c$ or, explicitly,

$$\phi_1 = \psi_{1L} + \epsilon_1\psi_{1R}^c = \cos\theta(\psi_L + \epsilon_1\psi_R^c) - \sin\theta(\psi_L^c + \epsilon_1\psi_R)\tag{2.55}$$
$$\phi_2 = \psi_{2L} + \epsilon_2\psi_{2R}^c = \sin\theta(\psi_L + \epsilon_2\psi_R^c) + \cos\theta(\psi_L^c + \epsilon_2\psi_R)\tag{2.56}$$

and, as required for Majorana fields,

$$\phi_k^c = (\psi_{kL})^c + \epsilon_k\psi_{kL} = \epsilon_k(\epsilon_k\psi_{kR}^c + \psi_{kL}) = \epsilon_k\phi_k\tag{2.57}$$

ϵ_k is the CP eigenvalue of the Majorana neutrino ϕ_k. So we finally get the analogous expression to (2.45):
$$2\mathcal{L} = m_1\bar{\phi}_1\phi_1 + m_2\bar{\phi}_2\phi_2.\tag{2.58}$$

From this general discussion one can take some interesting special aspects:

(1) $m_L = m_R = 0$ ($\theta = 45°$), resulting in $m_{1,2} = m_D$ and $\epsilon_{1,2} = \mp1$. As Majorana eigenstates, two degenerated states emerge:

$$\phi_1 = \frac{1}{\sqrt{2}}(\psi_L - \psi_R^c - \psi_L^c + \psi_R) = \frac{1}{\sqrt{2}}(\psi - \psi^c)\tag{2.59}$$

$$\phi_2 = \frac{1}{\sqrt{2}}(\psi_L + \psi_R^c + \psi_L^c + \psi_R) = \frac{1}{\sqrt{2}}(\psi + \psi^c).\tag{2.60}$$

These can be used to construct a Dirac field ψ:

$$\frac{1}{\sqrt{2}}(\phi_1 + \phi_2) = \psi_L + \psi_R = \psi.\tag{2.61}$$

The corresponding mass term (2.58) is (because $\bar{\phi}_1\phi_2 + \bar{\phi}_2\phi_1 = 0$)

$$\mathcal{L} = \frac{1}{2}m_D(\bar{\phi}_1 + \bar{\phi}_2)(\phi_1 + \phi_2) = m_D\bar{\psi}\psi.\tag{2.62}$$

We are left with a pure Dirac field. As a result, a Dirac field can be seen, using (2.61), to be composed of two degenerated Majorana fields; i.e., a Dirac ν looks like a pair of degenerated Majorana ν. The Dirac case is, therefore, a special solution of the more general Majorana case.

---

[4] An equivalent procedure for $\tilde{m}_k < 0$ would be a phase transformation $\psi_k \to i\psi_k$ resulting in a change of sign of the $\bar{\psi}^c\psi$ terms in (2.43). With $m_k = -\tilde{m}_k > 0$, positive m_k terms in (2.43) result.

(2) $m_D \gg m_L, m_R$ ($\theta \approx 45°$): In this case the states $\phi_{1,2}$ are almost degenerated with $m_{1,2} \approx m_D$ and such an object is called a pseudo-Dirac neutrino.

(3) $m_D = 0$ ($\theta = 0$): In this case $m_{1,2} = m_{L,R}$ and $\epsilon_{1,2} = 1$. So $\phi_1 = \psi_L + \psi_R^c$ and $\phi_2 = \psi_R + \psi_L^c$. This is the pure Majorana case.

(4) $m_R \gg m_D, m_L = 0$ ($\theta = (m_D/m_R) \ll 1$): One obtains two mass eigenvalues:

$$m_\nu = m_1 = \frac{m_D^2}{m_R} \qquad m_N = m_2 = m_R \left(1 + \frac{m_D^2}{m_R^2} \right) \approx m_R \qquad (2.63)$$

and

$$\epsilon_{1,2} = \mp 1.$$

The corresponding Majorana fields are

$$\phi_1 \approx \psi_L - \psi_R^c \qquad \phi_2 \approx \psi_L^c + \psi_R. \qquad (2.64)$$

The last scenario is especially popular within the seesaw model of neutrino mass generation and will be discussed in more detail in Chapter 5.

2.4.1   Generalization to n flavours

The discussion so far has been related to only one neutrino flavour. The generalization to n flavours will not be discussed in greater detail; only some general statements are made—see [Bil87, Kim93, Sch97] for a more complete discussion. A Weyl spinor is now an n-component vector in flavour space, given, for example, as

$$\nu_L = \begin{pmatrix} \nu_{1L} \\ . \\ . \\ . \\ \nu_{nL} \end{pmatrix} \qquad N_R = \begin{pmatrix} N_{1R} \\ . \\ . \\ . \\ N_{nR} \end{pmatrix} \qquad (2.65)$$

where every ν_{iL} and N_{iR} are normal Weyl spinors with flavour i. Correspondingly, the masses m_D, m_L, m_R are now $n \times n$ matrices M_D, M_L and M_R with complex elements and $M_L = M_L^T, M_R = M_R^T$. The general symmetric $2n \times 2n$ matrix is then, in analogy to (2.48),

$$M = \begin{pmatrix} M_L & M_D \\ M_D^T & M_R \end{pmatrix}. \qquad (2.66)$$

The most general mass term (2.47) is now

$$2\mathcal{L} = \bar{\Psi}_L M \Psi_R^c + \bar{\Psi}_R^c M^\dagger \Psi_L \qquad (2.67)$$

$$= \bar{\nu}_L M_D N_R + \bar{N}_L^c M_D^T \nu_R^c + \bar{\nu}_L M_L \nu_R^c + \bar{N}_L^c M_R N_R + h.c. \qquad (2.68)$$

where

$$\Psi_L = \begin{pmatrix} \nu_L \\ N_L^c \end{pmatrix} \qquad \text{and} \qquad \Psi_R^c = \begin{pmatrix} \nu_R^c \\ N_R \end{pmatrix}. \qquad (2.69)$$

Diagonalization of M results in $2n$ Majorana mass eigenstates with associated mass eigenvalues $\epsilon_i m_i (\epsilon_i = \pm 1, m_i \geq 0)$. In the previous discussion, an equal number of active and sterile flavours ($n_a = n_s = n$) is assumed. In the most general case with $n_a \neq n_s$, M_D is an $n_a \times n_s$, M_L an $n_a \times n_a$ and M_R an $n_s \times n_s$ matrix. So the full matrix M is an $(n_a + n_s) \times (n_a + n_s)$ matrix whose diagonalization results in $(n_a + n_s)$ mass eigenstates and eigenvalues.

In seesaw models, light neutrinos are given by the mass matrix (still to be diagonalized)

$$M_\nu = M_D M_R^{-1} M_D^T \qquad (2.70)$$

in analogy to m_ν in (2.63).

Having discussed the formal description of neutrinos in some detail, we now take a short look at the concept of lepton number.

2.5 Lepton number

Conserved quantum numbers arise from the invariance of the equation of motion under certain symmetry transformations. Continuous symmetries (e.g., translation) can be described by real numbers and lead to additive quantum numbers, while discrete symmetries (e.g., spatial reflections through the origin) are described by integers and lead to multiplicative quantum numbers. For some of them the underlying symmetry operations are known, as discussed in more detail in Chapter 3. Some quantum numbers, however, have not yet been associated with a fundamental symmetry such as baryon number B or lepton number L and their conservation is only motivated by experimental observations. The quantum numbers conserved in the individual interactions are shown in Table 2.1. Lepton number was introduced to characterize experimental observations of weak interactions. Each lepton is defined as having a lepton number $L = +1$, each antilepton $L = -1$. Moreover, each generation of leptons has its own lepton number L_e, L_μ, L_τ with $L = L_e + L_\mu + L_\tau$. An individual lepton number is not conserved, as has been established with the observation of neutrino oscillations (see Chapter 8). However, flavour changes of charged leptons (charged lepton flavour violation, CLFV) has not been observed yet. Classical examples for baryon and lepton number violation are proton decay, like $p \rightarrow e^+ \pi^0$ with $\Delta L = 1$ and $\Delta B = 1$ and neutron-antineutron oscillations which is a $\Delta B = 2$ process.

Consider the four Lorentz scalars discussed under a global phase transformation $e^{i\alpha}$:

$$\psi \rightarrow e^{i\alpha}\psi \qquad \bar{\psi} \rightarrow e^{-i\alpha}\bar{\psi} \qquad \text{so that} \qquad \bar{\psi}\psi \rightarrow \bar{\psi}\psi \qquad (2.71)$$

$$\psi^c \rightarrow (e^{i\alpha}\psi)^c = \eta_c C e^{i\bar{\alpha}}\bar{\psi}^T = e^{-i\alpha}\psi^c \qquad \bar{\psi}^c \rightarrow e^{i\alpha}\bar{\psi}^c. \qquad (2.72)$$

As can be seen, $\bar{\psi}\psi$ and $\bar{\psi}^c\psi^c$ are invariant under these transformations and are connected to a conserved quantum number, namely lepton number: ψ annihilates a lepton or creates an antilepton, $\bar{\psi}$ acts oppositely. $\bar{\psi}\psi$ and $\bar{\psi}^c\psi^c$ result in transitions $\ell \rightarrow \ell$ or $\bar{\ell} \rightarrow \bar{\ell}$ with $\Delta L = 0$. This does not relate to the other two Lorentz scalars

Table 2.1. Summary of conservation laws. B corresponds to baryon number and L to total lepton number.

Conservation law	Strong	Electromagnetic	Weak
Energy	yes	yes	yes
Momentum	yes	yes	yes
Angular momentum	yes	yes	yes
B, L	yes	yes	yes
P	yes	yes	no
C	yes	yes	no
CP	yes	yes	no
T	yes	yes	no
CPT	yes	yes	yes

$\bar{\psi}\psi^c$ and $\bar{\psi}^c\psi$ which force transitions of the form $\ell \to \bar{\ell}$ or $\bar{\ell} \to \ell$ corresponding to $\Delta L = \pm 2$ according to the assignment made earlier. For charged leptons such lepton-number-violating transitions are forbidden (i.e., $e^- \to e^+$) and they have to be Dirac particles (unless charged lepton flavour violation might be found in the future). But if one associates a mass to neutrinos both types of transitions are, in principle, possible.

If the lepton number is related to a global symmetry which has to be broken spontaneously, a Goldstone boson is associated with the symmetry breaking. In this case it is called a majoron (see [Moh86, 92, Kim93] for more details).

2.5.1 Experimental status of lepton flavour and number violation

As no underlying fundamental symmetry is known to conserve lepton number, one might think about observing lepton flavour violation at some level, which could also be searched for in mesonic decays. In the lepton sector the most sensitive searches for CLFV are linked to muons. The three classic processes are $\mu \to e\gamma$, $\mu \to 3e$ and coherent muon-electron conversion on nuclei:

$$
\begin{array}{ccccc}
 & \mu^- & + \,^A_Z\text{X} & \to \,^A_Z\text{X} & +e^- \\
L_e & 0 & +0 & \to 0 & +1 \\
L_\mu & 1 & +0 & \to 0 & +0
\end{array}
$$

They violate both L_e and L_μ conservation but would leave the total lepton number unchanged. None of these processes has been observed and some of the current upper limits are compiled in Table 2.2. For a comprehensive list see [PDG16].

A next generation of experiments or updates are about to appear. The MEG experiment improved the limit on the $\mu \to e\gamma$ decay [Bal16] and a planned upgrade MEGII will improve the sensitivity by about an order of magnitude. The Mu3e experiment at the Paul-Scherrer Institute (PSI) is preparing for a $\mu \to 3e$ experiment

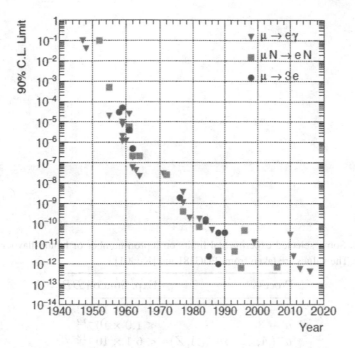

Figure 2.3. Time evolution of experimental limits of branching ratios for three rare LFV muon decays. Further improvements are expected by the next generation of experiments (from [Bal18]). Reproduced with permission of SNCSC.

[Wie17]. Furthermore, two experiments, Mu2e at Fermilab [Car08] and COMET at J-PARC [Kun13] aim to improve the muon-electron conversion down to 10^{-17} or even lower (for a general discussion see [Kun01]). Other processes like muonium–antimuonium conversion ($\mu^+ e^- \to \mu^- e^+$) have been studied as well. The time evolution of experimental progress of some of the searches is shown in Figure 2.3. Searches involving τ-leptons, e.g., $\tau \to \mu\gamma$, are also performed but are not as sensitive yet, although at the LHC and SuperKEKB significant improvements can be made. Another LFV process is neutrino oscillation, discussed in Chapter 8.

The 'gold-plated' process for total lepton number violation ($\Delta L = 2$) is neutrinoless double β-decay of a nucleus (A,Z)

$$(A, Z) \to (A, Z + 2) + 2e^-. \tag{2.73}$$

This process is only possible if neutrinos are massive Majorana particles and it will be discussed in more detail in Chapter 7. A compilation of various searches for $\Delta L = 2$ processes is given in Table 7.5. On the other hand, the observation of a static electric or magnetic moment would prove the Dirac character of the neutrino in case of CPT conservation (see Section 6.6).

Table 2.2. Some selected experimental limits on lepton-number or lepton-flavour violating processes. The values are taken from [PDG18] and [Kun01].

Process	Exp. limit on BR
$\mu \to e\gamma$	$< 4.2 \times 10^{-13}$
$\mu \to 3e$	$< 1.0 \times 10^{-12}$
$\mu^-(A, Z) \to e^-(A, Z)$	$< 6.1 \times 10^{-13}$
$\mu^-(A, Z) \to e^+(A, Z)$	$< 1.7 \times 10^{-12}$
$\tau \to \mu\gamma$	$< 4.4 \times 10^{-8}$
$\tau \to e\gamma$	$< 3.3 \times 10^{-8}$
$\tau \to 3e$	$< 2.7 \times 10^{-8}$
$\tau \to 3\mu$	$< 2.1 \times 10^{-8}$
$K^+ \to \pi^- e^+ e^+$	$< 6.4 \times 10^{-10}$
$K^+ \to \pi^- e^+ \mu^+$	$< 5.0 \times 10^{-10}$
$K^+ \to \pi^+ e^+ \mu^-$	$< 5.2 \times 10^{-10}$
$K^+ \to \pi^- \mu^+ \mu^+$	$< 8.6 \times 10^{-11}$

Chapter 3

The Standard Model of particle physics

DOI: 10.1201/9781315195612-3

In this chapter the basic features of the current Standard Model of elementary particle physics are discussed. As the main interest lies in neutrinos, the focus is on the weak or the more general electroweak interaction. For a more extensive introduction, see [Hal84, Kan87, Ait89, Nac90, Don92, Mar92, Lea96, Per00, Nar03, Bur06, Gri08, Lan09, Tho13].

3.1 The V–A theory of the weak interaction

Historically, the first theoretical description of the weak interaction for β-decay (see Chapter 6) after the discovery of the neutron by Chadwick, was given in the classical paper by Fermi [Fer34]. Nowadays, we rate this as a low-energy limit (the four-momentum transfer Q^2 (see Section 4.6) is much smaller than the W-mass) of the Glashow–Weinberg–Salam (GWS) model (see Section 3.3), but it is still valid to describe most of the low energy weak processes. Fermi chose a local coupling of four spin-$\frac{1}{2}$ fields (a four-point interaction) and took an ansatz quite similar to that of the later developed quantum electrodynamics (QED). In QED, the interaction of a charged particle (like the proton p) with an electromagnetic field A_μ is described by a Hamiltonian

$$H_{em} = e \int \mathrm{d}^3 \mathcal{H}_{em} = \int \mathrm{d}^3 x \bar{p}(x) \gamma^\mu p(x) A_\mu(x) \tag{3.1}$$

where $p(x)$ is the Dirac spinor of the proton. In analogy, Fermi introduced an interaction Hamiltonian for β-decay:

$$H_\beta = \frac{G_F}{\sqrt{2}} \int \mathrm{d}^3 x \, (\bar{p}(x) \gamma^\mu n(x))(\bar{e}(x) \gamma_\mu \nu(x)) + h.c. \tag{3.2}$$

with $p(x), n(x), e(x)$ and $\nu(x)$ being the proton, neutron, electron and neutrino spinor as described by the Dirac-equation (2.1), respectively. The new fundamental constant G_F is called the Fermi constant. It was soon realized that a generalization of (3.2) is necessary to describe all observed β-decays [Gam36]. In general this point-interaction can be written as current–current coupling in the form of two so-called

Table 3.1. Possible operators O_j and their transformation properties as well as their representation.

Operator	Transformation properties $(\Psi_f O_j \Psi_i)$	Representation with γ matrices
O_S (S)	scalar	$\mathbb{1}$
O_V (V)	vector	γ_μ
O_T (T)	tensor	$\frac{i}{2}(\gamma_\mu\gamma_\nu + \gamma_\nu\gamma_\mu)$
O_A (A)	axial vector	$\gamma_\mu\gamma_5$
O_P (P)	pseudo-scalar	γ_5

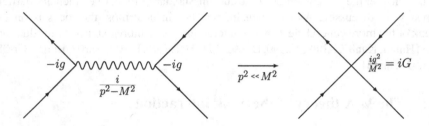

Figure 3.1. Left: Feynman diagram of a process like beta decay with an exchange particle (the W-boson) and a coupling constant g. Right: In case the 4-momentum is much smaller than the mass of the exchange particle (here about 80 GeV) this can equivalently be described by a point-like interaction with a coupling constant G (Fermi constant). This might become clearer during this chapter. With kind permission of S. Turkat.

currents J (either leptonic [Lan56, Sal57, Lee57] or hadronic [Fey58, The58]), in case of β-decay

$$\mathcal{L}(x) = \frac{G_F}{\sqrt{2}} J_L \cdot J_H \tag{3.3}$$

This coupling holds for the second and third generation of leptons and quarks as well. Dealing with this four-fermion interaction, the following question arises: Which is the correct structure of Lorentz-invariant combinations of the operators in the two currents? The weak Hamiltonian H_β can be deduced from a Lagrangian $\mathcal{L}$ by

$$H_\beta = \int \mathrm{d}^3 x\, \mathcal{L}(x). \tag{3.4}$$

The most general interaction Hamiltonian density describing nuclear beta decay is [Lee56, Jac57]

$$\mathcal{H}_\beta = (\bar{p}n)(\bar{e}(C_S + C'_S\gamma_5)\nu) \tag{3.5}$$
$$+ (\bar{p}\gamma_\mu n)(\bar{e}\gamma_\mu(C_V + C'_V\gamma_5)\nu)$$
$$+ \frac{1}{2}(\bar{p}\sigma_{\lambda\mu}n)(\bar{e}\sigma_{\lambda\mu}(C_T + C'_T\gamma_5)\nu)$$
$$- (\bar{p}\gamma_\mu\gamma_5 n)(\bar{e}\gamma_\mu\gamma_5(C_A + C'_A\gamma_5)\nu)$$
$$+ (\bar{p}\gamma_5 n)(\bar{e}\gamma_5(C_P + C'_P\gamma_5)\nu)$$
$$+ h.c$$

(for more details see [Sev06, Vos15]). The possible invariants j for the operators O are listed in Table 3.1. The kind of transformation property, called coupling, realized in nature was revealed by investigating allowed β-decay transitions (see Chapter 6). From the absence of Fierz interference terms (the Fierz transformation is an operation to change the order of the fermion fields in the four-fermion interaction Lagrangian, for more details see [Sch66, Wu66, Sev06] and Chapter 6) it could be concluded that Fermi transitions, nuclear transitions with no change of spin I and parity are either of S or V type, while Gamow–Teller transitions, nuclear transitions with change of $\Delta I = 0, 1$, excluding $0^+ \rightarrow 0^+$ transitions, and no parity change, could be only T- or A-type operators. P-type operators do not permit allowed transitions at all. After the discovery of parity violation, the measurements of electron–neutrino angular correlations in β-decay and the Goldhaber experiment (see Chapter 1), it became clear that the combination $\gamma_\mu(\mathbb{1} - \gamma_5)$ represented all the data accurately. Therefore, this is called the (V–A) structure of weak interactions. Current investigations of high precision measurements of nuclear and neutron β-decay especially in the form of angular distributions of the emitted electrons from polarised nuclei are used for searches of S- and T-type contributions. This is motivated by theories beyond the Standard Model and searches for a non-vanishing rest mass of the neutrino (see Chapter 6). Alternatively, searches are performed at the Large Hadron Collider (LHC) at CERN, a pp-collider with a center of mass energy of 14 TeV), looking for new heavy particles which predict such S and T contributions [Her95, Sev06, Dub11, Sev11, Vos15, Sev17, Gon19]. Both setups have similar sensitivities but very different experimental approaches. A compilation of current limits on T-type contributions is shown in Figure 3.2.

However, changing from quarks to nucleons, Equation (3.3) must be rewritten due to renormalization effects in strong interactions as [Fey58, The58]

$$J_H = \bar{p}(x)\gamma^\mu(g_V - g_A\gamma_5)n(x). \tag{3.6}$$

The coupling constants G_F, the vector- g_V and axial-vector g_A - constants have to be determined experimentally (see Section 3.4.1). Measurements of G_F in muon decay are in good agreement with those in nuclear β-decay (however, there is about a 2% difference which will be explained in Section 3.3.2) and lead to the concept of common current couplings (e–μ–τ universality, see Figure 3.3), also justified in

Table 3.2. Consequences to the coupling constants due to the violation of discrete symmetries (from [Sev06]):

Symmetry	Condition for violation
C	$(\mathrm{Re}(C_i) \neq 0$ and $\mathrm{Re}(C_i') \neq 0)$ or $(\mathrm{Im}(C_i) \neq 0$ and $\mathrm{Im}(C_i') \neq 0)$
P	$C_i \neq 0$ and $(C_i' \neq 0)$
T	$\mathrm{Im}(C_i/C_j) \neq 0$ or $\mathrm{Im}(C_i'/C_j') \neq 0$

measurements of τ-decays. The total leptonic current is then given by

$$J_L = J_e + J_\mu + J_\tau \tag{3.7}$$

each of them having the form of (3.3). Analogous arguments hold for the quark currents which can be extended to three families as well. Furthermore, the existence of a universal Fermi constant leads to the hypothesis of conserved vector currents (CVC) [Ger56, Fey58] showing that there are no renormalization effects in the vector current. Moreover, the observation that g_V and g_A are not too different (see Section 3.4.2) shows that renormalization effects in the axial vector current are small, leading to the concept of partially conserved axial vector currents (PCAC). For more details see [Gro90, Sev06, Gon19]. The formalism allows most of the observed low energy weak interactions to be described. It contains maximal parity violation and lepton universality, and it describes charged current interactions (see Chapter 4). How this picture is modified and embedded in the current understanding of gauge theories will be discussed now.

3.2   Gauge theories

All modern theories of elementary particles are gauge theories. Therefore, an attempt is made to indicate the fundamental characteristics of such theories without going into details. Theoretical aspects such as renormalization, the derivation of Feynman diagrams or triangle anomalies will not be discussed here and we refer to standard textbooks such as [Qui83, Hal84, Ait89, Don92, Lea96, Tho13]. However, it is important to realize that such topics do form part of the fundamentals of any such theory. One absolutely necessary requirement for such a theory is its *renormalizability*. Renormalization of the fundamental parameters is necessary to produce a relation between calculable and experimentally measurable quantities. The fact that gauge theories are *always* renormalizable, as long as the gauge bosons are massless, is of fundamental importance [t'Ho72, Lee72]. After this proof, gauge theories have become serious candidates for modelling interactions. One well-known non-renormalizable theory is general relativity.

A further aspect of the theory is its *freedom from anomalies*. The meaning of anomaly in this context is that the classical invariance of the equations of motion

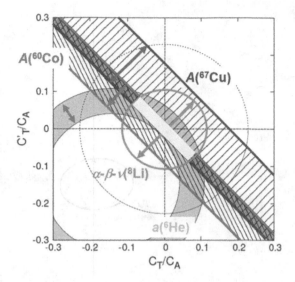

Figure 3.2. As an example, the limits on tensor coupling constants C_T and C_T' relative to the axial vector constant C_A. Four different $\beta - \nu$ correlation experiments have been studied on different isotopes and the overlap of all results is shown. The Standard Model predicts $C_T = C_T' = 0$ (from [Sev17]).

or, equivalently, the Lagrangian no longer exists in quantum field theoretical (QFT) perturbation theory. The reason for this arises from the fact that in such a case a consistent renormalization procedure cannot be found.

3.2.1 The gauge principle

The gauge principle can be explained by the example of classical electrodynamics. It is based on Maxwell's equations and the electric and magnetic fields - measurable quantities which can be represented as the components of the field-strength tensor $F_{\mu\nu} = \partial_\mu A_\nu - \partial_\nu A_\mu$. Here the four-potential A_μ is given by $A_\mu = (\phi, \mathbf{A})$, and the field strengths are derived from it as $\mathbf{E} = -\nabla\phi - \partial_t \mathbf{A}$ and $\mathbf{B} = \nabla \times \mathbf{A}$. If $\rho(t, \mathbf{x})$ is a well-behaved, differentiable real function, it can be shown that under a transformation of the potential such as

$$\phi'(t, \mathbf{x}) = \phi(t, \mathbf{x}) - \partial_t \rho(t, \mathbf{x}) \tag{3.8}$$

$$\mathbf{A}'(t, \mathbf{x}) = \mathbf{A}(t, \mathbf{x}) + \nabla\rho(t, \mathbf{x}) \tag{3.9}$$

all observable quantities remain invariant. The fixing of ϕ and $\mathbf{A}$ to particular values in order to, for example, simplify the equations of motion, is called *fixing the gauge*.

In gauge theories, this gauge freedom for certain quantities is raised to a fundamental principle. The existence and structure of interactions are determined

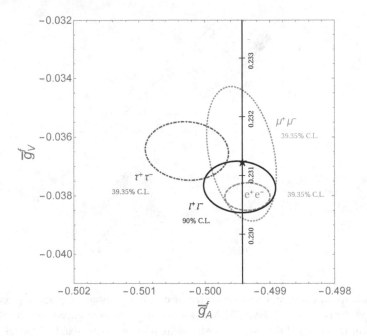

Figure 3.3. Lepton universality as probed in e^+e^- colliders at CERN and SLAC. Shown are 1σ contours (39.35 % for the Z-pole observables g_V^f and g_A^f for $f = e, \mu, \tau$. The best SM fit for the Weinberg angle (black line) is 0.23122. Also shown is the 90% CL under the assumption of universal coupling (from [PDG18]). With kind permission of M. Tanabashi *et al.* (Particle Data Group).

by the demand for such gauge-fixable but physically undetermined quantities. The inner structure of the gauge transformation is specified through a symmetry group.

As mentioned before, symmetries and behaviour under symmetry operations play a crucial role and will be considered next.

3.2.2   Global symmetries

Internal symmetries can be subdivided into discrete and continuous symmetries. We will concentrate on continuous symmetries. In quantum mechanics a physical state is described by a wavefunction $\psi(\boldsymbol{x}, t)$. However, only the modulus squared appears as a measurable quantity. This means that $\psi(\boldsymbol{x}, t)$ as well as the functions

$$\psi'(\boldsymbol{x}, t) = \mathrm{e}^{\mathrm{i}\alpha}\psi(\boldsymbol{x}, t) \tag{3.10}$$

are also solutions of the Schrödinger equation, where α is a real (space and time independent) function. This is called a *global symmetry* and relates to the space and time independence of α. Consider the wavefunction of a charged particle such as the electron. The relativistic equation of motion for a spin 1/2 object is the Dirac equation (Chapter 2.1):

$$i\gamma^\mu \partial_\mu \psi_e(\boldsymbol{x}, t) - m\psi_e(\boldsymbol{x}, t) = 0. \tag{3.11}$$

The invariance under the global transformation

$$\psi_e'(\boldsymbol{x}, t) = e^{ie\alpha} \psi_e(\boldsymbol{x}, t) \tag{3.12}$$

where e is a constant (in this case, the electric charge), is obvious:

$$e^{ie\alpha} i\gamma^\mu \partial_\mu \psi_e(\boldsymbol{x}, t) = e^{ie\alpha} m\psi_e(\boldsymbol{x}, t)$$
$$\Rightarrow i\gamma^\mu \partial_\mu e^{ie\alpha} \psi_e(\boldsymbol{x}, t) = m e^{ie\alpha} \psi_e(\boldsymbol{x}, t)$$
$$i\gamma^\mu \partial_\mu \psi_e'(\boldsymbol{x}, t) = m\psi_e'(\boldsymbol{x}, t). \tag{3.13}$$

Instead of discussing symmetries of the equations of motion, the Lagrangian $\mathcal{L}$ is often used. The equations of motion of a theory can be derived from the Lagrangian $\mathcal{L}(\phi, \partial_\mu \phi)$ with the help of the principle of least action (see, e.g., [Gol80]). For example, consider a real scalar field $\phi(x)$. Its free Lagrangian is

$$\mathcal{L}(\phi, \partial_\mu \phi) = \tfrac{1}{2}(\partial_\mu \phi \partial^\mu \phi - m^2 \phi^2). \tag{3.14}$$

From the requirement that the action integral S is stationary

$$\delta S[x] = 0 \quad \text{with } S[x] = \int \mathcal{L}(\phi, \partial_\mu \phi) \, dx^\mu \tag{3.15}$$

the equations of motion can be obtained:

$$\partial_\alpha \frac{\partial \mathcal{L}}{\partial(\partial_\alpha \phi)} - \frac{\partial \mathcal{L}}{\partial \phi} = 0. \tag{3.16}$$

The Lagrangian clearly displays certain symmetries of the theory. In general, it can be shown that the invariance of the field $\phi(x)$ under certain symmetry transformations results in the conservation of a four-current, given by

$$\partial_\alpha \left(\frac{\partial \mathcal{L}}{\partial(\partial_\alpha \phi)} \delta\phi \right) = 0. \tag{3.17}$$

This is known as *Noether's theorem* [Noe18]. Using this expression, time, translation and rotation invariance imply the conservation of energy, momentum and angular momentum, respectively. We now proceed to consider the differences introduced by local symmetries, in which α in Equation (3.10) is no longer a constant but shows a space and time dependence.

3.2.3 Local (= gauge) symmetries

If the requirement for space and time independence of α is neglected, the symmetry becomes a local symmetry. It is obvious that under transformations such as

$$\psi_e'(x) = e^{ie\alpha(x)} \psi_e(x) \tag{3.18}$$

the Dirac equation (3.11) does not remain invariant:

$$(i\gamma^\mu\partial_\mu - m)\psi'_e(x) = e^{ie\alpha(x)}[(i\gamma^\mu\partial_\mu - m)\psi_e(x) - e(\partial_\mu\alpha(x))\gamma^\mu\psi_e(x)]$$
$$= -e(\partial_\mu\alpha(x))\gamma^\mu\psi'_e(x) \neq 0. \tag{3.19}$$

The field $\psi'_e(x)$ is, therefore, not a solution of the free Dirac equation. If it would be somehow possible to compensate the additional term, the original invariance could be restored. This can be achieved by introducing a new auxiliary gauge field A_μ, which transforms itself in such a way that it compensates for the extra term. In order to achieve this, it is necessary to perform a transformation of

$$\partial_\mu \to D_\mu = \partial_\mu - ieA_\mu. \tag{3.20}$$

with D_μ being the covariant derivative. The invariance can be restored if all partial derivatives ∂_μ are replaced by the covariant derivative D_μ. The Dirac equation then becomes

$$i\gamma^\mu D_\mu\psi_e(x) = i\gamma^\mu(\partial_\mu - ieA_\mu)\psi_e(x) = m\psi_e(x). \tag{3.21}$$

If one now uses the transformed field $\psi'_e(x)$, it is straightforward to see that the original invariance of the Dirac equation can be restored if the gauge field transforms itself according to

$$A_\mu \to A_\mu + \partial_\mu\alpha(x). \tag{3.22}$$

Equations (3.18) and (3.22) describe the transformation of the wavefunction and the gauge field. They are, therefore, called *gauge transformations*. The whole electrodynamics can be described in this way as a consequence of the invariance of the Lagrangian $\mathcal{L}$ or, equivalently, the equations of motion, under phase transformations $e^{ie\alpha(x)}$. The resulting conserved quantity is the electric charge, e. The corresponding theory is called quantum electrodynamics (QED) and, as a result of its enormous success, it has become a paradigm of gauge theories. In the transition to classical physics, the gauge field A_μ becomes the classical vector potential of electrodynamics. The gauge field can be associated with the photon, which takes over the role of an exchange particle. It is found that generally in all gauge theories the gauge fields have to be massless. This is logical because, in case of a photon, a mass term would be proportional to $m_\gamma^2 A_\mu A^\mu$, which is not invariant. Hence, such a mass term cannot be used and any required masses have to be built in subsequently. The case discussed here corresponds to the gauge theoretical treatment of electrodynamics. In group-theory the multiplication with a phase factor (here $e^{i\alpha}$), which was used to explain gauge invariance, can be described by a unitary transformation, in this case the U(1) group. It has the unity operator as generator. The gauge principle can be generalized for Abelian gauge groups, i.e., groups whose generators commute with each other. It becomes somewhat more complex in the case of non-Abelian groups, as will be shown in the next section.

3.2.4   Non-Abelian gauge theories (= Yang–Mills theories)

Non-Abelian means that the generators of the groups no longer commute, but are subject to certain commutator relations and result in non-Abelian gauge theories

(Yang–Mills theories) [Yan54]. An example of operators with such commutator relations are the Pauli spin matrices σ_i,

$$[\sigma_i, \sigma_j] = i\hbar\sigma_k \tag{3.23}$$

which act as generators for the special unitary group SU(2). Generally SU(N) groups possess $N^2 - 1$ generators as S requires the determinant to be $+1$. A representation of the SU(2) group is all unitary 2×2 matrices with determinant $+1$. Consider the electron and neutrino as an example. Apart from their electric charge and their mass, these two particles behave identically with respect to the weak interaction, and one can imagine transformations such as

$$\left(\begin{array}{c} \psi_e(x) \\ \psi_\nu(x) \end{array} \right)' = U(x) \left(\begin{array}{c} \psi_e(x) \\ \psi_\nu(x) \end{array} \right) \tag{3.24}$$

where the transformation can be written as

$$U(a_1, a_2, a_3) = e^{i\frac{1}{2}(a_1\sigma_1 + a_2\sigma_2 + a_3\sigma_3)} = e^{i\frac{1}{2}a(x)\sigma}. \tag{3.25}$$

The particles are generally arranged in multiplets of the corresponding group (in (3.24) they are arranged as doublets). Using the Dirac equation and substituting the normal derivative for the covariant derivative by introducing a gauge field $W_\mu(x)$ and a quantum number g in analogy to (3.20)

$$D_\mu = \partial_\mu + \frac{ig}{2}W_\mu(x) \cdot \sigma \tag{3.26}$$

does *not* lead to gauge invariance. Rather, because of the non-commutation of the generators, an additional term results, an effect which did not appear in the electromagnetic interaction. Only transformations of the gauge fields such as

$$W_\mu \to W_\mu + \frac{1}{g}\partial_\mu a(x) - W_\mu \times a(x) \tag{3.27}$$

supply the desired invariance. (Note the difference compared with (3.22).) The non-commutation of the generators causes the exchange particles to carry 'charge' themselves (contrary to the case of the photon, which does not carry an electric charge) because of this additional term. Among other consequences, this results in a self-coupling of the exchange particles. We now discuss in more detail the non-Abelian gauge theories of the electroweak and strong interaction, which are unified in the *Standard Model of elementary particle physics*. As the main interest of this book lies in neutrinos, we will concentrate on the electroweak part of the Standard Model.

3.3   The Glashow–Weinberg–Salam model

We now consider a treatment of electroweak interactions in the framework of gauge theories. The exposition will be restricted to an outline; for a more detailed discussion

Table 3.3. (*a*) Properties of the quarks ordered with increasing mass: I, isospin and its third component I_3; S, strangeness; C, charm; Q, charge; B, baryon number; B^*, bottom; T, top. (*b*) Properties of leptons. L_i flavour-related lepton number, $L = \sum_{i=e,\mu,\tau} L_i$.

(*a*) Flavour	Spin	B	I	I_3	S	C	B^*	T	$Q[e]$
u	1/2	1/3	1/2	1/2	0	0	0	0	2/3
d	1/2	1/3	1/2	−1/2	0	0	0	0	−1/3
c	1/2	1/3	0	0	0	1	0	0	2/3
s	1/2	1/3	0	0	−1	0	0	0	−1/3
t	1/2	1/3	0	0	0	0	0	1	2/3
b	1/2	1/3	0	0	0	0	−1	0	−1/3

(*b*) Lepton	$Q[e]$	L_e	L_μ	L_τ	L
e^-	−1	1	0	0	1
ν_e	0	1	0	0	1
μ^-	−1	0	1	0	1
ν_μ	0	0	1	0	1
τ^-	−1	0	0	1	1
ν_τ	0	0	0	1	1

see the standard textbooks, for example [Hal84, Gre86, Ait89, Nac90, Don92, Mar92, Lea96, Per00, Lan09, Tho13].

Theoretically, the Standard Model group corresponds to a direct product of three groups, SU(3)⊗SU(2)⊗U(1), where SU(3) belongs to the colour group of quantum chromodynamics (QCD), SU(2) to the weak isospin and U(1) belongs to the hypercharge. The particle content with its corresponding quantum numbers is given in Table 3.3. The electroweak SU(2) ⊗ U(1) section, called the Glashow–Weinberg–Salam (GWS) model [Gla61, Sal64, Wei67, Sal68] or quantum flavour dynamics (QFD) consists of the weak isospin SU(2) and the hypercharge group U(1). The concept of weak isospin is in analogy to isospin in nuclear physics (see, e.g., [Kra88]). The elementary particles are arranged as doublets for chiral left-handed fields (see Chapter 2) and singlets for right-handed fields in the form

$$
\begin{pmatrix} u \\ d \end{pmatrix}_L, \quad \begin{pmatrix} c \\ s \end{pmatrix}_L, \quad \begin{pmatrix} t \\ b \end{pmatrix}_L, \quad \begin{pmatrix} e \\ \nu_e \end{pmatrix}_L, \quad \begin{pmatrix} \mu \\ \nu_\mu \end{pmatrix}_L, \quad \begin{pmatrix} \tau \\ \nu_\tau \end{pmatrix}_L
$$
$$
u_R \quad d_R \quad s_R \quad c_R \quad b_R \quad t_R \quad e_R \quad \mu_R \quad \tau_R. \tag{3.28}
$$

We want to discuss the theory along the line taken in [Nac90] taking the first generation of the three known chiral lepton fields e_R, e_L and ν_{eL} as an example. An extension to all three generations and quarks is straightforward. Neglecting any mass and switching off weak interactions and electromagnetism, the Lagrangian for the free Dirac fields can be written as

$$\mathcal{L}(x) = (\bar{\nu}_{eL}(x), \bar{e}_L(x))(i\gamma^\mu \partial_\mu)\begin{pmatrix} \nu_{eL}(x) \\ e_L(x) \end{pmatrix} + \bar{e}_R(x)i\gamma_\mu \partial_\mu e_R(x). \tag{3.29}$$

This Lagrangian is invariant with respect to global SU(2) transformations on the fields ν_{eL} and e_L. Going to a local SU(2) transformation, the Lagrangian is not invariant but we can compensate for this fact by introducing a corresponding number of gauge vector fields. In the case of SU(2) we have three generators and, therefore, we need three vector fields called $W^1_\mu, W^2_\mu, W^3_\mu$ (see Section 3.2.4). The Lagrangian including the W-fields can then be written as (see [Nac90])

$$\mathcal{L}(x) = -\tfrac{1}{2}\mathrm{Tr}(W_{\mu\rho}(x)W^{\mu\rho}(x)) + (\bar{\nu}_{eL}(x), \bar{e}_L(x))i\gamma_\mu(\partial_\mu + igW_\mu)\begin{pmatrix} \nu_{eL} \\ e_L \end{pmatrix}$$
$$+ \bar{e}_R(x)i\gamma_\mu \partial_\mu e_R(x). \tag{3.30}$$

The introduced gauge group SU(2) is called the weak isospin. Introducing the fields $W^\pm_\mu$ as

$$W^\pm_\mu = \frac{1}{\sqrt{2}}(W^1_\mu \mp iW^2_\mu) \tag{3.31}$$

from (3.30) the ν–e–W coupling term with a coupling constant g can be obtained as

$$\mathcal{L} = -g(\bar{\nu}_{eL}, \bar{e}_L)\gamma_\mu W_\mu \frac{\sigma}{2}\begin{pmatrix} \nu_{eL} \\ e_L \end{pmatrix}$$
$$= -g(\bar{\nu}_{eL}, \bar{e}_L)\gamma_\mu \frac{1}{2}\begin{pmatrix} W^3_\mu & \sqrt{2}W^+_\mu \\ \sqrt{2}W^-_\mu & -W^3_\mu \end{pmatrix}\begin{pmatrix} \nu_{eL} \\ e_L \end{pmatrix} \tag{3.32}$$
$$= -\frac{g}{2}\{W^3_\mu(\bar{\nu}_{eL}\gamma_\mu\nu_{eL} - \bar{e}_L\gamma_\mu e_L) + \sqrt{2}W^+_\mu \bar{\nu}_{eL}\gamma_\mu e_L + \sqrt{2}W^-_\mu \bar{e}_L\gamma_\mu\nu_{eL}\}$$

with σ as the Pauli matrices. This looks quite promising because the last two terms already have the $\gamma_\mu(\mathbb{1} - \gamma_5)$ structure as discussed in Section 2.1 and provides the coupling of an electron to a ν_e via a W-boson. Hence, by finding a method to make the W-boson very massive, at low energy the theory reduces to the Fermi four-point interaction mentioned at the beginning of this chapter.

Before discussing masses, we want to obtain electromagnetism. The easiest assumption for associating the remaining field W^3_μ with the photon field does not work because W^3_μ couples to neutrinos and not to e_R in contrast to the photon. Going back to (3.29) besides the SU(2) invariance, one can recognize an additional invariance under two further U(1) transformations with quantum numbers y_L, y_R:

$$\begin{pmatrix} \nu_{eL}(x) \\ e_L(x) \end{pmatrix} \to e^{+iy_L\chi}\begin{pmatrix} \nu_{eL}(x) \\ e_L(x) \end{pmatrix} \tag{3.33}$$

$$e_R(x) \to e^{+iy_R\chi}e_R(x). \tag{3.34}$$

However, this would result in two 'photon-like' gauge bosons in contrast to nature where we know there is only one. Therefore, we can restrict ourselves to one special combination of these phase transitions resulting in one U(1) transformation by choosing

$$y_L = -\tfrac{1}{2}. \tag{3.35}$$

y_R is fixed in (3.44). This U(1) group is called the weak hypercharge Y. We can make this U(1) into a gauge group as in QED, where the charge Q is replaced by the weak hypercharge Y. Between charge, hypercharge and the third component of the weak isospin, the following relation holds

$$Q = I_3 + \frac{Y}{2}. \tag{3.36}$$

The associated gauge vector field is called B_μ and the corresponding gauge coupling constant g'. Now we are left with two massless neutral vector fields W_μ^3, B_μ and the question arises as to whether we can combine them in a way to account for weak neutral currents (see Chapter 4) and electromagnetism. Let us define two orthogonal linear combinations resulting in normalized fields Z_μ and A_μ:

$$Z_\mu = \frac{1}{\sqrt{g^2 + g'^2}}(gW_\mu^3 - g'B_\mu) \tag{3.37}$$

$$A_\mu = \frac{1}{\sqrt{g^2 + g'^2}}(g'W_\mu^3 + gB_\mu). \tag{3.38}$$

By writing

$$\sin\theta_W = \frac{g'}{\sqrt{g^2 + g'^2}} \tag{3.39}$$

$$\cos\theta_W = \frac{g}{\sqrt{g^2 + g'^2}} \tag{3.40}$$

we can simplify the expressions to

$$Z_\mu = \cos\theta_W W_\mu^3 - \sin\theta_W B_\mu \tag{3.41}$$

$$A_\mu = \sin\theta_W W_\mu^3 + \cos\theta_W B_\mu. \tag{3.42}$$

The angle $\sin\theta_W$ is called the Weinberg angle (also sometimes called weak angle) and is one of the fundamental parameters of the Standard Model. Replacing the field W_μ^3 in (3.32) by Z_μ and A_μ results in

$$\begin{aligned}
\mathcal{L} = &- \frac{g}{\sqrt{2}}(W_\mu^+ \bar{\nu}_{eL}\gamma_\mu e_L + W_\mu^- \bar{e}_L\gamma_\mu \nu_{eL}) \\
&- \sqrt{g^2 + g'^2} Z_\mu \{ \tfrac{1}{2}\bar{\nu}_{eL}\gamma_\mu \nu_{eL} - \tfrac{1}{2}\bar{e}_L\gamma_\mu e_L \\
&- \sin^2\theta_W(-\bar{e}_L\gamma_\mu e_L + y_R \bar{e}_R\gamma_\mu e_R)\} \\
&- \frac{gg'}{\sqrt{g^2 + g'^2}} A_\mu(-\bar{e}_L\gamma_\mu e_L + y_R \bar{e}_R\gamma_\mu e_R).
\end{aligned} \tag{3.43}$$

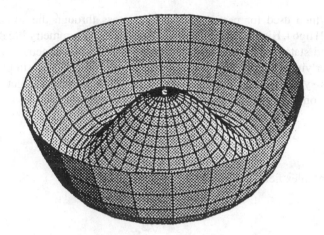

Figure 3.4. Schematic view of the Higgs potential ("Mexican hat") and its minimum for $\mu^2 < 0$ [Moh91] © 1991 World Scientific.

One can note that the Z_μ coupling results in neutral currents. Furthermore, A_μ no longer couples to neutrinos and is, therefore, a good candidate to be associated with the photon field. To reproduce electromagnetism, we have to choose the following

$$y_R = -1 \qquad \frac{gg'}{\sqrt{g^2 + g'^2}} = e \qquad (3.44)$$

which immediately yields another important relation by using (3.39)

$$\sin \theta_W = \frac{e}{g}. \qquad (3.45)$$

This finally allows us to write the Lagrangian using electromagnetic, weak charged and neutral currents:

$$\mathcal{L} = -e\left\{ A_\mu J_{em} + \frac{1}{\sqrt{2}\sin\theta_W}(W_\mu^+ \bar{\nu}_{eL}\gamma_\mu e_L + W_\mu^- \bar{e}_L \gamma_\mu \nu_{eL}) \right.$$
$$\left. + \frac{1}{\sin\theta_W \cos\theta_W} Z_\mu J_{NC}^\mu \right\} \qquad (3.46)$$

with the currents

$$J_{em}^\mu = -\bar{e}_L \gamma_\mu e_L - \bar{e}_R \gamma_\mu e_R = -\bar{e}\gamma_\mu e \qquad (3.47)$$
$$J_{NC}^\mu = \tfrac{1}{2}\bar{\nu}_{eL}\gamma_\mu \nu_{eL} - \tfrac{1}{2}\bar{e}_L \gamma_\mu e_L - \sin^2\theta_W J_{em}^\mu. \qquad (3.48)$$

3.3.1 Spontaneous symmetry breaking and the Higgs mechanism

In the formulation of the theory in the previous sections, all particles had to be massless to guarantee gauge invariance. The concept of spontaneous symmetry breaking is then used for particles to receive mass through the so-called *Higgs mechanism* [Hig64, Hig64a, Eng64, Kib67]. Spontaneous symmetry breaking results in the ground state of a system having no longer the full symmetry corresponding to the underlying Lagrangian. In the electroweak model, the simplest way of spontaneous symmetry breaking is achieved by introducing a doublet of complex scalar fields, one charged, one neutral:

$$\phi = \begin{pmatrix} \phi^\dagger \\ \phi^0 \end{pmatrix} \tag{3.49}$$

where the complex fields are given by

$$\phi^\dagger = \frac{\phi_1 + i\phi_2}{\sqrt{2}}$$

$$\phi^0 = \frac{\phi_3 + i\phi_4}{\sqrt{2}}. \tag{3.50}$$

Adding a kinetic term to the potential (3.49) leads to the following expression for the Lagrangian:

$$\mathcal{L}_{\text{Higgs}} = (\partial_\mu \phi)^\dagger (\partial^\mu \phi) - \mu^2 \phi^\dagger \phi - \lambda (\phi^\dagger \phi)^2. \tag{3.51}$$

Proceeding as before, the potential $V(\phi)$ has a minimum for $\mu^2 < 0$ at

$$\phi^\dagger \phi = \frac{-\mu^2}{2\lambda} = \frac{v^2}{2}. \tag{3.52}$$

Here again the minima, corresponding to the vacuum expectation values of ϕ, lie on a circle with $\langle \phi \rangle \equiv v/\sqrt{2} = \sqrt{-\mu^2/2\lambda}$. This ground state is degenerate and its orientation in two-dimensional isospin space is not defined. It can choose any value between $[0, 2\pi]$. From this infinite number of possible orientations we choose a particular field configuration which is defined as the vacuum state as

$$\phi_0 = \frac{1}{\sqrt{2}} \begin{pmatrix} 0 \\ v \end{pmatrix} \tag{3.53}$$

which is no longer invariant under SU(2) transformations. The upper component is motivated by the fact that the vacuum is electrically neutral. The field $\phi(x)$ can now be expanded around the vacuum

$$\phi = \frac{1}{\sqrt{2}} \begin{pmatrix} 0 \\ v + H(x) \end{pmatrix} \tag{3.54}$$

where a perturbation theory for $H(x)$ can be formulated as usual. Now consider the coupling of this field to fermions first. Fermions get their masses through coupling to the vacuum expectation value (vev) of the Higgs field. To conserve isospin invariance of the coupling, the Higgs doublet has to be combined with a fermion doublet and singlet. The resulting coupling is called Yukawa coupling and has the typical form (given here for the case of electrons)

$$
\begin{aligned}
\mathcal{L}_{\text{Yuk}} &= -c_e \left[\bar{e}_R \phi_0^\dagger \begin{pmatrix} \nu_{eL} \\ e_L \end{pmatrix} + h.c. \right] \\
&= -c_e \left[\bar{e}_R \phi_0^\dagger \begin{pmatrix} \nu_{eL} \\ e_L \end{pmatrix} + (\bar{\nu}_e, \bar{e}_L)\phi_0 e_R \right] \\
&= -c_e \left[\bar{e}_R \frac{1}{\sqrt{2}} v e_L + \bar{e}_L \frac{1}{\sqrt{2}} v e_R \right] \\
&= -c_e v \frac{1}{\sqrt{2}} (\bar{e}_R e_L + \bar{e}_L e_R) \\
&= -c_e \frac{v}{\sqrt{2}} \bar{e}e.
\end{aligned}
\tag{3.55}
$$

Here c_e is an adjustable coupling constant. The expression corresponds exactly to a mass term for the electron with an electron mass of

$$
m_e = c_e \frac{v}{\sqrt{2}}.
\tag{3.56}
$$

The same strategy holds for the other charged leptons and quarks with their corresponding coupling constant c_i, which are not predicted by theory but determined from experiment. In this way, fermions obtain their masses within the GWS model.

Neutrinos remain massless because with the currently accepted particle content there are no right-handed ν_R singlet states and one cannot write couplings like (3.55). With the evidence for massive neutrinos described later, one is forced to include right-handed neutrino singlets ν_R to the Standard Model particles or generate the masses in another way, for example using Higgs triplets (see Chapter 5).

Substituting the covariant derivative for the normal derivative in $\mathcal{L}$ as in (3.20) leads directly to the coupling of the Higgs field with the gauge fields. For details see [Nac90, Gun90]. The gauge bosons then acquire masses of

$$
m_W^2 = \frac{g^2 v^2}{4} = \frac{e^2 v^2}{4 \sin^2 \theta_W}
\tag{3.57}
$$

$$
m_Z^2 = \frac{(g^2 + g'^2)v^2}{4} = \frac{e^2 v^2}{4 \sin^2 \theta_W \cos^2 \theta_W}
\tag{3.58}
$$

resulting in

$$
\frac{m_W}{m_Z} = \cos \theta_W.
\tag{3.59}
$$

Hence the relation between charged current and neutral current couplings (see Chapter 4) is determined by the masses of the Z- and W-boson. An interesting quantity deduced from this relation is the ρ-parameter defined as

$$
\rho = \frac{m_W^2}{m_Z^2 \cos^2 \theta_W}.
\tag{3.60}
$$

which is determined by the Higgs structure. In the Standard Model with only one Higgs doublet $\rho = 1$. Thus, any experimental deviation from $\rho = 1$ would be a hint for new physics. The current value is given as $\rho = 1.00039 \pm 0.00019$ [PDG18].

An estimate for v can be given by (3.57) in lowest order perturbation theory resulting in

$$v = (\sqrt{2}G_F)^{-1/2} \approx 246 \text{ GeV}. \tag{3.61}$$

The inclusion of spontaneous symmetry breaking with the help of a complex scalar field doublet has another consequence, namely the existence of a new scalar particle called the Higgs boson, with a mass of m_H, such that

$$m_H^2 = 2\lambda v^2. \tag{3.62}$$

To obtain invariance under hypercharge transformations, we have to assign a hypercharge of $y_H = 1/2$ to the Higgs. In 2012 the Higgs boson has been discovered at the LHC and by now a mass of 125.09 GeV is obtained [PDG18], which would result in values of $\lambda \approx 0.13$ and $\mid m_H \mid \approx 88.8$ GeV (without loop corrections, these can be found in [But13]). For more details see Section 3.4.5.

3.3.2   The CKM mass matrix

It has been experimentally proved that the mass eigenstates for quarks are not identical to flavour eigenstates, which manifests itself by the difference in the Fermi-constant between neutron and muon decay. This is shown by the fact that transitions between the various generations are observed. Thus, the mass eigenstates of the d and s quarks are not identical to the flavour eigenstates d' and s', which take part in the weak interaction. They are connected via

$$\begin{pmatrix} d' \\ s' \end{pmatrix} = \begin{pmatrix} \cos\theta_C & \sin\theta_C \\ -\sin\theta_C & \cos\theta_C \end{pmatrix} \begin{pmatrix} d \\ s \end{pmatrix}. \tag{3.63}$$

The *Cabibbo angle* θ_C is about $13°$ ($\sin\theta_C = 0.222 \pm 0.003$). The extension to three generations leads to the so-called *Cabibbo–Kobayashi–Maskawa matrix* (CKM) [Kob73]

$$\begin{pmatrix} d' \\ s' \\ b' \end{pmatrix} = \begin{pmatrix} V_{ud} & V_{us} & V_{ub} \\ V_{cd} & V_{cs} & V_{cb} \\ V_{td} & V_{ts} & V_{tb} \end{pmatrix} \times \begin{pmatrix} d \\ s \\ b \end{pmatrix} = U \times \begin{pmatrix} d \\ s \\ b \end{pmatrix} \tag{3.64}$$

which can be parametrized with three mixing angles $\theta_{12}, \theta_{13}, \theta_{23}$ and a single complex phase $e^{i\delta}$:

$$U = \begin{pmatrix} c_{12}c_{13} & s_{12}c_{13} & s_{13}e^{-i\delta} \\ -s_{12}c_{23} - c_{12}s_{23}s_{13}e^{i\delta} & c_{12}c_{23} - s_{12}s_{23}s_{13}e^{i\delta} & s_{23}c_{13} \\ s_{12}s_{23} - c_{12}s_{23}s_{13}e^{i\delta} & -c_{12}s_{23} - s_{12}c_{23}s_{13}e^{i\delta} & c_{23}c_{13} \end{pmatrix} \tag{3.65}$$

where $s_{ij} = \sin\theta_{ij}, c_{ij} = \cos\theta_{ij}$ ($i, j = 1, 2, 3$). The individual matrix elements describe transitions between the different quark flavours and have to be determined experimentally. The current experimental results in combination with the constraint of unitarity of U give the values (90% CL) [PDG18]:

$$|U| = \begin{pmatrix} 0.97446 \pm 0.00010 & 0.22452 \pm 0.00044 & 3.65 \pm 0.12 \times 10^{-3} \\ 0.22438 \pm 0.00044 & 0.97359^{+0.00010}_{-0.00011} & 42.14 \pm 0.76 \times 10^{-3} \\ 8.96^{+0.00024}_{-0.00023} \times 10^{-3} & 41.33 \pm 0.74 \times 10^{-3} & 999.106 \pm 0.032 \times 10^{-3} \end{pmatrix}.$$

$$(3.66)$$

The Wolfenstein parametrization of U [Wol83] (an expansion with respect to $\lambda = \sin\theta_{12}$ accurate up to the third order in λ)

$$U = \begin{pmatrix} 1 - \frac{1}{2}\lambda^2 & \lambda & A\lambda^3(\rho - i\eta) \\ -\lambda & 1 - \frac{1}{2}\lambda^2 & A\lambda^2 \\ A\lambda^3(1 - \rho - i\eta) & -A\lambda^2 & 1 \end{pmatrix}, \qquad (3.67)$$

is useful. This parametrization assumes hierarchical matrix elements, with the diagonal terms being the strongest. This is fulfilled in the quark sector as can be seen in (3.66), but such a structure is not realized in the leptonic sector as we will see later. Useful concept are geometrical presentations in the complex (η, ρ) plane called unitarity triangles (Figure 3.5). The relations form triangles in the complex plane, with the feature that all triangles have the same area. The unitarity of the CKM matrix leads to various relations among the matrix elements where, in particular,

$$V_{ud}V_{ub}^* + V_{cd}V_{cb}^* + V_{td}V_{tb}^* = 0 \qquad (3.68)$$

is usually quoted as 'the unitarity triangle'. The third vertex is then given by the Wolfenstein parameters (ρ, η). A rescaled triangle is obtained by making $V_{cd}V_{cb}^*$ real (one side is then aligned to the real axis) and dividing the lengths of all sides by $V_{cd}V_{cb}^*$ (giving the side along the real axis length 1), which changes η, ρ to $\bar{\rho}, \bar{\eta}$, see [PDG16] for more information. Two vertices are then fixed at (0,0) and (1,0). With all the available data, one finds [PDG16] that

$$A = 0.836 \pm 0.015 \qquad \lambda = 0.22453 \pm 0.00044 \qquad (3.69)$$

$$\bar{\rho} = 0.122^{+0.018}_{-0.017} \qquad \bar{\eta} = 0.355 \pm^{+0.012}_{-0.011} \qquad (3.70)$$

$$\sin 2\beta = 0.691 \pm 0.017 \qquad \alpha = (84.5^{+5.9}_{-5.2})^o \qquad \gamma = (73.5^{+4.2}_{-5.1})^o$$

3.3.3 *CP* violation

The phase $e^{i\delta}$ in (3.65) can be linked to CP violation. The necessary condition for CP invariance of the Lagrangian is that the CKM matrix and its complex conjugate are identical; i.e., its elements are real. While this is always true for two families, for three families it is true in the previous parametrization only if $\delta = 0$ or $\delta = \pi$. This means that if δ does not equal one of those values, then CP violation occurs due to quark mixing (see, e.g., [Nac90]). The first evidence of CP violation has been observed in the kaon system [Chr64]. The experimentally observed particles K_S and K_L are only approximately identical to the CP eigenstates K_1 and K_2, so

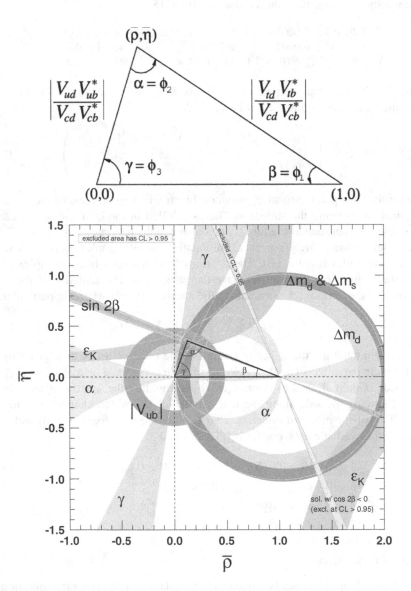

Figure 3.5. Top: Schematic picture of the unitarity triangle in the complex plane using the Wolfenstein parameters η, ρ. Bottom: Existing experimental limits constraining the apex of the triangle and the three angles α, β and γ. The shaded areas have 95 % CL (from [PDG18]). With kind permission of M. Tanabashi *et al.* (Particle Data Group).

that it is necessary to define the observed states K_L ($\simeq K_2$) and K_S ($\simeq K_1$) as (see, e.g., [Com83]):

$$|K_S\rangle = (1 + |\epsilon|^2)^{-1/2}(|K_1\rangle - \epsilon|K_2\rangle) \tag{3.71}$$
$$|K_L\rangle = (1 + |\epsilon|^2)^{-1/2}(|K_2\rangle + \epsilon|K_1\rangle). \tag{3.72}$$

CP violation caused by this mixing can be characterized by the parameter ϵ. The ratio of the amplitudes for the decay into two charged pions may be used as a measure of CP violation [Per00, PDG18]:

$$|\eta_{+-}| = \frac{A(K_L \to \pi^+\pi^-)}{A(K_S \to \pi^+\pi^-)} = (2.232 \pm 0.011) \times 10^{-3}. \tag{3.73}$$

A similar relation is obtained for the decay into two neutral pions, characterized in analogy as η_{00}. The ϵ appearing in Equations (3.71) and (3.72), together with a further parameter ϵ', can be connected with η via the relation

$$\eta_{+-} = \epsilon + \epsilon' \tag{3.74}$$
$$\eta_{00} = \epsilon - 2\epsilon' \tag{3.75}$$

from which it can be deduced (see, e.g., [Com83]) that

$$\left|\frac{\eta_{00}}{\eta_{+-}}\right| \approx 1 - 3\,\mathrm{Re}\left(\frac{\epsilon'}{\epsilon}\right). \tag{3.76}$$

Evidence for a non-zero ϵ' would show that CP is violated *directly* in the decay, i.e., in processes with change in strangeness $\Delta S = 1$, and does not depend only on the existence of mixing [Com83]. Indeed ϵ' is different from zero and its current value is [PDG18]

$$\mathrm{Re}(\epsilon'/\epsilon) = (1.67 \pm 0.23) \times 10^{-3}. \tag{3.77}$$

Other important kaon decays that will shed some light on CP violation are the decays $K^+ \to \pi^+\nu\bar{\nu}$ and $K_L \to \pi^0\nu\bar{\nu}$ which have small theoretical uncertainties. Several events of the first reaction have been observed in experiments at BNL [Adl00, Ani04, Art08] with large uncertainties. Also first data of the NA62 experiment have been released, showing one candidate event [Vel18].

CP violation has also been observed in B-meson decays. The gold-plated channel for investigation is $B_d \to J/\Psi + K_S$ because of the combination of the experimentally clean signature and exceedingly small theoretical uncertainties. It allows a measurement of $\sin(2\beta)$. The B factories at SLAC at Stanford (BaBar experiment [Bab95]) and at KEK in Japan (Belle experiment [Bel95]) have observed CP violation and $B^0 - \bar{B}^0$ oscillations in the B meson systems and provide very important results [Aub04, Abe05]. In 2010 anomalous CP violation in the B-meson systems has been observed [Aba10]. Also the equivalent of the long known $K^0 - \bar{K}^0$ oscillations have been observed in D^0- and B^0-mesons, whose amplitude is sensitive to new physics. With the observation of the up-type quarks D-meson oscillations [Aai18], complementary information with respect to strange and beauty oscillations will be achieved.

3.3.4   *CPT* and *T* violation

CP violation implies also T violation, as the application of the combination of the operators CPT in any ordering is invariant. This is a fundamental condition for quantum field theory (QFT). A classic test for T-violation is a potential static electric dipole moment of particles like the neutron or electron. Such a dipole moment would violate CP and thus also T if CPT is conserved. The current upper limits are $\mid d_n \mid \leq 3.0 \times 10^{-26}$ ecm (90% CL) [Pen15] while it is for the electron $\mid d_e \mid \leq 8.7 \times 10^{-29}$ ecm (90% CL) [Bar14]. The Standard Model predictions are about 5 or 10 orders of magnitude smaller respectively, but nevertheless these limits impose strong constraints on Beyond Standard Model physics. Furthermore, besides more sensitive experiments in the future also new experiments on the muon EDM are prepared at Fermilab and J-PARC.

T violation was directly observed for the first time in the kaon system by the CPLEAR experiment at CERN [Ang98], but the real T violation was observed in the $B^0 - \bar{B}^0$ system again [Lee12], see also [Ber15].

The newest developments of testing CPT violation are using anti-hydrogen. Here most tests rely either on optical transitions between different levels in atomic shells in comparison with normal hydrogen or on the comparison of nuclear properties like magnetic moments. These measurements using anti-hydrogen became feasible at the Antiproton Deccelerator (AD) at CERN. The BASE experiment measured the charge-to-mass ratio of proton and antiproton to agree to 69 parts per trillion [Ulm15]. A year later, the ASACUSA experiment has made a measurement of the antiproton-electron mass ratio which agrees on the level of 8×10^{-10} for the same ratio using protons. Strong tests could be performed by using optical precision spectroscopy using the classic atomic (1S-2S) hydrogen transition [Ahm16], which corresponds to an energy sensitivity of 10^{-23} GeV and the hyperfine structure transition with an agreement of 4 parts in 10000 with respect to the classic hydrogen hyperfine transition [Ahm17]. Meanwhile also the line shape has been measured, which improves the precision even more [Ahm18]. Last but not least, the magnetic moment of the antiproton by the BASE collaboration at CERN made a measurement which was better than the existing proton magnetic moment measurement and explores energy scales of less than 1.8×10^{-24} GeV [Smo17]. Far more results are expected in the near future.

In the leptonic sector, the issue could be similar: massive neutrinos will lead to a CKM-like matrix in the leptonic sector often called the Pontecorvo–Maki–Nakagawa–Sakata (PMNS) matrix [Pon60, Mak62] and, therefore, to CP violation. Furthermore, if neutrinos are Majorana particles, there would already be the possibility of CP violation with two families. For three flavours there will be three phases that will show up [Wol81] (see Chapter 5). A chance to probe one phase of CP violation in the leptonic sector exists with the planned neutrino factories (see Chapter 4). The Majorana phases have direct impact on the observables in neutrinoless double β-decay (see Chapter 7).

3.4 Experimental determination of fundamental parameters

Although it has been extraordinarily successful, not everything can be predicted by the Standard Model. In fact it has 18 free parameters as input, all of which have to be measured (see Chapter 5). This is excluding neutrino masses and mixing, which we know to exist as well. A few selected measurements are discussed now.

3.4.1 Measurement of the Fermi constant G_F

The Fermi constant G_F has been of fundamental importance in the history of weak interaction. Within the context of the current GWS model, it can be expressed as

$$\frac{G_F}{\sqrt{2}} = \frac{g^2}{8m_W^2}. \tag{3.78}$$

In the past the agreement of measurements of G_F in β-decay (now called G_β) and in μ-decay (now called G_μ) led to the hypothesis of conserved vector currents (CVC hypothesis, see Section 3.1); nowadays, the measurements can be used to test the universality of weak interactions. A small deviation between the two is expected because of the Cabibbo-mixing, which results in

$$\frac{G_\beta}{G_\mu} \simeq \cos\theta_C \approx 0.98. \tag{3.79}$$

In general, precision measurements of the fundamental constants including the Fermi constant, allow us to restrict the physics beyond the Standard Model [Her95, Mar99].

The best way to determine G_F which can also be seen as a definition of G_F ($G_F := G_\mu$) is the measurement of the muon lifetime τ:

$$\tau^{-1} = \Gamma(\mu \to e\nu_\mu\nu_e) = \frac{G_F^2 m_\mu^5}{192\pi^3}(1 + \Delta\rho) \tag{3.80}$$

where $\Delta\rho$ describes radiative corrections. Equation (3.80) can be expressed as [Rit00]

$$\tau^{-1} = \Gamma(\mu \to e\nu_\mu\nu_e) = \frac{G_F^2 m_\mu^5}{192\pi^3} F\left(\frac{m_e^2}{m_\mu^2}\right)\left(1 + \frac{3}{5}\frac{m_\mu^2}{m_W^2}\right)$$
$$\times \left(1 + \frac{\alpha(m_\mu)}{2\pi}\left(\frac{25}{4} - \pi^2\right)\right) \tag{3.81}$$

with ($x = m_e^2/m_\mu^2$)

$$F(x) = 1 - 8x - 12x^2 \ln x + 8x^3 - x^4 \tag{3.82}$$

and

$$\alpha(m_\mu)^{-1} = \alpha^{-1} - \frac{2}{3\pi}\left(\ln\frac{m_e}{m_\mu}\right) + \frac{1}{6\pi} \approx 136. \tag{3.83}$$

The second term in (3.81) is an effect of the W propagator and the last term is the leading contribution of the radiative corrections. The most precise experiment with an uncertainty of only 0.5 ppm has been performed at PSI by the MuLan collaboration, resulting in a value of [Web11a, Web11b, Tis13]

$$\frac{G_F}{(\hbar c)^3} = 1.166\,378\,7(6) \times 10^{-5} \text{ GeV}^{-2}. \tag{3.84}$$

This has a high sensitivity to new physics effects, especially when combined with other electroweak precision measurements [Alc07].

3.4.2 Neutrino–electron scattering and the coupling constants g_V and g_A

A fundamental electroweak process to study is νe-scattering, which can be of the form

$$\nu_e e \to \nu_e e \qquad \bar{\nu}_e e \to \bar{\nu}_e e \tag{3.85}$$

$$\nu_\mu e \to \nu_\mu e \qquad \bar{\nu}_\mu e \to \bar{\nu}_\mu e \tag{3.86}$$

$$\nu_\tau e \to \nu_\tau e \qquad \bar{\nu}_\tau e \to \bar{\nu}_\tau e. \tag{3.87}$$

While the first reaction can happen only via neutral current (NC) interactions, for the second both neutral current and charged current (CC) are possible (Figure 3.6); see also [Pan95].

3.4.2.1 Theoretical considerations

The Lagrangian for the first reaction (3.85) is

$$\mathcal{L} = -\frac{G_F}{\sqrt{2}}[\bar{\nu}_\mu\gamma_\alpha(\mathbb{1} - \gamma_5)\nu_\mu][\bar{e}\gamma_\alpha(g_V\mathbb{1} - g_A\gamma_5)e] \tag{3.88}$$

with the prediction from the GWS model of

$$g_V = -\tfrac{1}{2} + 2\sin^2\theta_W \qquad g_A = -\tfrac{1}{2}. \tag{3.89}$$

A similar term can be written for the other types of interaction. In addition, the CC contribution can be written as

$$\mathcal{L} = -\frac{G_F}{\sqrt{2}}[\bar{e}\gamma_\alpha(\mathbb{1} - \gamma_5)\nu_e][\bar{\nu}_e\gamma_\alpha(\mathbb{1} - \gamma_5)e] \tag{3.90}$$

$$= -\frac{G_F}{\sqrt{2}}[\bar{\nu}_e\gamma_\alpha(\mathbb{1} - \gamma_5)\nu_e][\bar{e}\gamma_\alpha(\mathbb{1} - \gamma_5)e] \tag{3.91}$$

where in the second step a Fierz transformation was applied (see [Bil94] for technical information on this). The predictions of the GWS model for the chiral couplings g_L and g_R are:

$$g_L = \tfrac{1}{2}(g_V + g_A) = -\tfrac{1}{2} + \sin^2\theta_W \qquad g_R = \tfrac{1}{2}(g_V - g_A) = \sin^2\theta_W. \tag{3.92}$$

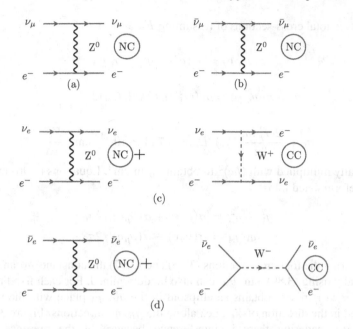

Figure 3.6. Feynman diagrams for neutrino–electron NC and CC reactions: $\nu_\mu e$ NC (*a*), $\bar{\nu}_\mu e$ NC (*b*), $\nu_e e$ NC + CC (*c*) and $\bar{\nu}_e e$ NC + CC scattering (*d*) (from [Sch97]). Reproduced with permission of SNCSC.

A detailed calculation [Sch97] leads to the expected cross-sections which are given by (see also Chapter 4)

$$\frac{d\sigma}{dy}(\overset{(-)}{\nu_\mu} e) = \frac{G_F^2 m_e}{2\pi} E_\nu \bigg[(g_V \pm g_A)^2$$

$$+ (g_V \mp g_A)^2 (1-y)^2 + \frac{m_e}{E_\nu}(g_A^2 - g_V^2)y \bigg] \qquad (3.93)$$

and

$$\frac{d\sigma}{dy}(\overset{(-)}{\nu_e} e) = \frac{G_F^2 m_e}{2\pi} E_\nu \bigg[(G_V \pm G_A)^2 + (G_V \mp G_A)^2 (1-y)^2$$

$$+ \frac{m_e}{E_\nu}(G_A^2 - G_V^2)y \bigg] \qquad (3.94)$$

with $G_V = g_V + 1$ and $G_A = g_A + 1$. The upper(lower) sign corresponds to $\nu e(\bar{\nu}e)$-scattering. The quantity y is called the inelasticity or the Bjorken y and is given by

$$y = \frac{T_e}{E_\nu} \approx \frac{E_e}{E_\nu} \qquad (3.95)$$

where T_e is the kinetic energy of the electron. Therefore, the value of y is restricted to $0 \leq y \leq 1$. The cross-sections are proportional to E_ν. An integration with respect to y leads to total cross-sections of (assuming $E_\nu \gg m_e$)

$$\sigma(\overset{(-)}{\nu_\mu} e) = \sigma_0(g_V^2 + g_A^2 \pm g_V g_A) \tag{3.96}$$

$$\sigma(\overset{(-)}{\nu_e} e) = \sigma_0(G_V^2 + G_A^2 \pm G_V G_A) \tag{3.97}$$

with

$$\sigma_0 = \frac{2G_F^2 m_e}{3\pi}(\hbar c)^2 E_\nu = 5.744 \times 10^{-42} \text{ cm}^2 \frac{E_\nu}{\text{GeV}}, \tag{3.98}$$

additionally multiplied with $(\hbar c)^2$ to obtain σ_0 in cm^2. Equations (3.96) and (3.97) can be reformulated into

$$g_V^2 + g_A^2 = [\sigma(\nu_\mu e) + \sigma(\bar{\nu}_\mu e)]/2\sigma_0 \tag{3.99}$$

$$g_V g_A = [\sigma(\nu_\mu e) - \sigma(\bar{\nu}_\mu e)]/2\sigma_0. \tag{3.100}$$

By measuring the four cross-sections (3.96) and (3.97) the constants g_V and g_A, and additionally using (3.89), $\sin^2 \theta_W$ can also be determined. For each fixed measured value of $\sigma(\nu e)/\sigma_0$ one obtains an ellipsoid in the g_V, g_A plane with the main axis orientated in the direction of 45°, i.e., along the g_R, g_L directions (Figure 3.7).

In $\nu_e e$-scattering there is interference because of the presence of both amplitudes (NC and CC) in the interactions. The cross-sections are given by

$$\sigma(\nu_e e) = (g_V^2 + g_A^2 + g_V g_A)\sigma_0 + 3\sigma_0 + 3(g_V + g_A)\sigma_0 \tag{3.101}$$

$$\sigma(\bar{\nu}_e e) = (g_V^2 + g_A^2 - g_V g_A)\sigma_0 + \sigma_0 + (g_V + g_A)\sigma_0 \tag{3.102}$$

where the interference term is given by

$$I(\nu_e e) = 3I(\bar{\nu}_e e) = 3(g_V + g_A)\sigma_0 = 3(2\sin^2 \theta_W - 1)\sigma_0. \tag{3.103}$$

The small cross-section requires experiments with a large mass and a high intensity neutrino beam. The signature of this type of event is a single electron in the final state. At high energies the electron is boosted in the forward direction and, besides a good energy resolution, a good angular resolution is required for efficient background suppression (see [Pan95] for details).

3.4.2.2 $\nu_\mu e$-scattering

The same experimental difficulties also occur in measuring $\nu_\mu e$-scattering cross-sections. Accelerators provide neutrino beams with energies in the MeV– GeV range (see Chapter 4). Experiments done in the 1980s consisted of calorimeters of more than 100 t mass (CHARM, E734 and CHARM-II), of which CHARM-II has, by far, the largest dataset [Gei93, Vil94]. With good spatial and energy resolution, an efficient background suppression was possible. The dominant background stems from ν_e CC - reactions due to beam contamination with ν_e - and NC- π^0 production, with $\pi^0 \to \gamma\gamma$ which could mimic electrons. The latter can be discriminated either by having a different shower profile in the calorimeter or by having a wider angular

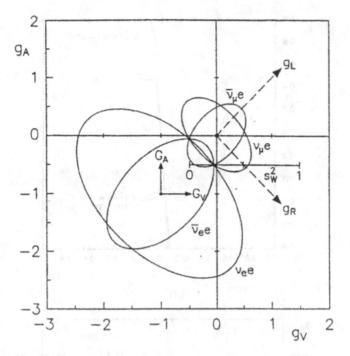

Figure 3.7. Schematic drawing of the four ellipses for fixed $\sigma(\nu e)/\sigma_0$ values in the (g_V, g_A) plane for the various νe-scattering processes. The directions of the g_L and g_R axis under $45°$ are shown as dashed lines and the GWS prediction $-\frac{1}{2} < g_V < \frac{3}{2}, g_A = -\frac{1}{2}$ for $0 < \sin^2 \theta_W < 1$ are also shown (from [Sch97]). Reproduced with permission of SNCSC.

distribution. The results are shown in Figure 3.8. However, there is still an ambiguity which fortunately can be solved by using data from forward–backward asymmetry measurements in elastic e^+e^--scattering (γZ interference) measured at LEP and SLC. The final solution is then [Sch97] (Figure 3.8)

$$g_V = -0.035 \pm 0.017 \qquad g_A = -0.503 \pm 0.017. \qquad (3.104)$$

This is in good agreement with GWS predictions (3.89) assuming $\sin^2 \theta_W = 0.23$.

3.4.2.3 $\nu_e e$ and $\bar{\nu}_e e$-scattering

Results on $\bar{\nu}_e e$-scattering rely on much smaller datasets. Using nuclear power plants as strong $\bar{\nu}_e$ sources, cross-sections of [Rei76]

$$\sigma(\bar{\nu}_e e) = (0.87 \pm 0.25) \times \sigma_0 \qquad 1.5 < E_e < 3.0 \text{ MeV} \qquad (3.105)$$
$$\sigma(\bar{\nu}_e e) = (1.70 \pm 0.44) \times \sigma_0 \qquad 3.0 < E_e < 4.5 \text{ MeV} \qquad (3.106)$$

were obtained, where σ_0 is the predicted integrated V–A cross-section (3.98) folded with the corresponding antineutrino flux.

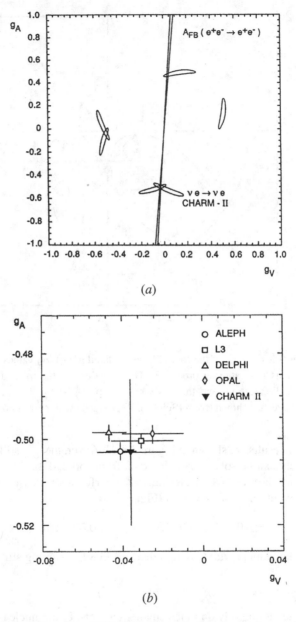

Figure 3.8. (*a*) Allowed regions (90% CL) of combinations in the (g_V, g_A) plane obtained with the CHARM-II data. Only statistical errors are considered. The small straight areas are the regions allowed by forward–backward asymmetry measurements in elastic e^+e^--scattering. Together they select a single solution consistent with $g_A = -\frac{1}{2}$. (*b*) Solution of the ambiguities. Together with the four LEP experiments a unique solution can be found. They are shown together with the CHARM-II result (from [Vil94]). © 1994 With permission from Elsevier.

Elastic $\nu_e e(\bar{\nu}_e e)$-scattering was investigated by E225 at LAMPF [All93]. Using muon-decay at rest, resulting in an average neutrino energy of $\langle E_\nu \rangle = 31.7$ MeV, 236 events were observed giving a cross-section of

$$\sigma(\nu_e e) = (3.18 \pm 0.56) \times 10^{-43} \text{ cm}^2. \tag{3.107}$$

By using $\langle E_\nu \rangle = 31.7$ MeV and the GWS prediction

$$\sigma(\nu_e e) = \sigma_0(\tfrac{3}{4} + 3\sin^2\theta_W + 4\sin^4\theta_W) = 9.49 \times 10^{-42} \text{ cm}^2 \frac{E_\nu}{\text{GeV}} \tag{3.108}$$

and $g_V, g_A = 0$ in (3.101) these are in good agreement with each other. The interference term was determined to be

$$I(\nu_e e) = (-2.91 \pm 0.57) \times 10^{-43} \text{cm}^2 = (-1.60 \pm 0.32)\sigma_0. \tag{3.109}$$

A new measurement was performed by LSND (see Chapter 8) resulting in [Aue01]

$$\sigma(\nu_e e) = [10.1 \pm 1.1(\text{stat.}) \pm 1.0(\text{sys.})] \times 10^{-45} \frac{E_\nu}{\text{MeV}} \tag{3.110}$$

also in good agreement with E225 data and the GWS prediction. With the continuous improvement on the precision of electroweak parameters, there might be valuable information available on physics beyond the Standard Model in the future [deG06].

3.4.2.4 Neutrino tridents

A chance to observe interference for the second generation is given by neutrino trident production (using ν_μ beams), the generation of a lepton pair in the Coulomb field of a nucleus

$$\nu_\mu N \rightarrow \nu_\mu \ell^+ \ell^- N. \tag{3.111}$$

A reduction in the cross-section of about 40% is predicted in the case of interference with respect to pure (V–A) interactions. Searches are done with high-energy neutrino beams (see Chapter 4) for events with low hadronic energy E_{had} and small invariant masses of the $\ell^+ \ell^-$ pair. Trident events were observed in several experiments [Gei90, Mis91]. Here also an interference effect could be observed.

3.4.3 Measurement of the Weinberg angle

One fundamental parameter of the GWS model is the Weinberg angle $\sin^2\theta_W$. In the language of higher-order terms, the definition for $\sin^2\theta_W$ has to be done very carefully [PDG16] because radiative corrections modify the mass and charge on different energy scales (see Chapter 5). The most popular ones are the on-shell and $\overline{\text{MS}}$ definitions (see [PDG16]). The on-shell definition relies on the tree level formula

$$\sin^2\theta_W = 1 - \frac{m_W^2}{m_Z^2} \tag{3.112}$$

obtained by dividing (3.57) and (3.58) so that it is also valid for the renormalized $\sin^2 \theta_W$ in all orders of perturbation theory. The modified minimal subtraction $\overline{\text{MS}}$ scheme (see [Lea96] for details) uses (see (3.39))

$$\sin^2 \theta_W(\mu) = \frac{g'^2(\mu)}{g^2(\mu) + g'^2(\mu)} \quad (3.113)$$

where the coupling constants are defined by modified minimal subtraction and the scale chosen, $\mu = m_Z$, which is convenient for electroweak processes.

The Weinberg angle can be measured in various ways. The measurement of the coupling constants g_V and g_A mentioned in (3.89) provide another way of determinating $\sin^2 \theta_W$. A different possibility is νN-scattering (for more details, see Chapter 4). Here, the NC *versus* CC ratios (see (4.121) and (4.122)) are measured, given by

$$R_\nu = \frac{\sigma_{NC}(\nu\text{N})}{\sigma_{CC}(\nu\text{N})} = \frac{1}{2} - \sin^2 \theta_W + \frac{20}{27} \sin^4 \theta_W \quad (3.114)$$

$$R_{\bar\nu} = \frac{\sigma_{NC}(\bar\nu\text{N})}{\sigma_{CC}(\bar\nu\text{N})} = \frac{1}{2} - \sin^2 \theta_W + \frac{20}{9} \sin^4 \theta_W. \quad (3.115)$$

The classic V-A interaction implies a small parity violation in low-energy measurements which is due to $\gamma - Z^0$ interference, first seen by the E122 experiment at SLAC [Pre78]. This effect is enhanced in atoms so the observation of atomic parity violation is searched for [Mas95, Blu95, Kum13]. From the V-A structure a weak charge Q_W can be deduced given as [Bou74]

$$Q_W(Z, N) = Z(1 - 4\sin^2 \theta_W) - N \quad (3.116)$$

with Z being the number of protons and N the number of neutrons in the atomic nucleus. In this way $\sin^2 \theta_W$ can be explored at different energy scales. The most precise measurements come from observables using the Z-pole, especially asymmetry measurements. These include the left–right asymmetry

$$A_{LR} = \frac{\sigma_L - \sigma_R}{\sigma_L + \sigma_R} \quad (3.117)$$

with $\sigma_L(\sigma_R)$ being the cross-section for left(right)-handed incident electrons. This has been measured precisely by SLD at SLAC. The left (L)–right (R) forward (F)–backward (B) asymmetry is defined as [Sch06]

$$A_{LR}^{FB}(f) = \frac{\sigma_{LF}^f - \sigma_{LB}^f - \sigma_{RF}^f + \sigma_{RB}^f}{\sigma_{LF}^f + \sigma_{LB}^f + \sigma_{RF}^f + \sigma_{RB}^f} = \frac{3}{4} A_f \quad (3.118)$$

where, e.g., σ_{LF}^f is the cross-section for a left-handed incident electron to produce a fermion f in the forward hemisphere. The Weinberg angle enters because A_f depends only on the couplings g_V and g_A:

$$A_f = \frac{2 g_V g_A}{g_V^2 g_A^2}. \quad (3.119)$$

A compilation of $\sin^2 \theta_W$ measurements is shown in Table 3.4.

Table 3.4. Compilation of measurements of the Weinberg angle $\sin^2 \theta_W$ (on-shell and in the $\overline{\text{MS}}$ scheme from various observables assuming global best-fit values. Taken into account are various masses and the width of the W- and Z-boson (after [PDG18]).

Dataset	$\sin^2 \theta_W$ $(\overline{MS})$	$\sin^2 \theta_W$ (on-shell)
All data	0.23129 ± 0.00005	0.22336 ± 0.00010
All data exc. m_H	0.23119 ± 0.00010	0.22312 ± 0.00022
All data exc. m_Z	0.23122 ± 0.00007	0.22332 ± 0.00011
All data exc. m_W	0.23132 ± 0.00006	0.22343 ± 0.00011
All data exc. m_t	0.23122 ± 0.00007	0.22303 ± 0.00024
m_H, m_Z, Γ_Z, m_t	0.23129 ± 0.00009	0.22342 ± 0.00016
LHC	0.23081 ± 0.00088	0.22298 ± 0.00088
Tevatron $+ m_Z$	0.23113 ± 0.00013	0.22307 ± 0.00030
LEP	0.23147 ± 0.00017	0.22346 ± 0.00047
SLD $+m_Z, \Gamma_Z, m_t$	0.23074 ± 0.00028	0.22229 ± 0.00055
$A_{FB}, m_Z, \Gamma_Z, m_t$	0.23299 ± 0.00029	0.22508 ± 0.00070
$m_{W,Z}, \Gamma_{W,Z}, m_t$	0.23106 ± 0.00014	0.22292 ± 0.00029
low energy $+ m_{H,Z}$	0.2328 ± 0.0014	0.2291 ± 0.0055

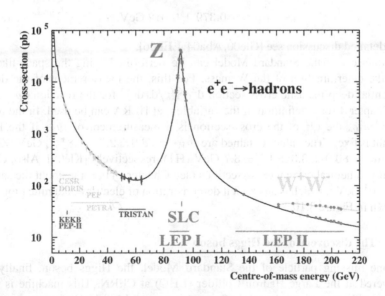

Figure 3.9. Cross-sections $(e^+e^- \to \text{hadrons})$, $(e^+e^- \to \mu^+\mu^-)$ and $(e^+e^- \to \gamma\gamma)$ as a function of the centre-of-mass energy. The sharp spike at the Z^0 resonance is clearly visible (from [Sch06]). © 2006 With permission from Elsevier.

3.4.4 Measurement of the gauge boson masses m_W and m_Z

The accurate determination of the mass of the Z-boson was one of the major goals of LEP and SLC. The Z^0 appears as a resonance in the cross-section in e^+e^--scattering

(Figure 3.9). With an accumulation of several million Z^0-bosons, the current world average is given by [PDG16]

$$m_Z = 91.1876 \pm 0.0021 \text{ GeV}. \tag{3.120}$$

Until 1996 the determination of the W-boson mass was the domain of $p\bar{p}$ machines like the $Sp\bar{p}S$ at CERN ($\sqrt{s} = 630$ GeV) and the Tevatron at Fermilab ($\sqrt{s} = 1.8$ TeV). With the start of LEP2, independent measurements at e^+e^- colliders became possible by W-pair production. Two effects could be used for an m_W measurement: the cross-sections near the threshold of W-pair production (Figure 3.11) and the shape of the invariant mass distribution of the W-pair. The combined LEP value is [Gle00, LEP06]

$$m_W = 80.350 \pm 0.056 \text{ GeV} \tag{3.121}$$

while the combined Tevatron measurements from the CDF and D0 experiments are [Aal13]

$$m_W = 80.387 \pm 0.016 \text{ GeV}. \tag{3.122}$$

Taking all these values and combining them with the first m_W measurement of ATLAS and CMS at LHC, a world average value is obtained to be [PDG18]

$$m_W = 80.379 \pm 0.012 \text{ GeV}. \tag{3.123}$$

For a detailed discussion see [Gle00, Aba04, LEP06].

A unique test of the Standard Model can be performed using the space-like and time-like determination of the W mass. For this, the measurement of the double differential deep inelastic cross-section $\text{d}^2\sigma^{CC}/\text{d}x\text{d}Q^2$ for the reaction $ep \rightarrow \nu X$ (see Chapter 4 for a definition of the variables) at HERA can be used. In the region $Q^2 \approx m_W^2$, the Q^2 of the cross-section is a measurement of m_W if the Fermi constant is fixed. The values obtained are $m_W = 78.9 \pm 2.0 \pm 1.8^{+2.0}_{-1.8}$ GeV (ZEUS) and $m_W = 80.9 \pm 3.3 \pm 1.7 \pm 3.7$ GeV (H1), respectively [Kle08]. Also, charge current and neutral current cross-sections (see Chapter 4) become about the same at $Q^2 \approx 10^4$ GeV^2, which is an explicit demonstration of electroweak unification. This is shown in Figure 3.10.

3.4.5 The discovery of the Higgs boson

The one missing particle of the Standard Model, the Higgs boson, finally was discovered at the Large Hadron Collider (LHC) at CERN. This machine is a pp-collider with an envisaged center of mass energy of 14 TeV. However, constraints on the Higgs mass could already be obtained before the LHC from electroweak precision measurements due to its contribution to radiative corrections. A best-fit value of 76^{+33}_{-24} GeV was determined by using all Z-pole data, m_t, m_W and Γ_W corresponding to an upper limit of 144 GeV with 95% CL [Alc07]. In the late phase of LEP2 a lower limit of the Higgs mass of $m_H > 114.4$ GeV could be obtained. As the Higgs boson coupling to the fermions is proportional to m_f^2, with f being the

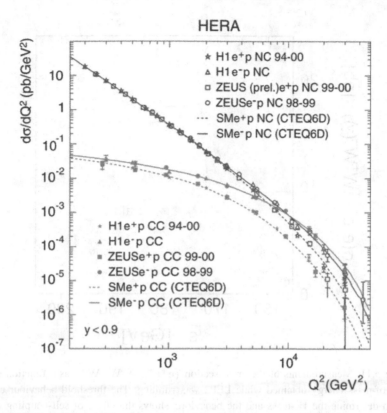

Figure 3.10. Comparison of the charged current and neutral current differential cross-sections as measured at HERA. The unification at high Q^2 is apparent (from [Kle08]). © 2008 With permission from Elsevier.

fermion mass, with the given bounds the most likely channel would be the decay H $\to b\bar{b}$, but the background from standard hadronic interactions is very high. Finally, the Higgs boson was discovered by the experiments ATLAS and CMS in the channels $H \to \gamma\gamma$ and $H \to WW$ (off-shell), where the W-bosons decay into charged leptons and neutrinos [Aad12, Cha12]. For a more lively account of the discovery period see [Dit13]. Currently the actual value of the Higgs boson mass is [PDG18]

$$m_H = 125.10 \pm 0.11 \text{ GeV}. \tag{3.124}$$

The Higgs sector might get more complicated as more Higgs particles could be involved. This is predicted in several extensions of the Standard Model like the minimal supersymmetric Standard Model discussed in Chapter 5.

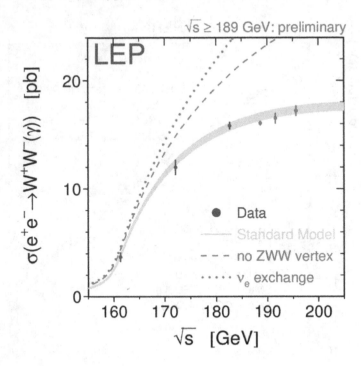

Figure 3.11. Measurements of the cross-section $(e^+e^- \rightarrow W^+W^-)$ as a function of the centre-of-mass energy obtained while LEP2 was running. The threshold behaviour can be used to determine the W-mass and the behaviour shows the effect of self-coupling of the gauge bosons. Scenarios with no ZWW vertex and pure ν_e exchange are clearly excluded. Predictions of Monte Carlo simulations are shown as lines (from [Pik02]). © 2002 With permission from Elsevier.

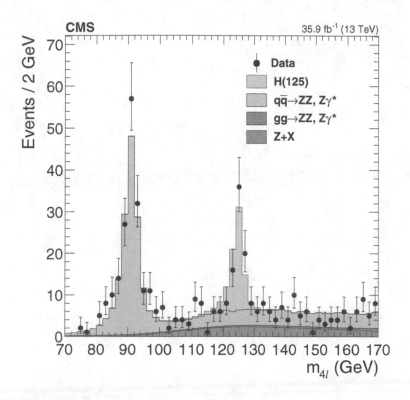

Figure 3.12. Invariant mass distribution of the 4 leptons from the Higgs decay $H \rightarrow ZZ^*$ as observed exemplarily by the CMS experiment at CERN. Clearly visible is the Z^0 resonance peak (left peak) and its associated background for the Higgs search and the Higgs boson peak (right peak). Data are shown for 13 TeV center of mass energy (from [PDG18]). With kind permission from M. Tanabashi *et al.* (Particle Data Group).

Chapter 4

Neutrinos as a probe of nuclear structure

DOI: 10.1201/9781315195612-4

Before exploring the intrinsic properties of neutrinos, we want to discuss how neutrinos can be used for measuring other important physical quantities. They allow a precise determination of various electroweak parameters and can be used to probe the structure of the nucleon via neutrino–nucleon scattering, which is a special case of lepton–nucleon scattering. On the other hand, the understanding of this process is mandatory for current and future long baseline, i.e., the distance from the neutrino source to the detector, neutrino experiments, which are discussed in Chapter 8. To perform systematic studies with enough statistics, artificial neutrino beams have to be created. Such sources are basically high-energy particle accelerators. Further information on this subject can be found in [Com83, Bil94, Lip95, Lea96, Sch97, Con98, Per00, Dev04, Dor08, Tho13].

4.1 Neutrino beams

Neutrino interactions have small cross-sections, hence to gain a reasonable event rate R (events per second) it is necessary that the target mass of the detector (expressed in numbers of nucleons in the target N_T) has to be quite large and the intensity I (ν cm^{-2}s^{-1}) of the beam should be as high as possible. An estimate of the expected event rate is then given by

$$R = N_T \sigma I \tag{4.1}$$

with σ being the appropriate cross-section (cm^2). First we discuss the potential beams.

4.1.1 Conventional beams

Neutrino beams have to be produced as secondary beams, because no direct, strongly focused, high-energy neutrino source is available. A schematic layout of a typical neutrino beam-line is shown in Figure 4.1. A proton synchrotron delivers bunches of high-energy protons (of the order of 10^{13} protons or more per bunch) on a fixed target (therefore, the commonly used luminosity unit is protons on target-"pot"), resulting in a high yield of secondary mesons, dominantly pions and kaons. By using beam

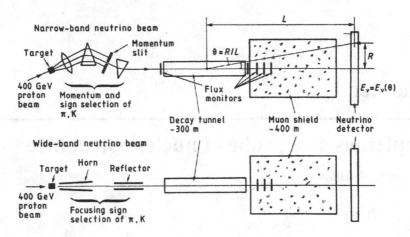

Figure 4.1. Schematic arrangements of neutrino beams: top, narrow-band beams; bottom, wide-band beams (from [Eis86]). © IOP Publishing. Reproduced with permission. All rights reserved.

optical devices (dipole or quadrupole magnets or magnetic horns) secondaries of a certain charge sign are focused into a long decay tunnel. There, the secondaries decay mostly via (assuming focusing of positive secondaries)

$$M^+ \rightarrow \mu^+ + \nu_\mu \qquad (M \equiv \pi, K) \qquad (4.2)$$

with a branching ratio of almost 100% for pions and 63.56% for kaons. As can be seen, a beam dominantly of ν_μ is produced (or, accordingly, a $\bar{\nu}_\mu$ beam if the oppositely signed charged mesons are focused). Only a fraction of the produced mesons decays in the tunnel with length L_D. The probability P for decay is given as

$$P = 1 - \exp(-L_D/L_0) \qquad (4.3)$$

with

$$L_0 = \beta c \times \gamma \tau_M = \frac{p_M}{m_M} \times c\tau_M = \begin{cases} 55.9 \text{ m} & \times & p_\pi/\text{GeV} \\ 7.51 \text{ m} & \times & p_K/\text{GeV} \end{cases} \qquad (4.4)$$

and τ_M being the relativistic boost factor. For $p_M = 200$ GeV and $L_D = 300$ m this implies: $L_0 = 11.2$ km, $P = 0.026$ (pions) and $L_0 = 1.50$ km, $P = 0.181$ (kaons). These probabilities have to be multiplied with the muonic branching ratios given before to get the number of neutrinos. To get a certain fraction of meson decays, L_D must increase proportional to momentum (energy) because of relativistic time dilation. At the end of the decay tunnel there is a long muon shield, to absorb the remaining pions and kaons via nuclear reactions and stop the muons by ionization and radiation losses. The experiments are located after this shielding. The neutrino spectrum can be determined from the kinematics of the two-body decay of the mesons. Energy (E_ν) and angle ($\cos\theta_\nu$) in the laboratory frame are related to the same quantities in the rest frame (marked with $*$) by

$$E_\nu = \bar{\gamma} E_\nu^* (1 + \bar{\beta} \cos \theta_\nu^*) \qquad \cos \theta_\nu = \frac{\cos \theta_\nu^* + \bar{\beta}}{1 + \bar{\beta} \cos \theta_\nu^*} \tag{4.5}$$

with

$$\bar{\beta} = \frac{p_M}{E_M} \qquad \bar{\gamma} = \frac{E_M}{m_M} \qquad \text{and} \qquad E_\nu^* = \frac{m_M^2 - m_\mu^2}{2 m_M}. \tag{4.6}$$

The two extreme values are given for $\cos \theta_\nu^* = \pm 1$ and result in

$$E_\nu^{\min} = \frac{m_M^2 - m_\mu^2}{2 m_M^2} (E_M - p_M) \approx \frac{m_M^2 - m_\mu^2}{4 E_M} \approx 0 \tag{4.7}$$

and

$$E_\nu^{\max} = \frac{m_M^2 - m_\mu^2}{2 m_M^2} (E_M + p_M) \approx \left(1 - \frac{m_\mu^2}{m_M^2}\right) \times E_M = \begin{cases} 0.427 \times E_\pi \\ 0.954 \times E_K \end{cases} \tag{4.8}$$

using $E_M \gg m_M$. With a meson energy spectrum $\phi_M(E_M)$ between $E_M^{\min}$ and $E_M^{\max}$ the resulting neutrino spectrum and flux is given by

$$\phi_\nu(E_\nu) \propto \int_{E_M^{\min}}^{E_M^{\max}} dE_M \, \phi_M(E_M) \frac{1}{p_M} \left(\frac{m_M^2 - m_\mu^2}{m_M^2} E_M - E_\nu\right). \tag{4.9}$$

Using (4.5) the following relation in the laboratory frame holds:

$$E_\nu(\theta_\nu) = \frac{m_M^2 - m_\mu^2}{2(E_M - p_M \cos \theta_\nu)} \approx E_M \frac{m_M^2 - m_\mu^2}{m_M^2 + E_M^2 \theta_\nu^2} \approx E_\nu^{\max} \frac{1}{1 + \bar{\gamma}_M^2 \theta_\nu^2}. \tag{4.10}$$

As can be seen, for typical configurations (the radius R of the detector is much smaller than the distance L to the source, meaning $\theta_\nu < R/L$) only the high-energy part of the neutrino spectrum hits the detector ($E_\nu(0) = E_\nu^{\max}$). Two types of beams can be produced which can be chosen by the physics goals and require the corresponding beam optical system. One is a narrow-band beam (NBB) using momentum selected secondaries, the other one is a wide-band beam (WBB) having a much higher intensity.

4.1.1.1 Narrow-band beams

An NBB collects the secondaries of interest coming from the target via quadrupole magnets. By using additional dipoles, it selects and focuses particles of a certain charge and momentum range (typically $\Delta p_M / p_M \approx 5\%$) that are leaving this area into the decay tunnel as a parallel secondary beam. Due to these two features (parallel and momentum selected), there is a unique relation between the radial distance with respect to the beam axis of a neutrino event in a detector and the neutrino energy for a given decay length (Figure 4.2). However, there is an ambiguity because two

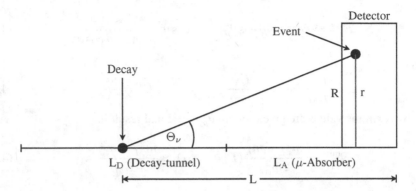

Figure 4.2. Geometric relation in an NBB between the position of meson decay (distance from the detector), decay angle θ_ν and radial position of the event in the detector (from [Sch97]). Reproduced with permission of SNCSC.

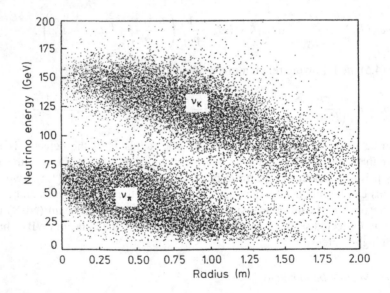

Figure 4.3. Scatter plot of E_ν with respect to radial event position for charged current (CC) events as obtained with the CDHSW detector at the CERN SPS. The dichromatic structure of the narrow-band beam (NBB) with $E_M = 160$ GeV is clearly visible and shows the neutrino events coming from pion and kaon decays (from [Ber87]). Reproduced with permission of SNCSC.

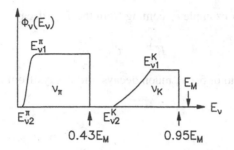

Figure 4.4. Schematic energy spectrum of neutrinos in an NBB hitting a detector. The contributions from pions and kaons are clearly separated (from [Sch97]). Reproduced with permission of SNCSC.

mesons (π, K) are present in the beam. Furthermore, the decay length is distributed along the decay tunnel, which results in a smearing into two bands. This is shown in Figure 4.3 for data obtained with the CDHSW experiment [Ber87]. For this reason NBBs are sometimes called dichromatic beams.

The main advantages of such a beam are a flat neutrino flux spectrum, the possibility of estimating E_ν from the radial position in the detector and a small contamination from other neutrino species. A schematic energy spectrum from an NBB is shown in Figure 4.4. However, the beam intensity is orders of magnitude smaller than in wide-band beams.

4.1.1.2 Wide-band beams

In a WBB the dipoles and quadrupoles are replaced by a system of so-called magnetic horns. They consist of two horn-like conductors which are pulsed with high currents synchronously with the accelerator pulse. This generates a magnetic field in the form of concentric circles around the beam axis, which focuses particles with the appropriate charge towards the beam axis. To increase this effect, a second horn, called the reflector, is often installed behind. In this case the prediction of the absolute neutrino energy spectrum and composition is a difficult task. Detailed Monte Carlo simulations are required to simulate the whole chain from meson production at the target all the way up to the neutrino flux at a detector. Instrumentation along the beam-line helps to determine accurate input parameters for the simulation. Particularly in the case of the West Area Neutrino Facility (WANF) at CERN, the SPY (secondary particle yields) experiment was performed to measure the secondary particle yield [Amb99], due to insufficient data from previous experiments [Ath80]. While in the NBB, because of the correlation of radial distance and neutrino energy, a reasonable estimation of E_ν can be deduced, in a WBB this is more difficult. In addition to beam-line simulations the observed event rates and distributions can be used to extract the neutrino flux by using known cross-sections ("empirical parametrization", see [Con98]). Furthermore, the beam can be polluted by other neutrino flavours, for example ν_e coming from the K_{e3}-decay

$$K^{\pm} \to \pi^0 e^{\pm} \overset{(-)}{\nu_e} \tag{4.11}$$

with a branching ratio of 5.1%, muon decays and decays from mesons produced in the absorber.

4.1.2 ν_τ beams

A completely different beam was necessary for the DONUT (E872) experiment at Fermilab (FNAL) [Kod01, Lun03a, Kod08]. Here the goal was to prove the existence of ν_τ by detecting τ - particles from CC reactions.

$$\nu_\tau + N \to \tau^- + X \tag{4.12}$$

and, therefore a ν_τ beam was needed. This was achieved by placing the detector only 36 m behind a 1 m long tungsten target, irradiated by an 800 GeV proton beam. The ν_τ beam results from the decay of the produced D_S-mesons via

$$D_S \to \tau \bar{\nu}_\tau (\text{BR} = 5.48 \pm 0.23\%) \quad \text{and} \quad \tau \to \nu_\tau + X. \tag{4.13}$$

They observed nine event candidates for process (4.13).

4.1.3   Off-axis superbeams

Conventional neutrino beams in the GeV range are dominated by systematics when investigating neutrino oscillations involving ν_μ and ν_e , because of the mentioned beam contaminations of ν_e from K_{e3} decays. To reduce this component, lower energy beams of less than a GeV ("Superbeams") are used. At these energies nuclear effects have to be considered as well. At lower energies quasi-elastic (QEL) interactions are dominant, which will be discussed in more detail in Section 4.5, and it is also convenient to use a narrow band beam. This can be achieved by installing detectors off-axis and using pion decay kinematics [McD01]. The most important kinematic property is, that by going off-axis the neutrino energy is in first order, independent of the energy of the parent pion. The ν_μ momentum in the laboratory frame is given for longitudinal and transverse momenta p_L and p_T by

$$p_L = \gamma(p^* \cos\theta^* + \beta p^*) \tag{4.14}$$
$$p_T = p^* \sin\theta^* \tag{4.15}$$

with $p^* = 0.03\,\text{GeV}/c$ as the neutrino momentum and θ^* as the polar angle of neutrino emission with respect to the pion direction of flight, both given in the pion rest frame. In the laboratory frame, θ is given by

$$\theta = \frac{R}{L} = \frac{1}{\gamma}\frac{\sin\theta^*}{1 + \cos\theta^*} \tag{4.16}$$

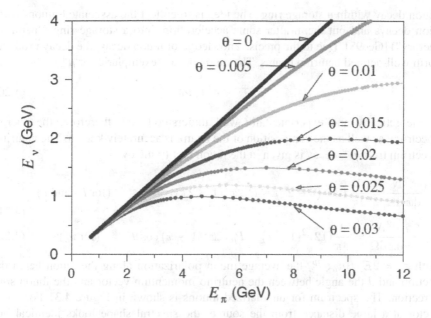

Figure 4.5. Neutrino energy as a function of pion energy. From kinematics it can be shown that on-axis pion and neutrino energies have a linear relation. However, if going to off-axis, for a large region of pion energies the neutrino energy remains almost constant (from [McD01]). With kind permission of K. T. McDonald.

with L as the baseline and R as the distance of the detector from the beam center. If the neutrino emission in the pion rest frame is perpendicular to the pion flight direction ($\theta^* = 90°$), then

$$\theta = \frac{1}{\gamma}. \tag{4.17}$$

The neutrino energy E_ν as a function of radial distance is given by

$$E_\nu(R) = \frac{2\gamma p^*}{1 + (\gamma R/L)^2} \tag{4.18}$$

which is half of the energy at beam centre for $\theta = 1/\gamma$. Hence, at this angle, the neutrino energy is, in first order, independent of the energy of the parent pion

$$\text{i.e.,} \quad \frac{\partial E_\nu}{\partial \gamma} = 0. \tag{4.19}$$

4.1.4 Alternative neutrino beams

Instead of using the neutrinos from the decay of secondary mesons as before, now there are two more ideas for getting a pure flavour beam. The first option is using muon decay within a storage ring. The idea is to collect the associated muons from pion decays and put them, after some acceleration, into a storage ring ("neutrino factory") [Gee98]. Due to the precise knowledge of muon decay, the decay products form well-defined neutrino beams. The muon decay (exemplaric for μ^+)

$$\mu^+ \to e^+ \nu_e \bar{\nu}_\mu \qquad (4.20)$$

is theoretically and experimentally well understood and, therefore, the energy spectrum, as well as the composition of the beam, is accurately known. The neutrino spectrum from μ^+-decay is given in the muon rest frame by

$$\frac{\mathrm{d}^2 N}{\mathrm{d}x\,\mathrm{d}\Omega} = \frac{1}{4\pi}\left(2x^2(3-2x) - P_\mu 2x^2(1-2x)\cos\theta\right) \qquad \text{(for } \bar{\nu}_\mu \text{ and } e) \tag{4.21}$$

$$\frac{\mathrm{d}^2 N}{\mathrm{d}x\,\mathrm{d}\Omega} = \frac{1}{4\pi}(12x^2(1-x) - P_\mu 12x^2(1-x)\cos\theta) \qquad \text{(for } \nu_e) \tag{4.22}$$

with $x = 2E_\nu/m_\mu$, P_μ the average muon polarization along the muon beam direction and θ the angle between the neutrino momentum vector and the muon spin direction. The spectrum for unpolarized muons is shown in Figure 4.31 For a detector at a large distance from the source, the spectral shape looks identical but the energy scale is multiplied by a Lorentz boost factor $2E_\mu/m_\mu$. The ν_e plays a special role because it is always emitted in the opposite direction to the muon polarization. Therefore 100% polarized muons with the right sign could produce a beam free of ν_e. Opposite-flavour beams are produced if the μ^--decay is used for the beam, resulting in a change of sign in (4.21) and (4.22). First experimental steps towards its realization have been performed; among them are the HARP experiment at CERN, which investigated the target material for optimal production of secondaries, the study of muon scattering (MUSCAT experiment) and muon cooling on hydrogen (MICE experiment) [Moh18]. For additional information see [Nuf01, Nuf02, Hub02, Gee07, Nuf08, Ban09, Gee09, Bog17]. A test storage ring called nuSTORM to experimentally study this idea is proposed [Ade15].

A second concept explored is the production of a pure beam of ν_e by accelerating β-unstable isotopes to a few hundred MeV ("beta beams") [Zuc02]. Research to realise such a beam has been performed at CERN investigating ^{6}He and ^{18}N but also ^{8}Li and ^{8}B [Ter04, Bur05, Vol07, Wil14]. Last but not least, a former idea of using a tagged beam has become under consideration as well [Dor18].

4.2   Neutrino detectors

A second important component is the detector. The small cross-sections involved in neutrino physics require detectors of large size and mass to get a reasonable event rate. As the focus is moving towards large distances between source and detector, high intensity beams are mandatory. Several requirements should be fulfilled by such a detector:

- identification of a charged lepton to distinguish CC and NC events,
- measurement of energy and scattering angle of the charged lepton to determine the kinematic variables of the event,
- measurement of the total hadronic energy, e.g., to reconstruct E_ν,
- identification of single secondary hadrons and their momenta to investigate in detail the hadronic final state,
- detection of short living particles (and optionally the usage of different target materials).

Some of these requirements are mutually exclusive of each other and there is no single detector to fulfil all of them. The actual design depends on the physics questions under exploration. For the long baseline beams a near (ND) and far (FD) detector are used. However, the relation between the two detectors is much more complex than just scaling with distance, as the original neutrino energy for a given event is not known and within the interaction nuclear effects have to be considered. Hence, the reconstruction of the neutrino energy is non-trivial. In the following sections, four examples of some detector concepts are presented.

4.2.1 OPERA

The OPERA experiment [Gul00, Aga18] in Europe was part of the long-baseline program using a neutrino beam from CERN to the Gran Sasso Laboratory (CNGS) [Els98]. The distance is 732 km. The aim of the experiment was to search directly for ν_τ appearance in a muon neutrino beam by using emulsions. The beam energy was optimised between cross-section and the prompt tau-background. The beam protons from the SPS at CERN are extracted with energies up to 450 GeV hitting a graphite target at a distance of 830 m from the SPS. After passing a magnetic horn system for focusing the pions (see Section 4.1), a decay pipe of 1000 m follows. Finally the average beam energy was around 17 GeV. The detection principle is to use lead as a massive target for neutrino interactions and, as the identification of a tau-lepton (average decay length $c\tau = 87 \,\mu$m) is non-trivial, thin emulsion sheets which were working conceptually as emulsion cloud chambers (ECC).

Finally a statistics of 17.97×10^{19} pot have been accumulated out of which 10 tau-candidate events could be extracted with an expected background of 2.0 ± 0.4 events [Aga18]. This is the only ν_τ-appearance oscillation measurement with a positive result, for more details see Chapter 9. Emulsion detectors have also been used in other experiments like DONUT at Fermilab to confirm the existence of ν_τ and CHORUS [Esk97], a neutrino oscillation experiment at CERN.

Alternative ways for detection by using active detectors like drift chambers combined with a muon spectrometer have been performed to study neutrino-nucleon cross-sections and neutrino oscillation searches. Here, the hadronic energy E_{had} and muon energy E_μ have to be reconstructed which is linked to the visible energy E_{vis} or neutrino energy by

$$E_\nu \approx E_{\text{vis}} = E_\mu + E_{\text{had}}. \tag{4.23}$$

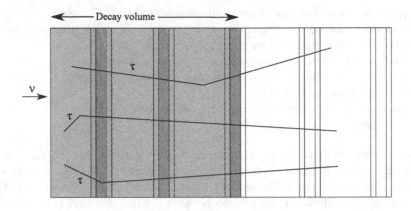

Figure 4.6. The emulsion cloud chamber (ECC) principle as used by OPERA. A τ-lepton produced in a charged current interaction can be detected via two mechanisms. If the τ-decay happens after the τ has traversed several emulsion sheets, there will be a kink between the various track segments (upper curve). In an early τ-decay an impact parameter analysis can be done because the interesting track is not pointing to the primary vertex (lower two curves). For simplicity the additional tracks from the primary vertex are not shown. With kind permission of S. Turkat.

The former CCFR [Sak90], NuTeV [Bol90], NOMAD [Alt98] and MINOS [Mic08] experiments have been working in this way.

4.2.2   NOVA

An alternative concept is used by NOVA, using liquid scintillators as trackers and calorimeters [Nov07, Muf15]. The experiment is located in the NuMi beam created at Fermilab [Ada16], with a near detector (ND) in about 1 km and a far detector (FD) about 810 km away. The FD contains cells of 3.9×6.6 cm in cross-section and 6.6 cm dimension along the beam direction and for a total length of 15.5 m. They are filled with mineral oil with a 5% pseudocumene (PC) admixture to act as a scintillator. The scintillation light collected in the cells is fed into wavelength-shifting optical fibers which are connected to avalanche photo diodes (APDs). These cells are organized into planes alternating in vertical and horizontal orientation. In total, 896 planes have been built, leading to a total mass of 14 ktons. As the experiment is off-axis, the experiment is about 14.6 mrad away from the central axis of the NuMi beam. The ND is much closer to the production region but has basically the same type of building blocks and has a size of 290 tons.

4.2.3   T2K

T2K (Tokai to Kamioka) at J-PARC is an off-axis beam experiment as well [Oya15]. The FD in this case is the Super-Kamiokande detector, a 50 kton water Cherenkov

Figure 4.7. Top view of a part of the NOVA far detector with the neutrino beam produced at Fermilab in a distance of 810 km away (from [Fer19]). With kind permission from Fermilab.

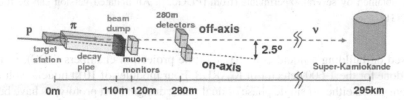

Figure 4.8. A schematical view of the T2K beamlines showing beamline components (target station, decay pipe and beam dump), as well as detector components (muon monitor, 280 m detectors and Super-Kamiokande) (from [Oya15]). With kind permission of the T2K Collaboration.

detector described in more detail in Chapter 9. The neutrino beam is produced by a 30 GeV proton beam hitting a graphite production target. The secondaries are monitored on- and off-axis (at 2.5°) in about 280 meters from the target, with Super-Kamiokande under the same angle in 295 km distance. Given the complexity of getting accurate neutrino beam predictions, T2K is upgrading constantly the area around the target region to improve the knowledge of the nuclear aspects of neutrino-nucleus scattering (Figure 4.8).

4.2.4 DUNE

A next generation long baseline neutrino experiment is DUNE (Deep Underground Neutrino Experiment) [Acc15]. The idea is to send a wide-band (60-120 GeV) high intensity neutrino beam from Fermilab to the Sanford Underground Facility in Homestake (see also Chapter 10), about 1300 km away. The detector technology

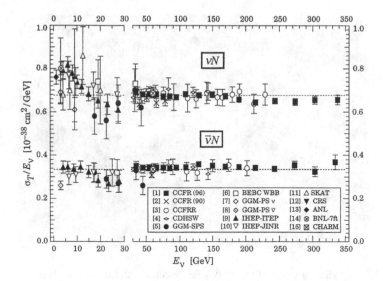

Figure 4.9. Compilation of total cross-sections σ_T/E_ν in νN and $\bar{\nu}N$ scattering as a function of E_ν obtained by several experiments (from [PDG02]). An updated version can be found in [PDG18].

is based on building a liquid argon (LAr) time projection chambers (TPC), as has been done for the T-600 detector of ICARUS. Four modules of 10 kt fiducial volume are considered either of single phase or dual phase design. First prototypes have been built at CERN.

Having discussed neutrino beams and detectors, we now proceed to physics results.

4.3   Total cross-section for neutrino–nucleon scattering

The total neutrino and antineutrino cross-sections for νN scattering have been measured in a large number of experiments. They can proceed (assuming ν_μ beams) via charged currents (CC) involving W-exchange and neutral current (NC) processes with Z-exchange

$$\nu_\mu N \to \mu^- X \qquad \bar{\nu}_\mu N \to \mu^+ X \quad (CC) \qquad (4.24)$$

$$\nu_\mu N \to \nu_\mu X \qquad \bar{\nu}_\mu N \to \bar{\nu}_\mu X \quad (NC) \qquad (4.25)$$

with $N \equiv p, n$ or an isoscalar target (average of neutrons and protons) and X as the hadronic final state. The total CC neutrino–nucleon cross-section on isoscalar targets[1] as a function of E_ν was determined dominantly by CCFR and CDHSW. Both were using NBB and an iron target. Except for small deviations at low energies ($E_\nu < 30\,\mathrm{GeV}$) a linear rise in the cross-section with E_ν was observed (Figure 4.9).

---

[1] A correction factor has to be applied for heavy nuclei because of a neutron excess there. For Fe it was determined to be -2.5% for $\sigma(\nu N)$ and $+2.3\%$ for $\sigma(\bar{\nu}N)$.

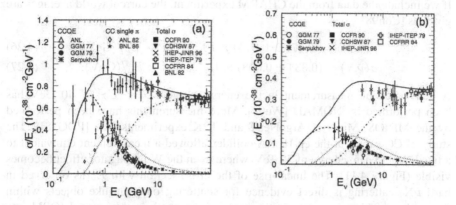

Figure 4.10. Compilation of total cross-section σ_T measurements as function of neutrino energy for neutrinos (left) and antineutrinos (right) according to the contributing processes (from [Ash05]). © 2005 by the American Physical Society. An updated version can be found in [PDG18].

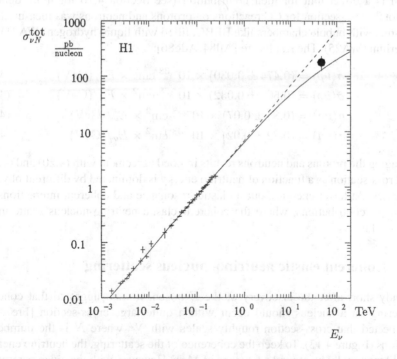

Figure 4.11. Compilation of $\sigma(E_\nu)$ from νN scattering (crosses) and from the H1 experiment (dot) at DESY. The dashed curve corresponds to a prediction without a W-propagator ($m_W = \infty$), the solid line is a prediction with W-propagator ($m_W = 80\,\text{GeV}$) (from [Ahm94]). © 1994 With permission from Elsevier.

If we include the data from the CHARM experiment, the current world averages are given as [Con98]

$$\sigma(\nu N) = (0.677 \pm 0.014) \times 10^{-38} \text{ cm}^2 \times E_\nu/(\text{GeV}) \qquad (4.26)$$

$$\sigma(\bar{\nu} N) = (0.334 \pm 0.008) \times 10^{-38} \text{ cm}^2 \times E_\nu/(\text{GeV}). \qquad (4.27)$$

A new precision measurement in the energy range $2.5 < E_\nu < 40$ GeV has been performed by NOMAD [Wu08]. More measurements have been performed by the MINOS, Minerva, ArgoNeuT and T2K experiments (see [PDG18]). The study of CC events at the ep-HERA collider allowed a measurement equivalent to a fixed target beam energy of 50 TeV, where even the W-propagator effect becomes visible (Figure 4.11). The linear rise of the cross-section with E_ν as observed in hard νN scattering is direct evidence for scattering on point-like objects within the nucleon. This assumption is the basis of the quark–parton–model (QPM, see Section 4.8), which predicts that deep-inelastic νN scattering can be seen as an incoherent superposition of quasi-elastic neutrino–(anti)quark scattering. At low energies ($E_\nu < 30$ GeV), the ratio $R = \sigma(\nu N)/\sigma(\bar{\nu} N) \approx 3$ agrees with the simple QPM prediction without sea-quark contributions. R is about 2 at higher energies which is a direct hint for their contribution (see Section 4.10 for more details). The total cross-section for CC reactions on protons and neutrons was measured, for example, with bubble chambers like BEBC, filled with liquid hydrogen (WA21) and deuterium (WA25). The results are [All84, Ade86]:

$$\sigma(\nu p) = (0.474 \pm 0.030) \times 10^{-38} \text{ cm}^2 \times E_\nu/(\text{GeV}) \qquad (4.28)$$

$$\sigma(\bar{\nu} p) = (0.500 \pm 0.032) \times 10^{-38} \text{ cm}^2 \times E_\nu/(\text{GeV}) \qquad (4.29)$$

$$\sigma(\nu n) = (0.84 \pm 0.07) \times 10^{-38} \text{ cm}^2 \times E_\nu/(\text{GeV}) \qquad (4.30)$$

$$\sigma(\bar{\nu} n) = (0.22 \pm 0.02) \times 10^{-38} \text{ cm}^2 \times E_\nu/(\text{GeV}) \qquad (4.31)$$

Averaging the protons and neutrons results in good agreement with (4.20) and (4.21). The cross-section as a function of neutrino energy is dominated by different physical processes. At lower energies, quasi-elastic, resonance and coherent interactions are the major contributions, where the coherent elastic neutrino-nucleus scattering is discussed first.

4.4   Coherent elastic neutrino–nucleus scattering

Already short after the discovery of the Z-boson, it was suggested that coherent scattering on a nucleus should occur with a quite large cross-section [Fre74]. It is expected that cross-section roughly scales with N^2 where N is the number of neutrons (Figure 4.12). To keep the coherence of the scattering, the neutrino energies should maximally be around a few tens of MeV. However, with the given parameters the measured recoil of the nucleus is only a few keV, which is hard to measure. However, finally this process could be measured at the Spallation Neutron Source (SNS) at Oak Ridge National Laboratory [Aki17]. A pulsed neutron beam of high intensity is used for this which also produces neutrinos due to charged pion decay

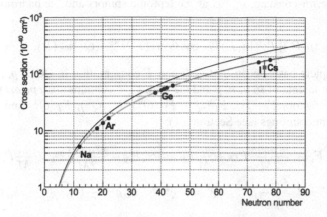

Figure 4.12. An illustration of the squared behaviour of the cross-section as a function of neutron number. The upper line assumes a unity-form factor, while the lower line assumes a Helm form factor (see [Hel56] for details) which covers an assumed 3% uncertainty of the rms neutron radius. The data points indicate the elements of the available detector materials used by the COHERENT experiment (from [Sch18]). With kind permission of K. Scholberg.

at rest. In this way a 14.6 kg CsI(Na) was exposed to the beam and lead to a measurement with high significance.

4.5 Quasi-elastic neutrino–nucleon scattering

Quasi-elastic (QEL) reactions are characterized by the fact that the nucleon does not break up and, therefore, $x \approx 1$. Reactions of the form $\nu + n \rightarrow \ell^- + p$ are quasi-elastic or, being more specific, in QEL $\nu_\mu N$ scattering the following reactions have to be considered [Gal11]:

$$\nu_\mu + n \rightarrow \mu^- + p \qquad E_{\text{Th}} = 110 \text{ MeV} \qquad (4.32)$$

$$\bar{\nu}_\mu + p \rightarrow \mu^+ + n \qquad E_{\text{Th}} = 113 \text{ MeV} \qquad (4.33)$$

$$\overset{(-)}{\nu_\mu} + p \rightarrow \overset{(-)}{\nu_\mu} + p. \qquad (4.34)$$

Corresponding reactions also hold for ν_e, but with different threshold energies. The quasi-elastic NC scattering on neutrons is, in practice, not measurable.

4.5.1 Quasi-elastic CC reactions

The most general matrix element in V–A theory for (4.32) is given by [Lle72,Com83, Str03]

$$ME = \frac{G_F}{\sqrt{2}} \times \bar{u}_\mu(p')\gamma_\alpha(1 - \gamma_5)u_\nu(p) \times \langle p(P')|J_\alpha^{CC}|n(P)\rangle \qquad (4.35)$$

with G_F as Fermi-constant, u_μ, u_ν as the leptonic spinors and the hadronic current is given as

$$\langle p(P')|J_\alpha^{CC}|n(P)\rangle = \cos\theta_C \bar{u}_p(P')\Gamma_\alpha^{CC}(Q^2)u_n(P). \tag{4.36}$$

This is a complete basis because of C, P and T conservation in QCD. p, p', P and P' are the 4-momenta of ν, μ, n, p and the term Γ_α^{CC} contains six *a priori* unknown complex form factors $F_S(Q^2)$, $F_P(Q^2)$, $F_V(Q^2)$, $F_A(Q^2)$, $F_T(Q^2)$ and $F_M(Q^2)$ for the different couplings (see Section 3.1):

$$\Gamma_\alpha^{CC} = \gamma_\alpha F_V + \frac{i\sigma_{\alpha\beta}q_\beta}{2M}F_M + \frac{q_\alpha}{M}F_S + \left[\gamma_\alpha F_A + \frac{i\sigma_{\alpha\beta}q_\beta}{2M}F_T + \frac{q_\alpha}{M}F_P\right]\gamma_5$$

$$q = P' - P = p - p' \qquad Q^2 = -q^2 \qquad \sigma_{\alpha\beta} = \frac{1}{2i}(\gamma_\alpha\gamma_\beta - \gamma_\beta\gamma_\alpha). \tag{4.37}$$

with M being the mass of the nucleon [Lea96]. The terms associated with F_T and F_S are called second class currents and F_M corresponds to weak magnetism. Assuming T-invariance and charge symmetry, the scalar and tensor form factors F_T and F_S have to vanish. Furthermore, terms in cross-sections containing pseudo-scalar interactions are always multiplied by m_l^2 for free nucleons [Lle72], with l being a lepton (here it is the μ) and can be neglected for high energies ($E_\nu \gg m_l$). Under these assumptions, Eq. (4.37) for V-A interactions is shortened to

$$\Gamma_\alpha^{CC} = \gamma_\alpha F_V + \frac{i\sigma_{\alpha\beta}q_\beta}{2M}F_M - \gamma_\alpha\gamma_5 F_A \tag{4.38}$$

containing vector and axial vector contributions as well as weak magnetism. Using the CVC hypothesis (see Section 3.1), F_V and F_M can be directly related to the electromagnetic form factors (G_E, G_M) of the nucleon, appearing in the Rosenbluth formula for the differential cross-section of elastic $eN \to eN$ ($N = p, n$) scattering via [Lea96]

$$F_V = \frac{G_E^V + \tau G_M^V}{1 + \tau} \tag{4.39}$$

$$F_M = \frac{G_M^V - \tau G_E^V}{1 + \tau} \tag{4.40}$$

with $\tau = Q^2/4M^2$. Experimentally, often an ansatz in form of a dipole is used

$$G_{E,M}(Q^2) = \frac{G_{E,M}(0)}{(1 + Q^2/M_V^2)^2} \qquad \text{with } M_V = 0.84\,\text{GeV} \tag{4.41}$$

with the normalization at $Q^2 = 0$:

$$G_E^p(0) = 1 \qquad G_E^n(0) = 0 \tag{4.42}$$

$$G_E^V(0) = \mu_p \qquad G_M^V(0) = \mu_n \tag{4.43}$$

with μ_p, μ_n as magnetic moments of proton and neutron in units of the nuclear magneton μ_N. Assuming the same dipole structure for F_A, using (4.41) by replacing

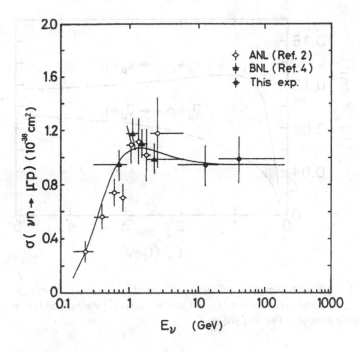

Figure 4.13. Compilation of results for the CC cross-section $\sigma(\nu_\mu n \to \mu p)$ of various experiments. The curve shows the prediction of V–A theory with $M_A = 1.05\,\text{GeV}$ (from [Kit83]). © 1983 by the American Physical Society.

M_V with M_A, and taking $F_A(0) = g_A/g_V = -1.2670 \pm 0.0030$ from neutron decay [PDG02], the only free parameter is M_A. It is measured in quasi-elastic νN scattering and has the average value of $M_A = (1.026 \pm 0.020)\,\text{GeV}$ [Ber02a] (Figure 4.13). New data from ep and eD scattering showed that (4.45) is accurate only to 10-20% and more sophisticated functions than the dipole approximation have to be used [Bos95, Bra02, Bud03]. Measurements of the axial vector mass from K2K and MiniBooNE came up with slightly higher values of $M_A = (1.144 \pm 0.077^{+0.078}_{-0.072})\,\text{GeV}$ [Gra06, Esp07] and $M_A = (1.23 \pm 0.20)\,\text{GeV}$ [Agu08], respectively. An accurate understanding of the quasi-elastic regime is essential for newly planned neutrino superbeams (see Section 8.10.4). A wider range in Q^2 and a more precise measurement of the actual shape of $F_A(Q^2)$ will be covered by the MINERνA experiment at Fermilab and by T2K at J-PARC. For a wide range of neutrino energies these experiments will explore cross-sections on He, C, Fe and Pb to study nuclear effects.

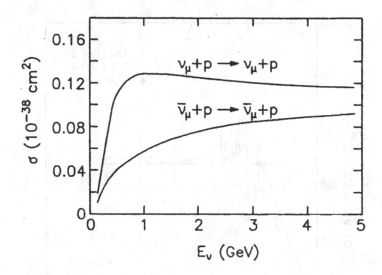

Figure 4.14. GWS prediction of the NC cross-sections $\sigma(\nu_\mu p \to \nu_\mu p)$ and $\sigma(\bar{\nu}_\mu p \to \bar{\nu}_\mu p)$ as a function of E_ν with the parameters $M_A = 1.00\,\mathrm{GeV}$ and $\sin^2\theta_W = 0.232$ (from [Hor82]). © 1982 by the American Physical Society.

Taking it all together, the quasi-elastic cross-sections are given by [Sch97]

$$\frac{\mathrm{d}\sigma_{QE}}{\mathrm{d}Q^2}\binom{\nu_\mu n \to \mu^- p}{\bar{\nu}_\mu p \to \mu^+ n} = \frac{M^2 G_F^2 \cos^2\theta_c}{8\pi E_\nu^2}\left(A_1(Q^2) \pm A_2(Q^2)\frac{s-u}{M^2}\right.$$
$$\left. + A_3(Q^2)\frac{(s-u)^2}{M^4}\right) \tag{4.44}$$

where $s - u = 4ME_\nu - Q^2$ and M is the mass of the nucleon and s, u are Mandelstam variables. The functions A_1, A_2 and A_3 depend on the form factors F_A, F_V, F_M and Q^2. The most generalized expressions are given in [Mar69]. Equation (4.44) is analogous to the Rosenbluth formula describing elastic eN scattering.

4.5.2 Quasi-elastic NC reactions

The matrix element for the NC nucleon current is analogous to (4.39) neglecting again S, P and T terms. Now, analog matrix elements exist for neutral currents, $\langle p(P')|J_\alpha^{NC}|p(P)\rangle$ and $\langle n(P')|J_\alpha^{NC}|n(P)\rangle$ with the corresponding form factors. For $\mathrm{d}\sigma/\mathrm{d}Q^2$ (4.44) holds but now with the corresponding NC form factors (Figure 4.14). Several experiments have measured the cross-section for this process (see [Man95]). $\sin^2\theta_W$ and M_A serve as fit parameters. Values obtained with the BNL experiment E734 result in [Ahr87]

$$M_A = (1.06 \pm 0.05)\,\mathrm{GeV} \qquad \sin^2\theta_W = 0.218^{+0.039}_{-0.047}. \tag{4.45}$$

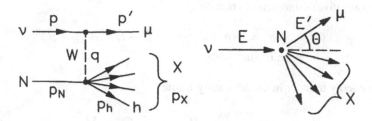

Figure 4.15. Kinematics of the CC reaction $\nu_\mu N \rightarrow \mu X$ via W-exchange. Left: The underlying Feynman graph. Right: Variables in the laboratory system (from [Sch97]). Reproduced with permission of SNCSC.

After discussing elastic and quasi-elastic processes, now inelastic scattering is discussed. To obtain more information about the structure of the nucleon, we have to look at deep inelastic scattering (DIS) dominating the cross-section for $E_\nu \geq 3$ GeV.

4.6 Kinematics of deep inelastic scattering

In deep inelastic lepton–nucleon scattering, leptons are used as point-like probes of nucleon structure. Reactions, especially those focusing on weak interaction properties, are done with neutrinos according to (4.18) and (4.19). In a similar fashion, the electromagnetic structure is explored via deep inelastic scattering with charged leptons

$$e^\pm + N \rightarrow e^\pm + X \qquad \mu^\pm + N \rightarrow \mu^\pm + X. \qquad (4.46)$$

Let us discuss the kinematics of CC interactions (4.18) on fixed targets as shown in Figure 4.15 using ν_μ as beam particle. The 4-momenta, $p, p', q = p - p', p_N, p_X$ and p_h of the incoming ν, the outgoing μ, the exchanged W, the incoming nucleon N, outgoing hadronic final state X and of a single outgoing hadron h are given in the laboratory frame as

$$p = (E_\nu, p_\nu) \qquad p' = (E_\mu, p_\mu) \qquad q = (\nu, q) \qquad (4.47)$$
$$p_N = (M, 0) \qquad p_X = (E_X, p_X) \qquad p_h = (E_h, p_h) \qquad (4.48)$$

with M being the nucleon mass. Measured observables in the laboratory frame are typically the energy $E' = E_\mu$ and the scattering angle $\theta = \theta_\mu$ of the outgoing muon (in analogy with the outgoing lepton in $eN/\mu N$ scattering) for a given neutrino energy $E = E_\nu$. These two quantities can be used to measure several important kinematic event variables.

- The total centre-of-mass energy $\sqrt{s}$:

$$s = (p + p_N)^2 = 2ME + M^2 \approx 2ME. \qquad (4.49)$$

- The (negative) 4-momentum transfer:

$$Q^2 = -q^2 = -(p - p')^2 = -(E - E')^2 + (p - p')^2$$
$$= 4EE' \sin^2 \tfrac{1}{2}\theta > 0 \tag{4.50}$$

- The energy transfer in the laboratory frame:

$$\nu = \frac{q \times p_N}{M} = E - E' = E_X - M. \tag{4.51}$$

- The Bjorken scaling variable x:

$$x = \frac{-q^2}{2q \times p_N} = \frac{Q^2}{2M\nu} \qquad \text{with } 0 \le x \le 1. \tag{4.52}$$

- The relative energy transfer (inelasticity) y (often called the Bjorken y)

$$y = \frac{q \times p_N}{p \times p_N} = \frac{\nu}{E} = 1 - \frac{E'}{E} = \frac{Q^2}{2MEx}. \tag{4.53}$$

- The total energy of the outgoing hadrons in their center-of-mass frame

$$W^2 = E_X^2 - p_X^2 = (E - E' + M)^2 - (p - p')^2 = -Q^2 + 2M\nu + M^2. \tag{4.54}$$

Equations (4.49) and (4.53) can be combined to give the useful relation

$$xy = \frac{Q^2}{2ME} = \frac{Q^2}{s - M^2}. \tag{4.55}$$

At a fixed energy E, inelastic reactions can, therefore, be characterized by two variables such as $(E', \theta), (Q^2, \nu), (x, Q^2)$ or (x, y). For quasi-elastic reactions $(x = 1)$, one variable $(E', \theta, Q^2$ or $\nu)$ is sufficient. Figure 4.16 shows the parameter space covered by current experiments. As can be seen, a wide range can be explored using various accelerators.

4.7   Coherent, resonant and diffractive production

Besides quasi-elastic and deep inelastic scattering, there are other mechanisms which can contribute to the neutrino cross-section. Among them are diffractive, resonance and coherent particle production, manifesting themselves dominantly in single pion production. Typical resonance reactions, in which intermediate resonance states like the $\Delta(1232)$ are produced, are

$$\nu_\mu p \to \mu^- p \pi^+ \tag{4.56}$$
$$\nu_\mu n \to \mu^- n \pi^+ \tag{4.57}$$
$$\nu_\mu n \to \mu^- p \pi^0 \tag{4.58}$$

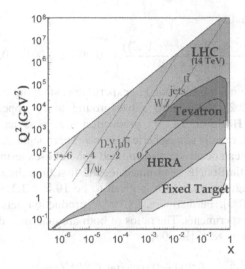

Figure 4.16. Allowed kinematic regions in the (x, Q^2) plane that can be explored by various experiments (from [PDG18]). With kind permission of M. Tanabashi et al. (Particle Data Group).

or NC reactions

$$\nu_\mu p \to \nu_\mu p \pi^0 \qquad \nu_\mu p \to \nu_\mu n \pi^+ \qquad (4.59)$$
$$\nu_\mu n \to \nu_\mu n \pi^0 \qquad \nu_\mu n \to \nu_\mu p \pi^- \qquad (4.60)$$

will not be discussed in more detail here (see [Pas00, Sor07]). As an example we briefly mention coherent π^0 production which directly probes the Lorentz structure of NC interactions. Helicity-conserving V, A interactions will result in a different angular distribution of the produced π^0 than the ones from helicity changing S, P, T interactions. For more extensive details see [Win00]. Coherent π^0 production

$$\nu + (A, Z) \to \nu + \pi^0 + (A, Z) \qquad (4.61)$$

leaves the nucleus intact. Because of helicity conservation in NC events, the π^0 is emitted at small angles in contrast to incoherent and resonant production. Several experiments have measured this process [Ama87, Cos88] and the results are compiled in Figure 4.17. The ratio of ν and $\bar{\nu}$ induced production is deduced to be

$$\frac{\sigma(\nu(A, Z) \to \nu \pi^0(A, Z))}{\sigma(\bar{\nu}(A, Z) \to \bar{\nu} \pi^0(A, Z))} = 1.22 \pm 0.33 \qquad (4.62)$$

still with a rather large uncertainty but they are in good agreement with theoretical expectations which predict a ratio of one [Rei81]. Improved measurements have been done by the K2K and MiniBooNE experiments (see Chapters 8 and 9). K2K measured the ratio of NC single π^0 production with respect to charged current interactions as [Nak05]

$$\frac{\sigma(\nu(A,Z) \to \nu\pi^0(A,Z))}{\sigma(\nu N \to \mu X)} = 0.064 \pm 0.001 \pm 0.007. \tag{4.63}$$

This process is the main background to experiments studying elastic $\nu_\mu e$-scattering (see Section 3.4.2.2) and is also a background to ν_e appearance experiments (see Chapter 8). However, NC π^0 production serves as an important tool for measuring total NC rates in atmospheric neutrino experiments (see Chapter 9). Single NC π^0 production can occur via resonant and coherent scattering, with the first one to be dominant. MiniBooNE determined for the first time the sum of the coherent and diffractive contribution, which is given to be 19.5 ± 1.1(stat.) ± 2.5(sys.) % of the total [Agu08a]. Furthermore, CC $\pi^{+,0}$ production acts as background for $\nu_\mu$disappearance experiments. The ratios of both are determined with respect to the QEL cross-section by K2K [Rod08]

$$\frac{\sigma_{CC\pi^0}}{\sigma_{QEL}} = 0.306 \pm 0.023(\text{stat.})^{+0.023}_{-0.021}(\text{sys.}) \tag{4.64}$$

$$\frac{\sigma_{CC\pi^+}}{\sigma_{QEL}} = 0.734 \pm 0.086(\text{fit})^{+0.076}_{-0.103}(\text{nucl.})^{+0.079}_{-0.073}(\text{sys.}). \tag{4.65}$$

In addition, K2K placed an upper limit on coherent charged pion production of 0.60×10^{-2} (90 % CL) with respect to the total CC cross-section [Has05]. The SciBooNE experiment, using the SciBar detector of K2K in the Fermilab neutrino beam ($\langle E_\nu \rangle \approx 0.8\,\text{GeV}$) for MiniBooNE (see Chapters 8 and 9 for more details) is measuring the CC single charged pion production with even greater sensitivity for neutrino and antineutrino beams, a recent theoretical calculation can be found in [Lei06].

Diffractive processes are characterized by leaving the nucleus intact, implying low momentum transfer. This can be described by a new kinematic variable t, being the square of the 4-momentum transferred to the target

$$t = (p - p')^2. \tag{4.66}$$

At low Q^2 and large ν, a virtual hadronic fluctuation of the gauge bosons, in the case of neutrinos the weak bosons W and Z, may interact with matter before being re-absorbed. Thus, diffractive production of mesons on a target might produce real mesons in the final state, e.g.,

$$\nu_\mu N \to \mu^- \rho^+ N. \tag{4.67}$$

In an analogous way the NC diffractive production of neutral vector mesons (V^0) such as $\rho^0, \omega, \Phi, J/\Psi$, etc. can also be considered (Figure 4.18). The elementary nature of the interaction is still not well understood. It can be described by the exchange of a colour singlet system, called Pomeron. In νN scattering diffractive production of π, $\rho^\pm$, a_1 and D_S^* mesons have been observed, while in lepto- and photo-production also ρ^0, ω, ϕ and J/Ψ have been seen due to the higher statistics.

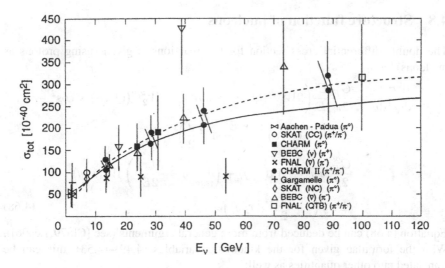

Figure 4.17. Compilation of results on coherent single-π production cross-sections in CC ν_μ and $\bar{\nu}_\mu$ interactions. The curve shows the prediction of the Rein–Sehgal model [Rei83] for $M_A = 1.3\,\text{GeV}/c^2$ (solid line) and the Bel'kov–Kopeliovich approach (dashed line). The results of the experiments are scaled according to both models to allow comparison (from [Win00]). © Cambridge University Press. In the low energy region new results from Minos [Ada16b] and NOVA [Ace18] have been produced (not shown here).

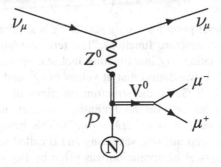

Figure 4.18. Feynman graph for diffractive vector meson (V^0) production via the exchange of a pomeron $\mathcal{P}$. With kind permission of S. Turkat.

A revival of interest in diffractive phenomena took place with the observation of 'rapidity gap' events at the ep collider HERA.

After discussing quasi-elastic scattering and giving a short insight of resonance and diffractive production, which dominate the cross section at low energies, we now want to focus on deep inelastic scattering which leads to the concept of structure functions.

4.8   Structure function of nucleons

The double differential cross-section for CC reactions is given (using protons as nucleons) by

$$
\frac{d^2\sigma^{\nu,\bar{\nu}}}{dQ^2\,d\nu} = \frac{G_F^2}{2\pi}\frac{E'}{E}\left(2W_1^{\nu,\bar{\nu}}(Q^2,\nu)\times\sin^2\frac{\theta}{2} + W_2^{\nu,\bar{\nu}}(Q^2,\nu)\times\cos^2\frac{\theta}{2}\right.
$$

$$
\left.\pm W_3^{\nu,\bar{\nu}}(Q^2,\nu)\frac{E+E'}{M}\sin^2\frac{\theta}{2}\right)
$$

$$
= \frac{G_F^2}{2\pi}\left(xy^2\frac{M}{\nu}W_1(x,y) + \left(1-y-\frac{Mxy}{2E}\right)W_2(x,y)\right.
$$

$$
\left.\pm xy\left(1-\frac{y}{2}W_3(x,y)\right)\right). \tag{4.68}
$$

Equation (4.68) can be deduced from more general arguments (see [Clo79, Lea96]). With the formulae given for the kinematic variables (4.49)–(4.53), this can be translated into other quantities as well:

$$
\frac{d^2\sigma}{dx\,dy} = 2ME\nu \times \frac{d^2\sigma}{dQ^2\,d\nu} = \frac{M\nu}{E'} \times \frac{d^2\sigma}{dE'\,d\cos\theta} = 2MEx \times \frac{d^2\sigma}{dx\,dQ^2}. \tag{4.69}
$$

The three structure functions W_i describe the internal structure of the proton as seen in neutrino–proton scattering. At very high energies, the W-propagator term can no longer be neglected and in (4.59) the replacement

$$
G_F^2 \to G_F^2 \left/ \left(1+\frac{Q^2}{m_W^2}\right)^2 \right. \tag{4.70}
$$

has to be made. The description for $ep/\mu p$ scattering is similar with the exception that there are only two structure functions. The term containing W_3 is missing because it is parity violating. By investigating inelastic ep scattering at SLAC in the late 1960s [Bre69], it was found that at values of Q^2 and ν are not too small ($Q^2 > 2$ GeV^2, $\nu > 2$ GeV), the structure functions did not depend on two variables independently but only on the dimensionless combination in the form of the Bjorken scaling variable $x = Q^2/2M\nu$ (4.52). This behaviour was predicted by Bjorken [Bjo67] for deep inelastic scattering and is called scaling invariance (or Bjorken scaling). A physical interpretation was given by Feynman as discussed in the next section. The same scaling behaviour is observed in high-energy neutrino scattering, leading to the replacements

$$
MW_1(Q^2,\nu) = F_1(x) \tag{4.71}
$$

$$
\nu W_2(Q^2,\nu) = F_2(x) \tag{4.72}
$$

$$
\nu W_3(Q^2,\nu) = F_3(x). \tag{4.73}
$$

4.9   The quark–parton model, parton distribution functions

The basic idea behind the parton model is the following [Fey69, Lea96, Sch97]: in elastic electromagnetic scattering of a point-like particle on an extended target, the

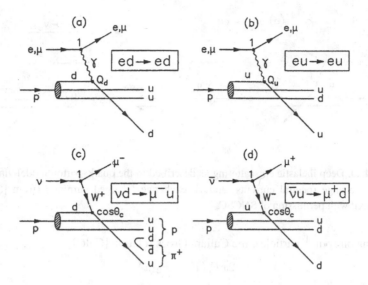

Figure 4.19. Graphs for the dominant processes in DIS in $ep/\mu p$ scattering (a, b), νp scattering (c) and $\bar{\nu}p$ scattering (d) (from [Sch97]). Reproduced with permission of SNCSC.

spatial extension can be described by a form factor $F(Q^2)$. This form factor can be seen as the Fourier transform of the spatial charge or magnetic moment distribution of the target. Form factors independent of Q^2 imply hard elastic scattering on point-like target objects, called partons. The SLAC results can be interpreted in such a way that the scaling invariance implies that deep inelastic ep-scattering can be seen as an incoherent superposition of hard elastic electron-parton scattering. The parton is kicked out of the proton, while the remaining partons (the proton remnant) act as spectators and are not involved in the interaction (Figure 4.19). After that the processes of fragmentation and hadronization follow, producing the particles observable in high-energy experiments. In this model, the variable x can be given an intuitive interpretation: assuming a proton with 4-momentum $p_p = (E_p, P_p)$, a parton has the 4-momentum $xp_p = (xE_p, xP_p)$ before its interaction. This means the variable x $(0 < x < 1)$ describes the fraction of the proton momentum and energy of the interacting parton (Figure 4.20). After several experiments on deep inelastic lepton–nucleon scattering, the result was that the partons are identical to the quarks proposed by Gell-Mann and Zweig in their SU(3) classification of hadrons [Gel64, Zwe64]. In addition to the valence quarks (a proton can be seen as a combination of uud-quarks, a neutron as of udd-quarks), the gluons also contribute, because, according to the Heisenberg uncertainty principle, they can fluctuate into quark–antiquark pairs for short times. These are known as the sea-(anti)quarks.

The picture described, called the quark–parton model (QPM), is today the basis for the description of deep inelastic lepton–nucleon scattering. For high Q^2 and the

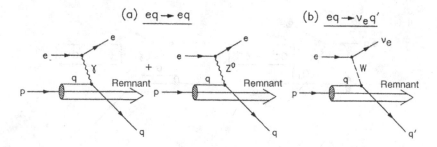

Figure 4.20. Deep inelastic ep scattering as described in the quark–parton model via photon and Z^0 exchange (neutral currents) and W-exchange (charged currents) (from [Sch97]). Reproduced with permission of SNCSC.

scattering on spin-$\frac{1}{2}$ particles, the Callan–Gross relation [Cal69]

$$2xF_1(x) = F_2(x) \tag{4.74}$$

holds between the first two structure functions. For a derivation see [Lea96].

4.9.1 Deep inelastic neutrino proton scattering

First, let us define the parton distribution functions (PDFs) within a proton. As an example, take the up-quark $u(x)$:

$$u(x)\,\mathrm{d}x = \text{Number of } u\text{-quarks in the proton with momentum}$$
$$\text{fraction between } x \text{ and } x + \mathrm{d}x \tag{4.75}$$

and corresponding definitions for the other quarks and antiquarks. They can be split into a valence- and a sea-quark contribution

$$u(x) = u_V(x) + u_S(x) \qquad d(x) = d_V(x) + d_S(x). \tag{4.76}$$

Symmetry of the $q\bar{q}$ sea requires

$$u_S(x) = \bar{u}(x) \qquad s(x) = \bar{s}(x)$$
$$d_S(x) = \bar{d}(x) \qquad c(x) = \bar{c}(x). \tag{4.77}$$

Because of the valence quark structure of the proton (uud), it follows that

$$\int_0^1 u_V(x)\,\mathrm{d}x = \int_0^1 [u(x) - \bar{u}(x)]\,\mathrm{d}x = 2 \tag{4.78}$$

$$\int_0^1 d_V(x)\,\mathrm{d}x = \int_0^1 [d(x) - \bar{d}(x)]\,\mathrm{d}x = 1. \tag{4.79}$$

The QPM predicts deep inelastic scattering as an incoherent sum of (quasi)-elastic lq or $l\bar{q}$ scattering on partons. The double differential cross-section can be written as

$$\frac{\mathrm{d}^2\sigma}{\mathrm{d}x\,\mathrm{d}y}(lp \to l'X) = \sum_{q,q'} q(x)\frac{\mathrm{d}\sigma}{\mathrm{d}y}(lq \to l'q') + \sum_{\bar{q},\bar{q}'} \bar{q}(x)\frac{\mathrm{d}\sigma}{\mathrm{d}y}(l\bar{q} \to l'\bar{q}'). \tag{4.80}$$

Using fundamental Feynman rules, one gets the following relations:

$$\frac{d\sigma}{dy}(eq \to eq) = \frac{d\sigma}{dy}(e\bar{q} \to e\bar{q}) = \frac{8\pi\alpha^2}{Q^4} m_q E q_q^2 \left(1 - y + \frac{y^2}{2}\right) \quad (4.81)$$

$$\frac{d\sigma}{dy}(\nu q \to \mu^- q') = \frac{d\sigma}{dy}(\bar{\nu}\bar{q} \to \mu^+ \bar{q}') = \frac{2G_F^2}{\pi} m_q E \quad (4.82)$$

$$\frac{d\sigma}{dy}(\nu\bar{q} \to \mu^- \bar{q}') = \frac{d\sigma}{dy}(\bar{\nu}q \to \mu^+ q') = \frac{2G_F^2}{\pi} m_q E(1 - y)^2 \quad (4.83)$$

where $y = 1 - E'/E = 1/2(1 - \cos\theta^*)$ (4.53) and q_q is the charge of the quark. Equation (4.72) describes electromagnetic interactions via photon exchange, while (4.82) and (4.83) follow from V-A theory ignoring the W-propagator. The additional term $(1 - y)^2$ for $\bar{\nu}$ scattering follows from angular momentum conservation because scattering with $\theta^* = 180°$ ($y = 1$) is not allowed. The corresponding cross-sections can then be written using the QPM formulae as

$$\frac{d\sigma}{dx\,dy}(\nu p) = \sigma_0 \times 2x[[d(x) + s(x)] + [\bar{u}(x) + \bar{c}(x)](1 - y)^2] \quad (4.84)$$

$$\frac{d\sigma}{dx\,dy}(\bar{\nu} p) = \sigma_0 \times 2x[[u(x) + c(x)](1 - y)^2 + [\bar{d}(x) + \bar{s}(x)]] \quad (4.85)$$

with (using (4.49))

$$\sigma_0 = \frac{G_F^2 ME}{\pi} = \frac{G_F^2 s}{2\pi} = 1.583 \times 10^{-38} \text{ cm}^2 \times E/\text{GeV}. \quad (4.86)$$

Equation (4.84) together with scaling invariance and the Callan-Gross relation (4.74) allows the derivation of the following relations:

$$F_2^{\nu p}(x) = 2x[d(x) + \bar{u}(x) + s(x) + \bar{c}(x)]$$
$$xF_3^{\nu p}(x) = 2x[d(x) - \bar{u}(x) + s(x) - \bar{c}(x)]$$
$$F_2^{\bar{\nu} p}(x) = 2x[u(x) + c(x) + \bar{d}(x) + \bar{s}(x)]$$
$$xF_3^{\bar{\nu} p}(x) = 2x[u(x) + c(x) - \bar{d}(x) - \bar{s}(x)]. \quad (4.87)$$

In a similar way, neutron structure functions can be written in terms of the proton PDFs by invoking isospin invariance:

$$u_n(x) = d_p(x) = d(x)$$
$$d_n(x) = u_p(x) = u(x)$$
$$s_n(x) = s_p(x) = s(x)$$
$$c_n(x) = c_p(x) = c(x). \quad (4.88)$$

The corresponding structure functions are then

$$F_2^{\nu n}(x) = 2x[u(x) + \bar{d}(x) + s(x) + \bar{c}(x)]$$
$$xF_3^{\nu n}(x) = 2x[u(x) - \bar{d}(x) + s(x) - \bar{c}(x)]. \quad (4.89)$$

Finally the cross-section for lepton scattering on an isoscalar target N is obtained by averaging

$$\frac{\mathrm{d}^2\sigma}{\mathrm{d}x\,\mathrm{d}y}(lN) = \frac{1}{2}\left(\frac{\mathrm{d}^2\sigma}{\mathrm{d}x\,\mathrm{d}y}(lp) + \frac{\mathrm{d}^2\sigma}{\mathrm{d}x\,\mathrm{d}y}(ln)\right) \qquad F_i^{lN} = \frac{1}{2}(F_i^{lp} + F_i^{ln}). \quad (4.90)$$

Combining (4.78), (4.79) and (4.81) and assuming $s = \bar{s}$, $c = \bar{c}$ results in

$$F_2^{e(\mu)N} = \frac{5}{18}x(u + d + \bar{u} + \bar{d}) + \frac{1}{9}x(s + \bar{s}) + \frac{4}{9}x(c + \bar{c})$$
$$F_2^{\nu N} = F_2^{\bar{\nu}N} = x[u + d + s + c + \bar{u} + \bar{d} + \bar{s} + \bar{c}] = x[q + \bar{q}]$$
$$xF_3^{\nu N} = x[u + d + 2s - \bar{u} - \bar{d} - 2\bar{c}] = x[q - \bar{q} + 2(s - c)]$$
$$xF_3^{\bar{\nu}N} = x[u + d + 2c - \bar{u} - \bar{d} - 2\bar{s}] = x[q - \bar{q} - 2(s - c)] \quad (4.91)$$
$$\text{with } q = u + d + s + c, \; \bar{q} = \bar{u} + \bar{d} + \bar{s} + \bar{c}.$$

As can be seen, the structure function $F_2^{\nu N}$ measures the density distribution of all quarks and antiquarks within the proton, while the $\nu/\bar{\nu}$ averaged structure function $F_3^{\nu N}$ measures the valence-quark distribution. Reordering (4.81) shows that F_2 and F_3 can be basically determined by the sum and difference of the differential cross-sections.

Experimentally the procedure is as follows (for details see [Die91, Con98]). Using the equations given earlier, the structure functions are determined from the differential cross-sections. From these, the single-quark distribution functions as well as the gluon structure function $xg(x)$ can be extracted. Figure 4.22 shows a compilation of such an analysis. As can be seen, the sea quarks are concentrated at low x ($x < 0.4$) values, while the valence quarks extend to higher values. It should be noted that the numbers are given for a fixed Q^2. Extensive measurements over a wide range of x and Q^2, increasing the explored parameter space by two orders of magnitude, have been performed at HERA (see [Abr99, Kle08]) and now at LHC [Ale17]. CCFR and CHORUS published new low x, low Q^2 analysis based on neutrino scattering data [Fle01, One06].

4.9.1.1   QCD effects

As already mentioned, measurements of structure functions over a wide range of Q^2 show a deviation from scaling invariance for fixed x:

$$F_i(x) \rightarrow F_i(x, Q^2). \quad (4.92)$$

For higher Q^2, $F_i(x, Q^2)$ rises at small x and gets smaller at high x (Figure 4.23). This can be understood by QCD. Higher Q^2 implies a better time and spatial resolution. Therefore, more and more partons from the sea with smaller and smaller momentum fractions can be observed, leading to a rise at small x. Quantitatively, this Q^2 evolution of the structure functions can be described by the DGLAP (named after Dokshitzer, Gribov, Lipatov, Altarelli and Parisi) equations [Gri72, Alt77, Dok77].

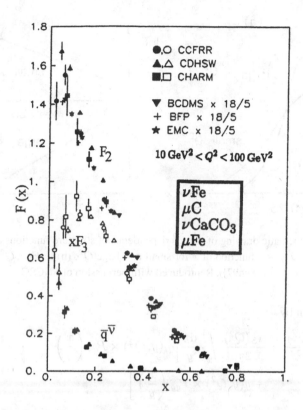

Figure 4.21. Compilation of the structure functions $F_2^{\nu N}$ and $xF_3^{\nu N}$ from $\nu/\bar{\nu}$ scattering as well as of (5/18) $F_2^{\mu N}$ from μ scattering on isoscalar targets and the distribution function $\bar{q}^{\bar{\nu}} = x(\bar{q} - \bar{s} - \bar{c})$ (from [Hik92]). © 1992 by the American Physical Society.

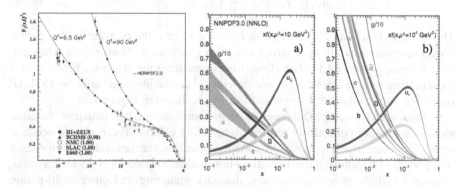

Figure 4.22. Left: Data of the structure functions $F_2^{\nu N}$ as a function of Bjorken x shown for two different Q^2. The closed lines are the fits using HERAPDF2.0. Right: Unpolarised parton distribution functions for individual quarks as function of x (from [PDG18]). With kind permission of M. Tanabashi et al. (Particle Data Group).

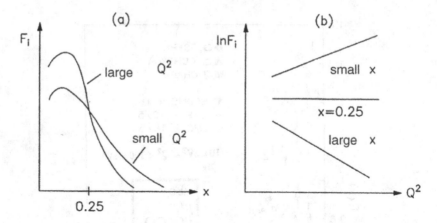

Figure 4.23. Schematic drawing of the Q^2 dependence of structure functions as predicted by QCD. (a) $F(x, Q^2)$ as a function of x for small and large Q^2; (b) $\ln F(x, Q^2)$ as a function of Q^2 for fixed x (from [Sch97]). Reproduced with permission of SNCSC.

They are given by

$$\frac{dq_i(x, Q^2)}{d\ln Q^2} = \frac{\alpha_S(Q^2)}{2\pi} \int_x^1 \frac{dy}{y} \left[q_i(y, Q^2) \times P_{qq}\left(\frac{x}{y}\right) \right.$$

$$\left. + g(y, Q^2) \times P_{qg}\left(\frac{x}{y}\right) \right] \tag{4.93}$$

$$\frac{dg_i(x, Q^2)}{d\ln Q^2} = \frac{\alpha_S(Q^2)}{2\pi} \int_x^1 \frac{dy}{y} \left[\sum_{j=1}^{N_f} [q_j(y, Q^2) + \bar{q}_j(y, Q^2)] \times P_{gq}\left(\frac{x}{y}\right) \right.$$

$$\left. + g(y, Q^2) \times P_{gg}\left(\frac{x}{y}\right) \right]. \tag{4.94}$$

The splitting functions $P_{ij}(x/y)$ (with $i, j = q, g$) give the probability that parton j with momentum y will be resolved as parton i with momentum $x < y$. They can be calculated within QCD. Therefore, from measuring the structure function at a fixed reference value Q_0^2, their behaviour with Q^2 can be predicted with the DGLAP equations. A compilation of structure functions is shown in Figure 4.24.

Non-perturbative QCD processes that contribute to the structure function measurements are collectively termed higher-twist effects. These effects occur at small Q^2 where the impulse approximation (treating the interacting parton as a free particle) of scattering from massless non-interacting quarks is no longer valid. Examples include target mass effects, di-quark scattering and other multi-parton effects. As neutrino experiments use heavy targets in order to obtain high interaction rates, nuclear effects must also be considered. They can be divided into three types. First of all, there is the Fermi motion and binding energy of the target nucleons (about 250 MeV) which change the interaction kinematics. In addition, there is the Pauli suppression of the available phase space to the final state nucleons leading to a

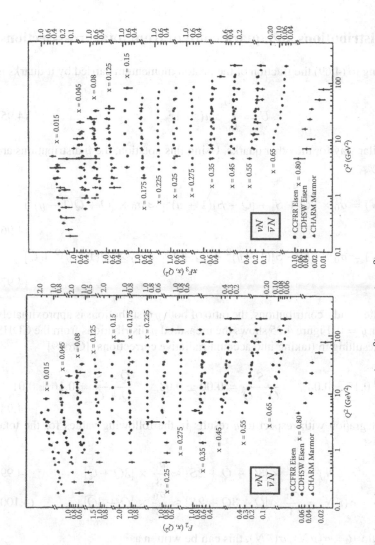

Figure 4.24. Compilation of the structure functions $F_2(x, Q^2)$ and $xF_3(x, Q^2)$ as obtained in νN and $\bar{\nu}N$ scattering with the CDHSW, CCFRR and CHARM experiments. The Q^2 dependence is plotted for fixed x (from [Hik92]). © 1992 by the American Physical Society.

Q^2 dependent cross-section reduction. Both can be described by a relativistic Fermi gas model [Smi72]. As a third effect the final state interactions inside the nucleus will change the composition and kinematics of the hadronic part of the final state. These nuclear effects have been observed in liquid argon detectors [Arn06]. For more detailed treatments see [Con98, Gal11, Mos16].

4.10   *y* distributions and quark content from total cross-sections

Corresponding to (4.69) the fraction of the proton momentum carried by u-quarks is defined by

$$U = \int_0^1 xu(x)\,\mathrm{d}x \qquad (4.95)$$

and in a similar way for the other quarks. Using this notation, the y distributions are then given by

$$\frac{\mathrm{d}\sigma}{\mathrm{d}y}(\nu N) = \sigma_0 \times [[Q+S] + [\bar{Q}-S](1-y)^2] \approx \sigma_0 \times [Q + \bar{Q}(1-y)^2] \qquad (4.96)$$

$$\frac{\mathrm{d}\sigma}{\mathrm{d}y}(\bar{\nu} N) = \sigma_0 \times [[Q-S](1-y)^2 + [\bar{Q}+S]] \approx \sigma_0 \times [Q(1-y)^2 + \bar{Q}] \qquad (4.97)$$

Neglecting the s and c contributions, the ratio of both y distributions is approximately about one for $y = 0$. Figure 4.25 shows the measured y distributions from the CDHS experiment resulting in (taking into account radiative corrections) [Gro79]

$$\frac{\bar{Q}}{Q+\bar{Q}} = 0.15 \pm 0.03 \qquad \frac{S}{Q+\bar{Q}} = 0.00 \pm 0.03 \qquad \frac{\bar{Q}+S}{Q+\bar{Q}} = 0.16 \pm 0.01. \qquad (4.98)$$

A further integration with respect to y results in the following values for the total cross sections:

$$\sigma(\nu N) = \frac{\sigma_0}{3} \times [3Q + \bar{Q} + 2S] \approx \frac{\sigma_0}{3} \times [3Q + \bar{Q}] \qquad (4.99)$$

$$\sigma(\bar{\nu} N) = \frac{\sigma_0}{3} \times [Q + 3\bar{Q} + 2S] \approx \frac{\sigma_0}{3} \times [Q + 3\bar{Q}]. \qquad (4.100)$$

Using the ratio $R = \sigma(\nu N)/\sigma(\bar{\nu} N)$, this can be written as

$$\frac{\bar{Q}}{Q} = \frac{3-R}{3R-1}. \qquad (4.101)$$

A measurement of $R < 3$ is a direct hint of the momentum contribution $\bar{Q}$ of the sea quarks (see Section 4.11). Using the measured values (4.89) resulting in $R = 2.02$,

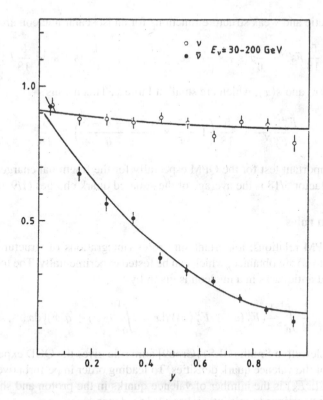

Figure 4.25. The differential cross-sections versus y as obtained by CDHS for νN and $\bar{\nu} N$ CC scattering. The dominant flat distribution for neutrinos and $(1-y)^2$ behaviour for antineutrinos shows that left- and right-handed couplings are different. The distributions are explained by dominant scattering from valence quarks with left-handed couplings (from [Eis86]). © IOP Publishing. Reproduced with permission. All rights reserved.

it follows that $Q \approx 0.41$ and $\bar{Q} \approx 0.08$. Therefore,

$$\int_0^1 F_2^{\nu N}(x)\,\mathrm{d}x = Q + \bar{Q} \approx 0.49$$

$$Q_V = Q - \bar{Q} \approx 0.33 \qquad Q_S = \bar{Q}_S = \bar{Q} \approx 0.08$$

$$\frac{\bar{Q}}{Q + \bar{Q}} \approx 0.16 \qquad \frac{\bar{Q}}{Q} \approx 0.19. \tag{4.102}$$

This shows that quarks and antiquarks carry about 49% of the proton momentum, whereas valence quarks contribute about 33% and sea quarks about 16%. Half of the proton spin has to be carried by the gluons. For more extensive reviews on nucleon structure see [Con98, Lam00].

The QPM equations allow predictions to be made about the different structure functions, which can serve as important tests for the model. As an example, the electromagnetic and weak structure functions for an isoscalar nucleon are related by

$$F_2^{\mu N, eN} = \frac{5}{18} F_2^{\nu N} - \frac{1}{6} x[s + \bar{s} - c - \bar{c}] \approx \frac{5}{18} F_2^{\nu N} - \frac{1}{6} x[s + \bar{s}] \approx \frac{5}{18} F_2^{\nu N} \quad (4.103)$$

neglecting $c(x)$ and $s(x)$, which are small at large x. This means

$$\frac{F_2^{\mu N, eN}}{F_2^{\nu N}} = \frac{5}{18} \left(1 - \frac{3}{5} \times \frac{s + \bar{s} - c - \bar{c}}{q + \bar{q}} \right) \approx \frac{5}{18}. \quad (4.104)$$

This is an important test for the QPM especially for the fractional charge of quarks, because the factor 5/18 is the average of the squared quark charges (1/9 and 4/9).

4.10.1   Sum rules

Using the QPM relations, important sum rules (integrations of structure functions with respect to x) are obtained, which can be tested experimentally. The total number of quarks and antiquarks in a nucleon is given by

$$\frac{1}{2} \int_0^1 \frac{1}{x} (F_2^\nu(x) + F_2^{\bar{\nu}}(x)) \, dx = \int_0^1 [q(x) + \bar{q}(x)] \, dx. \quad (4.105)$$

The Gross–Llewellyn Smith (GLS) [Gro69] sum rule gives the QCD expectation for the integral of the valence quark densities. To leading order in perturbative QCD, the integral $\int \frac{dx}{x} (x F_3)$ is the number of valence quarks in the proton and should equal three. QCD corrections to this integral result in a dependence on α_s

$$S_{\text{GLS}} = \frac{1}{2} \int_0^1 (F_3^\nu(x) + F_3^{\bar{\nu}}(x)) \, dx = \int_0^1 \bar{F}_3(x) \, dx = \int_0^1 [q(x) - \bar{q}(x)] \, dx$$

$$= 3 \left[1 - \frac{\alpha_s}{\pi} - a(n_f) \left(\frac{\alpha_s}{\pi} \right)^2 - b(n_f) \left(\frac{\alpha_s}{\pi} \right)^3 \right]. \quad (4.106)$$

In this equation, a and b are known functions of the number of quark flavours n_f which contribute to scattering at a given x and Q^2. This is one of the few QCD predictions that are available to the order of α_s^3. The world average is [Con98]

$$\int_0^1 F_3(x) \, dx = 2.64 \pm 0.06 \quad (4.107)$$

which is consistent with the next-to-next-to-leading order evaluation of (4.106) with the QCD parameter $\Lambda_{\text{QCD}} = 250 \pm 50$ MeV.

A further important sum rule is the Adler sum rule [Adl66]. This predicts the difference between the quark densities of the neutron and the proton, integrated over x (Figure 4.26). It is given at high energies (in all orders of QCD) by

$$S_A = \frac{1}{2} \int_0^1 \frac{1}{x} (F_2^{\nu n}(x) - F_2^{\nu p}(x)) \, dx = \int_0^1 [u_V(x) - d_V(x)] \, dx = 1. \quad (4.108)$$

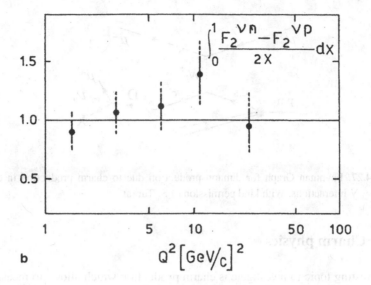

Figure 4.26. Test of the Adler sum rule. The estimated uncertainties (dashed lines) are shown separately (from [All85a]). Reproduced with permission of SNCSC.

Common to the determination of sum rules is the experimental difficulty of measuring them at very small x, the part dominating the integral.

For completeness, two more sum rules should be mentioned. The analogue to the Adler sum rule for charged-lepton scattering is the Gottfried sum rule [Got67]:

$$S_G = \int_0^1 \frac{1}{x} (F_2^{\mu p}(x) - F_2^{\mu n}(x))\, dx = \frac{1}{3} \int_0^1 [u(x) + \bar{u}(x) - d(x) - \bar{d}(x)]\, dx$$

$$= \frac{1}{3} \left(1 + 2 \int_0^1 [\bar{u}(x) - \bar{d}(x)]\, dx \right) = \frac{1}{3} \tag{4.109}$$

The experimental value is $S_G = 0.235 \pm 0.026$ [Arn94]. This is significantly different from expectation and might be explained by an isospin asymmetry of the sea, i.e., $\bar{u}(x) \neq \bar{d}(x)$, strongly supported by recent measurements [Ack98a]. Note that this assumption $\bar{u} = \bar{d}$ was not required in the Adler sum rule. Furthermore, there is the polarised Bjorken sum rule [Bjo67]

$$S_B = \int_0^1 [F_1^{\bar{\nu} p}(x) - F_1^{\nu p}(x)]\, dx = 1 - \frac{2\alpha_S(Q^2)}{3\pi}. \tag{4.110}$$

For more details see [Lea96]. We now continue to discuss a few more topics investigated in neutrino nucleon scattering. Because of the richness of possible observable quantities, we restrict ourselves to a few examples. For more details see [Sch97, Con98].

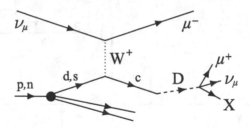

Figure 4.27. Feynman Graph for dimuon production due to charm production in charged current $\nu_\mu N$ interactions. With kind permission of S. Turkat.

4.11   Charm physics

An interesting topic to investigate is charm production which allows to measure the mass of the charm quark. In the case of neutrino scattering, the underlying process is a neutrino interacting with an s or d quark, producing a charm quark that fragments into a charmed hadron. The charmed hadrons decay semi-leptonically (BR $\approx 10\%$) and produce a second muon of opposite sign (the so-called opposite sign di-muon (OSDM) events) (Figure 4.27)

$$\nu_\mu + N \longrightarrow \mu^- + c + X \tag{4.111}$$
$$\hookrightarrow s + \mu^+ + \nu_\mu.$$

However, the relatively large mass m_c of the charm quark gives rise to a threshold behaviour in the dimuon production rate at low energies. This is effectively described by the slow rescaling model [Bar76, Geo76] in which x is replaced by the slow rescaling variable ξ given by

$$\xi = x \left(1 + \frac{m_c^2}{Q^2}\right). \tag{4.112}$$

The differential cross-section for dimuon production is then expressed generally as

$$\frac{\mathrm{d}^3\sigma(\nu_\mu N \to \mu^-\mu^+ X)}{\mathrm{d}\xi\,\mathrm{d}y\,\mathrm{d}z} = \frac{\mathrm{d}^2\sigma(\nu_\mu N \to cX)}{\mathrm{d}\xi\,\mathrm{d}y} D(z) B_c(c \to \mu^+ X) \tag{4.113}$$

where the function $D(z)$ describes the hadronization of charmed quarks and B_c is the weighted average of the semi-leptonic branching ratios of the charmed hadrons produced in neutrino interactions. As mentioned before, in leading order, charm is produced by direct scattering of the strange and down quarks in the nucleon. The leading order differential cross-section for an isoscalar target, neglecting target mass

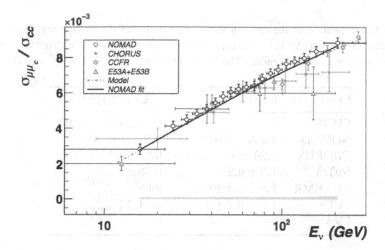

Figure 4.28. Compilation of observed dimuon *versus* single muon rates due to CC reactions as a function of energy E (shown in log scale) as obtained by E531, CDHS, CCFR, CHORUS and NOMAD (from [Sam13]). © 2013 With permission from Elsevier.

effects, is

$$\frac{d^3\sigma(\nu_\mu N \to cX)}{d\xi \, dy \, dz} = \frac{G_F^2 M E_\nu \xi}{\pi}[u(\xi, Q^2) + d(\xi, Q^2)]|V_{cd}|^2$$

$$+ 2s(\xi, Q^2)|V_{cs}|^2\left(1 - y + \frac{xy}{\xi}\right)D(z)B_c. \quad (4.114)$$

Therefore, by measuring the ratio of dimuon production *versus* single muon production as a function of neutrino energy, m_c can be determined from the threshold behaviour (Figure 4.28). The production of opposite-sign dimuons is also governed by the proportion of strange to non-strange quarks in the nucleon sea, $\kappa = 2\bar{s}/(\bar{u}+\bar{d})$, the CKM matrix elements V_{cd} and V_{cs} and B_c. Table 4.1 shows a compilation of such measurements. DIS data obtained at HERA have also been used for charm mass determinations [Ale13, Giz17].

The study of open charm production in the form of D-meson production is another important topic, especially to get some insight into the fragmentation process. CHORUS performed a search for D^0 production [Kay02]. In total, 283 candidates are observed, with an expected background of 9.2 events coming from K- and Λ-decay. The ratio $\sigma(D^0)/\sigma(\nu_\mu CC)$ is found to be $(1.99 \pm 0.13(\text{stat.})\pm0.17(\text{syst.})) \times 10^{-2}$ at 27 GeV average ν_μ energy. NOMAD performed a search for D^{*+}-production using the decay chain $D^{*+} \to D^0 + \pi^+$ followed by $D^0 \to K^- + \pi^+$. In total, 35 ± 7.2 events could be observed resulting in a D^{*+} yield in ν_μ CC interactions of $(0.79\pm0.17(\text{stat.})\pm0.10(\text{syst.}))\%$ [Ast02]. The same search was performed with CHORUS resulting in 22.1 ± 5.5 events and a yield with respect to ν_μ CC interactions of $(1.02\pm0.25(\text{stat.}) \pm 0.15(\text{syst.}))\%$ [One05]. Thus, within errors, both measurements are in good agreement. Another measurement

Table 4.1. Compilation of the mass of the charm quark and the strange sea parameter κ obtained by leading order fits in various experiments. The experiments are ordered with respect to increasing average neutrino energy. The NOMAD value for κ is given at $Q^2 = 20\ \text{GeV}^2/\text{c}^2$.

Experiment	m_c (GeV)	$\kappa \pm$ (stat.) $\pm$ (syst.)
CDHS	—	$0.47 \pm 0.08 \pm 0.05$
NOMAD	1.159 ± 0.075	0.591 ± 0.019
CHORUS	$1.26 \pm 0.16 \pm 0.09$	$0.33 \pm 0.05 \pm 0.05$
NuTeV	1.3 ± 0.2	0.38 ± 0.08
CHARMII	$1.8 \pm 0.3 \pm 0.3$	$0.39^{+0.07+0.07}_{-0.06-0.07}$
CCFR	$1.3 \pm 0.2 \pm 0.1$	$0.44^{+0.09+0.07}_{-0.07-0.02}$
FMMF	—	$0.41^{+0.08+0.103}_{-0.08-0.069}$

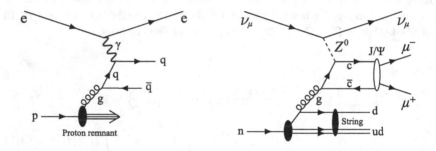

Figure 4.29. Feynman graph of boson-gluon fusion. Left: Photon-gluon fusion as obtained in $e, \mu N$ scattering producing J/Ψ mesons. This is a direct way to measure the gluon structure function $xg(x)$. Right: Z^0-gluon fusion responsible for neutral current J/Ψ production in νN scattering. With kind permission of S. Turkat.

related to charm is the production of bound charm–anticharm states like the J/Ψ. Due to the small cross-section, the expected number of events in current experiments is rather small. It can be produced via NC reactions by boson-gluon fusion as shown in Figure 4.29. They were investigated by three experiments (CDHS [Abr82], CHORUS [Esk01] and NuTeV [Ada00]) with rather inconclusive results. Their production in νN scattering can shed some light on the theoretical description of heavy quarkonium systems, which is not available in other processes [Pet99, Kni02].

The charm quark can be produced from strange quarks in the sea. This allows $s(x)$ to be measured by investigating dimuon production. It is not only possible to measure the strange sea of the nucleon, but also to get information about its polarization. This is done by measurements of the Λ-polarization. The polarization is measured by the asymmetry in the angular distributions of the protons in the parity-violating decay process $\Lambda \rightarrow p\pi^-$. In the Λ rest frame, the decay protons

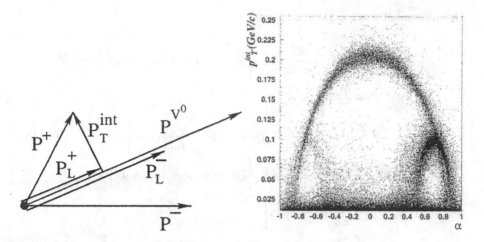

Figure 4.30. Left: Definition of kinematic variables. Right: Armenteros plot for neutral decaying particles V^0 as observed by the NOMAD experiment, showing clearly the distribution of kaons (big parabola), $\bar{\Lambda}$s (small parabola, left-hand corner) and Λs (small parabola, right-hand corner) (from [Ast00]). © 2000 With permission from Elsevier.

are distributed as follows

$$\frac{1}{N}\frac{dN}{d\Omega} = \frac{1}{4\pi}(1 + \alpha_\Lambda Pk) \tag{4.115}$$

where P is the Λ polarization vector, $\alpha_\Lambda = 0.642 \pm 0.013$ is the decay asymmetry parameter and k is the unit vector along the proton decay direction. Since NOMAD is unable to distinguish protons from pions in the range relevant for this search, any search for neutral strange particles (V^0) should rely on the kinematics of the V^0-decay. The definition of the kinematic variables and the so-called Armenteros plot are shown in Figure 4.30. Their results on Λ and $\bar{\Lambda}$ polarization can be found in [Ast00, Ast01].

4.12 Neutral current reactions

Inelastic neutral current (NC) reactions $\nu N \to \nu N$ are described by the QPM as elastic NC events such as

$$\nu q \to \nu q \qquad \nu \bar{q} \to \nu \bar{q} \tag{4.116}$$

$$\bar{\nu} q \to \bar{\nu} q \qquad \bar{\nu} \bar{q} \to \bar{\nu} \bar{q}. \tag{4.117}$$

The differential cross-sections are given by

$$
\frac{d\sigma}{dy}(\nu q) = \frac{d\sigma}{dy}(\bar{\nu}\bar{q}) = \frac{G_F^2 m_q}{2\pi} E_\nu \left[(g_V + g_A)^2 + (g_V - g_A)^2 (1-y)^2 \right.
$$
$$
\left. + \frac{m_q}{E_\nu}(g_A^2 - g_V^2)y \right]
$$
$$
= \frac{2G_F^2 m_q}{\pi} E_\nu \left[g_L^2 + g_R^2 (1-y)^2 - \frac{m_q}{E_\nu} g_L g_R y \right]
\tag{4.118}
$$
$$
\frac{d\sigma}{dy}(\bar{\nu}q) = \frac{d\sigma}{dy}(\nu\bar{q}) = \frac{G_F^2 m_q}{2\pi} E_\nu \left[(g_V - g_A)^2 + (g_V + g_A)^2 (1-y)^2 \right.
$$
$$
\left. + \frac{m_q}{E_\nu}(g_A^2 - g_V^2)y \right]
$$
$$
= \frac{2G_F^2 m_q}{\pi} E_\nu \left[g_R^2 + g_L^2 (1-y)^2 - \frac{m_q}{E_\nu} g_L g_R y \right].
\tag{4.119}
$$

For the following, the last term will be neglected because of $E_\nu \gg m_q$. The GWS predictions for the coupling constants are:

$$
g_V = \frac{1}{2} - \frac{4}{3}\sin^2\theta_W \qquad g_A = \frac{1}{2} \qquad \text{for } q \equiv u, c
$$
$$
g_V' = -\frac{1}{2} + \frac{2}{3}\sin^2\theta_W \qquad g_A' = -\frac{1}{2} \qquad \text{for } q \equiv d, s
\tag{4.120}
$$

and

$$
g_L = \frac{1}{2} - \frac{2}{3}\sin^2\theta_W \qquad g_R = -\frac{2}{3}\sin^2\theta_W \qquad \text{for } q \equiv u, c
$$
$$
g_L' = -\frac{1}{2} + \frac{1}{3}\sin^2\theta_W \qquad g_R' = \frac{1}{3}\sin^2\theta_W \qquad \text{for } q \equiv d, s.
\tag{4.121}
$$

According to the QPM, a similar relation holds as in CC events (4.71)

$$
\frac{d\sigma}{dx\,dy}\left(\overset{(-)}{\nu}p \to \overset{(-)}{\nu}X\right) = \sum_q q(x)\frac{d\sigma}{dy}(\overset{(-)}{\nu}q) + \sum_{\bar{q}} \bar{q}(x)\frac{d\sigma}{dy}(\overset{(-)}{\nu}\bar{q}).
\tag{4.122}
$$

The corresponding proton structure functions are then obtained:

$$
F_2^{\nu p, \bar{\nu}p} = 2x[(g_L^2 + g_R^2)[u + c + \bar{u} + \bar{c}] + (g_L'^2 + g_R'^2)[d + s + \bar{d} + \bar{s}]]
$$
$$
= x[(g_A^2 + g_V^2)[u + c + \bar{u} + \bar{c}] + (g_A'^2 + g_V'^2)[d + s + \bar{d} + \bar{s}]]
\tag{4.123}
$$
$$
xF_3^{\nu p, \bar{\nu}p} = 2x[(g_L^2 - g_R^2)[u + c - \bar{u} - \bar{c}] + (g_L'^2 - g_R'^2)[d + s - \bar{d} - \bar{s}]]
$$
$$
= 2x[g_V g_A[u + c - \bar{u} - \bar{c}] + g_V' g_A'[d + s - \bar{d} - \bar{s}]].
\tag{4.124}
$$

The neutron structure functions are obtained with the replacements given in (4.79) which leads to the structure functions for an isoscalar target:

$$F_2^{\nu N, \bar{\nu} N} = x[(g_L^2 + g_R^2)[u + d + 2c + \bar{u} + \bar{d} + 2\bar{c}]$$
$$+ (g_L'^2 + g_R'^2)[u + d + 2s + \bar{u} + \bar{d} + 2\bar{s}]]$$
$$xF_3^{\nu N, \bar{\nu} N} = x(g_L^2 - g_R^2)[u + d + 2c - \bar{u} - \bar{d} - 2\bar{c}]]$$
$$+ (g_L'^2 - g_R'^2)[u + d + 2s - \bar{u} - \bar{d} - 2\bar{s}]]. \quad (4.125)$$

Neglecting the s and c sea quarks, the corresponding cross-sections can be written as

$$\frac{d^2\sigma}{dx\,dy}(\nu N) = \sigma_0 \times x[(g_L^2 + g_L'^2)[q + \bar{q}(1-y)^2]$$
$$+ (g_R^2 + g_R'^2)[\bar{q} + q(1-y)^2]] \quad (4.126)$$

$$\frac{d^2\sigma}{dx\,dy}(\bar{\nu} N) = \sigma_0 \times x[(g_R^2 + g_R'^2)[q + \bar{q}(1-y)^2]$$
$$+ (g_L^2 + g_L'^2)[\bar{q} + q(1-y)^2]] \quad (4.127)$$

with $q = u + d$ and $\bar{q} = \bar{u} + \bar{d}$ and σ_0 given by (4.77). Comparing these cross-sections with the CC ones, integrating with respect to x and y and using the measureable ratios

$$R_\nu^N = \frac{\sigma_{NC}(\nu N)}{\sigma_{CC}(\nu N)} \qquad R_{\bar{\nu}}^N = \frac{\sigma_{NC}(\bar{\nu} N)}{\sigma_{CC}(\bar{\nu} N)} \qquad r = \frac{\sigma_{CC}(\bar{\nu} N)}{\sigma_{CC}(\nu N)} \quad (4.128)$$

leads to the following interesting relations for the couplings

$$g_L^2 + g_L'^2 = \frac{R_\nu^N - r^2 R_{\bar{\nu}}^N}{1 - r^2} \qquad g_R^2 + g_R'^2 = \frac{r(R_{\bar{\nu}}^N - R_\nu^N)}{1 - r^2}. \quad (4.129)$$

Using the GWS predictions for the couplings, the precise measurements of R_ν^N or $R_{\bar{\nu}}^N$ allows a measurement of the Weinberg angle ($r = 0.5$ and using (4.112))

$$R_\nu^N = (g_L^2 + g_L'^2) + r(g_R^2 + g_R'^2) = \frac{1}{2} - \sin^2\theta_W + (1+r)\frac{5}{9}\sin^4\theta_W \quad (4.130)$$

$$R_{\bar{\nu}}^N = (g_L^2 + g_L'^2) + \frac{1}{r}(g_R^2 + g_R'^2) = \frac{1}{2} - \sin^2\theta_W + \left(1 + \frac{1}{r}\right)\frac{5}{9}\sin^4\theta_W \quad (4.131)$$

These ratios were measured by several experiments, the most accurate ones being CHARM, CDHSW and CCFR [All87, Hai88, Blo90, Arr94]. The values obtained by CDHSW are:

$$R_\nu^N = 0.3072 \pm 0.0033 \qquad R_{\bar{\nu}}^N = 0.382 \pm 0.016. \quad (4.132)$$

For a precision measurement of $\sin^2\theta_W$ several correction factors have to be taken into account. The analyses for the three experiments result in values for $\sin^2\theta_W$ of 0.236 ± 0.006 ($m_c = 1.5$ GeV), 0.228 ± 0.006 ($m_c = 1.5$ GeV) and 0.2218 ± 0.0059 ($m_c = 1.3$ GeV). For a compilation of measurements of the Weinberg angle see Section 3.4.3.

As a general summary of all the observed results, it can be concluded that the GWS predictions are in good agreement with the experimental results.

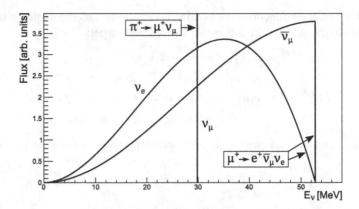

Figure 4.31. Energy spectrum of neutrinos coming from π^+ decay at rest. Besides a monoenergetic line of ν_μ at 29.8 MeV coming from pion decay, there are the continuous spectra of ν_e and $\bar{\nu}_\mu$ with equal intensity and energies up to 52.8 MeV from muon decay. With kind permission of S. Turkat.

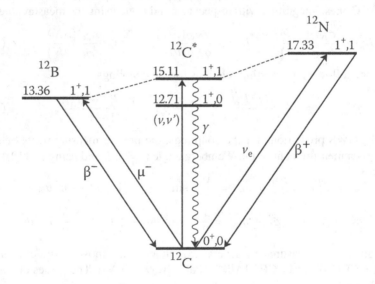

Figure 4.32. The $A = 12$ isobaric analogue triplet together with various possible transitions involving the ^{12}C ground state. With kind permission of G. Drexlin.

4.13   Neutrino cross-section on nuclei

After extensively discussing neutrino-nucleon scattering, it is worthwhile taking a short look at neutrino reactions with nuclei (for a recent review see [Eji19]). This is quite important not only for low-energy tests of electroweak physics, but also for neutrino astrophysics, either in the astrophysical process itself or in the detection of such neutrinos. This kind of neutrino spectroscopy has to be done with lower energy (a few MeV) neutrinos, typically coming from pion decay at rest (DAR) and subsequent muon decay (see Figure 4.31), giving rise to equal numbers of ν_e, ν_μ and $\bar{\nu}_\mu$. The study of such reactions allows important low-energy tests of NC and CC couplings and measurements of nuclear form factors. Consider, as an example, transitions between the ground state of ^{12}C and the isobaric analogue triplet states of the $A = 12$ system, i.e., ^{12}B, ^{12}C* and ^{12}N shown in Figure 4.32. It has well-defined quantum numbers and contains simultaneous spin and isospin flips $\Delta I = 1$, $\Delta S = 1$. Such neutrino reactions on carbon might be important for all experiments based on organic scintillators. The most stringent signature is the inverse β-reaction ^{12}C(ν_e, e^-) ^{12}N$_{gs}$, where ^{12}N$_{gs}$ refers to the ground state of ^{12}N. A coincidence signal can be formed by the prompt electron together with the positron from the ^{12}N$_{gs}$ β^+-decay with a lifetime of 15.9 ms. With appropriate spatial and time cuts, KARMEN (see Chapter 8) observed 536 such ν_e-induced CC events. The cross-section is dominated by the form factor F_A (see (4.41)), which is given using a dipole parametrization, the CVC hypothesis and scaling between F_M and F_A (see [Fuk88] for more details) by

$$\frac{F_A(Q^2)}{F_A(0)} = \frac{1}{(1 - \frac{1}{12}R_A^2 Q^2)^2}. \tag{4.133}$$

The radius of the weak axial charge distribution R_A has been determined by a fit as [Bod94]

$$R_A = (3.8^{+1.4}_{-1.8}) \text{ fm} \tag{4.134}$$

and the form factor at zero momentum transfer as

$$F_A(0) = 0.73 \pm 0.11 \tag{4.135}$$

in good agreement with values obtained from the ft-values (see Chapter 6) of ^{12}B and ^{12}N β-decay. For comparison, muon capture on ^{12}C is only able to measure the form factor at a fixed or zero momentum transfer.

Another reaction of interest is the NC inelastic scattering process ^{12}C(ν, ν') ^{12}C*$(1^+, 1; 15.1 \text{ MeV})$. The signal is a 15.1 MeV gamma ray. This peak is clearly visible in the KARMEN data (Figure 4.33). CC and NC reactions differ only by a Clebsch–Gordan coefficient of 1/2 and the fact that the ν_e and ν_μ spectra are almost identical allows the μ–e universality of the ν–Z^0 coupling at low energies to be tested. This can be done by looking at the ratio $R = \langle \sigma_{NC}(\nu_e + \bar{\nu}_\mu) \rangle / \langle \sigma_{CC}(\nu_e) \rangle$ which should be close to one. The measured value of KARMEN is

$$R = 1.17 \pm 0.11 \pm 0.12. \tag{4.136}$$

Using the NC inelastic scattering process, a test on the Lorentz structure of the weak interactions could also be performed. In the same way, the electron energy spectrum from muon decay is governed by the Michel parameter ρ; the ν_e energy spectrum depends on an analogous quantity ω_L. KARMEN measured

$$\omega_L = 2.7^{+3.8}_{-3.2} \pm 3.1 \times 10^{-2} \tag{4.137}$$

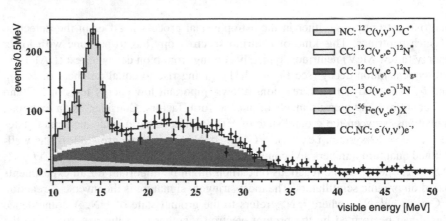

Figure 4.33. Energy spectrum of single prong events within the μ-decay time window (0.5–3.5 μs) as obtained by KARMEN. The peak corresponds to the reaction $^{12}\text{C}(\nu, \nu')\,^{12}\text{C}^*(1^+, 1; 15.1\,\text{MeV})$. The bump for energies larger than 16 MeV comes from a variety of ν_e-induced CC reactions on carbon and iron. The largest contribution is the CC reaction into excited states of ^{12}N (from [Eit08]). © IOP Publishing. Reproduced with permission. All rights reserved.

Table 4.2. Compilation of various nuclear cross-sections obtained by KARMEN and LSND in the $A = 12$ system averaged over the corresponding neutrino energies.

Reaction	σ (cm^2) KARMEN	σ (cm^2) LSND
$\langle\sigma(^{12}\text{C}(\nu_e, e^-)^{12}\text{N}_{gs})\rangle$	$9.3 \pm 0.4 \pm 0.8 \times 10^{-42}$	$9.1 \pm 0.4 \pm 0.9 \times 10^{-42}$
$\langle\sigma(^{12}\text{C}\,(\nu, \nu')\,^{12}\text{C}^*)\rangle$	$10.9 \pm 0.7 \pm 0.8 \times 10^{-42}$	—
$(\nu = \nu_e, \bar{\nu}_\mu)$		
$\langle\sigma(^{12}\text{C}(\nu_e, e^-)\,^{12}\text{N}^*)\rangle$	$5.1 \pm 0.6 \pm 0.5 \times 10^{-42}$	$5.7 \pm 0.6 \pm 0.6 \times 10^{-42}$
$\langle\sigma(^{12}\text{C}(\nu_\mu, \mu^-)^{12}\text{N}_{gs})\rangle$	—	$6.6 \pm 1.0 \pm 1.0 \times 10^{-41}$

in good agreement with the GWS prediction of $\omega_L = 0$ [Arm98]. A compilation of results from KARMEN and LSND (for both see Chapter 8) is shown in Table 4.2. Other examples will be discussed in the corresponding context.

After discussing neutrinos as probes of nuclear structure we now want to proceed to investigate neutrino properties especially in the case of non-vanishing neutrino masses. For that reason we start with a look at the physics beyond the standard model and the possibility of implementing neutrino masses in the Standard Model.

Chapter 5

Neutrino masses and physics beyond the Standard Model

DOI: 10.1201/9781315195612-5

Despite its enormous success in describing the available experimental data with high accuracy, the Standard Model discussed in Chapter 3 is generally not believed to be the last step in unification. In particular, there are several parameters that are not predicted as you would expect from theory. For example, the Standard Model contains 23 free parameters which have to be determined experimentally:

- the coupling constants e, α_S, $\sin^2 \theta_W$,
- the boson masses m_W, m_H,
- the lepton masses m_e, m_μ, m_τ, m_{ν_e}, m_{ν_μ}, m_{ν_τ}
- the quark masses m_u, m_d, m_s, m_c, m_b, m_t and
- the CKM and PMNS matrix parameters: each with three angles and a phase δ.

In addition, the mass hierarchy remains unexplained, left-handed and right-handed particles are treated very differently, and the quantization of the electric charge and the equality of the absolute values of proton and electron charge to a level better than 10^{-21} is not predicted. Furthermore, the SM does not provide a good particle candidate to act as dark matter (see Chapter 13).

However, what has undoubtedly succeeded is the unification of two of the fundamental forces at higher energies, namely weak interactions and electromagnetism. The question arises as to whether there is another more fundamental theory that will explain all these quantities and whether a further unification of forces at still higher energies can be achieved. The aim is to derive *all* interactions from the gauge transformations of *one simple* group G and, therefore, one coupling constant α (we will refrain here from discussing other, more specific solutions). Such theories are known as grand unified theories (GUTs). The grand unified group must contain the $SU(3) \otimes SU(2) \otimes U(1)$ group as a subgroup, i.e.,

$$G \supset SU(3) \otimes SU(2) \otimes U(1). \tag{5.1}$$

The gauge transformations of a simple group, which act on the particle multiplets characteristic for this group, result in an interaction between the elements within a multiplet which is mediated by a similarly characteristic number of gauge bosons. The three well-known and completely different coupling constants can be derived in the end from a single one only if the symmetry associated with the group G is broken in nature. The hope of achieving this goal is given by the experimental fact that it is known that the coupling constants are not really constants. For more extensive reviews on GUTs see [Lan81, Ros84, Moh86, 92, Fuk03a].

5.1   Running coupling constants

In quantum field theories like QED and QCD, dimensionless physical quantities $\mathcal{P}$ are expressed by a perturbation series in powers of the corresponding coupling constant α. Assume the dependence of $\mathcal{P}$ on a single coupling constant α and energy scale Q. Renormalization introduces another scale μ where the subtraction of the UV divergences is performed and, therefore, both $\mathcal{P}$ and α become functions of μ. Since $\mathcal{P}$ is dimensionless, it depends only on the ratio Q^2/μ^2 and on the renormalized coupling constant $\alpha(\mu^2)$. As the choice of μ is arbitrary, any explicit dependence of $\mathcal{P}$ on μ must be cancelled by an appropriate μ-dependence of α. It is natural to identify the renormalization scale with the physical energy scale of the process, $\mu^2 = Q^2$. In this case, α transforms into a running coupling constant $\alpha(Q^2)$ and the energy dependence of $\mathcal{P}$ enters only through the energy dependence of $\alpha(Q^2)$.

In general, there are equations in gauge theories that describe the behaviour of coupling constants α_i as a function of Q^2. These so-called 'renormalization group equations' have the general form

$$\frac{\partial \alpha_i(Q^2)}{\partial \ln Q^2} = \beta(\alpha_i(Q^2)). \tag{5.2}$$

The perturbative expansion of the beta function β depends on the group and the particle content of the theory. In lowest order, the coupling constants are given by

$$\alpha_i(Q^2) = \frac{\alpha_i(\mu^2)}{1 + \alpha_i(\mu^2)\beta_0 \ln(Q^2/\mu^2)}. \tag{5.3}$$

As an example in QCD, the lowest term is given by

$$\beta_0 = \frac{33 - 2N_f}{12\pi} \tag{5.4}$$

with N_f as the number of active quark flavors. Alternatively, quite often another parametrization is used in form of

$$\alpha_i(Q^2) = \frac{1}{\beta_0 \ln(Q^2/\Lambda^2)} \tag{5.5}$$

which is equivalent to (5.3) if

$$\Lambda^2 = \frac{\mu^2}{\exp(1/\beta_0 \alpha_i(\mu^2))}. \tag{5.6}$$

In the Standard Model (see Chapter 3) strong and weak interactions are described by Abelian and non-Abelian groups and, as a consequence, there is a decrease in the coupling constant with increasing energy, the so-called asymptotic freedom (Figure 5.1). This is due to the fact that the force-exchanging bosons like gluons and W, Z are carriers of the corresponding charge of the group itself, in contrast to QED, where photons have no electric charge. The starting points for the extrapolation are the values obtained at the Z^0 resonance from the world averages [PDG18]

$$\alpha_S(m_Z^2) = 0.1181 \pm 0.0011 \tag{5.7}$$
$$\alpha_{em}^{-1}(m_Z^2) = 127.906 \pm 0.019 \tag{5.8}$$

and $\sin^2 \theta_W$ as given in Table 3.4. These values are taken from [PDG18], see also discussion in [Kli17]. After the extrapolation is carried out, all three coupling constants should meet at a point roughly on a scale of 10^{16} GeV (see, however, Section 5.4.3) and from that point on an unbroken symmetry with a single coupling constant should exist. As previously mentioned, the particle contents also influence the details of the extrapolation and any new particles introduced as, e.g., in supersymmetry (SUSY) would modify the Q^2 dependence of the coupling constants.

The simplest group to realize unification is SU(5). We will, therefore, first discuss the minimal SU(5) model (Georgi–Glashow model) [Geo74], even if it is no longer experimentally preferred.

5.2   The minimal SU(5) model

For massless fermions, the gauge transformations fall into two independent classes for left- and right-handed fields, respectively. Let us assume the left-handed fields are the elementary fields (the right-handed transformations are equivalent and act on the corresponding charge conjugated fields). We simplify matters by considering only the first family, consisting of u, d, e and ν_e, giving 15 elementary fields, with c indicating antiparticles:

$$u_r, u_g, u_b, \nu_e$$
$$u_r^c, u_g^c, u_b^c, d_r^c, d_g^c, d_b^c \qquad e^+ \tag{5.9}$$
$$d_r, d_g, d_b, e^-$$

with r, g, b as the color index of QCD. The obvious step would be to arrange the particles in three 5-dimensional representations, which is the fundamental SU(5) representation. However, only particles within a multiplet can be transformed into each other and it is known that six of them, $u_r, u_g, u_b, d_r, d_g, d_b$, are transformed into each other via SU(2) and SU(3) transformations in the Standard Model. Therefore, the fields have to be arranged in higher representations as a 10- and a $\bar{5}$-dimensional representation ($\bar{5}$ is the representation complementary to the fundamental representation 5, although this is not significant for our current purposes). The actual arrangement of fields into the multiplets results from the just-mentioned quark transformations and the condition that the sum of the charges in

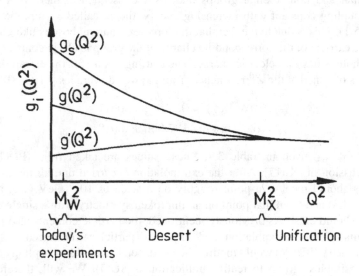

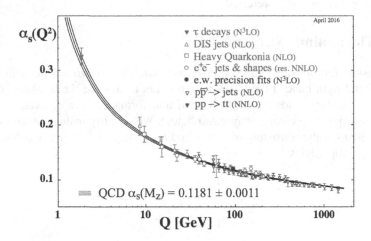

Figure 5.1. Top: Qualitative evolution with Q^2 of the three running coupling constants within the grand unification scale SU(5) (from [Gro90]) © 1990 CRC Press. Bottom: The clearest effect of running coupling with achievable energies is observed in the strong coupling α_S. Various experimental quantities can be used for its determination (from [PDG18]). With kind permission from M. Tanabashi et al. (Particle Data Group).

every multiplet has to be zero:

$$
\bar{5} = \begin{pmatrix} d_g^c \\ d_r^c \\ d_b^c \\ e^- \\ -\nu_e \end{pmatrix} \qquad 10 = \frac{1}{\sqrt{2}} \begin{pmatrix} 0 & -u_b^c & +u_r^c & +u_g & +d_g \\ +u_b^c & 0 & -u_g^c & +u_r & +d_r \\ -u_r^c & +u_g^c & 0 & +u_b & +d_b \\ -u_g & -u_r & -u_b & 0 & +e^+ \\ -d_g & -d_r & -d_b & -e^+ & 0 \end{pmatrix}. \qquad (5.10)
$$

The minus signs in these representations are conventional. SU(5) has 24 generators T_j (In general, SU(N) groups have $N^2 - 1$ generators), with the corresponding 24 gauge fields $\mathcal{B}_j$, which can be written in matrix form as

$$
\begin{pmatrix} G_{11} - \frac{2B}{\sqrt{30}} & G_{12} & G_{13} & X_1^c & Y_1^c \\ G_{21} & G_{22} - \frac{2B}{\sqrt{30}} & G_{23} & X_2^c & Y_2^c \\ G_{31} & G_{32} & G_{33} - \frac{2B}{\sqrt{30}} & X_3^c & Y_3^c \\ X_1 & X_2 & X_3 & \frac{W^3}{\sqrt{2}} + \frac{3B}{\sqrt{30}} & W^+ \\ Y_1 & Y_2 & Y_3 & W^- & -\frac{W^3}{\sqrt{2}} + \frac{3B}{\sqrt{30}} \end{pmatrix}.
$$

$$(5.11)$$

Here the 3×3 submatrix G characterizes the gluon fields of QCD, and the 2×2 submatrix W, B contains the gauge fields of the electroweak theory. In addition to the gauge bosons known to us, there are, however, a further 12 gauge bosons X, Y, which mediate transitions between quarks and leptons. The SU(5) symmetry has, however, to be broken down to the Standard Model. Here also the breaking occurs through the coupling to the Higgs fields which also has to be an SU(5) multiplet. SU(5) can be broken through a 24-dimensional Higgs multiplet with a vacuum expectation value (vev) of about 10^{15}–10^{16} GeV. This means that all particles receiving a mass via this breaking (e.g., the X, Y bosons) and its value is of the order of magnitude of the unification scale. By suitable SU(5) transformations we can ensure that only the X and Y bosons couple to the vacuum expectation value of the Higgs, while the other gauge bosons remain massless. An SU(5)-invariant mass term of the 24-dimensional Higgs field with the $\bar{5}$ and 10 representations of the fermions is not possible, so that the latter also remain massless. To break SU(2) ⊗ U(1) at about 100 GeV a further, independent 5-dimensional Higgs field is necessary, which gives the W, Z bosons and the fermions their mass.

We now leave this simplest unifying theory and consider its predictions. For a more detailed description see, e.g., [Lan81]. A few predictions can be drawn from (5.10):

(i) Since the sum of charges has to vanish in a multiplet, the quarks have to have 1/3 multiples of the electric charge. For the first time the appearance of non-integer charges is required.

(ii) From this immediately follows the equality of the absolute value of the electron and proton charge too.

(iii) The relation between the couplings of the $\mathcal{B}$-field to a SU(2) doublet (see Equation (3.28)) and that of the W^3-field is, according to Equation (5.11), given by $(3/\sqrt{15})$: 1. This gives a prediction for the value of the Weinberg angle $\sin^2 \theta_W$ [Lan81]:

$$\sin^2 \theta_W = \frac{g'^2}{g'^2 + g^2} = \frac{3}{8}. \tag{5.12}$$

This value is valid only for energies above the symmetry breaking. If renormalization effects are taken into consideration, at lower energies a slightly smaller value of

$$\sin^2 \theta_W = (0.218 \pm 0.006) \ln \left(\frac{100 \text{ MeV}}{\Lambda_{\text{QCD}}} \right) \tag{5.13}$$

results. This value is in agreement with the experimentally determined value (see Section 3.4.3).

(iv) Probably the most dramatic prediction is the transformation of quarks into leptons due to X, Y exchange. This would, among other things, permit the decay of the proton and with it ultimately the instability of all matter.

Because of the importance of the last process, it will be discussed in a little more detail.

5.2.1 Proton decay

As baryons and leptons are in the same multiplet, it is possible that bound protons and neutrons can decay. The main decay channels in accordance with the SU(5) model are [Lan81]:

$$p \rightarrow e^+ + \pi^0 \tag{5.14}$$

and

$$n \rightarrow \nu + \omega. \tag{5.15}$$

Here both, the baryon and lepton number, are violated by one unit while energetically possible. We specifically consider proton decay. The process $p \rightarrow e^+ + \pi^0$ should amount to about 30–50% of all decays. The proton decay can be calculated analogously to the muon decay, resulting in a lifetime [Lan81]

$$\tau_p \approx \frac{M_X^4}{\alpha_5^2 m_p^5} \tag{5.16}$$

with $\alpha_5 = g_5^2/4\pi$ as the SU(5) coupling constant. Using the renormalization group equations (5.2) with Standard Model particle contents, the two quantities M_X and α_5 can be estimated as [Lan81]

$$M_X \approx (1.3 \times 10^{14} \pm 50\%) \text{ GeV} \frac{\Lambda_{\text{QCD}}}{100 \text{ MeV}} \tag{5.17}$$

$$\alpha_5(M_X^2) = 0.0244 \pm 0.0002.$$

The minimal SU(5) model thus leads to the following prediction for the dominant decay channel [Lan86]:

$$\tau_p(p \to e^+\pi^0) = 6.6 \times 10^{28\pm0.7} \left[\frac{M_X}{1.3 \times 10^{14} \text{ GeV}}\right]^4 \text{ yr}$$

or equivalent

$$\tau_p(p \to e^+\pi^0) = 6.6 \times 10^{28\pm1.4} \left[\frac{\Lambda_{\text{QCD}}}{100 \text{ MeV}}\right]^4 \text{ yr.} \tag{5.18}$$

With $\Lambda_{\text{QCD}} = 200$ MeV the lifetime becomes $\tau_p = 1.0 \times 10^{30\pm1.4}$ yr. For reasonable assumptions on the value of Λ_{QCD}, the lifetime should, therefore, be smaller than 10^{32} yr. Besides the uncertainty in Λ_{QCD}, additional sources of error in the form of the quark wavefunctions in the proton must be considered. These are contained in the error on the exponent and a conservative upper value of $\tau_p = 1.0 \times 10^{32}$ yr can be assumed.

The experimental search for this decay channel is dominated by Super-Kamiokande, a giant water Cherenkov detector installed in the Kamioka mine in Japan (see Chapter 8). The decay should show the signature schematically shown in Figure 5.2. By not observing this decay a lower limit of $\tau_p/BR(p \to e^+\pi^0) > 1.6 \times 10^{34}$ yr (90% C.L.) for the decay $p \to e^+\pi^0$ could be deduced based on an exposure of 0.306 Mt × year [Abe17]. The disagreement with (5.18) rules out the minimal SU(5) model and other groups or extensions of SU(5) must be considered. The second reaction mentioned, $n \to \nu + \omega$, has also been searched for using the IMB-3 detector and a limit of $\tau_p/BR(n \to \nu\omega) > 1.08 \times 10^{32}$ yr (90% C.L.) is given [McG99]. A full list of all half-life limits of proton decay channels can be found in [PDG18].

5.3   The SO(10) model

One such alternative is the SO(10) model [Fri75, Geo75, Bab15] which contains the SU(5) group as a subgroup. The spinor representation is, in this case, 16-dimensional (see Figure 5.3):

$$16_{\text{SO(10)}} = 10_{\text{SU(5)}} \oplus \bar{5}_{\text{SU(5)}} \oplus 1_{\text{SU(5)}}. \tag{5.19}$$

The SU(5) singlet cannot take part in any renormalizable, i.e., gauge, SU(5) interaction. This new particle is, therefore, interpreted as the right-handed partner ν_R of the normal neutrino (more accurately, the field ν_L^C is incorporated into the multiplet). ν_R does not take part in any SU(5) interaction and, in particular, does not participate in the normal weak interaction of the GWS model. However, ν_R does participate in interactions mediated by the new SO(10) gauge bosons. Since the SO(10) symmetry contains the SU(5) symmetry, the possibility now exists that somewhere above M_X the SO(10) symmetry is broken down into the SU(5) symmetry and that it then breaks down further as already discussed:

$$\text{SO(10)} \to \text{SU(5)} \to \text{SU(3)} \otimes \text{SU(2)}_L \otimes \text{U(1)}. \tag{5.20}$$

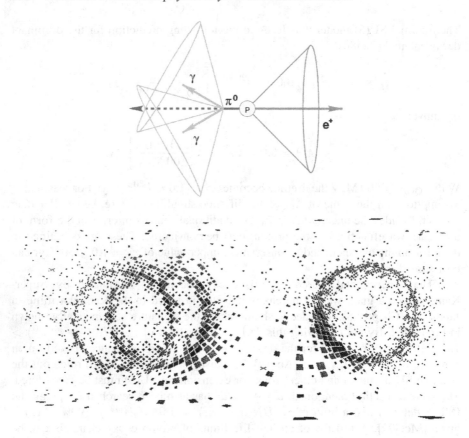

Figure 5.2. Top: Schematic picture of a proton decay $p \rightarrow e^+ \pi^0$ and the corresponding Cherenkov cones. Bottom: Monte Carlo simulation of such a proton decay for a water Cherenkov detector like Super-Kamiokande (from [Vir99]). With kind permission of Kamioka Observatory, ICRR (Institute for Cosmic Ray Research) and the University of Tokyo.

However, other breaking schemes for SO(10) do exist. For example, it can be broken down without any SU(5) phase and even below the breaking scale left–right symmetry remains. Thus, the SO(10) model does represent the simplest left–right symmetrical theory.

5.3.1 Left–right symmetric models

In this Pati–Salam model [Pat74] the symmetry breaking happens as follows:

$$\mathrm{SO}(10) \rightarrow \mathrm{SU}(4)_{\mathrm{EC}} \otimes \mathrm{SU}(2)_L \otimes \mathrm{SU}(2)_R \tag{5.21}$$

where the index EC stands for *extended color*, an extension of the strong interaction with the leptons as the fourth colour charge. The $\mathrm{SU}(2)_R$ factor can be seen as the right-handed equivalent of the left-handed $\mathrm{SU}(2)_L$. It describes a completely

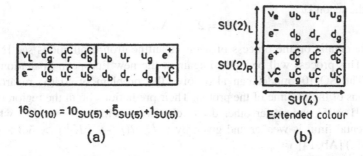

$$16_{SO(10)} = 10_{SU(5)} + \bar{5}_{SU(5)} + 1_{SU(5)}$$

(a)   **(b)**

Figure 5.3. All fermions of one family can be accommodated in *one* SO(10) multiplet. The 16th element is the as yet unseen right-handed neutrino ν_R or, equivalently, its CP conjugate ν_L^C. The illustrations correspond to different SO(10) breaking schemes. (*a*) A breaking scheme resulting in one SU(5) singlet and decouplet together with a $\bar{5}$ multiplet. (*b*) The breaking of the SO(10) multiplet according to the SU(4)$_{EC}\otimes$SU(2)$_L\otimes$SU(2)$_R$ structure (from [Gro90]). © 1990 CRC Press.

analogous right-handed weak interaction mediated by right-handed W bosons. Figure 5.3 shows the splitting of the multiplet according to the two symmetry-breaking schemes. The weak Hamiltonian in such a theory has to be extended by the corresponding terms involving right-handed currents:

$$H \sim G_F(j_L J_L^\dagger + \kappa j_R J_L^\dagger + \eta j_L J_R^\dagger + \lambda j_R J_R^\dagger) \tag{5.22}$$

with the leptonic currents j_i and hadronic currents J_i defined as in Chapter 3 and $\kappa, \eta, \lambda \ll 1$. The mass eigenstates of the vector bosons $W_{1,2}^\pm$ can be expressed as a mixture of the gauge bosons:

$$W_1^\pm = W_L^\pm \cos\theta + W_R^\pm \sin\theta \tag{5.23}$$
$$W_2^\pm = -W_L^\pm \sin\theta + W_R^\pm \cos\theta \tag{5.24}$$

with $\theta \ll 1$ and $m_2 \gg m_1$. This can be used to rewrite the parameters in (5.22):

$$\kappa \ll \eta \approx \tan\theta \qquad \lambda \approx (m_1/m_2)^2 + \tan^2\theta. \tag{5.25}$$

Lower bounds on the mass of right-handed W-boson exist [PDG18]:

$$m_{W_R} > 3 \text{ TeV}. \tag{5.26}$$

In contrast to the SU(5) model, which does not conserve B and L but does conserve $(B - L)$, $(B - L)$ does not necessarily have to be conserved in the SO(10) model. A baryon number as well as a lepton number violation of two units is possible and with that the possibility of not only neutrinoless double β-decay (see Chapter 7) but also of neutron–antineutron oscillations opens up. In the first case,

$$\Delta L = 2 \qquad \Delta B = 0 \tag{5.27}$$

and, in the second,

$$\Delta B = 2 \qquad \Delta L = 0. \tag{5.28}$$

For more details on the process of neutron–antineutron oscillations see [Moh96a, Phi16]. This process will be studied again at the new European Spallation Source (ESS). The SO(10) model can also solve the problem of SU(5) regarding the predictions of the lifetime of the proton. Their predictions lie in the region of 10^{32}–10^{38} yr [Lee95] and prefer other decay channels such as $p \rightarrow \nu K^+$ where the experimental limit is weaker and given by $\tau_p/BR(p \rightarrow \nu K^+) > 5.9 \times 10^{33}$ yr (90% C.L.) [Abe14].

It is convenient now to explore another extension of the Standard Model, which is given by supersymmetry (SUSY). This will also end with a short discussion of SUSY GUT theories.

5.4 Supersymmetry

A theoretical treatment of supersymmetry in all its aspects is far beyond the scope of this book. We restrict ourselves to some basic results and applications in particle physics. Several excellent textbooks and reviews exist on this topic for further reading [Dra87, Wes86, 90, Moh86, 92, Nil84, Hab85, Lop96, Tat97, Ell98, Oli99, Wei00, Wei05, Ait07, Bin07, Din07, Mar10, Lan18, All18].

Supersymmetry is a complete symmetry between fermions and bosons [Wes74]. This is a new symmetry and one as fundamental as that between particles and antiparticles. It expands the normal Poincaré algebra for the description of spacetime with extra generators, which changes fermions into bosons and *vice versa*. Let Q be a generator of supersymmetry such that

$$Q|(\text{Fermion})\rangle = |\text{Boson}\rangle \qquad \text{and} \qquad Q|(\text{Boson})\rangle = |\text{Fermion}\rangle.$$

In order to achieve this, Q itself has to have a fermionic character. In principle, there could be several supersymmetric generators Q, but we restrict ourselves to one ($N = 1$ supersymmetry). The algebra of the supersymmetry is then determined by the following relationships:

$$\{Q_\alpha, Q_\beta\} = 2\gamma^\mu_{\alpha\beta} p_\mu \tag{5.29}$$

$$[Q_\alpha, p_\mu] = 0 \tag{5.30}$$

Here p_μ is the 4-momentum operator. Note *that due to the anticommutator relation equation (5.29), internal particle degrees of freedom are connected to the external spacetime degrees of freedom*. This has the consequence that a *local* supersymmetry has to contain gravitation (supergravity theories, SUGRAs). A further generic feature of any supersymmetric theory is that the number of bosons equals that of fermions. A consequence for particle physics is then that the numbers of particles of the Standard Model are doubled. For every known fermion there is a boson and to each boson a fermion reduced by spin-$\frac{1}{2}$ exists. The nomenclature of the supersymmetric partners is as follows: the scalar partners of normal fermions are designated with a preceding

"S", so that, for example, the supersymmetric partner of the quark becomes the squark $\tilde{q}$. The super-partners of normal bosons receive the ending '-ino'. The partner of the photon, therefore, becomes the photino $\tilde{\gamma}$.

One of the most attractive features of supersymmetry with respect to particle physics is an elegant solution to the hierarchy problem. The problem here is to protect the electroweak scale (3.67) from the Planck scale (13.54) which arises from higher order corrections. This is especially dramatic for scalar particles like the Higgs boson. The Higgs mass receives a correction δm_H via higher orders where [Nil95]

$$\delta m_H^2 \sim g^2 \int^{\Lambda} \frac{\mathrm{d}^4 k}{(2\pi)^4 k^2} \sim g^2 \Lambda^2. \tag{5.31}$$

If the cut-off scale Λ is set at the GUT scale or even the Planck scale, the lighter Higgs particle would experience corrections of the order M_X or even M_{Pl}. In order to achieve a well-defined theory, it is then necessary to fine tune the parameters in all orders of perturbation theory. With supersymmetry, the problem is circumvented by postulating new particles with similar mass and equal couplings. Now corresponding to any boson with mass m_B in the loop there is a fermionic loop with a fermion mass m_F with a relative minus sign. So the total contribution to the 1-loop corrected Higgs mass is

$$\delta m_H^2 \simeq O\left(\frac{\alpha}{4\pi}\right)(\Lambda^2 + m_B^2) - O\left(\frac{\alpha}{4\pi}\right)(\Lambda^2 + m_F^2) = O\left(\frac{\alpha}{4\pi}\right)(m_B^2 - m_F^2). \tag{5.32}$$

When all bosons and fermions have the same mass, the radiative corrections vanish identically. The stability of the hierarchy requires only that the weak scale is preserved, meaning

$$|(m_B^2 - m_F^2)| \leq 1 \,\mathrm{TeV}^2. \tag{5.33}$$

Some remarks should be made. If this solution is correct, supersymmetric particles should be observed at the LHC, but they have not been observed yet. In addition, supersymmetry predicts that the masses of particles and their supersymmetric partners are identical. As the LHC has not yet observed supersymmetric particles, supersymmetry must be a broken symmetry. Furthermore, with the lower bounds from the LHC on the masses of SUSY particles compared to SM particles, the above mentioned cancellations in the loops (5.33) is not small enough anymore to solve the problem of the cosmological constant (see Chapter 13). In the following we restrict our discussion to the *minimal supersymmetric Standard Model* (MSSM); however, many more models with new parameters can be discussed.

5.4.1 The Minimal Supersymmetric Standard Model

As already stated, even in the minimal model we have to double the number of particles (introducing a superpartner to each particle) and we have to add another Higgs doublet (and its superpartner). The reason for the second Higgs doublet is given by the fact that there is no way to account for the up and down Yukawa couplings with only one Higgs field.

In the Higgs sector, both doublets obtain a vacuum expectation value (vev):

$$\langle H_1 \rangle = \begin{pmatrix} v_1 \\ 0 \end{pmatrix} \quad \langle H_2 \rangle = \begin{pmatrix} 0 \\ v_2 \end{pmatrix}. \tag{5.34}$$

Their ratio is often expressed as a parameter of the model:

$$\tan \beta = \frac{v_2}{v_1}. \tag{5.35}$$

Furthermore, in contrast to the SM, here one has eight degrees of freedom, three of which can be gauged away as in the SM. The net result is that there are five physical Higgs bosons: two CP-even (scalar) neutrals (h, H), one CP-odd (pseudo-scalar) neutral (A) and two charged Higgses ($H^\pm$).

There are four neutral fermions in the MSSM that receive mass but can mix as well. They are the gauge fermion partners of the B and W^3 gauge bosons (see Chapter 3), as well as the partners of the Higgs. They are, in general, called neutralinos or, more specifically, the bino $\tilde{B}$, the wino $\tilde{W}^3$ and the Higgsinos $\tilde{H}_1^0$ and $\tilde{H}_2^0$. The neutralino mass matrix can be written in the ($\tilde{B}, \tilde{W}^3, \tilde{H}_1^0, \tilde{H}_2^0$) basis as

$$\begin{pmatrix} M_1 & 0 & -M_Z s_{\theta_W} \cos\beta & M_Z s_{\theta_W} \sin\beta \\ 0 & M_2 & M_Z c_{\theta_W} \cos\beta & -M_Z c_{\theta_W} \sin\beta \\ -M_Z s_{\theta_W} \cos\beta & M_Z c_{\theta_W} \cos\beta & 0 & -\mu \\ M_Z s_{\theta_W} \sin\beta & -M_Z c_{\theta_W} \sin\beta & -\mu & 0 \end{pmatrix}$$

$$\tag{5.36}$$

where $s_{\theta_W} = \sin\theta_W$ and $c_{\theta_W} = \cos\theta_W$. The eigenstates are determined by diagonalizing the mass matrix. As can be seen, they depend on three parameters M_1 (coming from the bino mass term), M_2 (from the wino mass term) and μ (from the Higgsino mixing term $\frac{1}{2}\mu\tilde{H}_1\tilde{H}_2$). There are also four charginos coming from $\tilde{W}^\pm$ and $\tilde{H}^\pm$. The chargino mass matrix is composed similar to the neutralino mass matrix.

Using the universality hypothesis that, on the GUT scale, all the gaugino masses (spin-$\frac{1}{2}$ particles) are identical to a common mass $m_{1/2}$ and that all the spin-0 particle masses at this scale are identical to m_0, we end up with μ, $\tan\beta$, m_0, $m_{1/2}$ and A as free parameters. Here A is a soft supersymmetry-breaking parameter (for details see for example [Oli99]). In total in this Minimal Supersymmetric Standard Model (MSSM) five parameters remain which have to be explored experimentally.

5.4.2   R-parity

The MSSM is a model containing the minimal extension of the field contents of the Standard Model as well as minimal extensions of interactions. Only those required by the Standard Model and its supersymmetric generalization are considered. It is assumed that R-parity is conserved to guarantee the absence of lepton- and baryon-number-violating terms. R-parity is assigned as follows:

$$R_P = \quad 1 \qquad \text{for normal particles}$$
$$R_P = -1 \qquad \text{for supersymmetric particles.}$$

R_P is a multiplicative quantum number and is connected to the baryon number B, the lepton number L and the spin S of the particle by

$$R_P = (-1)^{3B+L+2S}. \tag{5.37}$$

Conservation of R-parity has two major consequences:

(i) Supersymmetric particles can be produced only in pairs.
(ii) The lightest supersymmetric particle (LSP) has to be stable.

However, even staying with the minimal particle content and being consistent with all symmetries of the theory, more terms can be written in the superpotential W which violate R-parity given as

$$W_{\rlap{/}R_p} = \lambda_{ijk}L_iL_j\bar{E}_k + \lambda'_{ijk}L_iQ_j\bar{D}_k + \lambda''_{ijk}U_i\bar{D}_j\bar{D}_k \tag{5.38}$$

where the indices i, j and k denote generations. L, Q denote lepton and quark doublet superfields and $\bar{E}$, $\bar{U}$ and $\bar{D}$ denote lepton and up, down quark singlet superfields, respectively. Terms proportional to λ, λ' violate lepton number; those proportional to λ'' violate baryon number. A compilation of existing bounds on the various coupling constants can be found in [Bed99, Bar05].

After shortly describing some basic features of the MSSM as one possible extension of the Standard Model and the possibility of R_P violation, it is obvious that one can also construct supersymmetric GUT theories, like SUSY SU(5), SUSY SO(10) and so on, with new experimental consequences. A schematic illustration of unification is shown in Figure 5.4, many more models exist. We now want to discuss briefly a few topics of the experimental search.

5.4.3 Experimental search for supersymmetry

Consider, first, the running coupling constants. As already mentioned, new particles change the parameters in the renormalization group equations (5.2). As can be seen in Figure 5.5, in contrast to the Standard Model extrapolation, the coupling constants including MSSM now unify and the unified value and scale are given by

$$M_{\text{GUT}} = 10^{15.8\pm0.3\pm0.1} \text{ GeV} \tag{5.39}$$

$$\alpha_{\text{GUT}}^{-1} = 26.3 \pm 1.9 \pm 1.0. \tag{5.40}$$

Even though this is not a proof that SUSY is correct, it at least gives a hint of its existence. The prediction of the Weinberg angle in supersymmetric models also corresponds better to the experimentally observed value (Chapter 3) than those of GUT theories without supersymmetry. The predictions of these theories are [Lan93b]:

$$\sin^2\theta_W(m_Z) = 0.2334 \pm 0.0050 \text{ (MSSM)} \tag{5.41}$$

$$\sin^2\theta_W(m_Z) = 0.2100 \pm 0.0032 \text{ (SM)}. \tag{5.42}$$

The experimental strategies to search for SUSY can be separated into four groups:

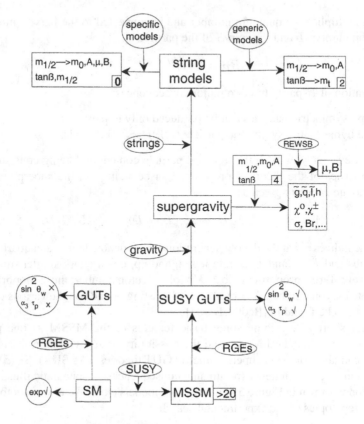

Figure 5.4. Schematic picture of the different steps in grand unification from the Fermi scale to the Planck scale. The numbers indicate the number of new parameters required to describe the corresponding model (from [Lop96]). © IOP Publishing. Reproduced with permission. All rights reserved.

- direct production of supersymmetric particles in high-energy accelerators,
- precision measurements,
- search for rare decays and
- dark matter searches.

For the accelerator searches, another constraint is applied to work with four free parameters (constrained MSSM, CMSSM). This requires gauge coupling unification at the GUT scale leading to the relation $M_1 = \frac{5}{3}\frac{\alpha_1}{\alpha_2}M_2$ and one can work only with the parameters μ, $\tan\beta$, m_0, $m_{1/2}$. Besides that, as long as R-parity is conserved, the LSP remains stable and acts as a good candidate for dark matter (see Chapter 13).

A good example for the second method is a search for electric dipole moments of particles like electrons and neutrons, where supersymmetry enters via loop corrections. The third one either uses existing stringent experimental bounds to restrict parameters like those coming from flavour violation, i.e., decays like

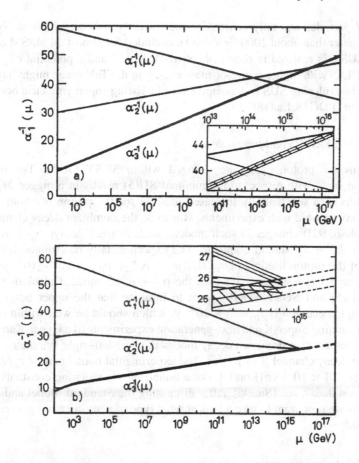

Figure 5.5. Running of the coupling constants. Top: Evolution assuming the SM particle content. Evidently the coupling constants do not meet at the unification scale (from [Ama91]). © 1991 With permission from Elsevier. Bottom: Unification is achieved by including the MSSM (from [Kla97a]). Reproduced with permission of SNCSC.

$b \rightarrow s + \gamma$, $\mu \rightarrow e + \gamma$, $\mu \rightarrow 3e$ and $\mu - e$ conversion. In general, various processes which might be enhanced or modified with respect to the Standard Model can be used. For more comprehensive reviews on SUSY see [Fen10, Mar10, Ath17, Rab17].

5.4.3.1   *SUSY signatures at high energy colliders*

SUSY particles can be pair produced at colliders. The obvious machine to look for it is the LHC with a current centre-of-mass energy of $\sqrt{s} = 13$ TeV with upgrades in the future. A common feature of all possible signals as long as we are working in the MSSM or CMSSM is a significant missing energy ($\not{E}_T$) and transverse momentum ($\not{p}_T$). The reason is that the produced stable LSPs escape detection. This signature is accompanied by either jets or leptons. So far all searches found no evidence of SUSY particles and only constraints and lower bounds are obtained. Typically, masses lighter than about 1000 GeV can be excluded for almost all MSSM particles.

If SUSY is realized in nature, the upgraded LHC and a potential e^+e^- linear collider (ILC) with higher centre-of-mass energy in the TeV range might have a rich program in exploring SUSY. A compilation of existing supersymmetric bounds can be found in [PDG18, Lan18].

5.4.3.2 SUSY GUTs and proton decay

Predictions for proton decay are changed within SUSY GUTs. The increased unification scale with respect to the minimal SU(5) results in a bigger M_X mass. This results in a substantially increased lifetime for the proton of about 10^{35} yr, which is compatible with experiments. However, the dominant decay channel (see, e.g., [Moh86, 92]) changes in such models, such that the decays $p \rightarrow K^+ + \bar{\nu}_\mu$ and $n \rightarrow K^0 + \bar{\nu}_\mu$ should dominate. The experimentally determined lower limit [Vir99] of the proton lifetime of $\tau_p/BR(p \rightarrow K^+ + \bar{\nu}_\mu) > 1.9 \times 10^{33}$ yr for the second channel is less restrictive than the $p \rightarrow e^+\pi^0$ mode. Calculations within SUSY SU(5) and SUSY SO(10) seem to indicate that the upper bound on the theoretical expectation is $\tau_p < 5 \times 10^{33}$ yr which should be well within the reach of longer running Super-K and next-generation experiments like Hyper-Kamiokande discussed later. Other dominant decay modes in some left–right symmetric models, prefer the decay channel $p \rightarrow \mu^+K^0$. The experimental bound here is $\tau_p/BR(p \rightarrow \mu^+K^0) > 1.3 \times 10^{32}$ yr [Kob05]. For a bound on R_p-violating constants coming from proton decay, see [Smi96]. After discussing the Standard Model and possible extensions we now want to take a look at what type of neutrino mass generation can be realized.

5.5   Neutrino masses

As already stated in Chapter 3, neutrino masses are set to zero in the Standard Model. Therefore, any evidence of a non-vanishing neutrino mass would indicate physics 'beyond the Standard Model'.[1] A general idea is to find some physics at very high energy which results in heavy particles with a certain degree of freedom and are unrelated to the electroweak symmetry breaking. A lot of model building has been performed to include neutrino masses in physics; for reviews see [Val03, Kin03, Str06, Moh07, Gon08, Lan12, deG16].

5.5.1   Neutrino masses in the electroweak theory

As will be seen in Chapter 6, neutrino masses are in the eV range or below. Additionally, measurements discussed later require that at least two neutrinos have

---

[1] It is a matter of taste what exactly 'beyond the Standard Model' means. Neutrino masses can be generated within the gauge structure of $SU(3) \otimes SU(2) \otimes U(1)$ by enlarging the particle content or adding non-renormalizable interactions. Even by adding new particles this sometimes is nevertheless still called 'Standard Model' because the gauge structure is unchanged.

a non-vanishing rest mass. Furthermore, neutrinos are pretty "isolated" compared to other fundamental fermions, the difference between neutrino and electron masses are almost 6 orders of magnitudes. Neutrino masses can be created in the Standard Model by extending the particle content of the theory. Dirac mass terms of the form (2.36) and the corresponding Yukawa couplings (3.61) can be written for neutrinos if one or more singlet ν_R are included in the theory as for all other fermions. This would result in (see (3.55))

$$\mathcal{L}_{\text{Yuk}} = -c_\nu \bar{\nu}_R \phi^\dagger \begin{pmatrix} \nu_{eL} \\ e_L \end{pmatrix} + h.c. \tag{5.43}$$

resulting in terms like (3.55)

$$= -c_\nu v \bar{\nu} \nu. \tag{5.44}$$

However, the smallness of the neutrino mass then requires a Yukawa coupling c_ν which is about 10^{-12} smaller compared to all other Standard Model particles. Furthermore, this is not trivial to generate, see also [Wei79, Moh07, Lan12, deG16]. If no additional fermions are included, the only possible mass terms are of Majorana type and, therefore, violate lepton number (equivalent to violating $B-L$, which is the only gauge-anomaly-free combination of these quantum numbers). Thus, we might introduce new Higgs bosons which can violate $B-L$ in their interactions. These Majorana type neutrinos must be incorporated into a Yukawa coupling, which could be done if other sources of electroweak symmetry breaking (EWSB) are existing, like another Higgs boson with a non-vanishing vev. Finally, also new sources independent of the EWSB scale given by $\langle v \rangle$ could exist, which would lead to two different mass scales. This is an often used assumption having a very high energy scale Λ (for example near the GUT scale) which can lead then to small masses.

The corresponding fermionic bilinears have a net $B-L$ number and the further requirement of gauge-invariant Yukawa couplings determine the possible Higgs multiplets, which can couple directly to the fermions:

- a triplet Δ and
- a singly charged singlet h^-.

The Higgs triplet is given by

$$\begin{pmatrix} \Delta^0 \\ \Delta^- \\ \Delta^{--} \end{pmatrix} \tag{5.45}$$

and its Yukawa coupling gives neutrinos their mass. The component Δ^0 requires a vacuum expectation value of v_3, which has to be much smaller than the one obtained by the standard Higgs doublet. As the Higgs potential now contains two multiplets ϕ (3.56) and Δ, both contribute to the mass of the gauge bosons. From that, an upper bound on v_3 can already be given:

$$\rho = \frac{m_W^2}{m_Z^2 \cos\theta_W} = \frac{1 + 2v_3^2/v^2}{1 + 4v_3^2/v^2} \rightarrow \frac{v_3}{v} < 0.07. \tag{5.46}$$

The second model introducing an SU(2) singlet Higgs h^- has been proposed by Zee [Zee80]. As h^- carries electric charge its vev must vanish and some other sources of $B - L$ violation must be found.

An independent possibility introducing neutrino masses in the Standard Model would be adding non-renormalizable operators. There is only one 5-dimensional operator (called Weinberg operator) [Wei79]. This operator is linked to a new scale Λ which is typically assumed to be very high (around the GUT scale) and the neutrino mass is proportional to $1/\Lambda$ as long as $v \ll \Lambda$. Higher dimensional operators have an even higher suppression so they are normally ignored (the 6-dimensional operators scale with $1/\Lambda^2$). From this, various see-saw models can be created, which are mentioned in Chapter 2. After discussing how by enlarging the particle content of the Standard Model neutrino masses can be generated, we now want to see what possibilities GUT and SUSY offer.

5.5.2   Neutrino masses in the minimal SU(5) model

In the multiplets given in (5.15) only ν_L with its known two degrees of freedom shows up, allowing only Majorana mass terms for neutrinos. The coupling to the Higgs field Φ has to be of the form $(\nu_L \otimes \nu_L^C)\Phi$. However, $5 \otimes 5$ results in combinations of $10 \oplus 15$ which does not allow us to write SU(5)-invariant mass terms, because with the Higgs, only couplings of 25 and 5 representations are possible. Therefore, in the minimal SU(5) neutrinos remain massless. But, as in the Standard Model, enlarging the Higgs sector allows us to introduce Majorana mass terms.

5.5.3   Neutrino masses in the SO(10) model and the seesaw mechanism

In the SO(10) model the free singlet can be identified with a right-handed neutrino (see Figure 5.3). It is, therefore, possible to produce Dirac mass terms. The corresponding Yukawa couplings have to be made with 10, 120 or 126 dimensional representations of the Higgs. However, as the neutrinos belong to the same multiplet as the remaining fermions, their mass generation is not independent from that of the other fermions and one finds, e.g., by using the 10-dimensional Higgs, that all Dirac mass terms are more or less identical, in strong contradiction to experiments where limits for neutrino masses are much smaller than the corresponding ones on charged leptons and quarks (see Chapter 6). This problem can be solved by adding the 126-dimensional representation of the Higgs field and assigning a vev to the SU(5) singlet component. This gives rise to a Majorana mass of the right-handed neutrino. This mass term can take on very large values up to M_X. Under these assumptions it is possible to obtain no Majorana mass term for ν_L and a very large term for ν_R. In this case the mass matrix (2.48) has the following form:

$$M = \begin{pmatrix} 0 & m_D \\ m_D & m_R \end{pmatrix} \tag{5.47}$$

where m_D is of the order of eV, while $m_R \gg m_D$. But this is exactly the requirement for a seesaw mechanism as discussed in Chapter 2. This means that it is possible for a suitably large Majorana mass m_R in Equation (5.47) to reduce the observable masses so far that they are compatible with experiment. This is the *seesaw* mechanism for the production of small neutrino masses [Gel78, Moh80]. If this is taken seriously, a quadratic scaling behaviour of the neutrino masses with the quark masses or charged lepton masses follows (2.63), i.e.,

$$m_{\nu_e} : m_{\nu_\mu} : m_{\nu_\tau} \sim m_u^2 : m_c^2 : m_t^2 \quad \text{or} \quad \sim m_e^2 : m_\mu^2 : m_\tau^2. \tag{5.48}$$

However, several remarks should be made. This relation holds on the GUT scale. By extrapolating down to the electroweak scale using the renormalization group equations, significant factors could disturb the relation. As an example, the ratio of the three neutrino masses for two different models is given by [Blu92]

$$m_1 : m_2 : m_3 = 0.05 m_u^2 : 0.09 m_c^2 : 0.38 m_t^2 \quad \text{SUSY–GUT} \tag{5.49}$$
$$m_1 : m_2 : m_3 = 0.05 m_u^2 : 0.07 m_c^2 : 0.18 m_t^2 \quad \text{SO(10)}. \tag{5.50}$$

Furthermore, it is assumed that the heavy Majorana mass shows no correlation with the Dirac masses. However, if this is the case, a linear seesaw mechanism arises. Of course many more models are existing.

5.5.3.1 *Almost degenerated neutrino masses*

If the upper left entry in (5.47) does not vanish exactly, the common seesaw formula might change. The common general seesaw term

$$m_\nu \approx -m_D^T m_R^{-1} m_D \tag{5.51}$$

is modified to

$$m_\nu \approx f \frac{v^2}{v_R} - m_D^T m_R^{-1} m_D \tag{5.52}$$

where the first term includes the vev of the Higgs fields. Clearly, if the first term dominates, there will be no hierarchical seesaw, but the neutrinos will be more or less degenerated in mass (sometimes called type II seesaw).

5.5.4 Neutrino masses in SUSY and beyond

Including SUSY in various forms like the MSSM, allowing R_p violation and SUSY GUT opens a variety of new possible neutrino mass generations. This can even be extended by including superstring-inspired models or those with extra dimensions. The neutrino mass schemes are driven here mainly by current experimental results such as those described in the following chapters. In the MSSM, neutrinos remain massless as in the Standard Model, because of lepton and baryon number conservation. However, an interesting feature is that the observed vacuum energy in the universe (see Chapter 13) can be written in natural units as $\Lambda \sim (0.003\,\text{eV})^4$. In this way a natural scale in the meV range would exist, which agrees nicely with discussed neutrino masses. For some recent models and further reviews, see [Die01, Moh01, Alt03, Hir02, Kin03, Str06, Moh06, Moh07, Gon08].

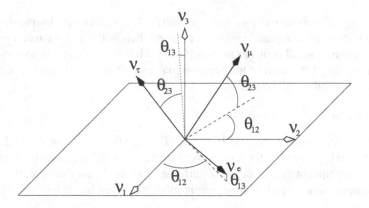

Figure 5.6. Graphical representation of the mixing matrix elements between flavour and mass eigenstates (from [Kin13]). With kind permission of S. King. © IOP Publishing. Reproduced with permission. All rights reserved.

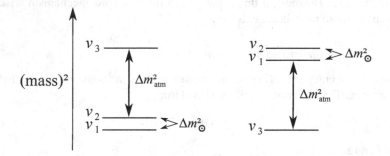

Figure 5.7. Normal and inverted mass hierarchies for three neutrinos. The inverted scheme is characterized by a $\Delta m_{23}^2 = m_3^2 - m_2^2 < 0$. With kind permission of S. Turkat.

5.6 Neutrino mixing

In the following chapters, it will be shown that neutrinos have a non-vanishing rest mass. Then the weak eigenstates ν_α do not need to be identical to the mass eigenstates ν_i. As in the quark sector the states could be connected by a unitary matrix U like the CKM matrix (see Chapter 3), called in the lepton sector PMNS-matrix (Pontecorvo–Maki–Nakagawa–Sakata) [Mak62]:

$$|\nu_\alpha\rangle = U_{\text{PMNS}}|\nu_i\rangle \qquad \alpha = e, \mu, \tau; \; i = 1 \ldots 3. \tag{5.53}$$

For three Dirac neutrinos U is given, in analogy to (3.65), as

$$U = \begin{pmatrix} c_{12}c_{13} & s_{12}c_{13} & s_{13}e^{-i\delta} \\ -s_{12}c_{23} - c_{12}s_{23}s_{13}e^{i\delta} & c_{12}c_{23} - s_{12}s_{23}s_{13}e^{i\delta} & s_{23}c_{13} \\ s_{12}s_{23} - c_{12}s_{23}s_{13}e^{i\delta} & -c_{12}s_{23} - s_{12}c_{23}s_{13}e^{i\delta} & c_{23}c_{13} \end{pmatrix} \tag{5.54}$$

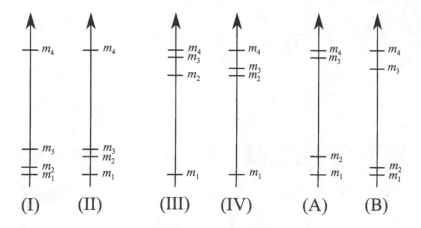

Figure 5.8. Various neutrino mass schemes can be built if there is the potential existence of a fourth neutrino state, often called sterile neutrino. The first four patterns shown are known as '3 + 1' schemes, because of the one isolated state m_4, while the remaining two are called '2 + 2' schemes. With kind permission of S. Turkat.

where $s_{ij} = \sin\theta_{ij}$, $c_{ij} = \cos\theta_{ij}$ $(i,j = 1,2,3)$. A graphical illustration of the mixing matrix elements ignoring the CP-phase is shown in Figure 5.6. In the Majorana case, the requirement of particle and antiparticle to be identical, restricts the freedom to redefine the fundamental fields. The net effect is the appearance of a CP-violating phase already in two flavours. For three flavours two additional phases have to be introduced resulting in a mixing matrix of the form

$$U = U_{\text{PMNS}} \, \text{diag}(1, e^{i\alpha}, e^{i\beta}). \tag{5.55}$$

In the three-flavour scenario several possible mass schemes can still be discussed which will become obvious in Chapters 8–10. In addition to normal and inverted mass schemes (Figure 5.7), almost degenerate neutrino masses $m_1 \approx m_2 \approx m_3$ are possible. Further common scenarios include a possible fourth neutrino as shown in Figure 5.8. Such a neutrino does not take part in weak interactions and is called a sterile neutrino. So far no indication for it is found. Having discussed the theoretical motivations and foundations for a possible neutrino mass, in the following we want to focus on experimental searches and evidence.

Chapter 6

Direct neutrino mass searches

DOI: 10.1201/9781315195612-6

In this chapter direct methods for neutrino mass determinations are discussed. The classical way to perform such searches for the rest mass of $\bar{\nu}_e$ is to investigate β-decay. From the historical point of view this process played a major role (see Chapter 1) because it was the motivation for W. Pauli to introduce the neutrino. Many fundamental properties of weak interactions were discovered by investigating β-decay. For an extensive discussion on weak interactions and β-decay see [Sch66, Sie68, Wu66, Kon66, Mor73, Rob88, Gro90, Wil01, Wei02, Ott08, Eji19].

6.1 Fundamentals of β-decay

Beta-decay is a nuclear transition, where the ordering number Z of the nucleus changes by one unit, while the atomic mass A remains the same.

This results in three possible decay modes:

$$(Z, A) \rightarrow (Z + 1, A) + e^- + \bar{\nu}_e \quad (\beta^--\text{decay}) \tag{6.1}$$

$$(Z, A) \rightarrow (Z - 1, A) + e^+ + \nu_e \quad (\beta^+-\text{decay}) \tag{6.2}$$

$$e^- + (Z, A) \rightarrow (Z - 1, A) + \nu_e \quad (\text{Electron capture}). \tag{6.3}$$

The basic underlying mechanism for (6.1) is given by

$$n \rightarrow p + e^- + \bar{\nu}_e \quad \text{or} \quad d \rightarrow u + e^- + \bar{\nu}_e \tag{6.4}$$

on the quark level respectively. The other decay modes can be understood in an analogous way. Free neutron decay into a proton can be observed, but the opposite is possible only in a nucleus. The corresponding decay energies are given by the following relations, where $m(Z, A)$ denotes the mass of the neutral atom (not the nucleus):

β^--decay:

$$Q_{\beta^-} = [m(Z, A) - Zm_e]c^2 - [(m(Z + 1, A) - (Z + 1)m_e) + m_e]c^2$$
$$= [m(Z, A) - m(Z + 1, A)]c^2. \tag{6.5}$$

The Q-value corresponds exactly to the mass difference between the mother and daughter atom.

β^+-decay:

$$
\begin{aligned}
Q_{\beta^+} &= [m(Z, A) - Zm_e]c^2 - [(m(Z - 1, A) - (Z - 1)m_e) + m_e]c^2 \\
&= [m(Z, A) - m(Z - 1, A) - 2m_e]c^2.
\end{aligned} \tag{6.6}
$$

As all masses are given for atoms, this decay requires the rest mass of two electrons. Therefore, the mass difference between both has to be larger than $2m_ec^2$ for β^+-decay to occur.

Electron capture:

$$
\begin{aligned}
Q_{\text{EC}} &= [m(Z, A) - Zm_e]c^2 + m_ec^2 - [m(Z - 1, A) - (Z - 1)m_e]c^2 \\
&= [m(Z, A) - m(Z - 1, A)]c^2.
\end{aligned} \tag{6.7}
$$

As can be expected the Q-values of the last two reactions are related by

$$
Q_{\beta^+} = Q_{\text{EC}} - 2m_ec^2. \tag{6.8}
$$

If Q is larger than $2m_ec^2$, both electron capture and β^+-decay are competitive processes, because they lead to the same daughter nucleus. For smaller Q-values only electron capture will occur. Obviously, for any of the modes to occur the corresponding Q-value has to be larger than zero.

As the way to determine the neutrino mass is related to β^--decay, this mode will be discussed in more detail. More accurately, this method measures the rest mass of $\bar{\nu}_e$, but CPT-conservation ensures that $m_{\bar{\nu}_e} \equiv m_{\nu_e}$.

The most important point is to understand the shape of the observed electron energy spectrum at the endpoint (see Chapter 1) and the impact of a non-vanishing neutrino mass which, for small neutrino masses, shows up only in the endpoint region of the energy spectrum. The following discussion is related to allowed and super-allowed transitions, meaning that the leptons do not carry away any orbital angular momentum ($l = 0$). The transition rate of β-decay to produce an electron in the energy interval between E and $E + \Delta E$ is given by Fermi's Golden Rule:

$$
\frac{d^2N}{dt\, dE} = \frac{2\pi}{\hbar} |\langle f|H_{if}|i\rangle|^2 \rho(E) \tag{6.9}
$$

where $|\langle f|H_{if}|i\rangle|$ describes the transition matrix element including the weak Hamilton operator H_{if}, $\rho(E)$ denotes the density of final states, and E_0 corresponds to the endpoint energy of the beta spectrum. In case of zero neutrino mass and ignoring any kind of final state excitation, i.e., looking at ground state transitions only, E_0 corresponds to the difference of the Q-value and the nuclear recoil energy. The latter is rather small; it gets its maximal value at the endpoint of the beta spectrum. For example, in case of molecular tritium decay, the center of mass kinetic energy of the daughter molecule ($^3H\,^3He^+$) is 1.72 eV. Neglecting nuclear recoil, the following relation is valid:

$$
E_0 = E_\nu + E_e. \tag{6.10}
$$

6.1.1   Matrix elements

Consider first the matrix element given by

$$|\langle f|H_{if}|i\rangle| = \int dV\,\psi_f^* H_{if}\psi_i. \tag{6.11}$$

The wavefunction ψ_i of the initial state is determined by the nucleons in the mother atom, while the final state wavefunction ψ_f has to be built by the wavefunction of the daughter as well as the wavefunction of the electron-neutrino field. The interaction between the nucleus and the leptons is weak, thus, in a first approximation wavefunctions normalized to a volume V can be treated as plane waves:

$$\phi_e(r) = \frac{1}{\sqrt{V}}e^{ik_e\cdot r} \tag{6.12}$$

$$\phi_\nu(r) = \frac{1}{\sqrt{V}}e^{ik_\nu\cdot r}. \tag{6.13}$$

These wavefunctions can be expanded in a Taylor series around the origin in the form

$$\phi_l(r) = \frac{1}{\sqrt{V}}(1 + ik_l\cdot r + \cdots) \qquad \text{with } l \equiv e, \nu. \tag{6.14}$$

The nuclear radius R can be estimated by $R = r_0 A^{1/3}$ with $r_0 \approx 1.2$ fm and the reduced Compton wavelength $\lambdabar$ of a 2 MeV electron is

$$\lambdabar = \frac{\hbar}{p} \simeq \frac{\hbar c}{E} = \frac{197\,\text{MeV fm}}{2\,\text{MeV}} \approx 10^{-11}\text{cm}. \tag{6.15}$$

Thus $k_l = 1/\lambdabar$ is about 10^{-2} fm^{-1} resulting in $k_l r \ll 1$. Therefore, in good approximation, the wavefunctions are

$$\phi_l(r) = \frac{1}{\sqrt{V}} \qquad \text{with } l \equiv e, \nu. \tag{6.16}$$

The electron wavefunction has to be modified taking into account the electromagnetic interaction of the emitted electron with the Coulomb field of the daughter nucleus $(A, Z + 1)$. For an electron the effect produces an attraction, while for positrons it results in a repulsion (Figure 6.4). The correction factor is called the Fermi function $F(Z + 1, E)$ and it is defined as

$$F(Z + 1, E) = \frac{|\phi_e(0)_{\text{Coul}}|^2}{|\phi_e(0)|^2}. \tag{6.17}$$

In the non-relativistic approach it can be approximated by [Pri68]

$$F(Z + 1, E) = \frac{z}{1 - e^{-z}} \tag{6.18}$$

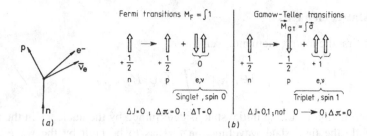

Figure 6.1. Neutron beta decay (*a*) and spin balance (*b*) for Fermi and Gamow–Teller transitions (from [May02]). Reproduced with permission of SNCSC.

with

$$z = \pm \frac{2\pi(Z+1)\alpha}{\beta} \qquad \text{for } \beta^{\mp}\text{-decay} \qquad (6.19)$$

and α as the fine structure constant and $\beta = v/c$. An accurate treatment has to take into account relativistic effects. A numerical compilation of Fermi functions can be found in [Beh69]. The lepton wavefunctions are practically constant all over the nuclear volume; thus using the first term in (6.14) is sufficient. As a consequence, the term $|\langle f|H_{if}|i\rangle|^2$ will contain a factor $|\phi_e(0)|^2|\phi_\nu(0)|^2 \simeq 1/V^2$. Introducing a coupling constant g to account for the strength of the interaction, the matrix element can be written as

$$|\langle f|H_{if}|i\rangle|^2 = g^2 F(E, Z+1)|\phi_e(0)|^2|\phi_\nu(0)|^2|M_{if}|^2$$
$$\simeq \frac{g^2}{V^2} F(E, Z+1)|M_{if}|^2 \qquad (6.20)$$

where the so-called nuclear matrix element M_{if} is given by

$$M_{if} = \int \mathrm{d}V \, \phi_f^* \mathcal{O} \phi_i. \qquad (6.21)$$

This expression now describes the transition between the two nuclear states, where $\mathcal{O}$ is the corresponding operator and, therefore, it is determined by the nuclear structure. Consider again only allowed transitions. In this case two kinds of nuclear transitions can be distinguished depending on whether the emitted leptons form a spin-singlet or spin-triplet state. Assume that the spins of electron and $\bar{\nu}_e$ are antiparallel with a total spin zero. Such transitions are called Fermi transitions (Figure 6.1). The transition operator corresponds to the isospin ladder operator τ^- and is given by

$$\mathcal{O}_F = I^- = \sum_{i=1}^{A} \tau^-(i) \qquad (6.22)$$

summing over all nucleons. The isospin I is introduced to account for the similar behaviour of protons and neutrons with respect to the nuclear force, defining the

Table 6.1. Characterization of β-decay transitions according to their angular momentum J and parity π. This leads to classification in form of allowed and forbidden decays as well as unique and non-unique decays. Shown are the changes in angular momentum and parity for X^{th}-fold forbidden unique transitions. $\Delta\pi = \pi_i\pi_f$ being the initial and final parity of the involved states (after [Suh07]).

X	1	2	3	4	5	6	7
ΔJ	2	3	4	5	6	7	8
$\Delta\pi$	-1	+1	-1	+1	-1	+1	-1

nucleon as an isospin $I = 1/2$ object with two projections ($I_3 = +1/2$ as proton and $I_3 = -1/2$ as neutron). As the transition does not change neither spin J and parity π nor isospin I, the following selection rules hold:

$$\Delta I = 0 \qquad \Delta J = 0 \qquad \Delta\pi = 0. \tag{6.23}$$

The second kind of transition is characterized by the fact that both leptons have parallel spins resulting in a total spin 1. Such transitions are called Gamow–Teller transitions and are described by

$$\mathcal{O}_{GT} = \sum_{i=1}^{A} \sigma(i)\tau^-(i) \tag{6.24}$$

where $\sigma(i)$ are the Pauli spin matrices, which account for the spin flip of the involved nucleons. Also here selection rules are valid:

$$\Delta I = 0, 1$$
$$\Delta J = 0, 1 \qquad \text{no } 0 \rightarrow 0 \text{ transition}$$
$$\Delta\pi = 0. \tag{6.25}$$

Beta decays, where the angular momentum changes by more than one unit or parity does not change with angular momentum are called forbidden transitions. They are characterised as shown in Table 6.1 and Table 6.2.

In summary, the nuclear matrix element for allowed transitions has the form

$$g^2|M_{if}|^2 = g_V^2|M_F|^2 + g_A^2|M_{GT}|^2 \tag{6.26}$$

already taking into account the different coupling strength of both transitions by using the vector- and axial vector coupling constants $g_V = G_\beta = G_F \cos\theta_C$ and g_A (see Chapter 3). The corresponding matrix elements have to be theoretically calculated. Under the assumptions made, M_{if} does not depend on energy. The matrix element is determined after summing over all spin states and averaging over the

Table 6.2. The same scheme as the one for Table 6.1, but for X^{th}-fold forbidden non-unique transitions.

X	1	2	3	4	5	6	7
ΔJ	0,1	2	3	4	5	6	7
$\Delta \pi$	-1	+1	-1	+1	-1	+1	-1

electron-neutrino correlation factor. For tritium this results in a matrix element of $|M_{if}|^2 = 5.55$ [Rob88]. The overlap between the initial and final wavefunction is especially large for mirror nuclei (the number of protons of one nucleus equals the number of neutrons in a second nucleus and vice versa); therefore, they have a large M_{if}. This becomes apparent for super-allowed $0^+ \rightarrow 0^+$ Fermi transitions. In this case $M_{GT} = 0$ and $M_F = \sqrt{2}$ and the ft-value (see Section 6.1.3) for such nuclei is constant and given by (see (6.40))

$$fT_{1/2} = f \frac{K}{2G_F^2 |V_{ud}|^2} \tag{6.27}$$

with V_{ud} as the CKM matrix element (see Section 3.3.2) and K given by (6.41) and (Figure 6.2). With the advent of Penning traps, this became the most precise way to measure V_{ud}. However, there are nuclei where electrons and neutrinos are emitted with $l \neq 0$ which means that the higher order terms of (6.14) have to be taken into account. The corresponding matrix elements are orders of magnitude smaller and the transitions are called forbidden. For a more extensive discussion on the classification and compilation of β-decays see [Wu66, Kon66, Sie68, Sin98, Suh07, Eji19]. From the discussion above it follows that the shape of the electron spectrum in allowed transitions is determined completely by the density of final states $\rho(E)$, which will be calculated next.

6.1.2   Phase space calculation

In general quantum mechanical treatment, the number of different states dn in phase space with momentum between p and $p + dp$ in a volume V is given by

$$\mathrm{d}n = \frac{4\pi V p^2 \, \mathrm{d}p}{h^3} = \frac{4\pi V p E \, \mathrm{d}E}{h^3}. \tag{6.28}$$

This translates into a density of states per energy interval of

$$\frac{\mathrm{d}n}{\mathrm{d}E} = \frac{4\pi V p E}{h^3} = \frac{V p E}{2\pi^2 \hbar^3}. \tag{6.29}$$

Dealing with a three-body decay and a heavy nucleus, the nucleus takes a negligible recoil energy but balances all momenta so the electron and neutrino momenta are

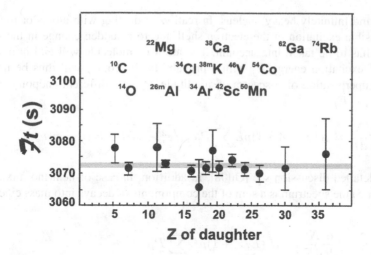

Figure 6.2. Experimental obtained and corrected ft-values observed in various super-allowed transitions. They result in an average value of 3072.08 ± 0.79 s allowing a precise determination of the mixing matrix element V_{ud} [Har15, Har16, Har18]. With kind permission of J.C. Hardy.

not directly correlated and can be treated independently. Thus, the two particle phase space density is given by

$$\rho(E) = \frac{\mathrm{d}n}{\mathrm{d}E_e} \cdot \frac{\mathrm{d}n}{\mathrm{d}E_\nu} = \frac{V^2 p_e E_e p_\nu E_\nu}{4\pi^4 \hbar^6}. \tag{6.30}$$

Using (6.10) and including a massive neutrino, the density of states can be expressed in terms of the kinetic energy of the electron E as (omitting subscript e)

$$\rho(E) = \frac{V^2 p E \sqrt{(E_0 - E)^2 - m^2(\nu_e)}(E_0 - E)}{4\pi^4 \hbar^6}. \tag{6.31}$$

Combining this together with (6.9) and (6.26) we get for the β-spectrum of electrons of allowed or super-allowed decays (with $\epsilon = E_0 - E$):

$$\frac{\mathrm{d}^2 N}{\mathrm{d}t\,\mathrm{d}E} = \underbrace{\frac{g_V^2 |M_F|^2 + g_A^2 |M_{GT}|^2}{2\pi^3 \hbar^7}}_{:=A} F(E, Z+1) p E$$

$$\times \sqrt{(E_0 - E)^2 - m^2(\nu_e)}(E_0 - E)\theta(E_0 - E - m(\nu_e))$$

$$= AF(E, Z+1) p E \epsilon \sqrt{\epsilon^2 - m^2(\nu_e)}\theta(\epsilon - m(\nu_e)). \tag{6.32}$$

with θ as the Heaviside function. As can be seen, the neutrino mass influences the spectral shape only at the upper end of the spectrum below E_0 leading to a change of the shape and a small constant offset proportional to $-m^2(\nu_e)$. Two important modifications might be necessary. First of all, (6.32) holds only for the decay of a bare and infinitely heavy nucleus. In reality, in dealing with atoms or molecules the possible excitation of the electron shell due to a sudden change in the nuclear charge has to be taken into account. The atom or molecule will end in a specific state of excitation energy E_j with a probability P_j. (6.32) will thus be modified into a superposition of β-spectra of amplitude P_j with different endpoint energies $\epsilon_j = E_0 - E_j$:

$$\frac{d^2N}{dt\,dE} = AF(E, Z+1)pE \sum_j P_j \epsilon_j \sqrt{\epsilon_j^2 - m^2(\nu_e)}\theta(\epsilon_j - m(\nu_e)). \quad (6.33)$$

For a detailed discussion see [Ott08]. In addition, in case of neutrino mixing (see Chapter 5) the spectrum is a sum of the components of decays into mass eigenstates ν_i:

$$\frac{d^2N}{dt\,dE} = AF(E, Z+1)pE \sum_j P_j \epsilon_j$$
$$\times \left(\sum_i |U_{ei}|^2 \sqrt{\epsilon_j^2 - m^2(\nu_i)}\theta(\epsilon_j - m(\nu_i)) \right). \quad (6.34)$$

As long as the experimental energy resolution is broader than the mass difference of involved neutrino states, the resulting spectrum can be analyzed in terms of a single observable—the electron neutrino mass:

$$m^2(\nu_e) = \sum_i |U_{ei}|^2 m^2(\nu_i) \quad (6.35)$$

by using (6.33). If the splitting of these states is larger than the resolution, it will lead to kinks in the energy spectrum (see Figure 6.11).

6.1.3 Kurie plot and *ft*-values

The decay constant λ for β-decays can be calculated from (6.32) by integration

$$\lambda = \frac{\ln 2}{T_{1/2}} = \int_0^{E_0} N(E)\,dE \quad (6.36)$$

This results in

$$\lambda = \int_0^{E_0} N(E)\,dE = (g_V^2|M_F|^2 + g_A^2|M_{GT}|^2)f(Z+1, \epsilon_0) \quad (6.37)$$

with

$$f(Z+1, \epsilon_0) = \int_1^{\epsilon_0} F(Z+1, \epsilon)\epsilon\sqrt{\epsilon^2 - 1}(\epsilon_0 - \epsilon)^2\,d\epsilon \quad (6.38)$$

as the so-called Fermi integral. ϵ, ϵ_0 are given by

$$\epsilon = \frac{E_e + m_e c^2}{m_e c^2} \qquad \epsilon_0 = \frac{Q}{m_e c^2}. \quad (6.39)$$

Table 6.3. Characterization of β-transitions according to their ft-values. Selection rules concerning spin J and parity π: $(+)$ means no parity change while $(-)$ implies parity change.

Transition	Selection rule	Log ft	Example	Half-life
Superallowed	$\Delta J = 0, \pm 1, (+)$	3.5 ± 0.2	^{1}n	11.7 min
Allowed	$\Delta J = 0, \pm 1, (+)$	5.7 ± 1.1	^{62}Zn	9.1 hr
First forbidden	$\Delta J = 0, \pm 1, (-)$	7.5 ± 1.5	^{198}Au	2.7 d
Unique first forbidden	$\Delta J = \pm 2, (-)$	8.5 ± 0.7	^{91}Y	58 d
Second forbidden	$\Delta J = \pm 2, (+)$	12.1 ± 1.0	^{137}Cs	30 yr
Third forbidden	$\Delta J = \pm 3, (-)$	18.2 ± 0.6	^{87}Rb	6×10^{10} yr
Fourth forbidden	$\Delta J = \pm 4, (+)$	22.7 ± 0.5	^{115}In	5×10^{14} yr

The product $fT_{1/2}$, given by

$$fT_{1/2} = \frac{K}{g_V^2 |M_F|^2 + g_A^2 |M_{GT}|^2} \tag{6.40}$$

is called the ft-value and can be used to characterize β-transitions (to be more accurate log ft is mostly used) as shown in Table 6.3. A compilation of ft-values of known β-emitters is shown in Figure 6.3. The constant K is given by

$$K = \frac{2 \ln 2 \pi^3 \hbar^7}{m_e^5 c^4}. \tag{6.41}$$

It is convenient in β-decay to plot the spectrum in the form of a so-called Kurie plot which is given by

$$\sqrt{\frac{N(E)}{p_e^2 F(Z+1, E)}} = A(E_0 - E_e) \left[1 - \left(\frac{m_\nu c^2}{E_0 - E_e} \right)^2 \right]^{1/4}. \tag{6.42}$$

Following from this, three important conclusions can be drawn:

(1) For massless neutrinos and allowed decays, the Kurie plot simplifies to

$$\sqrt{\frac{N(E)}{p_e^2 F(Z+1, E)}} = A(E_0 - E_e) \tag{6.43}$$

which is just a straight line intersecting the x-axis at the endpoint energy E_0.

(2) A light neutrino disturbs the Kurie plot in the region close to the endpoint, resulting in an endpoint at $E_0 - m_\nu c^2$ and the electron spectrum ends perpendicular to the x-axis.

(3) Assuming that there is a difference between the neutrino mass eigenstates and weak eigenstates as mentioned in Chapter 5 and discussed in more detail in

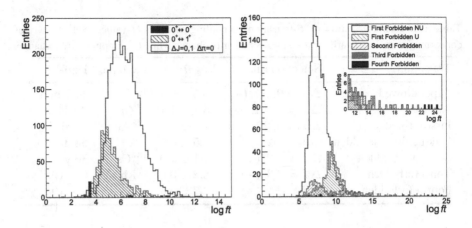

Figure 6.3. Compilation of all known $\log ft$ values for allowed and super-allowed decays (left) and forbidden decays (right) as provided by the IAEA. With kind permission of S. Turkat.

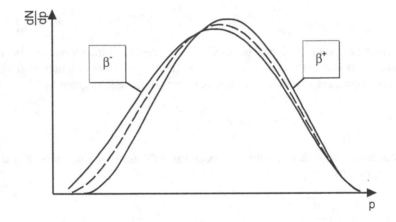

Figure 6.4. Schematic form of an electron beta spectrum. The phase space factor from (6.32) produces a spectrum with a parabolic fall at both ends for $m_\nu = 0$ (dotted line). This is modified by the interaction of the electron/positron with the Coulomb field of the final state nucleus (continuous lines). With kind permission of H. Wilsenach.

Chapter 8, the Kurie plot is modified to

$$\sqrt{\frac{N(p_e)}{p_e^2 F(Z+1, E)}} = A \sum_i |U_{ei}|^2 (E_0 - E_e) \left[1 - \left(\frac{m_i c^2}{E_0 - E_e} \right)^2 \right]^{1/4}. \quad (6.44)$$

The results are one or more kinks in the Kurie plot as discussed in Section 6.2.4.

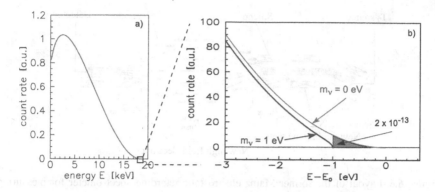

Figure 6.5. Endpoint region of a beta spectrum of tritium. The effect of a finite neutrino mass is a reduced endpoint at $E_0 - m_\nu c^2$. The black region indicates the last eV below the endpoint. Only about 10^{-13} electrons will fall in this region, hence a very intense source is needed (from [Kat01]). With kind permission of C. Weinheimer.

6.2   Searches for $m_{\bar{\nu}_e}$

Beta decay like all the following searches for neutrino masses is an example of kinematic searches. As the mass is given by the relativistic invariant total energy $m^2 = E^2 - p^2$ any uncertainty on the mass is given by

$$\Delta m_\nu^2 \approx \Delta E_\nu^2 + \Delta p_\nu^2 \approx 2E_\nu \Delta E_\nu + 2p_\nu \Delta p_\nu. \qquad (6.45)$$

Thus for a sensitive neutrino mass search the neutrino energy should be as small as possible; otherwise, relativity will hide any mass effect. On the other hand the decay rate shrinks with the phase space density and hence with the energy squared. In this way a compromise has to be found to use the most efficient way for a neutrino mass search and it is due to relativity that the mass limits by kinematical methods for the remaining two neutrino flavours are orders of magnitude worse.

6.2.1   General considerations

As already mentioned, a non-vanishing neutrino mass will reduce the phase space and leads to a change in the shape of the energy spectra, which for small masses can be investigated best near the endpoint of the energy spectrum (Figure 6.5). First measurements in search of neutrino masses have already been obtained in 1949 resulting in an upper bound of 100 keV [Han49]. A measurement done in 1952 gave a limit of less than 250 eV which led later to the general assumption of massless neutrinos [Lan52]. This was the motivation to implement neutrinos as massless particles in the Standard Model (see Chapter 3). Several aspects have to be considered before extracting a neutrino mass from a β-decay experiment [Hol92, Ott95, Wil01, Ott08]:

- the statistics of electrons with an energy close to the endpoint region is small (a small Q-value for the isotope under study is advantageous);

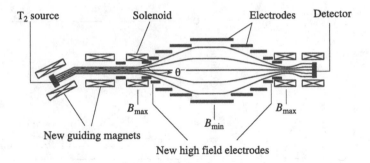

Figure 6.6. Layout of the former Mainz electrostatic retarding spectrometer for measuring tritium β-decay (from [Bon00]). Reproduced with permission of SNCSC.

- good energy resolution;
- energy loss within the source causing energy smearing;
- atomic and nuclear final state effects, excited state transitions; and
- a theoretical description of the involved wavefunctions.

From all isotopes, tritium is the most favoured one. But even in this case with the relatively low endpoint energy of about 18.6 keV, only a fraction of 2×10^{-13} of all electrons lies in a region of 1 eV below the endpoint (see Figure 6.5). A further advantage of tritium is $Z = 1$, making the distortion of the β-spectrum due to Coulomb interactions small and is allowing a sufficiently accurate quantum mechanical treatment. Furthermore, the half-life is relatively short ($T_{1/2} = 12.3$ yr) and the involved matrix element is energy independent (the decay is a super-allowed $\frac{1}{2}^{+} \rightarrow \frac{1}{2}^{+}$ transition between mirror nuclei). The underlying decay is

$$^{3}\mathrm{H} \rightarrow {}^{3}\mathrm{He}^{+} + e^{-} + \bar{\nu}_{e}. \tag{6.46}$$

In general, ^{3}H is not used in atomic form but rather in its molecular form H_2. In this case the molecular binding energies have to be considered as well and for an accurate determination, the small nuclear recoil E_R also has to be included (for details see [Ott08]). The newest available Penning trap measurement reports a value of $E_0 = (m(\mathrm{T}) - m(^{3}\mathrm{He}))c^2 = 18592.01(7)$ eV [Mye15]. Furthermore, only about 58% of the decays near the endpoint lead to the ground state of the $^{3}\mathrm{H}\,^{3}\mathrm{He}^{+}$ ion, making a detailed treatment of final states necessary. However, in the last 27 eV below the endpoint, there are no molecular excitations. An extensive discussion can be found in [Ott08].

6.2.2   Searches using spectrometers

While until 1990 magnetic spectrometers were mostly used for the measurements [Hol92, Ott95], the experiments performed afterwards in Mainz and Troitsk were using electrostatic filters with magnetic adiabatic collimation (MAC-E-Filters) [Lob85, Pic92]. The principle is that electrons emitted from the source spiral around magnetic field lines and will be guided into a spectrometer. The main advantage of such a spectrometer is the following: emitted electrons have a longitudinal kinetic energy T_L along the electric field lines, which is analysed by the spectrometer, and a transverse kinetic energy T_T in the cyclotron motion given by

$$T_T = -\mu \cdot \mathbf{B} \qquad \text{with } \mu = \frac{e}{2m_e}\mathbf{L}, \qquad (6.47)$$

with μ being the associated magnetic moment of the cyclotron motion. As angular momentum $\mathbf{L}$ is conserved and, therefore, μ is a constant of motion, showing that in an inhomogeneous magnetic field T_T changes proportional to $\mathbf{B}$. Thus, the energy in a decreasing magnetic field is transformed from $T_T \to T_L$ and *vice versa* in an increasing field. In the analyzing plane all cyclotron energy has been converted into analysable longitudinal energy T_L, except for a small rest between zero (emission under $\theta = 0°$, i.e., $T_T = 0$) and maximal ($\theta = 90°$, i.e., $T_T = T$). Therefore, the magnetic fields are very high (B_S), i.e., at the entrance and exit (detector) of the spectrometer and lowest ($B_{\min}$) in the middle, analysing plane. By a set of electrodes around the spectrometer a retarding electrostatic potential is created which has its maximum value (a barrier of eU_0 with $U_0 < 0$) in the analysing plane. The emitted electrons from the beta decay after entering the spectrometer will be decelerated by this potential: only those with sufficient energy can pass the potential barrier and will be accelerated and focused on the detector. The transmission function has a width of

$$\Delta T = \frac{B_{\min}}{B_S}T = \frac{1}{3000}T = 6 \text{ eV} \qquad (\text{if } T \approx 18 \text{ keV}). \qquad (6.48)$$

The Mainz spectrometer had a good energy resolution with filter width of only 4.8 eV. The major difference between the Mainz and the Troitsk spectrometer is the tritium source. While the Mainz experiment froze a thin film of T_2 onto a substrate, the Troitsk experiment used a gaseous tritium source. The obtained limits are [Lob03, Kra05]:

$$m_\nu^2 = -0.6 \pm 2.2(\text{stat.}) \pm 2.1(\text{sys.}) \text{ eV}^2/c^2 \qquad (6.49)$$
$$\to m_{\bar{\nu}_e} < 2.3 \text{ eV}/c^2 (95\% \text{ CL}) \qquad \text{Mainz}$$

$$m_\nu^2 = -2.3 \pm 2.5(\text{stat.}) \pm 2.0(\text{sys.}) \text{ eV}^2/c^2 \qquad (6.50)$$
$$\to m_{\bar{\nu}_e} < 2.05 \text{ eV}/c^2 (95\% \text{ CL}) \qquad \text{Troitsk}.$$

A long-standing problem of negative m_ν^2 values (m_ν^2 is a fit parameter to the spectrum and, therefore, can be negative) has finally disappeared, as a large amount of systematic uncertainties could be identified and reduced.The Troitsk number is obtained by including an observed anomaly in the analysis [Lob03].

6.2.2.1 The KATRIN experiment

For various physics arguments which will become clear throughout the book, it is important to improve the sensitivity of neutrino mass searches into a region

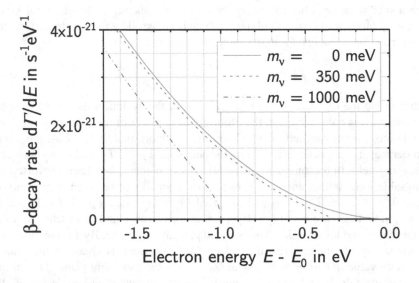

Figure 6.7. Differential energy spectra of electrons near the endpoint under the assumptions of various neutrino masses as could be seen by the KATRIN-experiment (from [Kle19]). Reproduced with permission of SNCSC.

Figure 6.8. Transport of the spectrometer towards its final destination at Karlsruhe. With kind permission of the KIT and the KATRIN collaboration.

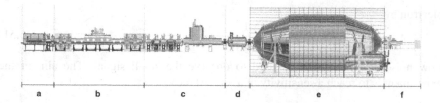

Figure 6.9. Schematic layout of the new KATRIN spectrometer. The KATRIN experimental setup, 70 m in length. The monitoring and calibration section (a) residing at the rear of the high-luminosity windowless source (b) provides stable and precise monitoring of tritium gas properties. The transport system (c) magnetically guides the electrons further downstream and prevents tritiated gas from entering the spectrometer section, which features two spectrometers operating as MAC-E-filters. The smaller pre-spectrometer (d) acts as a pre-filter for low energy electrons, and the larger main spectrometer (e) is used for the energy analysis in the endpoint region. A segmented semiconductor detector (f) acts as a counter for the transmitted signal electron (from [Kle19]). Reproduced with permission of SNCSC.

below 1 eV (Figure 6.7). However, this requires a very large spectrometer. The new KATRIN experiment [Wei03, Wei03a, Kat05] is designed to fulfil this need and probe neutrino masses down to 0.2 eV which is about an order of magnitude more sensitive than the Mainz experiment. For this to work an energy resolution at the transmission window of only 1 eV is necessary which corresponds to a ratio of $B_{\mathrm{min}}/B_S = 5 \times 10^{-5}$. A sketch of the layout is shown in Figure 6.9. The main features of the experiment are a windowless gaseous tritium source, minimizing the systematic uncertainties from the source itself, a pre-spectrometer, acting as an energy pre-filter to reject all lower energy electrons, except the ones in the region of interest close to the endpoint, and the main spectrometer. To obtain the required resolution the analysing plane has to have a spectrometer of diameter 10 m (Figure 6.8). The full spectrometer is 23 m long and is kept at an ultra-high vacuum below 10^{-11} mbar. The overall length of the experiment is 70 m. As mentioned, one difficulty is the fact that only 2×10^{-13} of all electrons from tritium decay fall into a region of 1 eV below the endpoint and thus a very intense tritium source is needed. In addition, the Si-detector at the end of the spectrometer has to be shielded, allowing only a background rate of 0.01 events s^{-1}. Also the detectors must have a good energy resolution (less than 600 eV at 18.6 keV). Recently, first results have been obtained, resulting in an upper mass limit of 1.1 eV (90 CL [Ake19].

6.2.2.2 Project 8

For a 0.2 eV sensitivity KATRIN has built a very large spectrometer. In case of a non-observation, new alternative techniques must be investigated if probing towards even lower neutrino masses is needed. One of the ideas is using frequencies which can be measured very accurately [Mon09, Esf17]. Here the cyclotron radiation emission spectroscopy (CRES) technique is used by measuring the cyclotron frequency of magnetically trapped ions. The frequency is linked to the kinetic energy of the electron by

$$f_c = \frac{1}{1} \frac{eB}{m_e + E_{\text{kin}}/c^2} \tag{6.51}$$

Low noise amplifiers will be used to observe the small signal. The aim of the experiment is to reach a sensitivity of about 40 meV.

6.2.3   Alternative searches

As mentioned, the number of electrons from beta decays close to the endpoint is very small and thus a deviation from almost a straight line is hard to detect (see Figure 6.7). Assume a small energy range ΔE close to the endpoint Q with $\Delta E \ll Q$, the number of decays in this range can be expanded as a Taylor series by [Moh91]

$$\int_{Q-\Delta E}^{Q} dE\, n(E) = \Delta E\, n(Q) - \frac{(\Delta E)^2}{2} n'(Q) + \frac{(\Delta E)^3}{6} n''(Q) + \dots \tag{6.52}$$

where the primes on n denote the derivatives with respect to the variable E. As shown in (6.40) for allowed beta decays the shape of the spectrum close to the endpoint is described by $(Q - E)^2$, hence the first two terms of 6.52 vanish. Thus, the fraction of decays with electron energies in the range $Q - \Delta E$ to Q is given by

$$\frac{\int_{Q-\Delta E}^{Q} dE n(E)}{\int_{0}^{Q} dE n(E)} \propto \left(\frac{\Delta E}{Q} \right)^3 . \tag{6.53}$$

Thus beta decay transitions with a very low Q-value are favourable. Various candidates exist, some of them are shown in Table 6.4, more can be found in [Eji19]. Ongoing measurements are performed with Penning traps to identify good candidates as they can provide the necessary atomic mass precision, see for example [Wel17].

A well known candidate for a ground state transition of low Q-value is ^{187}Re. The decay

$$^{187}\text{Re} \rightarrow {}^{187}\text{Os} + e^- + \bar{\nu}_e \tag{6.54}$$

has one of the lowest Q-values of all β-emitters as measured by Penning traps to be $Q = 2.492 \pm 20(\text{stat.}) \pm 15(\text{sys.})$ keV [Nes14]. The beta decay experiments are performed using cryogenic micro-calorimeters [Gat01, Fio01]. The idea behind this detector technology is a calorimetric energy measurement within an absorber, converting deposited energy into phonons which leads to a temperature rise. This will be detected by a sensitive thermometer. The same principle is also used in some double beta decay (see Chapter 7) and dark matter searches (see Chapter 13). For this to work, the device has to be cooled down into the mK region. The measurement of the electron energy is related to a temperature rise via the specific heat C_V by

$$\Delta T = \frac{\Delta E}{C_V} \tag{6.55}$$

Table 6.4. Some potential candidates for beta decay-transitions with low Q-value. This might include decays into excited states of the daughter nuclei. 'u' and 'nu' represent unique and non-unique transitions, respectively. Some Q-value are deduced from the Atomic Mass Evaluation [Wan17], potential Penning trap measurements are ongoing or planned (from [Eji19]).

Transition		$T_{1/2}$	E^*[keV]	Decay type	Q[keV]
^{77}As($3/2^-$)	$\rightarrow^{77}$ Se($5/2^+$)	38.3 h	680.1046(16)	1st nu β^-	2.8(18)
^{111}In($9/2^+$)	$\rightarrow^{111}$ Cd($3/2^+$)	2.805 d	864.8(3)	2nd u EC	- 2.8(50)
			866.60(6)	2nd u EC	- 4.6(50)
^{131}I($7/2^+$)	$\rightarrow^{131}$ Xe($9/2^+$)	8.025 d	971.22(13)	allowed β^-	-0.4(7)
^{146}Pm($3-$)	$\rightarrow^{146}$ Nd(2^+)	5.53 yr	1470.59	1st nu EC	1.4(40)
^{149}Gd($7/2^-$)	$\rightarrow^{149}$ Eu($5/2^+$)	9.28 d	1312(4)	1st nu EC	1(6)
^{155}Eu($5/2^+$)	$\rightarrow^{155}$ Gd($9/2^-$)	4.75 yr	251.7056(10)	1st u β^-	1.0(12)
^{159}Dy($3/2^-$)	$\rightarrow^{159}$ Tb($5/2^-$)	144 d	363.5449(14)	allowed EC	2.1(12)
^{161}Ho($7/2^-$)	$\rightarrow^{161}$ Dy($7/2^-$)	2.28 h	857.502(7)	allowed EC	1.4(27)
			858.7919(18)	2nd nu EC	0.1(27)
^{189}Ir($3/2^-$)	$\rightarrow^{189}$ Os($5/2^-$)	13.2 d	531.54(3)	1st nu EC	0.46(13)

where the specific heat is given in practical units as [Smi90]

$$C_V \approx 160 \left(\frac{T}{\Theta_D} \right)^3 \text{J cm}^{-3}\,\text{K}^{-1} \approx 1 \times 10^{18} \left(\frac{T}{\Theta_D} \right)^3 \text{keV cm}^{-3}\,\text{K}^{-1} \quad (6.56)$$

with Θ_D as material-dependent Debye temperature. This method allows the investigation of the β-decay of ^{187}Re without exploring final state effects. The associated half-life measurement of the decay is of the order of 10^{10} yr. The β-spectra (Figure 6.10) were measured successfully [Gat99, Ale99]. The deduced Q-values of 2481 ± 6 eV and 2460 ± 5(stat.) ± 10(sys.) eV are in agreement with the Penning trap measurement [Mou09]. A half-life for ^{187}Re of $T_{1/2} = 43 \pm 4$(stat.) ± 3(sys.) $\times 10^9$ yr has been obtained in agreement with measurements obtained by mass spectrometers which measured $T_{1/2} = 42.3 \pm 1.3 \times 10^9$ yr. A further opportunity is to study electrons from the 4-fold forbidden beta decay of ^{115}In into the first excited state of ^{115}Sn which has been measured by observing the 497.358 ± 0.024 keV γ-line [Cat05].

6.2.4 Kinks in β-decay

As already stated in Chapter 5, the existence of several neutrino mass eigenstates and their mixing and also atomic final state transitions in the daughter ion might lead to kinks in the Kurie plot of a β-spectrum. This is shown schematically in Figure 6.11. Assuming the energy range where the Kurie plot shows a kink (in the example here

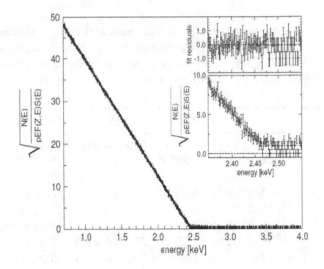

Figure 6.10. Kurie plot of the ^{187}Re spectrum obtained with a cryogenic AgReO$_4$ bolometer. (from [Sis04]). © 2004 With permission from Elsevier.

only two states are involved) is small, the spectrum depends on the involved mass eigenstates and the mixing angle θ (K given by (6.41)):

$$\frac{\Delta K}{K} \simeq \frac{\tan^2 \theta}{2} \left(1 - \frac{m_2^2 c^4}{(E_0 - E_e)^2} \right)^{1/2} \qquad \text{for } E_0 - E_e > m_2 c^2. \qquad (6.57)$$

Therefore, the position of the kink is determined by the heavier mass eigenstate m_2 and the size of the kink is related to the mixing angle θ between the neutrino states (see Chapter 8). Experimental searches were performed especially for heavier neutrino mass eigenstates in the keV range. A search for admixtures of keV neutrinos using the decay

$$^{63}\text{Ni} \rightarrow ^{63}\text{Cu} + e^- + \bar{\nu}_e \qquad (6.58)$$

with a Q-value of 67 keV has been performed [Hol99] and the limits on the admixture are shown in Figure 6.12. The discussion of mixtures with even heavier neutrinos states in the MeV, GeV region will be discussed in later.

6.3   Searches for m_{ν_e}

CPT invariance ensures that $m_{\bar{\nu}_e} = m_{\nu_e}$. However, some theories beyond the Standard Model offer the possibility of CPT violation [Kos11], which makes it worthwhile to measure m_{ν_e} directly as well. As stated in Section 6.2.3 decays with very low Q-values and therefore electron captures are considered. The isotope of most interest in the past and nowadays is ^{163}Ho. It has a very low Q-value of 2.833 ± 0.030(stat.) ± 0.015(sys.) [Eli15] and a half-life of $T_{1/2} = 4570 \pm 50$ years. The

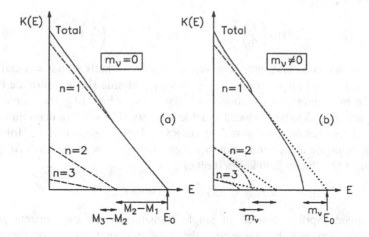

Figure 6.11. Schematic Kurie plot in the region of the beta decay endpoint. Shown are three different atomic final states of the daughter ion and the impact on the spectral shape. Similar features will be observed in the spectrum if (two) three neutrino mass eigenstates will exist. (from [Sch97]). Reproduced with permission of SNCSC.

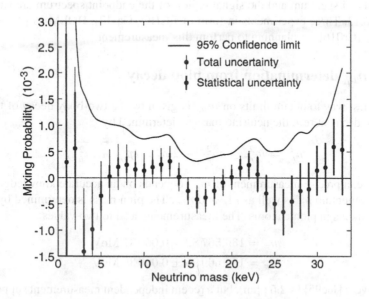

Figure 6.12. Best fit (points) of the mixing probability as a function of assumed neutrino mass in ^{63}Ni decay. The error bars combine statistical and systematic errors. The solid line is an upper limit at 95% CL (from [Hol99]). © 1999 With permission from Elsevier.

decay is characterized as

$$^{163}\text{Ho} \left(\frac{7}{2}\right)^- + e^- \rightarrow \nu_e + ^{163}\text{Dy}^* \left(\frac{5}{2}\right)^- \tag{6.59}$$

As the Q-value is so low, only electrons from atomic shells with n = 3 and higher can be used and - in first order - only s- and $p_{1/2}$ atomic shells contribute because of angular momentum conservation. As ^{163}Ho is very long-living, a source of kBq or MBq activity has to be produced to achieve reasonable neutrino mass limits. Such a measurement has been suggested by [deR81]. Using a source of $^{163}\text{HoF}_3$ and a Si(Li) detector the atomic transition between the 5p $\rightarrow$ 3s levels was investigated. Assuming a Q-value of 2.56 keV a limit of

$$m_{\nu_e} < 225 \text{ eV} \qquad (95\% \text{ CL}) \tag{6.60}$$

was obtained [Spr87]. Instead of single transitions, new experiments prefer a calorimetric approach by measuring the total endpoint energy of the internal bremsstrahlungs spectrum, which is well defined because of a new Q-value measurement [Eli15]. The spectrum can be described by

$$\frac{dN}{dE_C} = A(Q_{\text{EC}} - E_C)^2 \sqrt{1 - \frac{m_\nu^2}{(Q_{\text{EC}} - E_C)^2}} \sum_H \frac{C_H n_H B_H \phi_H^2(0)\Gamma_H/2\pi}{(E_C - E_H)^2 + \Gamma_H^2/4} \tag{6.61}$$

A calculated spectrum and the signal region of the endpoint spectrum are shown in Figure 6.13. Three experiments in form of ECHO [Gas17], HOLMES [Alp15] and NuMecs [Cro16] are planning to perfom this measurement.

6.4 m_{ν_μ} determination from pion decay

The easiest way to obtain limits on m_{ν_μ} is given by the two-body decay of the π^+. For pion decay at rest, the neutrino mass is determined by

$$m_{\nu_\mu}^2 = m_{\pi^+}^2 + m_{\mu^+}^2 - 2m_{\pi^+}\sqrt{p_{\mu^+}^2 + m_{\mu^+}^2}. \tag{6.62}$$

Therefore, a precise measurement of m_{ν_μ} depends on an accurate knowledge of the muon momentum p_μ as well as m_μ and m_π. The pion mass is determined by X-ray measurements in pionic atoms. The measurements lead to two values

$$m_\pi = 139.567\,82 \pm 0.000\,37 \text{ MeV}$$
$$m_\pi = 139.569\,95 \pm 0.000\,35 \text{ MeV} \tag{6.63}$$

respectively [Jec95] ($\approx$2.5 ppm), but a recent independent measurement supports the higher value by measuring $m_\pi = 139.570\,71 \pm 0.000\,53$ MeV [Len98]. The muon mass is determined by measuring the ratio of the magnetic moments of muons and protons. This results in [PDG08]

$$m_\mu = (105.658\,3668 \pm 0.000\,0038) \text{ MeV} \qquad (\approx 0.04 \text{ ppm}). \tag{6.64}$$

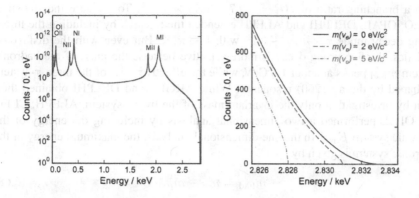

Figure 6.13. Left: Calculated ^{163}Ho EC spectrum for a total number of 10^{14} events with $Q_{EC} = 2.833\,\text{keV}$, considering only first order excitations for the daughter ^{163}Dy atom using the parameters given in [Fae15] and assuming zero neutrino mass. Right: Shape of the Bremsstrahlung spectrum near the endpoint calculated for neutrino masses of 0, 2 and $5\,\text{eV/c}^2$, respectively. The effect of a finite electron neutrino mass is shown on a linear scale. (from [Gas17]). Reproduced with permission of SNCSC.

Latest π-decay measurements were performed at the Paul-Scherrer Institute (PSI) resulting in a muon momentum of [Ass96]

$$p_\mu = (29.792\,00 \pm 0.000\,11)\,\text{MeV} \qquad (\approx 4\,\text{ppm}). \tag{6.65}$$

Combining all numbers, a limit of

$$m_{\nu_\mu}^2 = (-0.016 \pm 0.023)\,\text{MeV}^2 \rightarrow m_{\nu_\mu} < 190\,\text{keV} \qquad (90\%\,\text{CL}) \tag{6.66}$$

could be achieved.

6.5   Mass of the ν_τ from tau decay

Before discussing the mass of ν_τ it should be mentioned that the direct detection of ν_τ via CC reactions has been observed only recently [Kod01,Lun03a,Kod08]. It was the goal of E872 (DONUT) at Fermilab to detect exactly this reaction (see Chapter 4) and they came up with nine candidate events expecting 1.5 background events.

The present knowledge of the mass of ν_τ stems from measurements with ARGUS (DORIS II) [Alb92], CLEO(CESR) [Cin98], OPAL [Ack98], DELPHI [Pas97] and ALEPH [Bar98] (LEP) all using the reaction $e^+e^- \rightarrow \tau^+\tau^-$. The energy E_τ is given by the different collider centre-of-mass energies $E_\tau = \sqrt{s}/2$. Practically all experiments use the τ-decay into five charged pions:

$$\tau \rightarrow \nu_\tau + 5\pi^\pm(\pi^0) \tag{6.67}$$

with a branching ratio of $BR = (9.7 \pm 0.7) \times 10^{-4}$. To increase the statistics, CLEO, OPAL, DELPHI and ALEPH extended their search by including the three-prong decay mode $\tau \rightarrow \nu_\tau + 3h^\pm$ with $h \equiv \pi, K$. But even with the disfavored statistics, the five-prong decay is more sensitive because the mass of the hadronic system m_{had} peaks at about 1.6 GeV, while the effective mass of the three π-system is shaped by the $a_1(1260)$ resonance. While ARGUS and DELPHI obtained their limit by investigating only the invariant mass of the five π-system, ALEPH, CLEO and OPAL performed a two-dimensional analysis by including the energy of the hadronic system E_{had}. In the one-dimensional analysis, the maximum energy of the hadronic system is given by

$$m_{had} = m_\tau - m_\nu \tag{6.68}$$

and, therefore, results in an upper bound on m_ν. A bound can also be obtained from the hadronic energy coming from

$$m_\nu < E_\nu = E_\tau - E_{had} \tag{6.69}$$

where E_{had} is given in the rest frame of the τ by

$$E_{had} = \frac{(m_\tau^2 + m_{had}^2 - m_\nu^2)}{2m_\tau} \tag{6.70}$$

which will be boosted in the laboratory frame. A finite neutrino mass leads to a distortion of the edge of the triangle of a plot of the E_{had}–m_{had} plane as shown in Figure 6.14.

The most stringent one is given by ALEPH to be [Bar98]

$$m_{\nu_\tau} < 18.2 \, \text{MeV} \quad (95\% \, \text{CL}) \tag{6.71}$$

A combined limit for all four LEP experiments improves this limit only slightly to 15.5 MeV. A chance for improvement might be offered by an investigation of leptonic D_S^+-decays [Pak03].

6.6 Electromagnetic properties of neutrinos

A further experimental aspect where a non-vanishing neutrino mass could show up is the search for electromagnetic properties of neutrinos such as electromagnetic moments. Even with charge neutrality, neutrinos can participate in electromagnetic interactions by coupling with photons via loop diagrams (see Figure 6.16). As for other particles the electromagnetic properties can be described by form factors (see Chapter 4). The Lorentz and gauge invariance of the electromagnetic current j_μ allows four independent form factors for Dirac neutrinos, the charge and axial charge form factors $F(Q^2)$ and $G(Q^2)$ and the electric and magnetic dipole moment form factors $D(Q^2)$ and $M(Q^2)$. $F(Q^2)$ and $G(Q^2)$ have to vanish for $Q^2 \rightarrow 0$ because of electric charge neutrality. The values of $D(Q^2)$ and $M(Q^2)$ for $Q^2 = 0$ are the

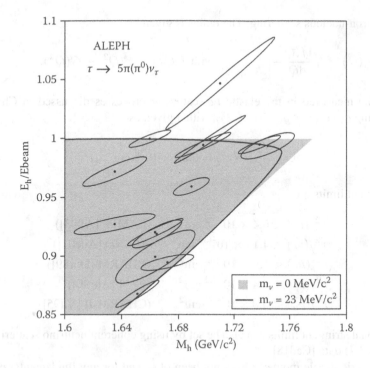

Figure 6.14. Two-dimensional plot of the hadronic energy *versus* the invariant mass of the 5(6)π-system. The uncertainty ellipses are positively correlated because both, the hadronic mass and the hadronic energy, are determined from the momenta of the particles composing the hadronic system (from [Bar98]). Reproduced with permission of SNCSC.

electric $D(0) = d_\nu$ and magnetic dipole moment $M(0) = \mu_\nu$ of the Dirac neutrinos. CPT and CP invariance make the electric dipole moment vanish. The previously mentioned static moments correspond to the diagonal elements of a "moment" matrix. The off-diagonal elements, which not only lead to a spin flip in a magnetic field but also include a change in flavour, are called transition moments. For more details see [Kim93, Fuk03a, Giu09, Giu15].

For Majorana neutrinos $F(Q^2)$, $D(Q^2)$ and $M(Q^2)$ vanish, because of their self-conjugate properties. Only $G(Q^2)$ and transition moments are possible.

6.6.1   Electric dipole moments

The Fourier transforms of the previously mentioned form factors in general are the spatial distributions of charges and dipole moments (EDM). This allows a possible spatial extension of neutrinos to be defined via an effective mean charge radius $\langle r^2 \rangle$ ('effective size of the neutrino') in the same spirit as radii of nuclei are determined by electron-nucleus scattering. The radius is given by

$$\langle r^2 \rangle = 6 \frac{\mathrm{d}f(Q^2)}{\mathrm{d}Q^2} \bigg|_{Q^2=0} \qquad \text{with } f(Q^2) = F(Q^2) + G(Q^2). \tag{6.72}$$

It can be measured in the elastic νe-scattering processes discussed in Chapter 4 (replacing $g_V, G_V \rightarrow g_V, G_V + 2\delta$), with δ given as

$$\delta = \frac{\sqrt{2}\pi\alpha}{3G_F} \langle r^2 \rangle = 2.38 \times 10^{30} \text{ cm}^{-2} \langle r^2 \rangle. \tag{6.73}$$

The current limits are:

$$
\begin{aligned}
\langle r^2 \rangle(\nu_e) &< 5.4 \times 10^{-32} \text{ cm}^2 && \text{(LAMPF [All93])} \\
\langle r^2 \rangle(\nu_e) &< 4.14 \times 10^{-32} \text{ cm}^2 && \text{(LSND [Aue01])} \\
\langle r^2 \rangle(\nu_\mu) &< 1.0 \times 10^{-32} \text{ cm}^2 && \text{(CHARM [Dor89])} \\
\langle r^2 \rangle(\nu_\mu) &< 2.4 \times 10^{-32} \text{ cm}^2 && \text{(E734 [Ahr90])} \\
\langle r^2 \rangle(\nu_\mu) &< 6.0 \times 10^{-32} \text{ cm}^2 && \text{(CHARM-II [Vil95])}
\end{aligned}
\tag{6.74}
$$

Additional stringent limits can also be set by using coherent neutrino scattering (see Section 4.4) data [Cad18].

Electric dipole moments have not been observed for any fundamental particle. They always vanish as long as CP or, equivalently, T is conserved as this implies $d_\nu = 0$. Nevertheless, very stringent bounds exist on the EDM of the electron [And18], muon [Ben09] and neutron [Pen15]. However, very little is known about CP violation in the leptonic sector (see Chapter 8). This might change with running and planned experiments. Until then we can use the limits on magnetic dipole moments from νe-scattering as bounds, because for not-too-small energies a contribution from an electric dipole moment to the cross-section is identical. Then bounds of the order of $d_\nu < 10^{-20} \, e \, \mathrm{cm}$ (ν_e, ν_μ) and $d_\nu < 10^{-17} \, e \, \mathrm{cm}$ (ν_τ) result. For Majorana neutrinos, CPT invariance ensures that $d_\nu = 0$.

6.6.2 Magnetic dipole moments

A further option to probe a non-vanishing neutrino mass and the neutrino character is the search for its magnetic moment. In the Standard Model neutrinos have no magnetic moment because they are massless and a magnetic moment would require a coupling of a left-handed state with a right-handed one–the latter does not exist. A simple extension by including right-handed singlets allows for Dirac masses. In this case, it can be shown that due to loop diagrams neutrinos can obtain a magnetic moment (see Figure 6.16) which is proportional to their mass and is given by [Lee77, Mar77]

$$\mu_\nu = \frac{3eG_F}{8\sqrt{2}\pi^2} m_\nu = 3.2 \times 10^{-19} \left(\frac{m_\nu}{\mathrm{eV}}\right) \mu_B \tag{6.75}$$

with μ_B as Bohr magneton. For neutrino masses in the eV range, this is either far too small to be observed or to have any significant effect in astrophysics. Nevertheless, there exist models, which are able to increase the expected magnetic moment [Fuk87, Bab87, Pal92]. However, Majorana neutrinos still have a vanishing static moment because of CPT invariance. This can be seen from the following argument (a more theoretical treatment can be found in [Kim93]). The electromagnetic energy E_{em} of a neutrino with spin direction σ in an electromagnetic field is given by

$$E_{em} = -\mu_\nu \sigma \cdot \mathbf{B} - d_\nu \sigma \cdot \mathbf{E}. \tag{6.76}$$

Applying CPT results in $B \rightarrow B$, $E \rightarrow E$ and $\sigma \rightarrow -\sigma$ which results in $E_{em} \rightarrow -E_{em}$. However, CPT transforms a Majorana neutrino into itself ($\bar{\nu} = \nu$) which allows no change in E_{em}. Therefore, $E_{em} = 0$ which is possible only if $\mu_\nu = d_\nu = 0$.

Experimental limits on magnetic moments can be probed by searching for modifications in $\nu_e e$-scattering experiments and astrophysical considerations. The differential cross-section for $\nu_e e$-scattering in the presence of a magnetic moment is given by [Dom71, Vog89]

$$\frac{d\sigma}{dT} = \frac{G_F^2 m_e}{2\pi} \left[(g_V + x + g_A)^2 + (g_V + x - g_A)^2 \left(1 - \frac{T}{E_\nu} \right)^2 \right.$$
$$\left. + (g_A^2 - (x + g_V)^2) \frac{m_e T}{E_\nu^2} \right] + \frac{\pi \alpha^2 \mu_\nu^2}{m_e^2} \frac{1 - T/E_\nu}{T} \tag{6.77}$$

where T is the kinetic energy of the recoiling electron and x is related to the charge radius $\langle r^2 \rangle$:

$$x = \frac{2m_W^2}{3} \langle r^2 \rangle \sin^2 \theta_W \qquad x \rightarrow -x \qquad \text{for } \bar{\nu}_e. \tag{6.78}$$

The contribution associated with the charge radius can be neglected if $\mu_\nu \gtrsim 10^{-11} \mu_B$. As can be seen, the largest effect of a magnetic moment can be observed in the low-energy region and because of destructive interference with the electroweak terms, searches with antineutrinos would be preferred. The obvious sources are, therefore, nuclear reactors (see Figure 6.15).

One of the latest dedicated experiments was the MUNU experiment [Ams97] performed at the Bugey reactor. It consisted of a 1 m³ time projection chamber (TPC) loaded with CF_4 under a pressure of 5 bar. The usage of a TPC allowed not only the electron energy to be measured but also, for the first time in such experiments, the scattering angle, making the reconstruction of the neutrino energy possible. To suppress the background, the TPC was surrounded by 50 cm anti-Compton scintillation detectors as well as a passive shield of lead and polyethylene. The neutrino energy spectrum in reactors in the energy region 1.5 MeV$< E_\nu <$ 8 MeV is known at the 3% level. If there is no magnetic moment the expected count rate is 9.5 events per day increasing to 13.4 events per day if $\mu_\nu = 10^{-10} \mu_B$ for an energy threshold of 500 keV. The experiment did not see any hint for a magnetic moment and derived an upper limit of [Dar05] $\mu_{\bar{\nu}_e} < 9 \times 10^{-11} \mu_B$(90% CL).

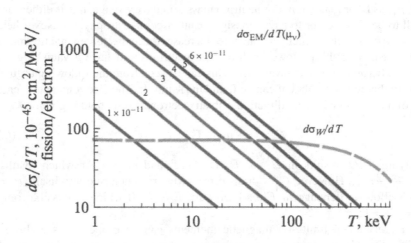

Figure 6.15. Differerential cross-section for electron-neutrino scattering as a function of the electron recoil energy T for different values of the magnetic moment. The dashed cross-section assumes no magnetic moment while the other lines show exemplaric modifications of the cross-sections due to a magnetic moment in the region from 1-6 $\times 10^{-11}$ μ_B (from [Bed13]). Reproduced with permission of SNCSC.

Another approach is placing Ge semiconductor detectors close to a reactor core. A further approach is TEXONO [Won07], using a 1 kg Ge-detector in combination with 46 kg of CsI(Tl) scintillators. The use of a low background Ge–NaI spectrometer in a shallow depth near a reactor has also been considered. A first search resulted in a limit of $\mu_{\bar{\nu}_e} < 7.4 \times 10^{-11} \mu_B$ [Won07]. The same idea is used by the GEMMA experiment using a Ge-detector only. The limit obtained so far is $\mu_{\bar{\nu}_e} < 3 \times 10^{-11} \mu_B$ [Bed13]. A search can also be performed with large scale detectors designed to study other neutrino issues like solar neutrinos. Two of them, Superkamiokande (see Chapter 9) and Borexino (see Chapter 10) have placed upper limits on the magnetic moment of $\mu_{\bar{\nu}_e} < 1.1 \times 10^{-10} \mu_B$ [Liu04] and $\mu_{\bar{\nu}_e} < 3 \times 10^{-11} \mu_B$ ([Ago17]), respectively.

Astrophysical limits exist and are somewhat more stringent but also more model dependent that sometimes its understanding is limited. Bounds from supernovae will be discussed in Chapter 11. The major constraint on magnetic moments arises from stellar energy-loss arguments. Transverse and longitudinal excitations in a stellar plasma ('plasmons') are both kinematically able to decay into neutrino pairs of sufficiently small mass, namely $2m_\nu < K^2$, where K is the plasmon 4-momentum. In addition, an effective ν–γ coupling is introduced. For $\mu_\nu > 10^{-12} \mu_B$ this process can compete with standard energy-loss mechanisms of stars if the plasma frequency is around 10 keV. The cooling of the hottest white dwarfs will be faster if plasmon decay into neutrinos occurs and, therefore, a suppression of the hottest white dwarfs in the luminosity function might occur. From observations, bounds of the order

$\mu_\nu < 10^{-11}\mu_B$ could be obtained [Raf99]. More reliable are globular cluster stars. Here asymptotic giant branch (AGB) stars and low mass red giants before the He flash would be affected if there is an additional energy loss in the form of neutrinos. To prevent the core mass at He ignition from exceeding its standard value by less than 5%, a bound of $\mu_\nu < 2 \times 10^{-12}\mu_B$ has been obtained [Raf99, Arc15].

Accelerator measurements based on $\nu_e e \rightarrow \nu_e e$ and $\nu_\mu e \rightarrow \nu_\mu e$ scattering were done at LAMPF and BNL yielding bounds for ν_e and ν_μ of [Kra90] (see also [Dor89, Ahr90, Vil95])

$$\mu_{\nu_e} < 10.8 \times 10^{-10}\mu_B \qquad (\text{if } \mu_{\nu_\mu} = 0) \qquad (6.79)$$

$$\mu_{\nu_\mu} < 7.4 \times 10^{-10}\mu_B \qquad (\text{if } \mu_{\nu_e} = 0). \qquad (6.80)$$

Combining these scattering results and Super-Kamiokande observations (see Chapter 9), a limit for the magnetic moment of ν_τ was obtained [Gni00]:

$$\mu_{\nu_\tau} < 1.9 \times 10^{-9}\mu_B. \qquad (6.81)$$

As can be seen, the experimental limits are still orders of magnitude away from the predictions (6.75).

6.7   Neutrino decay

Another physical process which is possible if neutrinos have a non-vanishing rest mass is neutrino decay. Depending on the mass of the heavy neutrino ν_H various decay modes into a light neutrino ν_L can be considered, the most common are:

$$
\begin{aligned}
&\nu_H \rightarrow \nu_L + \gamma \\
&\nu_H \rightarrow \nu_L + \ell^+ + \ell^- \qquad \cdot (\ell \equiv e, \mu) \\
&\nu_H \rightarrow \nu_L + \nu + \bar{\nu} \\
&\nu_H \rightarrow \nu_L + \chi.
\end{aligned}
\qquad (6.82)
$$

The first mode is called radiative neutrino decay and the fourth process is a decay with the emission of a majoron χ, the Goldstone boson of lepton symmetry breaking (see Chapter 7). Because of the non-detectable majoron the last two modes are often called invisible decays. Note that it is always a mass eigenstate that decays, meaning e.g., the decay $\nu_\mu \rightarrow \nu_e + \gamma$ is in a two-neutrino mixing scheme caused by the decay $\nu_2 \rightarrow \nu_1 + \gamma$.

6.7.1   Radiative decay $\nu_H \rightarrow \nu_L + \gamma$

The two simplest Feynman graphs for radiative neutrino decay are shown in Figure 6.16. The decay rate is given as [Fei88]

$$\Gamma(\nu_H \rightarrow \nu_L + \gamma) = \frac{1}{8\pi}\left[\frac{m_H^2 - m_L^2}{m_H}\right]^3 (|a|^2 + |b|^2) \qquad (6.83)$$

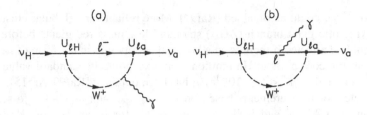

Figure 6.16. Feynman diagrams describing radiative neutrino decay $\nu_H \to \nu_L + \gamma$. For virtual photons these graphs determine the magnetic moment of the neutrino (from [Sch97]). Reproduced with permission of SNCSC.

where for Dirac neutrinos the amplitudes a, b are

$$a_D = -\frac{eG_F}{8\sqrt{2}\pi^2}(m_H + m_L)\sum_l U_{lH}U_{lL}^* F(r_l) \tag{6.84}$$

$$b_D = -\frac{eG_F}{8\sqrt{2}\pi^2}(m_H - m_L)\sum_l U_{lH}U_{lL}^* F(r_l) \tag{6.85}$$

with U as the corresponding mixing matrix elements and $F(r_l)$ as a smooth function of $r_l = (m_l/m_W)^2$ with $F(r_l) \approx 3r/4$ if $r_l \ll 1$. For Majorana neutrinos $a_M = 0$, $b_M = 2b_D$ or $a_M = 2a_D$, $b_M = 0$ depending on the relative CP-phase of the neutrinos ν_H and ν_L. Taking only tau-leptons which dominate the sum in (6.84), one obtains for $m_L \ll m_H$ a decay rate of

$$\Gamma \approx \frac{m_H^5}{30 \text{ eV}}|U_{\tau H}U_{\tau L}^*|^2 \times 10^{-29} \text{ yr}^{-1}. \tag{6.86}$$

This implies very long lifetimes against radiative decays of the order $\tau > 10^{30}$ yr. However, in certain models, like the left–right symmetric models (see Section 5.3.1), this can be reduced drastically.

Experimentally, the following searches have been performed:

- Search for photons at nuclear reactors by using liquid scintillators. This probes the admixture of ν_H to $\bar{\nu}_e$; therefore, it is proportional to $|U_{eH}|^2$. At the Goesgen reactor no difference was observed in the on/off phases of the reactor resulting in [Obe87]

$$\frac{\tau_H}{m_H} > 22(59)\frac{s}{\text{eV}} \qquad \text{for } a = -1(+1) \qquad (68\% \text{ CL}). \tag{6.87}$$

- At LAMPF, using pion and muon decays at rest (therefore looking for $|U_{\mu H}|^2$). No signal was observed and a limit of [Kra91]

$$\frac{\tau_H}{m_H} > 15.4\frac{s}{\text{eV}} \qquad (90\% \text{ CL}) \tag{6.88}$$

was obtained.

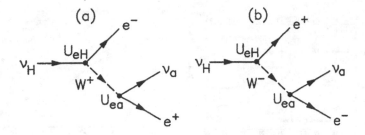

Figure 6.17. Feynman diagrams describing radiative neutrino decay $\nu_H \to \nu_L + e^+ + e^-$ (from [Sch97]). Reproduced with permission of SNCSC.

- From the experimental solar x-ray and γ-flux a lower bound was derived as [Raf85]

$$\frac{\tau_H}{m_H} > 7 \times 10^9 \, \frac{s}{eV}. \tag{6.89}$$

  Observations performed during a solar eclipse to measure only decays between the moon and the Earth have also been performed [Bir97, Cec11].

- Maybe the most stringent limits come from supernova SN1987A (see Chapter 11). There was no excess of the γ-flux measured by the gamma-ray spectrometer (GRS) on the solar maximum mission (SMM) satellite during the time when the neutrino events were detected, which can be converted in lower bounds of [Blu92a, Obe93]

$$\tau_H > 2.8 \times 10^{15} B_\gamma \frac{m_H}{eV} \qquad m_M < 50 \, eV$$

$$\tau_H > 1.4 \times 10^{17} B_\gamma \qquad 50 \, eV < m_M < 250 \, eV \tag{6.90}$$

$$\tau_H > 6.0 \times 10^{18} B_\gamma \frac{eV}{m_H} \qquad m_M > 250 \, eV$$

where B_γ is the radiative branching ratio.

6.7.2 The decay $\nu_H \to \nu_L + e^+ + e^-$

The Feynman graphs for this decay are shown in Figure 6.19. Clearly this decay is possible only if $m_H > 2m_e \approx 1$ MeV. The decay rate is given by

$$\Gamma(\nu_H \to \nu_L + e^+ + e^-) = \frac{G_F^2 m_H^5}{192\pi^3} |U_{eH}^2|. \tag{6.91}$$

Here Dirac and Majorana neutrinos result in the same decay rate. Searches are performed with nuclear reactors and high-energy accelerators. The obtained limits on the mixing U_{eH}^2 as well as such on $U_{\mu H}^2$ are shown in Figure 6.18.

6.7.3 The decay $\nu_H \to \nu_L + \chi$

To avoid several astrophysical and cosmological problems associated with radiative decays, the invisible decay into a majoron is often considered. Its decay rate is given

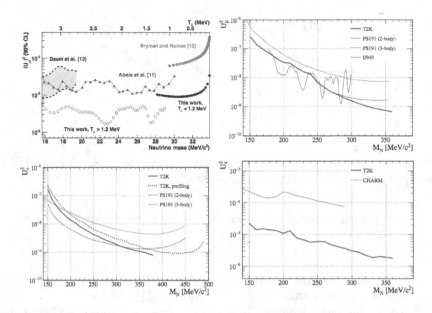

Figure 6.18. Limits on U_{eH}^2, $U_{\mu H}^2$ and $U_{\tau H}^2$ as a function of a heavy neutrino mass m_H. Top left: Examplaric graph on limits on U_{eH}^2 at low energies (< 50 MeV). The curve 'this work' refers to reference [Agu19], also shown are data from [Abe81, Dau87, Bry96] (from [Agu19]). © 2019 With permission from Elsevier. Top right: A continuation of limits for U_{eH}^2 in a higher energy range between 100-400 MeV obtained by the PS191 and T2K experiments. Bottom left and right: For completeness also $U_{\mu H}^2$ and $U_{\tau H}^2$ are shown from the PS191 [Ber86, Ber88], E949 [Art15], CHARM [Orl02] and T2K experiments in the range of 100-400 MeV. ((b), (c) and (d) from [Abe19]). © 2019 by the American Physical Society. For a review see [Dre17].

for highly relativistic neutrinos as [Kim93]

$$\Gamma(\nu_H \to \nu_L + \chi) = \frac{g^2 m_L m_H}{16\pi E_H}\left(\frac{x}{2} - 2 - \frac{2}{x}\ln x + \frac{2}{x^2} - \frac{1}{2x^3}\right) \qquad (6.92)$$

with g being an effective coupling constant and $x = m_H/m_L$. Little is known experimentally about this invisible decay. Matter can enhance the decay rates as discussed in [Kim93]. However, no neutrino decay has yet been observed.

6.8 Heavy neutrinos

Also searches for heavy neutrinos in the GeV and TeV range can be performed at accelerators like LHC. As an example, a search for right-handed heavy neutrino and W-boson has been conducted [Aab19a]. As can be seen, strong lower limits of about 1 TeV can be implied on both particles. Similar results have been obtained by ATLAS and the CMS experiment.

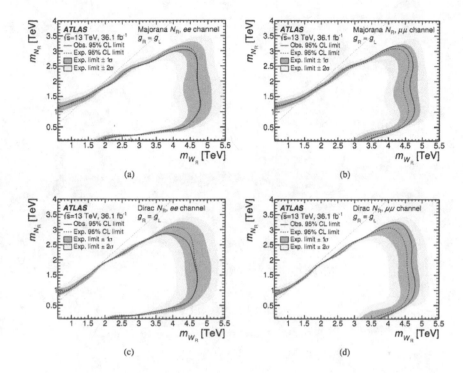

Figure 6.19. Exclusion plots of right-handed heavy neutrinos N_R versus a right-handed W-boson W_R in the electron (left) and muon (right) channel for both options Majorana (upper) and Dirac (lower) neutrinos. The couplings g_L and g_R are assumed to be the same. The data shown are obtained by the ATLAS experiment at CERN at a center of mass energy of 13 TeV (from [Aab19a]). Reproduced with permission of SNCSC.

We now proceed to a further process where neutrino masses can show up and which is generally considered as the gold-plated channel for probing the fundamental character of neutrinos, discussed in Chapter 2.

Chapter 7

Double beta decay

DOI: 10.1201/9781315195612-7

A further nuclear decay which is extremely important for neutrino physics is neutrinoless double β-decay. This lepton-number-violating process requires, in addition to a non-vanishing neutrino mass, that neutrinos are Majorana particles. It is, therefore, often regarded as the gold-plated process for probing the fundamental character of neutrinos. For additional literature see [Doi83, Hax84, Doi85, Gro90, Boe92, Kla95, Rod11, Saa13, Dol19, Eji19].

7.1 Introduction

Double β-decay is characterized by a nuclear process changing the nuclear charge Z by two units while leaving the atomic mass A unchanged. It is a transition among isobaric isotopes. Using the semi-empirical mass formula of Weizsäcker [Wei35], isobars can be described as a function of the ordering number Z as

$$m(Z, A = \text{const.}) \propto \text{constant} + \alpha Z + \beta Z^2 + \delta_P \tag{7.1}$$

with δ_P as the pairing energy, empirically parametrized as [Boh75]

$$\delta_P = \begin{cases} -a_P A^{-1/2} & \text{even–even nuclei} \\ 0 & \text{even–odd and odd–even nuclei} \\ +a_P A^{-1/2} & \text{odd–odd nuclei} \end{cases} \tag{7.2}$$

with $a_P \approx 12\,\text{MeV}$. For odd A the pairing energy vanishes resulting in one parabola with one stable isobar, while for even A two parabolae separated by $2\delta_P$ exist (Figure 7.1). The second case allows double β-decay and, therefore, all double β-decay emitters are even–even nuclei. This process can be understood as two subsequent β-decays via a virtual intermediate state. Thus, a necessary requirement for double β-decay to occur is

$$m(Z, A) > m(Z + 2, A) \tag{7.3}$$

and, for practical purposes, β-decay has to be forbidden

$$m(Z, A) < m(Z + 1, A) \tag{7.4}$$

165

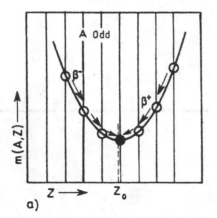 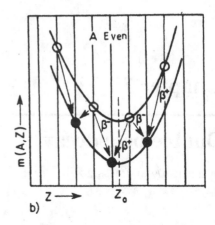

Figure 7.1. Mass parabola as a function of ordering number Z for nuclei with the same mass number A (isobars): stable nuclei are denoted by bold circles. Left: nuclei with odd mass number A. Right: nuclei with even mass number A. A splitting of the mass parabola due to the pairing energy δ_P is apparent. As is shown, some isotopes on the left (right) side are blocked but can be transformed to the second next neighbour. This allows for double beta (left side) of the parabola and double electron capture or double positron decay (right side) (from [Kla95]). © Taylor & Francis Group.

or at least strongly suppressed. Such a strong suppression of β-transitions between the involved nuclear states is caused by a large difference ΔL in spin (see Chapter 6), as in the cases of ^{96}Zr and ^{48}Ca (ΔL equal to 5 or 6). It turns out that in this case highly forbidden single beta decays compete with double beta decays. As ground states of even–even nuclei have spin 0 and parity $(+)$, the double beta decay ground state transitions are characterized as $(0^+ \rightarrow 0^+)$. We know 35 possible double β-decay emitters on both sides of the parabola; the most important double beta isotopes are listed in Table 7.1. A full list can be found in [Boe92].

In the following, the two-nucleon mechanism (2n mechanism) is explored in more detail. Discussions of other mechanisms (Δ, π^-) where the same nucleon experiences two successive β-decays can be found in [Mut88]. For $(0^+ \rightarrow 0^+)$ transitions they are forbidden by angular momentum selection rules [Boe92].

Double β-decay was first discussed by M. Goeppert-Mayer [Goe35] in the form of

$$(Z, A) \rightarrow (Z + 2, A) + 2e^- + 2\bar{\nu}_e \qquad (2\nu\beta\beta\text{-decay}). \tag{7.5}$$

This process can be seen as two simultaneous neutron decays in a nucleus (Figure 7.2). This decay mode conserves lepton number and is allowed within the Standard Model, independently of the nature of the neutrino. This mode is of second-order Fermi theory and, therefore, the lifetime is proportional to $(G_F \cos \theta_C)^{-4}$. Within the GWS model (see Chapter 3), this corresponds to a fourth-order process. As double β-decay is a higher-order effect, expected half-lives are long compared to β-decay: rough estimates illustrated in [Wu66, Kla95] result in half-lives of the order

Table 7.1. Compilation of $\beta^-\beta^-$-emitters with a Q-value of at least $2\,\text{MeV}$. Q-values are determined from AME 16 [Wan17], all Q-values are based on precision measurements with Penning traps. Natural abundances are taken from [Boe92] and phase space factors G from [Mir15]. $G^{0\nu}$ is given in units of $10^{-15}\,\text{yr}^{-1}$ and $G^{2\nu}$ in units of $10^{-21}\,\text{yr}^{-1}$. All transitions are ground state transitions.

Transition	Q-value (keV)	Nat. ab. (%)	$G^{0\nu}$	$G^{2\nu}$
$^{48}_{20}\text{Ca}\rightarrow {}^{48}_{22}\text{Ti}$	4262.96 ± 0.84	0.187	24.65	15536
$^{76}_{32}\text{Ge}\rightarrow {}^{76}_{34}\text{Se}$	2039.006 ± 0.050	7.8	2.372	46.47
$^{82}_{34}\text{Se}\rightarrow {}^{82}_{36}\text{Kr}$	2997.9 ± 0.3	9.2	10.14	1573
$^{96}_{40}\text{Zr}\rightarrow {}^{96}_{42}\text{Mo}$	3356.097 ± 0.086	2.8	20.48	6744
$^{100}_{42}\text{Mo}\rightarrow {}^{100}_{44}\text{Ru}$	3034.40 ± 0.17	9.6	15.84	3231
$^{110}_{46}\text{Pd}\rightarrow {}^{110}_{48}\text{Cd}$	2017.85 ± 0.64	11.8	4.915	132.5
$^{116}_{48}\text{Cd}\rightarrow {}^{116}_{50}\text{Sn}$	2813.50 ± 0.13	7.5	16.62	2688
$^{124}_{50}\text{Sn}\rightarrow {}^{124}_{52}\text{Te}$	2292.64 ± 0.39	5.64	9.047	551.4
$^{130}_{52}\text{Te}\rightarrow {}^{130}_{54}\text{Xe}$	2527.518 ± 0.013	34.5	14.25	1442
$^{136}_{54}\text{Xe}\rightarrow {}^{136}_{56}\text{Ba}$	2457.83 ± 0.37	8.9	14.54	1332
$^{150}_{60}\text{Nd}\rightarrow {}^{150}_{62}\text{Sm}$	3371.38 ± 0.20	5.6	61.94	35397

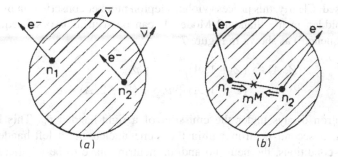

Figure 7.2. Schematic view of double β-decay. Left: The expected process of simultaneous decay of two neutrons in the same nucleus. Right: The lepton-number violating neutrinoless decay. It requires neutrinos to be Majorana particles and a process for helicity matching, mostly done by introducing a non-vanishing neutrino rest mass (from [Gro90]. © CRC Press.

of 10^{20} yr and higher. Together with proton decay, this is among the rarest processes envisaged and, therefore, special experimental care has to be taken to observe this process. In contrast to proton decay which can be searched for in water detectors with several kilotons, it is not easy to build detectors of the same size for double β-decay, because here one is restricted to the isotope of interest which currently implies typical sample sizes of several kilograms to hundreds of kilograms.

Shortly after the classical paper by Majorana [Maj37] discussing a two-

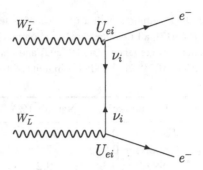

Figure 7.3. Feynman diagram of neutrinoless double β-decay. The initial particles being neutrons (d-quarks) are not shown. Further, the coupling via the PNMS matrix element U_{ei} and the corresponding mass term m_i are part of the diagram. With kind permission of S. Turkat.

component neutrino, another decay mode in form of [Rac37, Fur39]

$$(Z, A) \rightarrow (Z + 2, A) + 2e^- \qquad (0\nu\beta\beta\text{-decay}) \qquad (7.6)$$

was discussed. Clearly, this process violates lepton number conservation by two units and is forbidden in the Standard Model. It can be seen as two subsequent steps ('Racah sequence') as shown in Figure 7.2:

$$(Z, A) \rightarrow (Z + 1, A) + e^- + \bar{\nu}_e$$
$$(Z + 1, A) + \nu_e \rightarrow (Z + 2, A) + e^-. \qquad (7.7)$$

First a neutron decays under the emission of a right-handed $\bar{\nu}_e$. This has to be absorbed at a second neutron within the same nucleus as a left-handed ν_e. To fulfil these conditions, the neutrino and antineutrino have to be identical; i.e., the neutrinos have to be Majorana particles (see Chapter 2). Moreover, to allow for helicity matching, a neutrino mass is required. The reason is that for a massive neutrino, helicity is not a good quantum number anymore (see Chapter 2) and V-A interactions allow for only left-handed charged current reactions. The wavefunction describing neutrino mass eigenstates for $m_\nu > 0$ has no fixed helicity and, therefore, besides the dominant left-handed contribution, has an admixture of a right-handed one with an amplitude proportional to m_ν / E. Another method to account for helicity matching could be new interactions like V + A allowing right-handed charged currents. This would allow a coupling of the other helicity states to right-handed W-bosons. Such an interaction could result from left–right symmetric theories like SO(10) (see Chapter 5). The left–right symmetry is broken at low energies because the right-handed vector mesons $W_R^\pm$ and Z_R^0 have not yet been observed. Then, in addition to the neutrino mass mechanism, right-handed leptonic and hadronic currents can also contribute. The general Hamiltonian used for $0\nu\beta\beta$-decay rates is then given by

$$H = \frac{G_F \cos\theta_C}{\sqrt{2}}(j_L J_L^\dagger + \kappa j_L J_R^\dagger + \eta j_R J_L^\dagger + \lambda j_R J_R^\dagger) \qquad (7.8)$$

with the left- and right-handed leptonic currents as

$$j_L^\mu = \bar{e}\gamma^\mu(1 - \gamma_5)\nu_{eL} \qquad j_R^\mu = \bar{e}\gamma^\mu(1 + \gamma_5)\nu_{eR}. \qquad (7.9)$$

The hadronic currents J (converting neutrons into protons) can be expressed in an analogous way by quark currents taking into account the corresponding coupling constants (see Chapter 3). Often nucleon currents are used in a non-relativistic approximation treating nucleons within the nucleus as free particles (impulse approximation). The mass eigenstates of the vector bosons $W_{1,2}^\pm$ are mixtures of the left- and right-handed gauge bosons

$$W_1^\pm = W_L^\pm \cos\theta + W_R^\pm \sin\theta \qquad (7.10)$$
$$W_2^\pm = -W_L^\pm \sin\theta + W_R^\pm \cos\theta \qquad (7.11)$$

with $\theta \ll 1$ and $M_2 \gg M_1$. Thus, the parameters can be expressed in left–right symmetric GUT models as

$$\eta < \kappa \approx \tan\theta \qquad \lambda \approx (M_1/M_2)^2 + \tan^2\theta. \qquad (7.12)$$

The coupling constants κ, η, λ vanish in the GWS model. It can be shown that in gauge theories a positive observation of $0\nu\beta\beta$-decay would prove a finite Majorana mass term [Sch82, Tak84]. The reason is that, regardless of the mechanism causing $0\nu\beta\beta$-decay, the two emitted electrons together with the two u-, d-quarks that are involved in the $n \rightarrow p$ transition can be coupled to the two ν_e at some loop level in such a way that a neutrino–antineutrino transition as in the Majorana mass term occurs (Figure 7.4). However a calculation results in a very tiny mass of about 10^{-28} eV [Due11] which has nothing to do with those neutrino masses discussed later. For an illustrative deduction see [Kay89]. This statement can be generalized to a wider field of lepton number violating processes [Hir06].

An equivalent process to the one discussed is $\beta^+\beta^+$-decay also in combination with electron capture (EC). There are three different variants possible depending on the Q-value:

$$(Z, A) \rightarrow (Z - 2, A) + 2e^+(+2\nu_e) \qquad (\beta^+\beta^+) \qquad (7.13)$$
$$e_B^- + (Z, A) \rightarrow (Z - 2, A) + e^+(+2\nu_e) \qquad (\beta^+/\text{EC}) \qquad (7.14)$$
$$2e_B^- + (Z, A) \rightarrow (Z - 2, A)(+2\nu_e) \qquad (\text{EC/EC}). \qquad (7.15)$$

$\beta^+\beta^+$ is always accompanied by EC/EC or β^+/EC-decay. The $\beta^+\beta^+$ Q-value is reduced by $4m_e c^2$ to account for the two emitted positrons. The rate for $\beta^+\beta^+$ is, therefore, small and energetically, possible only for six nuclides (Table 7.2). Predicted half-lives for $2\nu\beta^+\beta^+$ are of the order 10^{28} yr and higher, while for

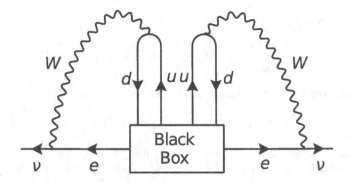

Figure 7.4. Graphical representation of the Schechter-Valle theorem (from [Val15]). With kind permission of José W. F. Valle. © John Wiley & Sons.

Table 7.2. Compilation of the six known $\beta^+\beta^+$ emitters in nature. The Q-values after subtracting $4m_ec^2$, natural abundances and phase space factors (taken from [Mir15]) are given. $G^{0\nu}$ is given in units of 10^{-20} yr^{-1} and $G^{2\nu}$ in units of 10^{-29} yr^{-1}.

Transition	Q-value (keV)	Nat. ab. (%)	$G^{0\nu}$	$G^{2\nu}$
^{78}Kr $\rightarrow$ ^{78}Se	838	0.35	243.2	9159
^{96}Ru $\rightarrow$ ^{96}Mo	676	5.5	80.98	942.3
^{106}Cd $\rightarrow$ ^{106}Pd	738	1.25	91.75	1794
^{124}Xe $\rightarrow$ ^{124}Te	822	0.10	107.8	4261
^{130}Ba $\rightarrow$ ^{130}Xe	534	0.11	23.82	91.54
^{136}Ce $\rightarrow$ $^{136}_{48}$Ba	362	0.19	2.126	0.2053

β^+/EC (reduction by $Q - 2m_ec^2$) this can be reduced by orders of magnitude down to 10^{22-23} yr, making an experimental detection more realistic. The lowest expected half-life is for the 2νEC/EC process which is also the hardest to detect experimentally. A possible 0νEC/EC needs additional particles in the final state because of energy–momentum conservation. Double K-shell capture forbids the emission of a real photon in $0^+ \rightarrow 0^+$ transitions because of angular momentum conservation [Doi92, Doi93]. Therefore other possibilities like KL-shell capture, the emission of internal bremsstrahlung (often called radiative decay) or pair production in the field of the nucleus are considered. However, atomic effects might interfere.

Currently there is a revived interest in the $\beta^+\beta^+$-decay. If $0\nu\beta\beta$-decay is ever observed, it will be very important to clarify the underlying physics mechanism and β^+/EC modes show an enhanced sensitivity to right-handed weak currents [Hir94]. Furthermore, it has been shown that if 0νEC/EC decay occurs into an excited state of the daughter, which is degenerate in energy with the initial state, a resonant enhancement of the decay rate might be expected [Suj04]. Such systems have been found by Penning trap measurements [Eli11a, Eli11b].

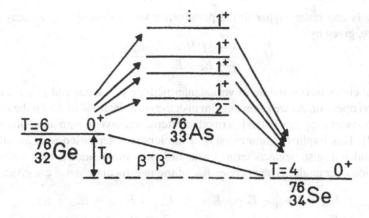

Figure 7.5. Principle of a transition via an intermediate state for $2\nu\beta\beta$-decay. Shown is the transition ^{76}Ge $\rightarrow$ ^{76}Se, which can occur via intermediate 1^+ states ("left" leg and "right" leg) into ^{76}As. For the neutrinoless mode, intermediate states of all multipolarities can contribute (from [Kla95]). © Taylor & Francis Group.

To summarize, the observation of $0\nu\beta\beta$-decay would prove the Majorana character of neutrinos. This would be a step beyond the Standard Model and violates lepton number by two units.

7.2 Decay rates

Decay rates can be described analogously to β-decay starting from Fermi's Golden Rule, but now the processes under discussion are of second-order perturbation theory. The details of the calculations are rather complex. We refer to the existing literature [Kon66,Doi83,Hax84,Doi85,Mut88,Tom88,Gro90,Boe92,Kla95,Suh98,Eji19] and will give only a brief discussion.

7.2.1 The $2\nu\beta\beta$ decay rates

As ground-state transitions are of the type $(0^+ \rightarrow 0^+)$, they can be seen as two subsequent Gamow–Teller transitions, and selection rules then require the intermediate states to be 1^+. Fermi transitions are forbidden or at least strongly suppressed by isospin selection rules (Figure 7.5) [Hax84].

Using time-dependent perturbation theory, the transition probability W per time from an initial state i to a final state f is given by Fermi's golden rule (see Chapter 6)

$$\frac{dW}{dt} = \frac{2\pi}{\hbar} |\langle f|H_{if}|i\rangle|^2 \delta(E_f - E_i) \qquad (7.16)$$

where the δ-function illustrates the fact that we are dealing with discrete energy levels instead of a density of final states. The corresponding matrix element for double

β-decay is one order higher in the pertubation series than single β-decay and is, therefore, given by

$$M_{if} = \sum_m \frac{\langle f|H_{if}|m\rangle\langle m|H_{if}|i\rangle}{E_i - E_m - E_\nu - E_e} \qquad (7.17)$$

where m characterizes the set of virtual intermediate 1^+- states and H_{if} is the weak Hamilton operator. As we cannot distinguish the combinations in which the electron-neutrino system appears in the intermediate steps, we have to sum all configurations in (7.17). This implies a summation over the lepton polarisation and, as only rates are considered, also neglects terms linear in lepton momentum which vanish after integration over angles. The energies E_m of the intermediate states are given as

$$E_m = E_{Nm} + E_{e1} + E_{\nu 1} \qquad E_m = E_{Nm} + E_{e2} + E_{\nu 2} \qquad (7.18)$$
$$E_m = E_{Nm} + E_{e1} + E_{\nu 2} \qquad E_m = E_{Nm} + E_{e2} + E_{\nu 1} \qquad (7.19)$$

where E_{Nm} is the energy of the intermediate nucleus. Without an explicit derivation (see [Kon66, Gro90, Boe92] for details), the obtained decay rate is given by

$$\lambda_{2\nu} = \frac{G_F^4 \cos^4 \theta_C}{8\pi^7} \int_{m_e}^{Q+m_e} F(Z, E_{e1})p_{e1}E_{e1}\, dE_{e1}$$
$$\times \int_{m_e}^{Q+2m_e - E_{e1}} F(Z, E_{e2})p_{e2}E_{e2}\, dE_{e2}$$
$$\times \int_0^{Q+2m_e - E_{e1} - E_{e2}} E_{\nu 1}^2 E_{\nu 2}^2\, dE_{\nu 1} \sum_{m,m'} A_{mm'} \qquad (7.20)$$

with Q as the nuclear transition energy available to the leptons

$$Q = E_{e1} + E_{e2} + E_{\nu 1} + E_{\nu 2} - 2m_e \qquad (7.21)$$

and $F(Z, E)$ the Fermi function (see Chapter 6). The quantity $A_{mm'}$ contains the Gamow-Teller nuclear matrix elements and the typical energy denominators from the perturbative calculations

$$A_{mm'} = \langle 0_f^+\|t_\sigma\|1_j^+\rangle\langle 1_j^+\|t_\sigma\|0_i^+\rangle\langle 0_f^+\|t_\sigma\|1_j^+\rangle\langle 1_j^+\|t_\sigma\|0_i^+\rangle$$
$$\times \tfrac{1}{3}(K_m K_{m'} + L_m L_{m'} + \tfrac{1}{2}K_m L_{m'} + \tfrac{1}{2}L_m K_{m'}) \qquad (7.22)$$

with t_- as the isospin ladder operator converting a neutron into a proton, σ as spin operator, as already introduced in Chapter 6, and

$$K_m = \frac{1}{E_{Nm} + E_{e1} + E_{\nu 1} - E_i} + \frac{1}{E_{Nm} + E_{e2} + E_{\nu 2} - E_i} \qquad (7.23)$$
$$L_m = \frac{1}{E_{Nm} + E_{e1} + E_{\nu 2} - E_i} + \frac{1}{E_{Nm} + E_{e2} + E_{\nu 1} - E_i}. \qquad (7.24)$$

For a definition of the reduced matrix elements in (7.22) averaging over spin states see [Doi85]. The double bars indicate an average over spin states. Two more assumptions are good approximations in the case of $0^+ \to 0^+$ transitions. First of all, the lepton energies can be replaced by their corresponding average value, $E_e + E_\nu \approx Q/2 + m_e$ in the denominator of (7.23) and (7.24). This implies that

$$K_m \approx L_m \approx \frac{1}{E_{Nm} - E_i + Q/2 + m_e} = \frac{1}{E_{Nm} - (M_i + M_f)/2}. \qquad (7.25)$$

With this approximation the nuclear physics and kinematical parts separate. The second approach is a simplified Fermi function, often called the Primakoff-Rosen approximation [Pri68], given in (6.19). The single-electron spectrum can then be obtained by integrating over $\mathrm{d}E_{\nu 1}$ and $\mathrm{d}E_{e2}$ in Equation (7.20). Using the Primakoff-Rosen approximation allows us an analytic integration and results in a single electron spectrum [Boe92]:

$$\frac{\mathrm{d}N}{\mathrm{d}T_e} \approx (T_e + 1)^2 (Q - T_e)^6 [(Q - T_e)^2 + 8(Q - T_e) + 28] \qquad (7.26)$$

where T_e is the electron kinetic energy in units of the electron mass. Most experiments measure the sum energy K (also in units of m_e) of both electrons. Here, the spectral form can be obtained by changing to the variables $E_{e1} + E_{e2}$ and $E_{e1} - E_{e2}$ in (7.20) and performing an integration with respect to the latter, resulting in

$$\frac{\mathrm{d}N}{\mathrm{d}K} \approx K(Q - K)^5 \left(1 + 2K + \frac{4K^2}{3} + \frac{K^3}{3} + \frac{K^4}{30}\right) \qquad (7.27)$$

which shows a maximum at about one third of the Q-value. The total rate $\lambda_{2\nu}$ is obtained by integrating over Equations (7.20) and (7.27)

$$\lambda_{2\nu} \approx Q^7 \left(1 + \frac{Q}{2} + \frac{Q^2}{9} + \frac{Q^3}{90} + \frac{Q^4}{1980}\right). \qquad (7.28)$$

The total decay rate scales with Q^{11}. The decay rate can then be transformed into a half-life which, in its commonly used form, is written as

$$\lambda_{2\nu}/\ln 2 = (T_{1/2}^{2\nu})^{-1} = G^{2\nu}(Q, Z) \left|M_{GT}^{2\nu} + \frac{g_V^2}{g_A^2} M_F^{2\nu}\right|^2 \qquad (7.29)$$

with $G^{2\nu}$ as the phase space and the matrix elements given by

$$M_{GT}^{2\nu} = \sum_j \frac{\langle 0_f^+ \|t_-\sigma\| 1_j^+ \rangle \langle 1_j^+ \|t_-\sigma\| 0_i^+ \rangle}{E_j + Q/2 + m_e - E_i} \qquad (7.30)$$

$$M_F^{2\nu} = \sum_j \frac{\langle 0_f^+ \|t_-\| 1_j^+ \rangle \langle 1_j^+ \|t_-\| 0_i^+ \rangle}{E_j + Q/2 + m_e - E_i}. \qquad (7.31)$$

As already mentioned, Fermi transitions are strongly suppressed.

In earlier times the virtual energies of the intermediate states E_m were replaced by an average energy $\langle E_m \rangle$ and the sum of the intermediate states was taken using $\sum_m |1_m^+\rangle\langle 1_m^+| = 1$ (closure approximation). The advantage was that only the wavefunctions of the initial and final state were required and the complex calculations of the intermediate states could be avoided. However, interference between the different individual terms of the matrix element (7.22) is important and must be considered. Thus, the amplitude of each intermediate state has to be weighted with the corresponding energy E_m of the state and the closure approximation is not appropriate for estimating $2\nu\beta\beta$-decay rates.

7.2.2   The $0\nu\beta\beta$ decay rates

Now let us consider the neutrinoless case. As stated, beside requiring neutrinos to be Majorana particles, we further have to assume a non-vanishing mass or other lepton number violating ($\Delta L = 2$) processes to account for the helicity mismatch. Different physics mechanisms require different nuclear matrix elements [Doi85,Mut88,Suh07, Eji19]. A general formulation of the problem can be found in [Pae99]. First consider the case of the light Majorana neutrino only. The decay rate is then given by [Boe92]

$$\lambda_{0\nu} = 2\pi \sum_{spin} |R_{0\nu}|^2 \delta(E_{e1} + E_{e2} + E_f - M_i)\, \mathrm{d}^3 p_{e1}\, \mathrm{d}^3 p_{e2} \qquad (7.32)$$

where $R_{0\nu}$ is the transition amplitude containing leptonic and hadronic parts. Because of the complexity, we concentrate on the leptonic part (for details see [Doi85,Gro90,Suh07]). The two electron phase space integral is

$$G^{0\nu} \propto \int_{m_e}^{Q+m_e} F(Z, E_{e1})F(Z, E_{e2})p_{e1}p_{e2}E_{e1}E_{e2}\delta(Q - E_{e1} - E_{e2})\, \mathrm{d}E_{e1}\, \mathrm{d}E_{e2}$$
$$(7.33)$$

with $Q = E_{e1} + E_{e2} - 2m_e$. Using the Primakoff–Rosen approximation (6.18), the decay rate is

$$\lambda_{0\nu} \propto \left(\frac{Q^5}{30} - \frac{2Q^2}{3} + Q - \frac{2}{5} \right). \qquad (7.34)$$

Here, the total rate scales with Q^5 compared to the Q^{11} dependence of $2\nu\beta\beta$-decay.

The phase space for neutrinoless double β-decay is about a factor 10^6 larger than for $2\nu\beta\beta$-decay because of a correspondingly larger number of final states. The reason is that the existence of the virtual neutrino in the process (7.6) is restricted to the radius r of the nucleus which, according to Heisenberg's uncertainty principle ($\Delta p \times \Delta r > \hbar$), requires taking states up to about $100\,\mathrm{MeV}$ into account. Furthermore, all multipole states contribute here. This is unlike the $2\nu\beta\beta$-decay case where only 1^+ states contribute. As real neutrinos are emitted, the number of final states is restricted by the Q-value, which is below $5\,\mathrm{MeV}$.

The signature for the sum energy spectrum of both electrons in $0\nu\beta\beta$-decay is outstanding, namely a peak at the Q-value of the transition. The single electron spectrum is given in the used approximation by

$$\frac{dN}{dT_e} \propto (T_e + 1)^2 (Q + 1 - T_e)^2. \tag{7.35}$$

It should be noted that almost all kinds of double β-decay transitions could also occur into excited states of the daughter, dominantly into excited 0^+ and 2^+ states, if the states are below the Q-value. The phase space will be reduced and different nuclear matrix elements will be involved. However, the emitted gamma rays from the de-excitation serve as a good experimental signature.

The total decay rate for $0\nu\beta\beta$-decay is then

$$(T^{0\nu}_{1/2})^{-1} = G^{0\nu}(Q, Z)|M^{0\nu}_{GT} - M^{0\nu}_F - M_T|^2 \left(\frac{\langle m_{\nu_e}\rangle}{m_e}\right)^2 \tag{7.36}$$

with the matrix elements

$$M^{0\nu}_{GT} = \sum_{m,n} \langle 0^+_f \| t_{-m} t_{-n} H(r) \sigma_m \sigma_n \| 0^+_i \rangle \tag{7.37}$$

$$M^{0\nu}_T = \sum_{m,n} \langle 0^+_f \| t_{-m} t_{-n} H(r) \sigma_m \sigma_n S_{mn} \| 0^+_i \rangle \tag{7.38}$$

$$M^{0\nu}_F = \sum_{m,n} \langle 0^+_f \| t_{-m} t_{-n} H(r) \| 0^+_i \rangle \left(\frac{g_V}{g_A}\right)^2 \tag{7.39}$$

with $\mathbf{r} = |\mathbf{r}_m - \mathbf{r}_n|$ as the distance between two nucleons and S_{mn} as spin tensor operator. The dependence of the lifetime on the neutrino mass arises from the leptonic part of $|R_{0\nu}|$. By integrating the neutrino propagator with respect to the exchanged neutrino momentum, a term proportional to the neutrino mass remains. The effect of the propagator can be described by a neutrino potential $H(r)$ acting on the nuclear wavefunctions allowing Fermi-transitions as well [Mut88]. The neutrino potential introduces a dependence of the transition operator on the coordinates of the two nucleons.

The quantity $\langle m_{\nu_e}\rangle$, called the effective Majorana neutrino mass, which can be deduced from the half-life measurement, is of course of great interest for neutrino physics. Taking neutrino mixing (see Chapter 8) into account at each vertex, it is given by

$$\langle m_{\nu_e}\rangle = \left|\sum_i U^2_{ei} m_i\right| \tag{7.40}$$

with U_{ei} as the mixing matrix elements ((5.53) and (5.6)) and m_i as the corresponding mass eigenvalues. However, the various CP-phases have to be considered as discussed in Section 5.6. For three generations of particles this includes one Dirac CP-phase. Hence, $\langle m_{\nu_e}\rangle$ can be described in terms of the PMNS-matrix elements as

$$\langle m_{\nu_e}\rangle = \left|\cos\theta^2_{12}\cos\theta^2_{13}m_1 + \sin\theta^2_{12}\cos\theta^2_{13}e^{i2\alpha_1}m_2 + \sin\theta^2_{13}e^{i2(\alpha_2-\delta)}m_3\right|. \tag{7.41}$$

Note the fact of a possible interference among the different terms contributing to the sum in (7.40) in contrast to single β-decay. Furthermore, mass from double β-decay is valid only for Majorana neutrinos, while β-decay is not sensitive on the character of the neutrino. Hence, both measurements deliver quite complementary information.

If right-handed currents are included, expression (7.36) can be generalized to

$$(T_{1/2}^{0\nu})^{-1} = C_{mm} \left(\frac{\langle m_{\nu_e} \rangle}{m_e} \right)^2 + C_{\eta\eta} \langle \eta \rangle^2 + C_{\lambda\lambda} \langle \lambda \rangle^2 \tag{7.42}$$

$$+ C_{m\eta}(\frac{\langle m_{\nu_e} \rangle}{m_e})\langle \eta \rangle + C_{m\lambda} \left(\frac{\langle m_{\nu_e} \rangle}{m_e} \right) \langle \lambda \rangle + C_{\eta\lambda} \langle \eta \rangle \langle \lambda \rangle \tag{7.43}$$

where the coefficients C contain the phase space factors and the matrix elements and the effective quantities are

$$\langle \eta \rangle = \eta \sum_j U_{ej} V_{ej} \qquad \langle \lambda \rangle = \lambda \sum_j U_{ej} V_{ej} \tag{7.44}$$

with V_{ej} as the mixing matrix elements among the right-handed neutrino states. Equation (7.42) reduces to (7.36) when $\langle \eta \rangle, \langle \lambda \rangle = 0$. For example the element C_{mm} is given by

$$C_{mm} = |M_{GT}^{0\nu} - M_F^{0\nu} - M_T|^2 G^{0\nu}(Q, Z). \tag{7.45}$$

The ratio $R = \langle \lambda \rangle / \langle \eta \rangle$, being independent of V_{ej}, is, under certain assumptions, a simple function of $K = (m_{W_L}/m_{W_R})^2$ and of the mixing angle θ introduced in (7.10) [Suh93].

7.3   Nuclear structure effects on matrix elements

A severe uncertainty in extracting a bound or a value on $\langle m_{\nu_e} \rangle$ from experimental half-life limits is the involved nuclear matrix element. Thus, due to its importance, a few additional measurements to improve the issue will be briefly discussed [Zub05, Suh07, Eji19]. Different nuclear matrix elements are associated with the various decay modes. $2\nu\beta\beta$-decay basically requires an understanding of the Gamow–Teller (GT) strength distribution B(GT), as only intermediate 1^+ states are involved. The fact that $2\nu\beta\beta$-decay has been observed experimentally and calculations can be compared with measurements is beneficial. $0\nu\beta\beta$-decay with the exchange of light Majorana neutrinos does not have selection rules on multipoles and states up to 100 MeV have to be considered [Zub05]. Here, some of the nuclear issues involved in the matrix elements are mentioned shortly.

- **Charge exchange reactions:** The measurement of the "left leg" and the "right leg" for the 1^+ states (Figure 7.5) can be explored in several ways. If only the ground state of the intermediate nucleus is of interest, the log ft-values from the beta decay and the electron capture of the intermediate ground state nuclide would be sufficient. Otherwise, for excited state contributions nuclear reactions like (p,n) or $(^3\mathrm{He},t)$ or (n,p) or $(d,^2\mathrm{He})$ reactions at accelerators have been done. Performing measurements under 0 degrees, i.e., the linear momentum transfer q is going towards 0, allows to use a simple relation between the B(GT) strength and the differential cross-section

$$\frac{d\sigma}{d\Omega}\big|_{q=0} = \hat{\sigma} B(GT) \qquad (7.46)$$

with $\hat{\sigma}$ as unit cross-section. To reveal the individual states involved, the energy resolution is very important. Currently the Grand Raiden Spectrometer is the best machine to measure B(GT) using the $(^3\mathrm{He},t)$ reaction. Furthermore, under larger angles (about 2-3 degrees) spin-dipole transitions $(0^+ \rightarrow 2^-)$ can be excited as well as higher multipoles. However, in these cases there is no simple relation between the differential cross-sections and the strength functions. Alternatively the NUMEN project is trying to extract informations by performing double charge exchange reactions [Cap18].

- **Nucleon transfer reactions:** Double β-decay results in the replacement of two nucleons in the neutron shell and the addition of two protons in the final shell. The vacancies and occupancies of the included shells are important. This can be studied by using the combination of the reactions $(^3\mathrm{He},n)$ and (p,t), which also transforms two neutrons into protons, see for example [Sch08].

- **Ordinary muon capture:** A muon capture on a nucleus releases the corresponding ν_μ and leaves the daughter nucleus in a highly excited state of about the muon mass (105 MeV). These nuclei will de-excite by emission of protons, neutrons and gamma rays. The proton and neutron emission channels produce new isotopes in highly excited states as well. Hence several nuclei de-excite at the same time leaving a very complex pattern for the used gamma spectroscopy. Furthermore, also other processes like 2p-2h excitations have to be considered, see for example [Has18, Zin19].

- **Deformation:** Nuclei can have different shapes, which might have an impact on the nuclear matrix element. Theoretical studies have been performed which have shown that the nuclear matrix element is relatively high if the mother and daughter nuclide have the same shape as the overlap of the initial and final wave-functions is best. The shapes of the initial and final nuclei is likely different especially at heavier nuclei. This results in a smaller overlap of the wave-functions, which leads to a smaller matrix element [Rod11].

The precise description of short range correlations of nucleons and nuclear deformations are among the most sensitive nuclear model parameters to be investigated. Three basic strategies are followed in the calculations: either the nuclear shell model approach, the quasi-random phase approximation (QRPA) or the interacting boson model have been used [Bar09a]. All calculations are quite complex and beyond the scope of this book. Detailed treatments can be found in [Hax84, Doi85, Sta90, Mut88, Gro90, Suh98, Suh07, Eji19].

$2\nu\beta\beta$-decay is a standard weak process and does not involve any uncertainty from particle physics aspects. Its rate is governed by (7.17). The first factor in the numerator is identical to the β^+, (n,p) or $(d,^2\mathrm{He})$ amplitude for the final state nucleus; the second factor is equivalent to the β^-, (p,n) or $(^3\text{He},t)$ amplitude of the initial nucleus. In principle, all GT amplitudes including their signs have to be used. The difficulty is that the 2ν matrix elements exhaust only a small fraction (10^{-5}–10^{-7}) of the double GT sum rule [Vog88, Mut92] and, hence, it is sensitive to details of the nuclear structure. Various approaches have been done, a compilation is given in [Suh98, Eji19]. The main ingredients are a repulsive particle–hole spin–isospin interaction and an attractive particle-particle interaction. They play a decisive role in concentrating the β^- - strength in the GT resonance and for the relative suppression of β^+ - strength and its concentration at low excitation energies. The calculations typically show a strong dependence on the strength of a particle–particle force g_{PP}, which for realistic values is often close to its critical value ('collapse'). This indicates a rearrangement of the nuclear ground state but QRPA is meant to describe small deviations from the unperturbed ground state and, thus, is not fully applicable near the point of collapse. QRPA and its various extensions are trying to explain the experimental values by adjusting only one parameter. It seems that in some isotopes it looks like one low-lying 1^+-state accounts for the whole matrix element (single state dominance). The increase of computing power allows the nuclear shell model methods to become capable of handling much larger configuration spaces than before and can be used for descriptions as well [Cau08]. They avoid the above difficulties of QRPA and can also be tested with other data from nuclear spectroscopy.

In $0\nu\beta\beta$-decay mediated by light virtual Majorana neutrinos, several new features arise. According to Heisenberg's uncertainty relation, the virtual neutrino can have a momentum up to $q \simeq 1/r_{mn} \simeq 50$–$100\,\text{MeV}$ where r_{mn} is the distance between the decaying nucleons. Therefore, the dependence on the energy of the intermediate state is small and the closure approximation can be applied. Also, because $qR > 1$ (R being the radius of the nucleus), the expansion in multipoles does not converge and all multipoles contribute by comparable amounts (Figure 7.6). Finally the neutrino propagator results in a long-range neutrino potential. A compilation of representative matrix element calculations is shown in Figure 7.7. For a more detailed discussion and the treatment of heavy Majorana neutrinos, see [Mut88, Boe92, Suh98, Eji19].

7.4   Experiments

Typical energies for double β-decay are in the region of a few MeV distributed among the two (or four, depending on the channel) leptons which are emitted as s-waves. The signal for neutrinoless double β-decay is a peak in the sum energy spectrum of both electrons at the Q-value of the transition, while for the $2\nu\beta\beta$-decay a continuous spectrum with the form given in (7.28) can be expected (Figure 7.9). In tracking experiments where the energy of both electrons can be measured separately, angular distributions can also be used to distinguish among the various transitions and underlying processes. In the 2n mechanism, the individual electrons can be

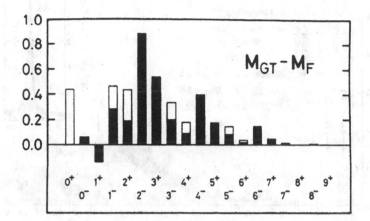

Figure 7.6. Decomposition of the nuclear matrix element $M_{GT} - M_F$ into contributions of the intermediate states with spin and parity I^π for the $0\nu\beta\beta$-decay of ^{76}Ge. Open and filled histograms describe the contributions of M_F and M_{GT}, respectively (from [Kla95]). © Taylor & Francis Group.

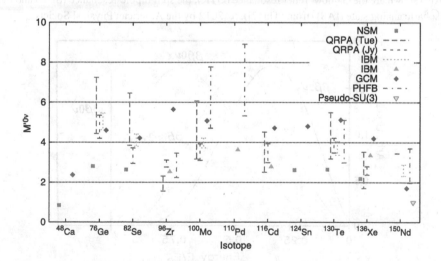

Figure 7.7. Comparison of representative nuclear matrix element calculations for double beta isotopes with a Q-value above 2 MeV. Shown are the various theoretical approaches (from [Due11a]). © 2011 by the American Physical Society.

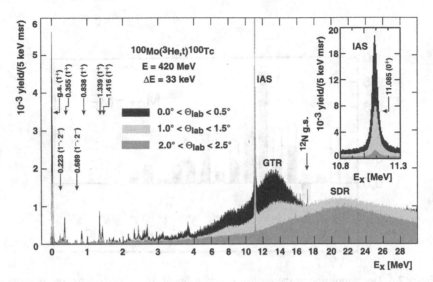

Figure 7.8. Examplaric plot of charge exchange measurements using the (^{3}He,t) reaction on ^{100}Mo, one of the double beta nuclides, at RCNP Osaka. Shown is the GT strength of individual excited nuclear levels as a function of excitation energy. In this case the dominant 1^+-strength is via the ground state. A few excited 1^+-states shown add a little bit to the strength. The analysis of these 1^+-states will provide the strength of the "left" leg of the transition. In case of the shown ^{100}Mo, single state dominance (SSD) is a good approximation. Also shown are the Gamow-Teller resonance (GTR), the spin-dipole-resonance (SDR) and the isobaric analog state (IAS) (from [Thi12]). © 2012 by the American Physical Society.

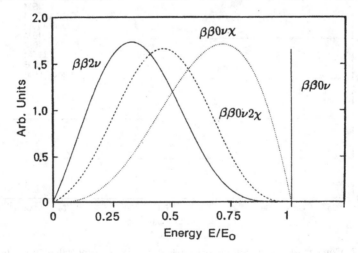

Figure 7.9. Different spectral shapes of observable sum energy spectra of emitted electrons in double β-decay. The $n = 1,3$ forms (dashed lines) correspond to different majoron accompanied modes, $n = 5$ (solid line) is the $2\nu\beta\beta$-decay and the $0\nu\beta\beta$-decay results in a peak (from [Kla97]). © Taylor & Francis Group.

described by the following angular distributions:

$$P(\theta_{12}) \propto 1 - \beta_1 \beta_2 \cos\theta_{12} \qquad (0^+ \to 0^+) \tag{7.47}$$

$$P(\theta_{12}) \propto 1 + \tfrac{1}{3}\beta_1 \beta_2 \cos\theta_{12} \qquad (0^+ \to 2^+) \tag{7.48}$$

with θ_{12} as the angle between both electrons and $\beta_{1,2} = p_{1,2}/E_{1,2}$ their velocity. For a compilation of angular distributions of additional decay modes, see [Tre95]. Being a nuclear decay, the actual measured quantity is a half-life, whose value can be determined from the radioactive decay law assuming $T_{1/2} \gg t$:

$$T_{1/2}^{0\nu} = \ln 2 \times m \times a \times t \times N_A/N_{\beta\beta} \tag{7.49}$$

with m the used mass, a the isotopical abundance, t the measuring time, N_A the Avogadro constant and $N_{\beta\beta}$ the number of double β events, which has to be taken from the experiment. If no peak is observed and a constant background is assumed scaling linearly with time, $N_{\beta\beta}$ is often estimated at the 1σ level as a possible fluctuation of the background events N_B as $N_{\beta\beta} = \sqrt{N_B}$. With this simple assumption the $0\nu\beta\beta$ half-life limit can then be estimated from experimental quantities to be

$$T_{1/2}^{0\nu} \propto a \times \epsilon \sqrt{\frac{m \times t}{B \times \Delta E}} \tag{7.50}$$

where ϵ is the efficiency for detection, M is the mass of the active detector volume, ΔE is the energy resolution at the peak position and B the background index normally given in counts/keV/kg/year. In this case $\langle m_{\nu_e} \rangle$ scales with the square root of the measuring time (7.40). However, in case of no background, the half-life measurement depends linearly on the measuring time, hence the $\langle m_{\nu_e} \rangle$ sensitivity itself already scales with $\sqrt{Mt}$ [Moe91a].

7.4.1 Practical considerations in low-level counting

For a fair chance of detection, isotopes with large phase space factors (high Q-value as the rate scales with Q^5) and favourable large nuclear matrix elements should be considered. For that reason only isotopes with a Q-value of at least $2\,\text{MeV}$ are considered for experiments; they are compiled in Table 7.1. A significant amount of source material should be available, which is acquired in most second-generation double β-decay experiments by using isotopical enriched materials. Being an extremely rare process, the use of low-level counting techniques is necessary. The main concern is, therefore, background, i.e., reactions which deposit the same energy in a detector at the Q-value as a double beta event. Some of the most common background sources follow:

- Cosmic ray interactions in the atmosphere producing secondary particles like muons and hadrons. To a large extent they can be shielded by going underground (Figure 7.10). To compare the overburden of different underground locations, the rock shielding above the experiment is converted

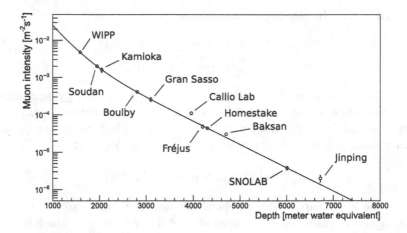

Figure 7.10. Muon intensity *versus* depth of some of the most important underground laboratories. Their shielding depth is given in metres of water equivalent (m.w.e.) together with the attenuation of the atmospheric muon flux. With kind permission of S. Turkat.

into an equivalent column of water with the same shielding (metre water equivalent (m.w.e.)).

- Neutrons either produced by the interactions of the remaining muons in the surrounding rock, the experiment or by (α, n) reactions in the rock due to fission. Thus, a neutron shielding might be required.

- Natural radioactivity in the form of decay chains (^{235,238}U, ^{232}Th). The most energetic natural γ-line with some significant intensity is at 2.614 MeV (from ^{208}Tl decay). Only six nuclides have a Q-value beyond this energy, hence the experiments do not suffer from this background. Further prominent background components coming from these chains are ^{210}Pb producing electrons with energies up to 1.1 MeV, ^{214}Bi β-decay up to 3 MeV and α-particles. In addition, ^{222}Rn is typically disturbing, being parent to ^{214}Bi and ^{210}Pb.

- Anthropogenic activities. In particular, ^{137}Cs (prominent γ-line at 662 keV) should be mentioned.

- α-activities close to the detectors which cannot be shielded.

- Cosmogenic activation. Production of radio-isotopes by cosmic ray spallation in the materials during their stay on the Earth's surface. It depends on the materials and isotope of interest to determine potential dangerous nuclides.

- Natural ^{40}K (γ-line at 1.461 MeV).

These background components influence not only double β-decay experiments but some other underground neutrino experiments in general. However, there might be additional background components that are more specific to a certain experiment. Even in the cleanest environment an irreducible background for $0\nu\beta\beta$-decay will be $2\nu\beta\beta$-decay. Typically the half-life for $2\nu\beta\beta$-decay is at least 5-6 orders of magnitude shorter, thus resulting in a background in the $0\nu\beta\beta$-decay peak range.

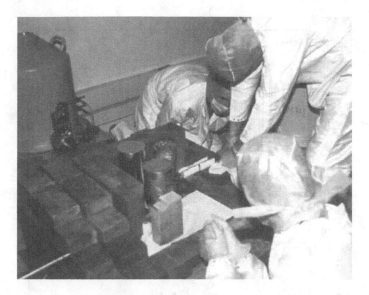

Figure 7.11. Photograph of a "classical" installation (detectors surrounded by shielding materials like copper and lead) of enriched detectors here exemplarily shown from the former Heidelberg-Moscow experiment. With kind permission of H. V. Klapdor-Kleingrothaus.

At this point energy resolution of the used detectors becomes a crucial parameter. For more details on low-level counting techniques see [Heu95].

All direct experiments focus on electron detection and can be either active or passive. The advantage of active detectors is that the source and detector are identical but therefore they often measure only the sum energy of both electrons. However, passive detectors allow us to get more information (e.g., they measure energy and tracks of both electrons separately by using foils in a time projection chamber), but they usually have a smaller source strength. Some experiments will now be described in a little more detail.

7.4.2 Direct counting experiments

7.4.2.1 Semiconductor experiments

In this type of experiment, first done by a group from Italy [Fio67], germanium diodes have been used. The source and detector are identical; the isotope under investigation is ^{76}Ge with a Q-value of 2039.00 keV. The big advantage is the excellent energy resolution of Ge semiconductor detectors (typically about 3–4 keV at 2 MeV). However, this technique allows the measurement of only the sum energy of the two electrons. A big step forward was taken by using enriched germanium detectors containing more than 85% ^{76}Ge (the natural abundance of ^{76}Ge is 7.8%).

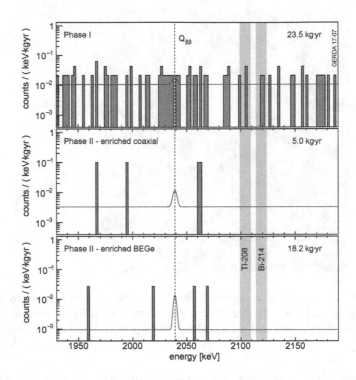

Figure 7.12. Measured events (histogram) around the position of the expected peak region (dashed line). The average background and maximal signal for different data phases of the GERDA experiment is shown as solid line. The top part shows the data of phase 1. The two lower plots show data for phase 2 with two different kinds of detectors. The grey vertical bars indicate the position of two gamma lines from the natural decay chains. No signal is apparent in the peak region (from [Ago18a]). © 2018 by the American Physical Society.

The GERDA and MAJORANA experiments   After a series of Ge-experiments in the last decades the current ones are the GERDA- experiment [Ack13] at Gran Sasso Laboratory and the MAJORANA-experiment [Abg14] at Sanford Underground Laboratory. The MAJORANA demonstrator follows the "classic" setup as exemplarily shown already in Figure 7.11. All materials are selected for high radiopurity and the detectors are shielded with copper, lead and a muon veto. The GERDA approach is to minimize the material around the Ge-diodes. For that, an 80 ton liquid argon tank was installed for shielding and cooling. The bare Ge-diodes are mounted on strings which are implemented into the tank. After a first phase of data-taking, a veto system and more enriched detectors were implemented. Both experiments are using pulse-shape discrimination to distinguish single energy deposition (possible signal) against multiple ones (background event).

After 26 kg × yr of data-taking with the MAJORANA demonstrator, the peak

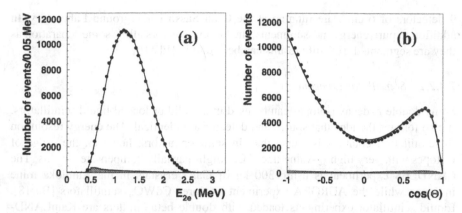

Figure 7.13. Measured two neutrino double beta decay spectrum of ^{100}Mo as obtained by NEMO-3. Shown are the sum energy spectrum (left) and the opening angle among both electrons (right) (from [Arn05]). © 2005 by the American Physical Society.

region reveals no signal and a half-life limit of

$$T^{0\nu}_{1/2} > 2.7 \times 10^{25} \text{ yr} \qquad (90\% \text{ CL}) \tag{7.51}$$

has been deduced [Alv19]. GERDA has released data for two phases and an exposure of 48 kg × yr (see Figure 7.12) with a half-life limit of [Ago19]

$$T^{0\nu}_{1/2} > 9.0 \times 10^{25} \text{ yr} \qquad (90\% \text{ CL}) \tag{7.52}$$

With the given uncertainties of the nuclear matrix elements, the estimated half-life leads to an upper mass bound of about $\lesssim 0.1$ eV. Both experiments have merged and are now preparing for a 200 kg enriched experiment in the GERDA facility (LEGEND) [Abg17].

The $2\nu\beta\beta$-decay half-life was measured by GERDA too and results in [Ago15]

$$T^{2\nu}_{1/2} = (1.926 \pm 0.095) \times 10^{21} \text{ yr}. \tag{7.53}$$

Moreover, there is always the possibility of depositing a double β-decay emitter near a semiconductor detector to study its decay but then only transitions to excited states can be observed by detecting the corresponding gamma rays. Searches for $\beta^+\beta^+$-decay and β^+/EC-decay were also done in this way searching for the 511 keV photons. This has been widely used in the past.

COBRA. A further approach to take advantage of the good energy resolution of semiconductors is COBRA [Zub01]. The idea here is to use room temperature CdTe or CdZnTe detectors, mainly to explore ^{116}Cd and ^{130}Te decays. In total, there are seven (nine in the case of CdZnTe) double β-emitters within the detector including those of EC/EC-decay. Two prototypes, one of 64 CdZnTe detectors of 1 cm^3 and

9 detectors of 6 cm³, are running at the Gran Sasso Underground Laboratory. In addition to pure energy measurements, the larger detectors allow some separation as they are segmented. Half-life results can be found in [Ebe16].

7.4.2.2   Scintillator experiments

Some double β-decay isotopes can be used in as solid or loaded liquid scintillators. It also follows the idea that source and detector are identical. The energy resolution in scintillation counters is worse than in semiconductors; however, the usage of isotopes with very high Q-value like ^{48}Ca might partially compensate for this. The CANDLES experiment is using 300 kg of CaF$_2$ crystals in the Kamioka mine in Japan while the AURORA experiment explores CdWO$_4$ scintillators [Bar18]. Liquid scintillator experiments loaded with double beta emitters are KamLAND-Zen using enriched ^{136}Xe (see Chapter 8) and SNO+ [And16]. Here about 800 tons of liquid scintillators are filled in the infrastructure of the former SNO experiment (see Chapter 9) which can be loaded with double beta emitters, in the SNO+ case it will be ^{130}Te. Obtained limits (both 90% CL) for these two solid scintillators are

$$T_{1/2}^{0\nu}(^{48}\text{Ca}) > 5.8 \times 10^{22} \text{ yr} \qquad \text{(CANDLES)} \qquad (7.54)$$

$$T_{1/2}^{0\nu}(^{116}\text{Cd}) > 2.2 \times 10^{23} \text{ yr} \qquad \text{(AURORA)}. \qquad (7.55)$$

7.4.2.3   Cryogenic detectors

Another important technique which allows to study a variety of double beta nuclei is the usage of cryogenic bolometers. The experiments are running in a cryostat at very low temperature (mK) (see Chapter 6). In dielectric materials the specific heat $C(T)$ at such temperatures scales according to (6.55). Therefore, the energy deposition, ΔE, of double β-decay would lead to a temperature rise ΔT of

$$\Delta T = \frac{\Delta E}{C(T)M}. \qquad (7.56)$$

Such detectors normally have a very good energy resolution of a few keV at 2 MeV. The CUORE experiment has started data taking using 988 crystals with a total mass of about 730 kg of TeO$_2$ crystals at about 10 mK to search for the ^{130}Te decay. As detectors NTD Ge thermistors are used. The current obtained half-life limit for ^{130}Te corresponds to [Ald18]

$$T_{1/2}^{0\nu}(^{130}\text{Te}) > 1.5 \times 10^{25} \text{ yr} \qquad \text{(90\% CL)}. \qquad (7.57)$$

As mentioned, several other double beta isotopes can be used as cryo-detectors as well; this is a large ongoing worldwide effort.

7.4.2.4   Ionization experiments

These passive experiments are mostly built in the form of time projection chambers (TPCs) where the emitter is either the filling gas (e.g., ^{136}Xe) or is included in thin

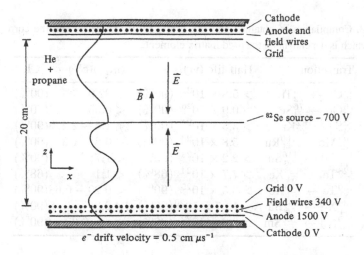

Figure 7.14. Schematic view of the setup of the TPC at UC Irvine, showing the wires, direction of the fields and the ^{82}Se source. A sample electron trajectory is shown on the left-hand side (from [Ell88]). © 1988 With permission from Elsevier.

foils. The advantage is that energy measurements as well as tracking of the two electrons is possible. The disadvantages are the worse energy resolution and, in the case of thin foils, the limited source strength. It was such a device that first gave evidence for $2\nu\beta\beta$-decay in a direct counting experiment using ^{82}Se [Ell87]. The experiment used a 14 g selenium source, enriched to 97% in ^{82}Se, in the form of a thin foil installed in the centre of a TPC (Figure 7.14). The TPC was shielded against cosmic rays by a veto system. After 7960 hr of measuring time and background subtraction, 36 events remained which, if attributed to $2\nu\beta\beta$-decay, resulted in a half-life of

$$T_{1/2}^{2\nu}(^{82}\text{Se}) = (1.1^{+0.8}_{-0.3}) \times 10^{20} \text{ yr}. \tag{7.58}$$

Another experiment of this type is NEMO-3 in the Fréjus Underground Laboratory. It is a passive source detector using thin foils made out of double beta elements. It consists of a tracking (wire chambers) and a calorimetric (plastic scintillators) device put into a 3 mT magnetic field. The total source strength is about 10 kg which is dominated by using enriched ^{100}Mo foils. The experiment provided a half-life limit of ^{100}Mo to be [Arn15]

$$T_{1/2}^{0\nu}(^{100}\text{Mo}) > 1.1 \times 10^{24} \text{ yr} \qquad (90\% \text{ CL}). \tag{7.59}$$

In addition, a major step forward due to NEMO-3 was a variety of new $2\nu\beta\beta$-decay detections with high statistics. A compilation of some obtained double β results is shown in Table 7.3. Last but not least, there is the EXO-200 experiment as a Xenon-filled TPC which is planning an upgrade to nEXO. One unique feature which is explored is barium tagging as ^{136}Ba is the daughter of the ^{136}Xe decay. The latest

Table 7.3. Compilation of obtained half-life limits for $0\nu\beta\beta$-decay and the corresponding $\langle m_{\nu_e} \rangle$, which is dependent on the used matrix element.

Transition	Half-life (yr)	$\langle m_{\nu_e} \rangle$(eV)	CL
$^{48}_{20}\text{Ca} \rightarrow {}^{48}_{22}\text{Ti}$	$> 5.8 \times 10^{22}$ (90%)	< 22.9	(90%)
$^{76}_{32}\text{Ge} \rightarrow {}^{76}_{34}\text{Se}$	$> 0.9 \times 10^{26}$ (90%)	< 0.35	(90%)
$^{82}_{34}\text{Se} \rightarrow {}^{82}_{36}\text{Kr}$	$> 3.5 \times 10^{24}$ (90%)	$< 1.4 - 2.2$	(90%)
$^{100}_{42}\text{Mo} \rightarrow {}^{100}_{44}\text{Ru}$	$> 5.8 \times 10^{23}$ (90%)	$< 0.8 - 1.3$	(90%)
$^{116}_{48}\text{Cd} \rightarrow {}^{116}_{50}\text{Sn}$	$> 2.2 \times 10^{23}$ (90%)	< 1.7	(90%)
$^{128}_{52}\text{Te} \rightarrow {}^{128}_{54}\text{Xe}$	$> 7.7 \times 10^{24}$ (68%)	< 1.1	(68%)
$^{130}_{52}\text{Te} \rightarrow {}^{130}_{54}\text{Xe}$	$> 3.5 \times 10^{25}$ (90%)	$< 0.20 - 0.68$	(90%)
$^{136}_{54}\text{Xe} \rightarrow {}^{136}_{56}\text{Ba}$	$> 1.07 \times 10^{26}$ (90%)	< 2.3	(90%)
$^{150}_{60}\text{Nd} \rightarrow {}^{150}_{62}\text{Sm}$	$> 2.0 \times 10^{22}$ (90%)	$< 4 - 6.3$	(90%)

result by EXO-200 is [Alb18]

$$T^{0\nu}_{1/2}(^{136}\text{Xe}) > 1.5 \times 10^{25} \text{ yr} \quad (90\% \text{ CL}). \tag{7.60}$$

7.4.3 Geochemical experiments

The geochemical approach is to use old ores containing double beta emitters, which could produce a significant amount of daughter nuclei over geological time scales. The decay would lead to an isotopical enhancement of the daughter isotope in the natural abundance element which could be measured by mass spectrometry. Clearly the advantage of such experiments is the long exposure time of up to billions of years. Using the age T of the ore, and measuring the abundance of the mother $N(Z, A)$ and daughter $N(Z \pm 2, A)$ isotopes, the decay rate can be determined from the exponential decay law ($t \ll T_{1/2}$)

$$\lambda \simeq \frac{N(Z \pm 2, A)}{N(Z, A)} \times \frac{1}{T}. \tag{7.61}$$

As only the total amount of the daughter is observed, this type of measurement does not allow us to differentiate between the production mechanisms; therefore, the measured decay rate is

$$\lambda = \lambda_{2\nu} + \lambda_{0\nu}. \tag{7.62}$$

To be useful, several requirements and uncertainties have to be taken into account if applying this method. The isotope of interest should be present in a high concentration within the ore. In addition, a high initial concentration of the daughter should be avoided if possible. Other external effects which could influence the daughter concentration should be excluded. Last but not least, an accurate age determination of the ore is necessary. From all these considerations, only Se and Te ores are usable in practice. ^{82}Se, ^{128}Te and ^{130}Te decay to inert noble gases (^{82}Kr, ^{128,130}Xe). The noble gas concentration during crystallization and ore formation is considered to be small. The detection of the small expected isotopical anomaly is made possible due to the large sensitivity of noble gas mass spectrometry [Kir86]. Although experiments of this type were initially already performed in 1949, real convincing evidence for double β-decay was observed later in experiments using selenium and tellurium ores [Kir67, Kir68, Kir86]. More measurements can be found in [Lin88, Ber93, Mes08, Tho08]. Comparing the decay rates of the two Te isotopes, phase space arguments ($2\nu\beta\beta$-decay scales with Q^{11}, while $0\nu\beta\beta$-decay scales with Q^5) and the assumption of almost identical matrix elements show that the observed half-life for ^{130}Te can be attributed to $2\nu\beta\beta$-decay [Mut88].

A different approach using thermal ionization mass spectrometry allowed the determination of the $2\nu\beta\beta$-decay half-life of ^{96}Zr to be about 10^{19} yr [Kaw93, Wie01], a measurement also performed by NEMO-3 [Arg10].

7.4.4 Radiochemical experiments

This method takes advantage of the radioactive decay of the daughter nuclei, allowing a shorter "measuring" time than geochemical experiments ("milking experiments"). It is also independent of some uncertainties in the latter, e.g., the geological age of the sample, original concentration of the daughter and possible diffusion effects of noble gases in geochemical samples. No information on the decay mode can be obtained—only the total concentration of daughter nuclei is measured.

Two possible candidates are the decays ^{232}Th $\rightarrow$ ^{232}U and ^{238}U $\rightarrow$ ^{238}Pu with Q-values of 850 keV (^{232}Th) and 1.15 MeV (^{238}U), respectively. Both daughters are unstable against α-decay with half-lives of 70 yr (^{232}Th) and 87.7 yr (^{238}U), respectively. For the detection of the ^{238}U $\rightarrow$ ^{238}Pu decay, the emission of a 5.5 MeV α-particle from the ^{238}Pu decay is used as a signal. The first such experiment was originally performed in 1950 using a six-year-old UO$_3$ sample. From the non-observation of the 5.51 MeV α-particles, a lower limit of

$$T_{1/2}^{0\nu}(^{238}\text{U}) > 6 \times 10^{18} \text{ yr} \tag{7.63}$$

was deduced. A sample of 8.47 kg of uranium nitrate, which was purified in 1956 and analysed in 1989, was investigated, and a half-life of

$$T_{1/2}^{2\nu}(^{238}\text{U}) = (2.0 \pm 0.6) \times 10^{21} \text{ yr} \tag{7.64}$$

was obtained [Tur92]. Both geo- and radio-chemical methods measure only the total decay rate by examining the concentration of the daughter nuclei. As it is not possible to distinguish between the different decay modes, their sensitivity is finally limited by $2\nu\beta\beta$-decay. This makes it almost impossible to establish a real positive evidence for the neutrinoless mode by these methods.

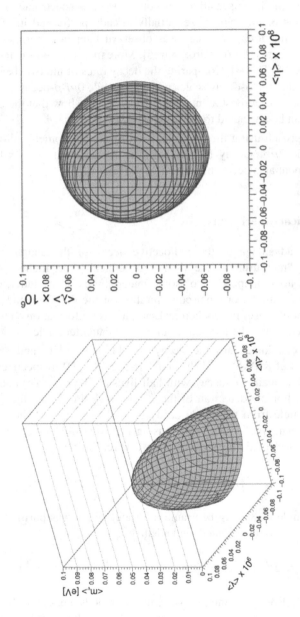

Figure 7.15. Obtained neutrino mass $\langle m_{\nu_e} \rangle$ as a function of the two right-handed current parameters $\langle \lambda \rangle$, $\langle \eta \rangle$. The half life of ^{136}Xe has been used. Left: Three-dimensional ellipsoid. Right: Projection on the $\langle \eta \rangle - \langle \lambda \rangle$ plane. Usually limits on the neutrino mass are presented assuming $\langle \eta \rangle = \langle \lambda \rangle = 0$. With kind permission of J. Volkmer.

Table 7.4. Compilation of obtained half-lives for $2\nu\beta\beta$-decay.

Isotope	Experiment	$T_{1/2}$ (10^{20} yr)
^{48}Ca	Calt.-KIAE	$0.43^{+0.24}_{-0.11} \pm 0.14$
^{48}Ca	NEMO-3	$0.64^{+0.07}_{-0.06}\ ^{+0.12}_{-0.09}$
^{76}Ge	GERDA	$18.4^{+1.4}_{-1.0}$
^{82}Se	NEMO-3	$0.939 \pm 0.017 \pm 0.058$
^{96}Zr	NEMO 3	$0.235 \pm 0.014 \pm 0.016$
^{100}Mo	CUPID	$0.0712^{+0.0018}_{-0.0014} \pm 0.0010$
^{100}Mo	ELEGANT V	$0.115^{+0.03}_{-0.02}$
^{100}Mo	NEMO 3	$0.0681 \pm 0.0001^{+0.0038}_{-0.0040}$
^{100}Mo	UCI	$0.0675^{+0.0037}_{-0.0042} \pm 0.0068$
^{116}Cd	NEMO-3	$0.274 \pm 0.004 \pm 0.018$
^{116}Cd	ELEGANT V	$0.26^{+0.09}_{-0.05}$
^{116}Cd	AURORA	$0.263^{+0.011}_{-0.012}$
^{128}Te*	Wash. Uni-Tata	77000 ± 4000
^{130}Te	NEMO 3	$6.9 \pm 0.9^{+1.0}_{-0.7}$
^{130}Te	CUORE	8.2 ± 0.8
^{136}Xe	EXO-200	$2.165 \pm 0.016 \pm 0.059$
^{136}Xe	KL-Zen	2.38 ± 0.16
^{150}Nd	ITEP/INR	$0.188^{+0.066}_{-0.039} \pm 0.019$
^{150}Nd	UCI	$0.0675^{+0.0037}_{-0.0042} \pm 0.0068$
^{150}Nd	NEMO-3	$0.0934 \pm 0.0022^{+0.0062}_{-0.0060}$
^{238}U		20 ± 6

7.5 Interpretation of the obtained results

The current best half-life limit for $0\nu\beta\beta$-decay has been obtained with ^{76}Ge by the GERDA experiment giving an upper bound of about 0.1 eV for $\langle m_{\nu_e} \rangle$. If right-handed currents are included, $\langle m_{\nu_e} \rangle$ is fixed by an ellipsoid which is shown in Figure 7.15. The weakest mass limit allowed occurs for $\langle \lambda \rangle, \langle \eta \rangle \neq 0$. In this case the half life of (7.53) corresponds to limits of

$$\langle m_{\nu_e} \rangle < 0.56 \, \text{eV} \tag{7.65}$$

$$\langle \eta \rangle < 6.5 \times 10^{-9} \tag{7.66}$$

$$\langle \lambda \rangle < 8.2 \times 10^{-7}. \tag{7.67}$$

With the given bounds this will result in a lower limit of a potential W_R boson mass of 1.5 TeV.

7.5.1 Effects of MeV neutrinos

Equation (7.40) has to be modified for heavy neutrinos ($m_\nu \geq 1 \, \text{MeV}$). Now the neutrino mass in the propagator can no longer be neglected with respect to the neutrino momentum. This results in a change in the radial shape of the used neutrino potential $H(r)$ from a Coulomb type to a Yukawa type form

$$H(r) \propto \frac{1}{r} (\text{light neutrinos}) \rightarrow H(r) \propto \frac{\exp(-m_h r)}{r} (\text{heavy neutrinos}). \tag{7.68}$$

The change in $H(r)$ can be accommodated by introducing an additional factor $F(m_h, A)$ into (7.40) resulting in an atomic mass A dependent contribution:

$$\langle m_{\nu_e} \rangle = \left| \sum_{i=1,\text{light}}^{N} U_{ei}^2 m_i + \sum_{h=1,\text{heavy}}^{M} F(m_h, A) U_{eh}^2 m_h \right|. \tag{7.69}$$

By comparing the $\langle m_{\nu_e} \rangle$ obtained for different isotopes, interesting limits on the mixing angles for an MeV neutrino can be deduced [Hal83, Zub97].

7.5.2 Transitions to excited states

In addition to the discussed ground-state transition, decays into 0^+ and 2^+ excited states of the daughter are also possible. The phase space for these transitions is smaller (it is now given by $Q - E_\gamma$) but the de-excitation photon might allow a good experimental signal. So far the neutrino accompanied decay into the first excited 0^+ state has been observed in two systems. The half life observed for the first excited 0^+-state for ^{100}Mo is given as [Arn14]

$$T_{1/2} = 7.5 \pm 0.6(\text{stat.}) \pm 0.6(\text{sys.}) \times 10^{20} \text{yr} \tag{7.70}$$

and for ^{150}Nd the half-life is [Kid14]

$$T_{1/2} = 1.07^{+0.45}_{-0.25}(\text{stat.}) \pm 0.07(\text{sys.}) \times 10^{20} \text{yr} \tag{7.71}$$

From the point of view of right-handed currents, observations of transitions to the first excited 2^+ state were long thought to be clear evidence for this process because here the contribution of the mass term vanishes in first order. However, by taking recoil corrections into account, it could be shown that $0^+ \rightarrow 2^+$ transitions have the same relative sensitivity to $\langle m_{\nu_e} \rangle$ and $\langle \eta \rangle$ $0^+ \rightarrow 0^+$ transitions, but the $0^+ \rightarrow 2^+$ transitions are relatively more sensitive to $\langle \lambda \rangle$ [Tom00]. As long as no signal is seen, bounds on $\langle \eta \rangle$ and $\langle \lambda \rangle$ from ground-state transitions are much more stringent.

7.5.3 Majoron accompanied double β-decay

A completely new class of decays emerge in connection with the emission of a majoron χ [Doi88]

$$(Z, A) \to (Z + 2, A) + 2e^- + \chi. \tag{7.72}$$

Majorana mass terms violate lepton number by two units and, therefore, also $(B - L)$ symmetry, which is the only anomaly-free combination of both quantum numbers. A breaking can be achieved in basically three ways:

- explicit $(B - L)$ breaking, meaning the Lagrangian contains $(B - L)$ breaking terms,
- spontaneous breaking of a local $(B - L)$ symmetry and
- spontaneous breaking of a global $(B - L)$ symmetry.

Associated with the last method is the existence of a Goldstone boson, which is called the majoron χ. Depending on its transformation properties under weak isospin, singlet [Chi80], doublet [San88] and triplet [Gel81] models exist. The triplet and pure doublet model are excluded by the measurements of the Z-width at LEP because such majorons would contribute the analogue of 2 (triplet) or 0.5 (doublet) neutrino flavours to the Z-width (see Chapter 3).

A consequence for experiments is a different sum energy spectrum of the electrons due to three particles in the final state. The predicted spectral shapes are analogous to (7.27) as

$$\frac{\mathrm{d}N}{\mathrm{d}K} \propto (Q - K)^n \left(1 + 2K + \frac{4K^2}{3} + \frac{K^3}{3} + \frac{K^4}{30} \right) \tag{7.73}$$

where the spectral index n is now 1 for the triplet majoron, 3 for lepton-number-carrying majorons and 7 for various other majoron models. The different shape allows discrimination with respect to $2\nu\beta\beta$-decay, where $n = 5$. In the $n = 1$ model, the effective neutrino–majoron coupling $\langle g_{\nu\chi} \rangle$ can be deduced from

$$(T_{1/2}^{0\nu\chi})^{-1} = |M_{GT}^{0\nu\chi} - M_F^{0\nu\chi}|^2 G^{0\nu\chi} |\langle g_{\nu\chi} \rangle|^2 \tag{7.74}$$

where $\langle g_{\nu\chi} \rangle$ is given by

$$\langle g_{\nu\chi} \rangle = \sum_{i,j} g_{\nu\chi} U_{ei} U_{ej}. \tag{7.75}$$

Present half-life limits for the decay mode ($n = 1$) are of the order 10^{21}–10^{24} yr resulting in a deduced upper limit on the coupling constant (7.75) of about

$$\langle g_{\nu\chi} \rangle \lesssim 10^{-4}. \tag{7.76}$$

A first half-life limit for the $n = 3$ mode was obtained with ^{76}Ge [Zub92], the current value can be found in [Ago15].

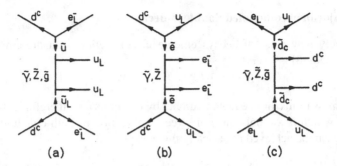

Figure 7.16. Dominant Feynman graphs from R-parity-violating SUSY contributing to double β-decay (from [Moh86]). © 1986 by the American Physical Society.

7.5.4 Decay rates for SUSY-induced $0\nu\beta\beta$ decay

The obtained half-life limit also sets bounds on other physical quantities because the intermediate transition can be realized by other $\Delta L = 2$ mechanisms. Among these are double charged Higgs bosons, the already mentioned right-handed weak currents, R-parity-violating SUSY, leptoquarks and others (see [Kla99]).

Double β-decay can also proceed via R_P-violating SUSY graphs [Moh86, Hir95, Hir96]: the dominant ones are shown in Figure 7.16. The obtainable half-life is given by

$$(T^{0\nu}_{1/2}(0^+ \to 0^+))^{-1} \propto G \left(\frac{\lambda'_{111}}{m^4_{\tilde{q},\tilde{e}} m_{\tilde{g},\chi}} M \right)^2 \qquad (7.77)$$

with G and M the corresponding phase space factor and nuclear matrix element, λ'_{111} the strength of the R-parity violation (see (5.38)) and $m_{\tilde{q},\tilde{e},\tilde{g},\chi}$ as the mass of the involved squarks, selectrons, gluinos and neutralinos (see Chapter 5).

7.6 Positron decay and electron capture decay modes

Relatively mild interest has been shown in another form of the decay, namely the modes with electron capture and positron decay (see (7.13)). The reasons for that are firstly, for every positron produced the available phase-space is reduced by $m_e c^2$, and secondly that the double EC modes require the detection of X-rays and Auger electrons, which makes the experimental detection more challenging. There is a revived interest in these processes because it has been shown that β^+/EC modes show an enhanced sensitivity to right-handed weak currents [Hir94]. In addition, in neutrinoless EC/EC decay a resonant enhancement of the decay rate can occur if there exists an excited state in the daughter nucleus which is energetically degenerate with the initial ground state. As mentioned before, this has been experimentally studied with Penning traps. The experimental signatures of the decay modes involving positrons, (7.13) and (7.14), in the final state are promising because of two or four 511 keV photons. Experimentally more challenging is the EC/EC mode. In an excited state transition, characteristic gamma rays can be used in association with X-ray emission. In the 0ν mode, because of energy and momentum conservation, additional particles must be emitted such as an e^+e^- pair or internal bremsstrahlung photons, often called radiative decay [Doi93]. The emission of a real photon requires that one of the electrons has to be captured from the L-shell resulting in a reduced decay rate. Various half-life limits have been obtained and are of the order of 10^{21} yr for several isotopes; however, the first observation of the 2ν double EC has been observed in ^{124}Xe [Bar14a, Apr19]. A compilation of limits can be found in [Tre02]. The COBRA experiment has the chance of simultaneously measuring five different isotopes for this decay channel [Zub01]. As the decay is intrinsic to the CdZnTe detectors, there is a good chance of observing the 2ν double EC and the positron-emitting modes coincidences among the crystals can be used.

7.7 CP phases and double beta decay

As already mentioned, two additional CP-phases exist in the case of Majorana neutrinos. The neutrino mixing matrix (5.54) for three flavours can be written in the form

$$U = U_{\text{PMNS}} \, \text{diag}(1, e^{i\alpha}, e^{i\beta}) \qquad (7.78)$$

where U_{PMNS} is given in Chapter 5 and α, β are the new phases associated with Majorana neutrinos. Neutrino oscillations (see Chapter 8) can probe only δ because it violates flavour lepton number but conserves total lepton number. Double β-decay is unique in a sense for having these additional Majorana phases. The effective Majorana mass can be written in the three flavour scenario as given in (7.40). If CP is conserved, which is the case for $\alpha, \beta = k\pi$ with $k = 0, 1, 2, \ldots$, the equation changes to

$$\langle m_{\nu_e} \rangle = | \, m_1 U_{e1}^2 \pm m_2 U_{e2}^2 \pm m_3 U_{e3}^2 \, | \quad . \qquad (7.79)$$

Depending on the signs, destructive interference among the individual terms can happen. Taking the individual terms of the PMNS matrix into account, an alternative way of expressing $\langle m_{\nu_e} \rangle$ is

$$\langle m_{\nu_e} \rangle = \cos^2_{12} \cos^2_{13} \, m_1 + \sin^2_{12} \cos^2_{13} \, e^{i\alpha} \sqrt{m_1^2 + \Delta m_{12}^2}$$
$$+ \sin^2_{13} \, e^{i\beta} \sqrt{m_1^2 + \Delta m_{12}^2 + \Delta m_{23}^2} \qquad (7.80)$$

This form is convenient when investigating the link between double beta decay and neutrino oscillation results, which are discussed in the next chapter. Investigations on the effect of Majorana phases can be found in [Rod01a, Pas02]. They might play a crucial role in creating a baryon asymmetry in the early Universe via leptogenesis (see Chapter 13).

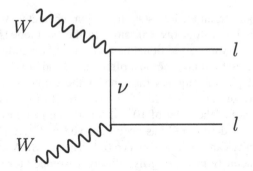

Figure 7.17. General Feynman diagram for $\Delta L = 2$ processes mediated by a virtual Majorana neutrino. With kind permission of S. Turkat.

7.8   Generalization to three flavours

In general, there is a 3×3 matrix of effective Majorana masses, the elements being

$$\langle m_{\alpha\beta} \rangle = \left| \sum m_m \eta_m^{CP} U_{\alpha m} U_{\beta m} \right| \qquad \text{with } \alpha, \beta \equiv e, \mu, \tau. \qquad (7.81)$$

Double β-decay measures the element $\langle m_{\nu_e} \rangle = \langle m_{ee} \rangle$. In contrast to $0\nu\beta\beta$-decay, little is known about the other matrix elements.

7.8.1   General considerations

The underlying Feynman graph for all these $\Delta L = 2$ processes mediated by a virtual massive Majorana neutrino is shown in Figure 7.17. The general behaviour can be described by

$$\sigma \propto \frac{m_i^2}{(q^2 - m_i^2)^2} \to \begin{cases} m_i^2 & \text{for } m_i^2 \ll q^2 \\ m_i^{-2} & \text{for } m_i^2 \gg q^2. \end{cases} \qquad (7.82)$$

with q^2 as four-momentum transfer and m_i as neutrino mass eigenstates. As long as an experimental bound does not intersect the cross-section prediction, a limit on $\langle m_{\alpha\beta} \rangle$ can, in principle, be obtained by linearly extrapolating the low-energy part. However, such a limit is unphysical and should give only a rough estimate of how far away from actually becoming meaningful the result still is. What physical processes can explore the remaining eight terms? It should already be mentioned here that all following bounds are unphysical because the experimental limits are currently not strong enough.

7.8.1.1   *Muon–positron conversion on nuclei*

$$\mu^- + (A, Z) \to e^+ + (A, Z - 2) \qquad (7.83)$$

is a process closely related to double β-decay and, within the context discussed here measures $\langle m_{e\mu} \rangle$. The current best bound comes from SINDRUMII and is given by [Kau98]

$$\frac{\Gamma(\text{Ti} + \mu^- \to \text{Ca}^{GS} + e^+)}{\Gamma(\text{Ti} + \mu^- \to \text{Sc} + \nu_\mu)} < 1.7 \times 10^{-12} \quad (90\% \text{ CL}) \quad (7.84)$$

which can be converted into a new limit of $\langle m_{e\mu} \rangle < 17(82)\,\text{MeV}$ depending on whether the proton pairs in the final state are in a spin singlet or triplet state [Doi85]. A calculation [Sim01] comes to a cross-section ten orders of magnitude smaller, which will worsen the bound by five orders of magnitude. Clearly this has to be better understood. Note that a process like $\mu \to e\gamma$ does not give direct bounds on the quantities discussed here, because it measures $m_{e\mu} = \sqrt{\sum U_{ei} U_{\mu i} m_i^2}$. Therefore, without specifying a neutrino-mixing and mass scheme, the quantities are rather difficult to compare. However, if this can be done, these indirect bounds are more stringent.

7.8.1.2 *Processes investigating* $\langle m_{\mu\mu} \rangle$

Three different kinds of search can be considered. One process under study is muon lepton-number-violating ($\Delta L_\mu = 2$) trimuon production in neutrino–nucleon scattering via charged current (CC) reactions

$$\nu_\mu N \to \mu^- \mu^+ \mu^+ X \quad (7.85)$$

where X denotes the hadronic final state. Detailed calculations can be found in [Fla00]. Taking the fact that, in past experiments, no excess events of this type were observed on the level of 10^{-5} of CC events, a limit of $\langle m_{\mu\mu} \rangle \lesssim 10^4\,\text{GeV}$ can be deduced.

A further possibility for probing $\langle m_{\mu\mu} \rangle$ is to explore rare meson decays such as the rare kaon decay [Zub00a]

$$K^+ \to \pi^- \mu^+ \mu^+. \quad (7.86)$$

A new upper limit on the branching ratio of

$$\frac{\Gamma(K^+ \to \pi^- \mu^+ \mu^+)}{\Gamma(K^+ \to \text{all})} < 8.6 \times 10^{-11} \quad (90\% \text{ CL}) \quad (7.87)$$

could be deduced [PDG18] resulting in a bound of $\langle m_{\mu\mu} \rangle \lesssim 100\,\text{MeV}$. Other rare meson decays can be envisaged; the current status of some decays is shown in Table 7.5. A full compilation is given in [PDG18].

A realistic chance to bring $\langle m_{\mu\mu} \rangle$ at least into the physical region by improving both methods and especially using trimuon production will be given by a neutrino factory [Rod01]. However, this would require a muon beam energy of at least 500 GeV, which is currently not a favored option.

Table 7.5. Branching ratios of $\Delta L = 2$ decays of rare mesons, which can be described by the same Feynman graph as double β-decay (from [PDG18]).

Decay mode	Limit on branching ratio
$K^+ \to \pi^- e^+ e^+$	6.4×10^{-10}
$K^+ \to \pi^- \mu^+ \mu^+$	8.6×10^{-11}
$K^+ \to \pi^- e^+ \mu^+$	5.0×10^{-10}
$D^+ \to \pi^- e^+ e^+$	1.1×10^{-6}
$D^+ \to \pi^- \mu^+ \mu^+$	2.2×10^{-8}
$D^+ \to \pi^- e^+ \mu^+$	2.0×10^{-6}
$D_s^+ \to \pi^- e^+ e^+$	4.1×10^{-6}
$D_s^+ \to \pi^- \mu^+ \mu^+$	1.2×10^{-7}
$D_s^+ \to \pi^- e^+ \mu^+$	8.4×10^{-6}
$B^+ \to \pi^- e^+ e^+$	2.3×10^{-8}
$B^+ \to \pi^- \mu^+ \mu^+$	4.0×10^{-9}
$B^+ \to \pi^- e^+ \mu^+$	1.5×10^{-7}

Probably the closest analogy for performing a measurement on nuclear scales would be μ^- capture by nuclei with a μ^+ in the final state as discussed in [Mis94]. No such experiment has yet been performed, probably because of the requirement to use radioactive targets due to energy conservation arguments. The ratio with respect to standard muon capture can be given in the case of the favored ^{44}Ti and a light neutrino exchange ($m_i \ll q^2$) as

$$R = \frac{\Gamma(\mu^- + \text{Ti} \to \mu^+ + \text{Ca})}{\Gamma(\mu^- + \text{Ti} \to \nu_\mu + \text{Sc})} \simeq 5 \times 10^{-24} \left(\frac{\langle m_{\mu\mu} \rangle}{250 \text{ keV}} \right)^2. \tag{7.88}$$

many orders of magnitude smaller than current μ-e conversion experiments.

7.8.1.3 *Limits on $\langle m_{\tau\tau} \rangle$ from CC events at HERA*

Limits for mass terms involving the τ-sector were obtained by using HERA data [Fla00a]. The process studied is

$$e^\pm p \to \overset{(-)}{\nu_e} l^\pm l'^\pm X \quad \text{with } (ll') = (e\tau), (\mu\tau), (\mu\mu) \text{ and } (\tau\tau). \tag{7.89}$$

Such a process has a spectacular signature with large missing transverse momentum ($\not{p}_T$) and two like-sign leptons, isolated from the hadronic remnants [Rod00].

Unfortunately, all the bounds given except for $\langle m_{ee} \rangle$ are still without physical meaning and currently only the advent of a neutrino factory might change the situation. Nevertheless, it is worthwhile considering these additional processes because, as in the case of $0\nu\beta\beta$-decay, they might provide stringent bounds on other quantities such as those coming from R-parity-violating SUSY.

After discussing only the limits for a possible neutrino mass, we now come to neutrino oscillations where evidence for a non-vanishing rest mass are found.

Chapter 8

Neutrino oscillations

DOI: 10.1201/9781315195612-8

In the case of a non-vanishing rest mass of the neutrino, the weak and mass eigenstates are not necessarily identical, a fact well known in the quark sector where both types of states are connected by the CKM matrix (see Section 3.3.2). This allows for the phenomenon of neutrino oscillations, a flavour oscillation which is already known in other particle systems, e.g., $K^0\overline{K^0}$ oscillation. It can be described by pure quantum field theory. Oscillations are observable as long as the neutrino wave packets form a coherent superposition of states. Such oscillations among the different neutrino flavours do not conserve individual flavour lepton numbers, only a total lepton number. We start with the most general case first, before turning to the more common two- and three-flavour scenarios. For additional literature see [Bil78, Bil87, Kay81, Kay89, Boe92, Kim93, Gri96, Gro97, Bil99, Lip99, Giu07, Dor08, Akh09].

8.1 General formalism

The following discussion is based on simplified arguments, nevertheless, resulting in correct equations. A sophisticated derivation can be done within quantum field theory; see [Kay81, Gri96, Akh09].

Let us assume that there is an arbitrary number of n orthonormal eigenstates. The n flavour eigenstates $|\nu_\alpha\rangle$ with $\langle\nu_\beta|\nu_\alpha\rangle = \delta_{\alpha\beta}$ are connected to the n mass eigenstates $|\nu_i\rangle$ with $\langle\nu_i|\nu_j\rangle = \delta_{ij}$ via a unitary mixing matrix U:

$$|\nu_\alpha\rangle = \sum_i U_{\alpha i}|\nu_i\rangle \qquad |\nu_i\rangle = \sum_\alpha (U^\dagger)_{i\alpha}|\nu_\alpha\rangle = \sum_\alpha U^*_{\alpha i}|\nu_\alpha\rangle \qquad (8.1)$$

with

$$U^\dagger U = 1 \qquad \sum_i U_{\alpha i}U^*_{\beta i} = \delta_{\alpha\beta} \qquad \sum_\alpha U_{\alpha i}U^*_{\alpha j} = \delta_{ij}. \qquad (8.2)$$

In the case of antineutrinos, i.e., $U_{\alpha i}$ has to be replaced by $U^*_{\alpha i}$:

$$|\bar\nu_\alpha\rangle = \sum_i U^*_{\alpha i}|\bar\nu_i\rangle. \qquad (8.3)$$

The number of parameters in an $n \times n$ unitary matrix is n^2. The $2n - 1$ relative phases of the $2n$ neutrino states can be fixed in such a way that $(n-1)^2$ independent parameters remain. It is convenient to write them as $\frac{1}{2}n(n-1)$ weak mixing angles of an n-dimensional rotational matrix together with $\frac{1}{2}(n-1)(n-2)$ CP-violating phases.

The mass eigenstates $|\nu_i\rangle$ are stationary states and show a time dependence according to

$$|\nu_i(x,t)\rangle = e^{-iE_i t}|\nu_i(x,0)\rangle \tag{8.4}$$

assuming neutrinos with momentum p emitted by a source positioned at $x = 0$ ($t = 0$)

$$|\nu_i(x,0)\rangle = e^{ipx}|\nu_i\rangle \tag{8.5}$$

and being relativistic

$$E_i = \sqrt{m_i^2 + p_i^2} \simeq p_i + \frac{m_i^2}{2p_i} \simeq E + \frac{m_i^2}{2E} \tag{8.6}$$

for $p \gg m_i$ and $E \approx p$ as neutrino energy. Assume that the difference in mass between two neutrino states with different mass $\Delta m_{ij}^2 = m_i^2 - m_j^2$ cannot be resolved. Then the flavour neutrino is a coherent superposition of neutrino states with definite mass.[1] Neutrinos are produced and detected as flavour states. Therefore, neutrinos with flavour $|\nu_\alpha\rangle$ emitted by a source at $t = 0$ propagate with time into a state

$$|\nu(x,t)\rangle = \sum_i U_{\alpha i} e^{-iE_i t}|\nu_i\rangle = \sum_{i,\beta} U_{\alpha i} U_{\beta i}^* e^{ipx} e^{-iE_i t}|\nu_\beta\rangle. \tag{8.7}$$

Different neutrino masses imply that the phase factor in (8.7) is different. This means that the flavour content of the final state differs from the initial one. At macroscopic distances this effect can be large in spite of small differences in neutrino masses. The time-dependent transition amplitude for a flavour conversion $\nu_\alpha \to \nu_\beta$ is then given by

$$A(\alpha \to \beta)(t) = \langle \nu_\beta | \nu(x,t)\rangle = \sum_i U_{\beta i}^* U_{\alpha i} e^{ipx} e^{-iE_i t}. \tag{8.8}$$

Using (8.6) this can be written as

$$A(\alpha \to \beta)(t) = \langle \nu_\beta | \nu(x,t)\rangle = \sum_i U_{\beta i}^* U_{\alpha i} \exp\left(-i\frac{m_i^2}{2}\frac{L}{E}\right)$$
$$= A(\alpha \to \beta)(L) \tag{8.9}$$

with $L = x = ct$ being the distance between source and detector. In an analogous way, the amplitude for antineutrino transitions can be derived (8.8):

$$A(\bar{\alpha} \to \bar{\beta})(t) = \sum_i U_{\beta i} U_{\alpha i}^* e^{-iE_i t}. \tag{8.10}$$

---

[1] This is identical to the kaon system. The states K^0 and $\bar{K}^0$ are states of definite strangeness which are related to K_S^0 and K_L^0 as states with definite masses and widths.

The transition probability P can be obtained from the transition amplitude A:

$$P(\alpha \to \beta)(t) = |A(\alpha \to \beta)|^2 = \sum_i \sum_j U_{\alpha i} U_{\alpha j}^* U_{\beta i}^* U_{\beta j} e^{-i(E_i - E_j)t}$$

$$= \sum_i |U_{\alpha i} U_{\beta i}^*|^2 + 2 \operatorname{Re} \sum_{j>i} U_{\alpha i} U_{\alpha j}^* U_{\beta i}^* U_{\beta j} \exp\left(-i\frac{\Delta m_{ij}^2}{2}\right)\frac{L}{E}$$

$$(8.11)$$

The second term in (8.11) describes the time- (or spatial-) dependent neutrino oscillations. The first one is an average transition probability, which also can be written as

$$\langle P_{\alpha \to \beta} \rangle = \sum_i |U_{\alpha i} U_{\beta i}^*|^2 = \sum_i |U_{\alpha i}^* U_{\beta i}|^2 = \langle P_{\beta \to \alpha} \rangle. \qquad (8.12)$$

Using CP invariance ($U_{\alpha i}$ real), this can be simplified to

$$P(\alpha \to \beta)(t) = \sum_i U_{\alpha i}^2 U_{\beta i}^2 + 2\sum_{j>i} U_{\alpha i} U_{\alpha j} U_{\beta i} U_{\beta j} \cos\left(\frac{\Delta m_{ij}^2}{2}\frac{L}{E}\right)$$

$$= \delta_{\alpha\beta} - 4\sum_{j>i} U_{\alpha i} U_{\alpha j} U_{\beta i} U_{\beta j} \sin^2\left(\frac{\Delta m_{ij}^2}{4}\frac{L}{E}\right). \qquad (8.13)$$

Evidently, the probability of finding the original flavour is given by

$$P(\alpha \to \alpha) = 1 - \sum_{\alpha \neq \beta} P(\alpha \to \beta). \qquad (8.14)$$

As can be seen from (8.11) there will be oscillatory behaviour as long as at least one neutrino mass eigenstate is different from zero and if there is a mixing (non-diagonal terms in U) among the flavours. In addition, the observation of oscillations allows no absolute mass measurement; oscillations are sensitive to only Δm^2. Last but not least, neutrino masses should not be exactly degenerated. Another important feature is the dependence of the oscillation probability on L/E. Majorana phases as described in Chapter 7 are unobservable in oscillations because the form given in (7.28) and implemented in (8.11) shows that the diagonal matrix containing these phases always results in the identity matrix [Bil80]. The same results for oscillation probabilities are also obtained by performing a more sophisticated quantum field theoretical treatment or using wave packets [Kay81, Gri96, Akh09].

The result can also be obtained from very general arguments [Gro97, Lip99], which show that such flavour oscillations are completely determined by the propagation dynamics and the boundary condition that the probability of observing the wrong flavour at the position of the source at any time must vanish. The propagation in free space for each state is given in (8.4). The expansion of the neutrino wavefunction in energy eigenstates is

$$\psi = \int g(E)\,\mathrm{d}E\,\mathrm{e}^{-iEt} \sum_{i=1}^3 c_i \mathrm{e}^{ipx} |\nu_i\rangle \qquad (8.15)$$

with the energy independent coefficients c_i. The function $g(E)$ describing the exact form of the energy wave packet is irrelevant at this stage. Each energy eigenstate has three terms, one for each mass eigenstate, if three generations are assumed. The boundary condition for creating a ν_e and only a ν_e at the source (or at $t = 0$) then requires

$$\sum_{i=1}^{3} c_i \langle \nu_i | \nu_\mu \rangle = \sum_{i=1}^{3} c_i \langle \nu_i | \nu_\tau \rangle = 0. \tag{8.16}$$

The momentum of each of the three components is determined by the energy and the neutrino masses. The propagation of this energy eigenstate, the relative phases of its three mass components and its flavour mixture at the detector are completely determined by the energy–momentum kinematics of the three mass eigenstates and lead to the same oscillation formula as described before.

8.2 CP and T violation in neutrino oscillations

Comparison of (8.8) with (8.10) yields a relation between neutrino and antineutrino transitions:

$$A(\bar{\alpha} \to \bar{\beta})(t) = A(\alpha \to \beta)(t) \neq A(\beta \to \alpha)(t). \tag{8.17}$$

This relation is a direct consequence of the CPT theorem. CP violation manifests itself if the oscillation probabilities of $\nu_\alpha \to \nu_\beta$ are different from its CP conjugate process $\bar{\nu}_\alpha \to \bar{\nu}_\beta$. So one observable for detection could be

$$\Delta P_{\alpha\beta}^{CP} = P(\nu_\alpha \to \nu_\beta) - P(\bar{\nu}_\alpha \to \bar{\nu}_\beta) \neq 0 \qquad \alpha \neq \beta. \tag{8.18}$$

This might be done with the proposed neutrino superbeams and neutrino factories (see Section 8.10). Similarly, T violation can be tested if the probabilities of $\nu_\alpha \to \nu_\beta$ are different from the T conjugate process $\nu_\beta \to \nu_\alpha$. Here, the observable is

$$\Delta P_{\alpha\beta}^{T} = P(\nu_\alpha \to \nu_\beta) - P(\nu_\beta \to \nu_\alpha) \neq 0 \qquad \alpha \neq \beta. \tag{8.19}$$

If CPT conservation holds, which is the case for neutrino oscillations in vacuum, violation of T is equivalent to violation of CP. Using U_{PMNS} it can be shown explicitly that in vacuum $\Delta P_{\alpha\beta}^{CP}$ and $\Delta P_{\alpha\beta}^{T}$ are equal and given by

$$\Delta P_{\alpha\beta}^{CP} = \Delta P_{\alpha\beta}^{T}$$
$$= -16 J_{\alpha\beta} \sin\left(\frac{\Delta m_{12}^2}{4E} L\right) \sin\left(\frac{\Delta m_{23}^2}{4E} L\right) \sin\left(\frac{\Delta m_{13}^2}{4E} L\right) \tag{8.20}$$

where

$$J_{\alpha\beta} \equiv \text{Im}[U_{\alpha 1} U_{\alpha 2}^* U_{\beta 1}^* U_{\beta 2}] = \pm c_{12} s_{12} c_{23} s_{23} c_{13}^2 s_{13} \sin\delta \tag{8.21}$$

corresponds to the Jarlskog invariant as in the quark sector, and the $+(-)$ sign is denoting cyclic (anticyclic) permutation of $(\alpha, \beta) = (e, \mu), (\mu, \tau), (\tau, e)$. Note that for CP or T violation effects to be present, all the angles must be non-zero and, therefore, three-flavour mixing is essential. To be more specific, first we now consider the case of two-flavour oscillations.

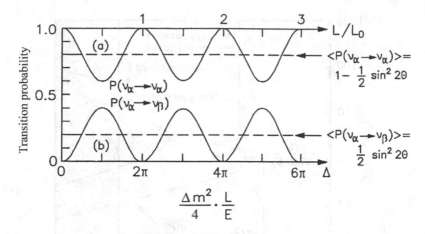

Figure 8.1. Example of neutrino oscillations in the two-flavour scheme. (a) $P(\nu_\alpha \to \nu_\alpha)$ (disappearance, reduction of the flavour); (b) $P(\nu_\alpha \to \nu_\beta)$ (appearance, a flavour shows up not seen before) as a function of $L/L_0 = \Delta m^2/4\pi$ for $\sin^2 2\theta = 0.4$. The dashed lines show the average oscillation probabilities (from [Sch97]). Reproduced with permission of SNCSC.

8.3 Oscillations with two neutrino flavours

The simplest case is the two flavour discussion without any matter effects which might be sufficient for some oberservations. However, in general a full three-flavour analysis including matter effects has to be done. In the first case the relation between the neutrino states is described by one mixing angle θ and one mass difference, for example $\Delta m_{12}^2 = m_2^2 - m_1^2$. The unitary transformation (8.1) is then analogous to the Cabibbo matrix in the quark sector and given by (taking ν_e and ν_μ as flavour eigenstates):

$$\begin{pmatrix} \nu_e \\ \nu_\mu \end{pmatrix} = \begin{pmatrix} \cos\theta & \sin\theta \\ -\sin\theta & \cos\theta \end{pmatrix} \begin{pmatrix} \nu_1 \\ \nu_2 \end{pmatrix}. \tag{8.22}$$

Using the formulae from the previous section, the corresponding two-flavour transition probability is given because there is no CP violating phase by

$$P(\nu_e \to \nu_\mu) = P(\nu_\mu \to \nu_e) = P(\bar\nu_e \to \bar\nu_\mu) = P(\bar\nu_\mu \to \bar\nu_e)$$

$$= \sin^2 2\theta \times \sin^2 \left(\frac{\Delta m^2}{4} \times \frac{L}{E} \right) = 1 - P(\nu_e \to \nu_e). \tag{8.23}$$

This formula explicitly shows that oscillations occur only if both θ and Δm^2 are non-vanishing. All two-flavour oscillation probabilities can be characterized by these two quantities because $P(\nu_\alpha \to \nu_\beta) = P(\nu_\beta \to \nu_\alpha)$. The phase factor can be rewritten as

$$\frac{E_i - E_j}{\hbar} t = \frac{1}{2\hbar c} \Delta m_{ij}^2 \frac{L}{E} = 2.534 \frac{\Delta m_{ij}^2}{\text{eV}^2} \frac{L/\text{m}}{E/\text{MeV}} \tag{8.24}$$

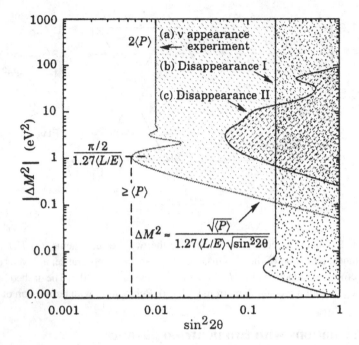

Figure 8.2. Standard double logarithmic plot of Δm^2 *versus* $\sin^2 2\theta$. The excluded parameter ranges and average oscillation probability $\langle P \rangle$ of hypothetical appearance and disappearance experiments are shown. At low Δm^2 the experiment loses sensitivity being too close to the source, so the oscillation barely develops. This implies a slope of -2 until one reaches maximal sensitivity in the first oscillation maximum. At very high Δm^2 the oscillation itself can no longer be observed, only an average transition probability $\langle P \rangle$. In case of non-observation the excluded parameter regions are always on the right side of the curve (from [PDG00]).

where in the last step some practical units were used. The oscillatory term can then be expressed as

$$\sin^2 \left(\frac{\Delta m_{ij}^2}{4} \frac{L}{E} \right) = \sin^2 \pi \frac{L}{L_0} \tag{8.25}$$

$$\text{with } L_0 = 4\pi\hbar c \frac{E}{\Delta m^2} = 2.48 \frac{E/\text{MeV}}{\Delta m^2/\text{eV}^2} \text{m}.$$

In the last step the oscillation length L_0, describing the period of one full oscillation cycle, is introduced (Figure 8.1). It becomes larger with higher energies and smaller Δm^2. The mixing angle $\sin^2 2\theta$ determines the amplitude of the oscillation while Δm^2 influences the oscillation length. Both unknown parameters are typically drawn in a double logarithmic plot as shown in Figure 8.2.

Phrasing it slightly different, the relative phase of the two neutrino states at a position x is (see (8.15))

$$\delta\phi(x) = (p_1 - p_2)x + \frac{(p_1^2 - p_2^2)}{(p_1 + p_2)}x = \frac{\Delta m^2}{(p_1 + p_2)}x. \tag{8.26}$$

Since the neutrino mass difference is small compared to all momenta $|m_1 - m_2| \ll p \equiv (1/2)(p_1 + p_2)$, this can be rewritten in first order in Δm^2 as

$$\delta\phi(x) = \frac{\Delta m^2}{2p}x \tag{8.27}$$

identical to (8.23) with $x = L$ and $p = E$.

8.4 The case for three flavours

A more realistic scenario to consider is that of three known neutrino flavours. The mixing matrix U_{PMNS} is given in Chapter 5. Note that now more Δm^2 quantities are involved both in magnitude and sign; although in a two-flavour oscillation in vacuum the sign does not enter, in three-flavour oscillation, which includes both matter effects (see Section 8.8) and CP violation, the signs of the Δm^2 quantities enter and can, in principle, be measured. In the absence of any matter effect, the probability is given by

$$P(\nu_\alpha \to \nu_\beta) = \delta_{\alpha\beta} - 4 \sum_{i>j=1}^{3} \text{Re}(K_{\alpha\beta,ij}) \sin^2\left(\frac{\Delta m_{ij}^2 L}{4E}\right)$$
$$+ 4 \sum_{i>j=1}^{3} \text{Im}(K_{\alpha\beta,ij}) \sin\left(\frac{\Delta m_{ij}^2 L}{4E}\right) \cos\left(\frac{\Delta m_{ij}^2 L}{4E}\right) \tag{8.28}$$

where

$$K_{\alpha\beta,ij} = U_{\alpha i} U_{\beta i}^* U_{\alpha j}^* U_{\beta j}. \tag{8.29}$$

The general formulae in the three-flavour scenario are quite complex; therefore, the following assumption is made: in most cases only one mass scale is relevant, i.e., $\Delta m_{\text{atm}}^2 \sim 10^{-3}$ eV^2, which is discussed in more detail in Chapter 9. Furthermore, one possible neutrino mass spectrum such as the hierarchical one is taken:

$$\Delta m_{21}^2 = \Delta m_{\text{sol}}^2 \ll \Delta m_{31}^2 \approx \Delta m_{32}^2 = \Delta m_{\text{atm}}^2. \tag{8.30}$$

Then the expressions for specific oscillation transitions are:

$$P(\nu_\mu \to \nu_\tau) = 4|U_{33}|^2|U_{23}|^2 \sin^2\left(\frac{\Delta m_{\text{atm}}^2 L}{4E}\right)$$
$$= \sin^2(2\theta_{23}) \cos^4(\theta_{13}) \sin^2\left(\frac{\Delta m_{\text{atm}}^2 L}{4E}\right) \tag{8.31}$$

$$P(\nu_e \to \nu_\mu) = 4|U_{13}|^2 |U_{23}|^2 \sin^2\left(\frac{\Delta m_{\text{atm}}^2 L}{4E}\right)$$

$$= \sin^2(2\theta_{13}) \sin^2(\theta_{23}) \sin^2\left(\frac{\Delta m_{\text{atm}}^2 L}{4E}\right) \tag{8.32}$$

$$P(\nu_e \to \nu_\tau) = 4|U_{33}|^2 |U_{13}|^2 \sin^2\left(\frac{\Delta m_{\text{atm}}^2 L}{4E}\right)$$

$$= \sin^2(2\theta_{13}) \cos^2(\theta_{23}) \sin^2\left(\frac{\Delta m_{\text{atm}}^2 L}{4E}\right). \tag{8.33}$$

8.5 Experimental considerations

The search for neutrino oscillations has to be performed in different ways: an appearance or disappearance mode. In the latter case, one explores whether the number of neutrinos of a produced flavour arrive at a detector with less than the expected R^{-2} flux reduction or whether the spectral shape changes if observed at various distances from a source. This method is not able to determine the new neutrino flavour. An appearance experiment searches for possible new flavours, which do not exist in the original beam or produce an enhancement of an existing neutrino flavour. The identification of the various flavours relies on the detection of the corresponding charged lepton produced in their charged current interactions

$$\nu_l + \text{N} \to l^- + \text{X} \qquad \text{with } l \equiv e, \mu, \tau \tag{8.34}$$

where X denotes the hadronic final state.

Several neutrino sources can be used to search for oscillations which will be discussed in this and the following chapters more extensively. The most important ones are:

- nuclear power plants ($\bar{\nu}_e$),
- accelerators ($\nu_e, \nu_\mu, \bar{\nu}_e, \bar{\nu}_\mu$),
- the atmosphere ($\nu_e, \nu_\mu, \bar{\nu}_e, \bar{\nu}_\mu$) and
- the Sun (ν_e)

which shows that the various mentioned sources sometimes cannot probe each other; i.e., high-energy accelerators ($E \approx 1$–100 GeV, $L \approx 1$ km) are not able to check the solar neutrino data ($E \approx 1$ MeV, $L \approx 10^8$ km). Equation (8.35) also defines the minimal Δm^2 which can be explored. Three cases have to be considered with respect to a possible observation of oscillations (Figure 8.3):

- $L/E \ll \frac{4}{\Delta m^2}$, i.e., $L \ll L_0$. Here, the experiment is too close to the source and the oscillations have no time to develop.
- $L/E \approx \frac{4}{\Delta m^2}$, i.e., $L/E \approx \frac{1}{\Delta m^2}$. This is a necessary condition to observe oscillations and it is the most sensitive region.

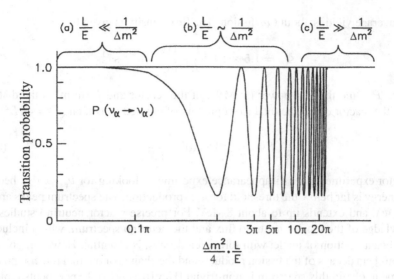

Figure 8.3. Logarithmic plot of the oscillation probability $P(\nu_\alpha \to \nu_\alpha)$ as a function of L/E for $\sin^2 2\theta = 0.83$. The brackets denote three possible cases: (a) no oscillations ($L/E \ll 1/\Delta m^2$); (b) maximal sensitivity to oscillations $L/E \approx 1/\Delta m^2$; and (c) only average oscillation measurement due to finite resolution for $L/E \gg 1/\Delta m^2$ (from [Sch97]). Reproduced with permission of SNCSC.

- $L/E \gg \frac{4}{\Delta m^2}$, i.e., $L \gg L_0$. Several oscillations occurred between the source and the detector. Normally, experiments then measure L/E not precisely enough to resolve the oscillation pattern but measure only an average transition probability.

Thus, the part of the Δm^2–$\sin^2 2\theta$ parameter space explored depends on the ratio L/E. The most sensitive range of an experiment is at

$$\Delta m^2 \approx E/L. \tag{8.35}$$

Two more points which influence the experimental sensitivity and the observation of oscillations have to be considered. First of all, L is often not well defined. This is the case when dealing with an extended source (Sun, atmosphere, decay tunnels). Alternatively, E might not be known exactly. This might be the case if the neutrino source has an energy spectrum $N(E)$ and E will not be measured in a detector. Last but not least, for some experiments there is no chance to vary L and/or E because it is fixed (e.g., in the case of the Sun); therefore, the explorable Δm^2 region is constrained by nature.

8.6 Nuclear reactor experiments

Nuclear reactors are the strongest terrestrial antineutrino sources, stemming from the β-decays of unstable neutron-rich fission products of ^{235}U, ^{238}U, ^{239}Pu and ^{241}Pu.

The average yield is about $6\bar{\nu}_e$/fission. The flux density is given by

$$\Phi_\nu = 1.5 \times 10^{12} \frac{P/\text{MW}}{L^2/\text{m}^2} \text{ cm}^{-2} \text{ s}^{-1} \quad (8.36)$$

where P is the thermal power (in MW) of the reactor and L (in m) is the distance from the reactor core. The total isotropic flux of emitted $\bar{\nu}_e$ is then ($F = 4\pi L^2$)

$$F\Phi_\nu = 1.9 \times 10^{17} \frac{P}{\text{MW}} \text{ s}^{-1}. \quad (8.37)$$

Reactor experiments are disappearance experiments looking for $\bar{\nu}_e \rightarrow \bar{\nu}_X$, because the energy is far below the threshold for μ, τ-production. The spectrum peaks around 2-3 MeV and extends up to about 8 MeV. For precise reactor neutrino studies, the knowledge of the reactor neutrino flux and the neutrino spectrum, which includes a significant fraction of nuclei with forbidden decays, is essential. However, given the various beta decays of the fission products and the change of the fuel rod composition in time to obtain this spectrum is non-trivial [Hay16, Hay19]. Experiments typically try to measure the positron energy spectrum which can be deduced from the $\bar{\nu}_e$ spectrum and either compare it directly to the theoretical predictions or measure it at several distances from the reactor and search for spectral changes. Both types of experiments have been performed in the past. However, the first approach requires a detailed theoretical understanding of the fission processes as well as a good knowledge of the operational parameters of the reactor during a duty cycle which changes the relative contributions of the fission products.

The detection reaction used is inverse beta decay

$$\bar{\nu}_e + p \rightarrow e^+ + n \quad (8.38)$$

with an energy threshold of 1.806 MeV. The $\bar{\nu}_e$ energy can be obtained by measuring the positron energy spectrum as

$$E_{\bar{\nu}_e} = E_{e^+} + m_n - m_p = E_{e^+} + 1.293 \text{ MeV} = T_{e^+} + 1.806 \text{ MeV} \quad (8.39)$$

with T_{e^+} as kinetic energy of the positron and neglecting the small neutron recoil energy (≈ 20 keV). The cross-section for (8.38) is given by

$$\sigma(\bar{\nu}_e + p \rightarrow e^+ + n) = \sigma(\nu_e + n \rightarrow e^- + p)$$

$$= \frac{G_F^2 E_\nu^2}{\pi} |\cos\theta_c|^2 \left(1 + 3\left(\frac{g_A}{g_V}\right)^2\right)$$

$$= 9.23 \times 10^{-42} \left(\frac{E_\nu}{10 \text{ MeV}}\right)^2 \text{ cm}^2 \quad (8.40)$$

with $\cos\theta_c$ being the Cabibbo-angle (see Section 3.3). All experiments are using liquid scintillators for the measurements. Hence, the same signature is used as for the discovery of the neutrino (see Chapter 1). Therefore, coincidence techniques are used

Table 8.1. List of finished 'short-baseline' ($\lesssim 1000$ m) reactor experiments. The power of the reactors and the distance of the experiments with respect to the reactor are given.

Reactor	Thermal power [MW]	Distance [m]
ILL-Grenoble (F)	57	8.75
Bugey (F)	2800	13.6, 18.3
Rovno (USSR)	1400	18.0, 25.0
Savannah River (USA)	2300	18.5, 23.8
Gösgen (CH)	2800	37.9, 45.9, 64.7
Krasnojarsk (Russia)	-	57.0, 57.6, 231.4
Bugey III (F)	2800	15.0, 40.0, 95.0
CHOOZ (F)	2×4200	998, 1115
Palo Verde (USA)	11600	890, 750

for detection between the annihilation photons of the positron and the neutrons which diffuse and thermalize within 10–200 μs. The neutrons will be captured by (n, γ)-reactions resulting in a second pulse of gamma rays for detection, some experiments add ^{157}Gd to the scintillator to increase the sensitivity because of its very high thermal neutron capture cross-section (about $\sigma = 253000$ barn).

Sometimes in the past experiments have used heavy water reactors allowing reactions of

$$\bar{\nu}_e + \mathrm{D} \to e^+ + n + n \qquad (E_{\mathrm{Thr}} = 4.0\,\mathrm{MeV}) \qquad (\mathrm{CC}) \qquad (8.41)$$

$$\bar{\nu}_e + \mathrm{D} \to \bar{\nu}_e + p + n \qquad (E_{\mathrm{Thr}} = 2.2\,\mathrm{MeV}) \qquad (\mathrm{NC}). \qquad (8.42)$$

The main backgrounds in reactor neutrino experiments originate from uncorrelated cosmic-ray hits in coincidence with natural radioactivity and correlated events from cosmic-ray muons and induced neutrons.

8.6.1 Experimental status

Several reactor experiments have been performed in the past (see Table 8.1). More recent experiments which will be discussed now lead to the measurement of the two mixing angles θ_{12} and θ_{13}.

8.6.1.1 *KamLAND–Measurement of θ_{12}*

From the solar neutrino data (see Chapter 10), one of the preferred oscillation solutions has been indicating a parameter region of $\Delta m_{12}^2 \approx 8 \times 10^{-5}$ eV^2 with a large mixing angle $\sin^2 2\theta_{12}$ and it also included matter effects (see Section 8.8) [Egu03, Ara05]. As reactor and solar neutrinos have about the same energy, an oscillation result of solar neutrinos including matter effects would be equivalent to a baseline in vacuum (identical to air) of several hundred kilometres to prove the solar solution region. An experiment designed for this goal called KamLAND [Egu03] in Japan has been installed in the Kamioka mine. Using a large number of reactors nearby and also from South Korea resulted in a total amount of 55 commercial nuclear power plants which are together delivering a power of about 155 GW. Thus, the total flux of $\bar{\nu}_e$ at Kamioka is about 4×10^6 cm^{-2} s^{-1} (or 1.3×10^6 cm^{-2} s^{-1} for $E_{\bar{\nu}} > 1.8$ MeV). Almost half of the power (≈ 70 GW) is produced from reactors in a distance of 175 ± 35 km. The detector itself consists of 1000 t of liquid scintillator contained within a nylon sphere read out by photomultiplier tubes. The scintillator is based on mineral oil and pseudocumene to achieve a sufficiently high light yield and n–γ discrimination by pulse shape analysis. This inner sphere is surrounded by 2.5 m of non-scintillating fluid as shielding. Both parts are contained and mechanically supported by a spherical stainless steel vessel. On this vessel 1325 phototubes for readout of the fiducial volume are also mounted.

Impressive results based on three measuring periods have been obtained [Egu03, Ara05, Abe08] with a total exposure of 2881 ton×years. 1609 $\bar{\nu}_e$ events were observed while the expectation had been 2179 ± 89 events (Figure 8.4). The obtained ratio is

$$\frac{N_{\text{obs}} - N_{\text{BG}}}{N_{\text{exp}}} = 0.593 \pm 0.020(\text{stat.}) \pm 0.026(\text{sys.}). \tag{8.43}$$

This is clear evidence for neutrino oscillations in form of a $\bar{\nu}_e$ disappearance. The best fit parameters are given as

$$\Delta m_{12}^2 = 7.58^{+0.21}_{-0.2} \times 10^{-5} \text{eV}^2 \quad \text{and} \quad \tan^2\theta_{12} = 0.56^{+0.14}_{-0.09} \tag{8.44}$$

Especially the mass difference Δm^2 is well defined. An improved measurement could be done in a phase when the reactors were off, which allowed a background measurement [Gan15] and led to the new values of

$$\Delta m_{12}^2 = 7.53 \pm 0.18 \times 10^{-5} \text{eV}^2 \quad \text{and} \quad \tan^2\theta_{12} = 0.436^{+0.029}_{-0.025}. \tag{8.45}$$

The implications of this result with respect to mixing angles and the solar neutrino problem will be discussed in Chapter 10.

8.6.1.2   Double Chooz, RENO and Daya Bay–Measurement of θ_{13}

Within the framework of the next generation experiments, a full three-flavour analysis of oscillation data including various effects like CP violation and matter effects must be done. The oscillation probably shows a degeneracy among different quantities, which has to be disentangled. As all angles θ_{13}, θ_{23} and θ_{12} show up within the product of the Jarlskog invariant (8.21), these angles have to be non-zero to allow a search for CP violation. Furthermore, the value of $\sin^2\theta_{13}$ should be larger than about 0.01 because otherwise there is a drastic change in the CP sensitivity which might prohibit its measurement. Reactor experiments provide a

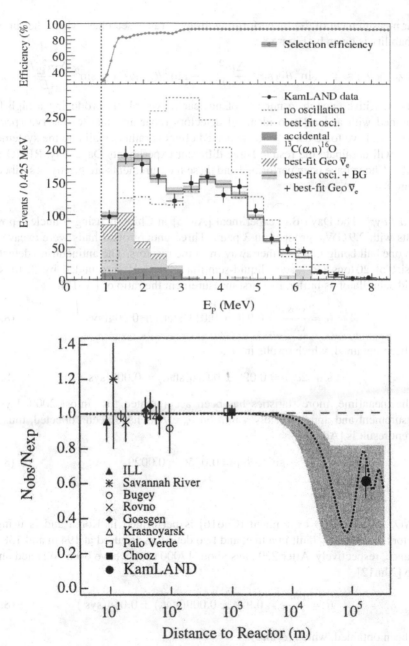

Figure 8.4. Oscillation results from KamLAND. Top: The measured positron spectrum. The deviation from the expected spectral shape can be clearly seen (from [Abe08]). © 2008 by the American Physical Society. Bottom: Ratio between observed and expected events as a function of L/E. A clear reduction with respect to short baseline reactor experiments is seen. For comparison, a theoretical oscillation curve is included (from [Egu03]). © 2003 by the American Physical Society.

clear measurement of θ_{13} by performing disappearance searches, where the survival probability is given by Figure 8.5

$$P(\bar{\nu}_e \to \bar{\nu}_e) \approx 1 - \sin^2 \theta_{13} \sin^2 \frac{\Delta m_{13}^2 L}{4E} - \sin^2 \theta_{12} \cos^4 \theta_{13} \sin^2 \frac{\Delta m_{12}^2 L}{4E}. \quad (8.46)$$

To be sensitive enough, a number of nuclear power plants producing a high flux combined with at least two identical detectors (near and far) to observe spectral distortions have to be used, as the expected effect is rather small and the systematic errors will dominate the result. Three different experiments Daya Bay, RENO and Double Chooz have been performed and these measurements are briefly described in Figure 8.4.

Daya Bay: The Daya Bay experiment [An16] in China is using 6 nuclear power plants with 2.9 GW$_{th}$ grouped in 3 pairs. Three underground halls were excavated, with one hall being much farther away from the reactors. The antineutrino detectors consist of 20 tons Gd-loaded liquid-scintillator vessels surrounded by 20 tons of liquid scintillator as buffer. In a first measurement the ratio of [An12]

$$R = \frac{\text{obs.}}{\text{exp.}} = 0.940 \pm 0.011(\text{stat.}) \pm 0.004(\text{sys.}) \quad (8.47)$$

has been obtained, which results in

$$\sin^2 2\theta_{13} = 0.092 \pm 0.016(\text{stat.}) \pm 0.005(\text{sys.}) \quad (8.48)$$

In the meantime, more statistics has been accumulated for almost 2000 days of measurement and approximately 4 million $\bar{\nu}_e$ events have been collected, thus the current result is [Ade18]

$$\sin^2 2\theta_{13} = 0.0856 \pm 0.0029 \quad (8.49)$$

RENO: The RENO experiment [Cho16] is performed in Korea and is using 6 reactors of 2.8 GW$_{th}$ built in a line, and two detectors are used at 194 m and 1383 m distance, respectively. After 229 days about 17000 events have been collected and a ratio [Ahn12]

$$R = \frac{obs.}{exp.} = 0.920 \pm 0.009(\text{stat.}) \pm 0.014(\text{sys.}) \quad (8.50)$$

has been obtained, which results in

$$\sin^2 2\theta_{13} = 0.113 \pm 0.013(\text{stat.}) \pm 0.019(\text{sys.}) \quad (8.51)$$

In the meantime much more statistics has been accumulated and the current value is given as [Bak18]

$$\sin^2 2\theta_{13} = 0.0896 \pm 0.0048(\text{stat.}) \pm 0.0047(\text{sys.}) \quad (8.52)$$

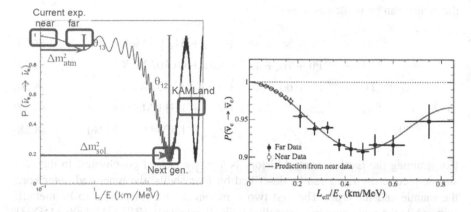

Figure 8.5. Left: Oscillation probabilities as a function of distances. For reactor neutrinos and the measurement of $\sin^2 2\theta_{13}$ the detector should be very close to the power plant while the measurement of $\sin^2 2\theta_{12}$ has a larger amplitude and is further away from it. The small amplitude suggests a measurement of at least two reactors (from [Sta15]). Right: Oscillation probability as a function of L/E exemplaric shown for the RENO experiment (from [Bak18]). © 2018 by the American Physical Society.

Double Chooz: The Double Chooz experiment in France [Abe12] is using two detectors at about 400 m and 1050 m distance from the reactor core. Their latest result is a value of [Abe16a], see also [deK19]

$$\sin^2 2\theta_{13} = 0.095^{+0.038}_{-0.039}(\text{sys.}) \tag{8.53}$$

In summary, within a relatively short time period three experiments have determined a value of $\sin^2 2\theta_{13}$ producing the most precise angle within the PMNS matrix.

As an exciting byproduct of the reactor oscillation search, for the first time neutrinos from the radioactive decays within the Earth have been observed, which are a low energy background contribution to the reactor spectra as seen in Figure 8.4, but also offer the opportunity to learn about the radioactive contribution to the heat budget of the Earth.

8.6.2   Geoneutrinos

As a very remarkable side effect to the reactor neutrino measurements, the first detection of geoneutrinos was done by KamLAND [Ara05a] and data with more statistics have been released [Abe08]. In the meantime also the Borexino experiment (see Chapter 10) has observed geoneutrinos [Bel10, Bel15]. These neutrinos are produced by the radioactive decays of very long-living isotopes, dominantly the natural decay chains of ^{238}U, ^{232}Th and ^{235}U as well as ^{40}K and other smaller contributions within the Earth (Figure 8.7). The summation of the energy within the chains can be written as [Fio07]

$$^{238}U \rightarrow {}^{206}Pb + 8\alpha + 6e^- + 6\bar{\nu}_e + 51.698 \text{ MeV} \tag{8.54}$$

$$^{235}U \rightarrow {}^{207}Pb + 7\alpha + 4e^- + 4\bar{\nu}_e + 46.402 \text{ MeV} \tag{8.55}$$

$$^{232}Th \rightarrow {}^{208}Pb + 6\alpha + 4e^- + 4\bar{\nu}_e + 42.652 \text{ MeV} \tag{8.56}$$

$$^{40}K \rightarrow {}^{40}Ca \quad + e^- \quad + \bar{\nu}_e \quad + 1.311 \text{ MeV } (89\%) \tag{8.57}$$

$$^{40}K + e^- \rightarrow {}^{40}Ar \quad\quad + \nu_e \quad + 1.505 \text{ MeV } (11\%) \tag{8.58}$$

Constraining the Earth's thermal history is a major task of geophysics. In the very simple picture the Earth can be described by four shells: the inner and outer core, the mantle and the crust. The first two components are considered to be metallic iron, while the remaining two are called bulk silicate earth (BSE) [McD99, McD03]. This model links the elemental abundance of the Earth with the one of CI chondritic meteorites due to the common origin in the formation of the solar system. The heat loss of the Earth is a balance between its secular cooling and the radiogenic heat production. The thermal heat conduction between the core and the BSE is weak, hence the major contribution comes from the BSE. The estimated total heat loss is about 47 ± 2 TW, split between the loss in the core (about 10 TW), mantle heat production (12 TW), mantle cooling (17 TW) and crustal heat production (8 TW). Out of the Bulk Silicate Earth Model various reference geoneutrino models have been created to compare and predict event rates at different locations on the Earth [Man04, Fog05, Eno05]. The typical antineutrino flux of U,Th is approximately 2×10^6 cm^{-2}s^{-1} which leads to a new unit called TNU (terrestrial neutrino unit) to be 1 event per year and 10^{32} free protons. Hence, the flux is in the same order as the solar ^{8}B neutrino flux (see Chapter 10). Experimental locations far from continents (oceans) have a higher sensitivity to the mantle, as the Earth crust is relatively thin; the opposite is valid for experiments in the middle of continents. With KamLAND and Borexino two locations have been measured, where Borexino has measured $21.2^{+9.5}_{-9.0}$(stat.)$^{+1.1}_{-0.9}$(sys.) TNU for the mantle and a total value of $47.0^{+8.4}_{-7.7}$(stat.)$^{+2.4}_{-1.9}$(sys.) TNU [Bel15, Ago20]. More measurements and thus data points will come from new experiments like SNO+, JUNO and Jingping. For more details see [Fio07, Sra12, Lud13].

8.7   Accelerator-based oscillation experiments

High-energy accelerators offer the chance for both appearance and disappearance searches. Both were and are still commonly used. Having typically much higher beam energies than reactors, they probe normally higher Δm^2 regions. However, because of the intensity of the beam, the event rate can be much higher, allowing smaller mixing angles $\sin^2 2\theta$ to be probed. Long-baseline ($L \gg 100$ km) experiments are able to extend the accelerator searches down to Δm^2 regions relevant for atmospheric neutrino studies and will be discussed in Chapter 9.

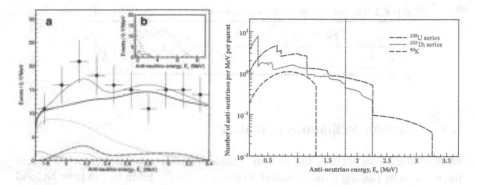

Figure 8.6. Left: Antineutrino energy spectrum as observed by KamLAND. The data points can be fitted (solid line) with five contributions. The dominant ones are reactor antineutrinos (above about 2.3 MeV) and potential fake events resulting from the $^{13}C(\alpha,n)^{16}O$ reaction (below 2.3 MeV). Additionally, two contributions from the U (dash-dotted line) and Th (dotted line) are needed to describe the bump around 2.1 MeV. At very low energy, random coincidences have to be taken into account (from [Ara05a]). Reproduced with permission of SNCSC. Right: Expected geoneutrino energy spectrum from various radioactive sources within the Earth. The vertical dotted line corresponds to the energy threshold of inverse beta decay (1.806 MeV). Using this reaction only ^{238}U and ^{232}Th can be observed (from [Ara05a]). Reproduced with permission of SNCSC.

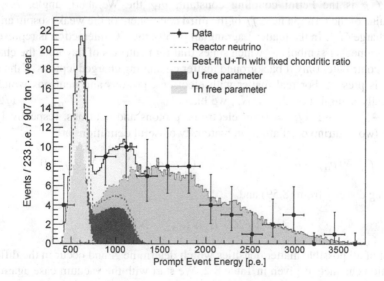

Figure 8.7. Decomposition of the various components of the spectrum of the latest Borexino measurement on geoneutrinos at Gran Sasso Laboratory in central Italy (from [Ago15a]). © 2015 by the American Physical Society, see also [Ago20].

Table 8.2. The list of matter densities relevant for two-neutrino oscillations.

	$\nu_e \to \nu_{\mu,\tau}$	$\nu_e \to \nu_s$	$\nu_\mu \to \nu_\tau$	$\nu_{\mu,\tau} \to \nu_s$
$\dfrac{A}{2\sqrt{2}EG_F}$	N_e	$N_e - \frac{1}{2}N_n$	0	$-\frac{1}{2}N_n$

8.8   Neutrino oscillations in matter

Matter effects can occur if the neutrinos under consideration experience different interactions by passing through matter. In the Sun and the Earth ν_e can have NC and CC interactions with leptons because of the existence of electrons, while for ν_μ and ν_τ only NC reactions are possible. In addition, for a ν_μ beam traversing the Earth, in the case of the existence of sterile neutrinos ν_S, there is a difference between weak reactions (ν_μ) and no weak interactions at all (ν_S), see also [Kuo89, Kim93, Sch97, Bil99].

Starting from the weak interaction Lagrangian (3.48) one gets for low-energy neutrino interactions of flavour ℓ with the background matter

$$-\mathcal{L}_{\nu_\ell} = \frac{G_F}{\sqrt{2}} \nu_\ell^\dagger (1 - \gamma_5) \nu_\ell \sum_f N_f (\delta_{\ell f} + I_{3f_L} - 2\sin^2 \theta_W Q_f) \qquad (8.59)$$

where G_F is the Fermi coupling constant, θ_W the Weinberg angle, I_{3f_L} the eigenvalue of the fermion field f_L of the third component of the weak isospin and Q_f is the charge of f. In the matter Lagrangian (8.59), the CC interaction is represented by the Kronecker symbol $\delta_{\ell f}$ which states that for neutrinos of flavour ℓ the charged current contributes only if background matter containing charged leptons of the same flavour is present. For real matter with electrons, protons and neutrons which are electrically neutral, i.e., $N_e = N_p$, we have $I_{3e_L} = -I_{3p_L} = I_{3n_L} = -1/2$ and $Q_e = -Q_p = -1$, $Q_n = 0$ for electrons, protons and neutrons, respectively. To discuss two-neutrino oscillations in matter, two useful definitions are:

$$N(\nu_\alpha) \equiv \delta_{\alpha e} N_e - \tfrac{1}{2}N_n \qquad (\alpha \equiv e, \mu, \tau) \qquad N(\nu_s) \equiv 0 \qquad (8.60)$$

following directly from (8.59) and

$$A \equiv 2\sqrt{2}G_F E(N(\nu_\alpha) - N(\nu_\beta)). \qquad (8.61)$$

The list of all possible matter densities which determine A and occur in the different oscillation channels is given in Table 8.2. We start with the vacuum case again. The time dependence of mass eigenstates is given by (8.4). Neglecting the common phase by differentiation, we obtain the equation of motion (Schrödinger equation)

$$i\frac{d\nu_i(t)}{dt} = \frac{m_i^2}{2E}\nu_i(t) \qquad (8.62)$$

which can be written in matrix notation as follows:

$$i\frac{d\nu(t)}{dt} = H^i\nu(t)$$

with

$$\nu = \begin{pmatrix} \nu_1 \\ \cdot \\ \cdot \\ \cdot \\ \nu_n \end{pmatrix} \tag{8.63}$$

and

$$H_{ij}^i = \frac{m_i^2}{2E}\delta_{ij}.$$

H^i is the Hamilton matrix ('mass matrix') in the ν_i representation and it is diagonal, i.e., the mass eigenstates in vacuum are eigenstates of H. By applying the unitary transformation

$$\nu = U^\dagger\nu' \qquad \text{with } \nu' = \begin{pmatrix} \nu_\alpha \\ \cdot \\ \cdot \\ \cdot \end{pmatrix} \tag{8.64}$$

and the mixing matrix U, the equation of motion and the Hamilton matrix H^α can be written in the representation of flavour eigenstates ν_α:

$$i\frac{d\nu'(t)}{dt} = H^\alpha\nu'(t) \qquad \text{with } H^\alpha = UH^iU^\dagger. \tag{8.65}$$

Consider the case of two neutrinos (ν_e, ν_μ): the Hamilton matrix can be written in both representations as

$$H^i = \frac{1}{2E}\begin{pmatrix} m_1^2 & 0 \\ 0 & m_2^2 \end{pmatrix}$$

$$H^\alpha = \frac{1}{2E}\begin{pmatrix} m_{ee}^2 & m_{e\mu}^2 \\ m_{e\mu}^2 & m_{\mu\mu}^2 \end{pmatrix}$$

$$= \frac{1}{2E}\begin{pmatrix} m_1^2\cos^2\theta + m_2^2\sin^2\theta & (m_2^2 - m_1^2)\sin\theta\cos\theta \\ (m_2^2 - m_1^2)\sin\theta\cos\theta & m_1^2\sin^2\theta + m_2^2\cos^2\theta \end{pmatrix}$$

$$= \frac{1}{4E}\Lambda\begin{pmatrix} 1 & 0 \\ 0 & 1 \end{pmatrix} + \frac{1}{4E}\Delta m^2\begin{pmatrix} -\cos 2\theta & \sin 2\theta \\ \sin 2\theta & \cos 2\theta \end{pmatrix} \tag{8.66}$$

with $\Lambda = m_2^2 + m_1^2$ and $\Delta m^2 = m_2^2 - m_1^2$. How does the behaviour change in matter? As already stated, the ν_e mass is modified in matter according to (using ν_e and ν_μ as examples)

$$m_{ee}^2 \to m_{eem}^2 = m_{ee}^2 + A \qquad \text{with } A = 2\sqrt{2}G_FEN_e \tag{8.67}$$

the latter following directly from (8.61). The Hamilton matrix H_m^α in matter is, therefore, given in the flavour representation as

$$H_m^\alpha = H^\alpha + \frac{1}{2E}\begin{pmatrix} A & 0 \\ 0 & 0 \end{pmatrix} = \frac{1}{2E}\begin{pmatrix} m_{ee}^2 + A & m_{e\mu}^2 \\ m_{e\mu}^2 & m_{\mu\mu}^2 \end{pmatrix}$$

$$= \frac{1}{4E}(\Lambda + A)\begin{pmatrix} 1 & 0 \\ 0 & 1 \end{pmatrix}$$

$$+ \frac{1}{4E}\begin{pmatrix} A - \Delta m^2\cos 2\theta & \Delta m^2\sin 2\theta \\ \Delta m^2\sin 2\theta & -A + \Delta m^2\cos 2\theta \end{pmatrix}. \quad (8.68)$$

The same relations hold for antineutrinos with the exchange $A \to -A$. Transforming this matrix back into the (ν_1, ν_2) representation results in

$$H_m^i = U^\dagger H_m^\alpha U = U^\dagger H^\alpha U + \frac{1}{2E}U^\dagger\begin{pmatrix} A & 0 \\ 0 & 0 \end{pmatrix}U$$

$$= H^i + \frac{1}{2E}U^\dagger\begin{pmatrix} A & 0 \\ 0 & 0 \end{pmatrix}U$$

$$= \frac{1}{2E}\begin{pmatrix} m_1^2 + A\cos^2\theta & A\cos\theta\sin\theta \\ A\cos\theta\sin\theta & m_2^2 + A\sin^2\theta \end{pmatrix}. \quad (8.69)$$

The matrix now contains nondiagonal terms, meaning that the mass eigenstates of the vacuum are no longer eigenstates in matter. To obtain the mass eigenstates (ν_{1m}, ν_{2m}) in matter and the corresponding mass eigenvalues (m_{1m}^2, m_{2m}^2) (effective masses) H_m^i must be diagonalized. This results in mass eigenstates of

$$m_{1m,2m}^2 = \frac{1}{2}\left[(\Lambda + A) \mp \sqrt{(A - \Delta m^2\cos 2\theta)^2 + (\Delta m^2)^2\sin^2 2\theta}\right]. \quad (8.70)$$

For $A \to 0$, it follows that $m_{1m,2m}^2 \to m_{1,2}^2$. Considering now a mixing matrix U_m connecting the mass eigenstates in matter $m_{1m,2m}$ with the flavour eigenstates (ν_e, ν_μ) the corresponding mixing angle θ_m is given by

$$\tan 2\theta_m = \frac{\sin 2\theta}{\cos 2\theta - A/\Delta m^2}$$

$$\sin 2\theta_m = \frac{\sin 2\theta}{\sqrt{(A/\Delta m^2 - \cos 2\theta)^2 + \sin^2 2\theta}}. \quad (8.71)$$

Here again, for $A \to 0$, it follows that $\theta_m \to \theta$. Using the relation

$$\Delta m_m^2 = m_{2m}^2 - m_{1m}^2 = \Delta m^2\sqrt{\left(\frac{A}{\Delta m^2} - \cos 2\theta\right)^2 + \sin^2 2\theta} \quad (8.72)$$

the oscillation probabilities in matter can be written analogously to those of the vacuum:

$$P_m(\nu_e \to \nu_\mu) = \sin^2 2\theta_m \times \sin^2\frac{\Delta m_m^2}{4} \times \frac{L}{E} \quad (8.73)$$

$$P_m(\nu_e \to \nu_e) = 1 - P_m(\nu_e \to \nu_\mu) \quad (8.74)$$

with a corresponding oscillation length in matter:

$$L_m = \frac{4\pi E}{\Delta m_m^2} = \frac{L_0}{\sqrt{\left(\frac{A}{\Delta m^2} - \cos 2\theta\right)^2 + \sin^2 2\theta}} = \frac{\sin 2\theta_m}{\sin 2\theta} L_0. \tag{8.75}$$

Note already here that (8.68) allows the possibility of maximal mixing in matter, $\sin 2\theta_m \approx 1$, even for small $\sin \theta$ because of the resonance type form.

8.9 Future activities – Determination of the PMNS matrix elements

Having established neutrino oscillations, one of the major goals now is to determine the PMNS matrix elements more precisely and search for a possible CP violation in the lepton sector. Obviously, this requires a full three-flavour analysis of all the available data. The expressions for the three-flavour oscillation probabilities including matter effects are quite complex, as an example $P(\nu_\mu \rightarrow \nu_e)$ can be expressed as

$$P(\nu_\mu \rightarrow \nu_e) = P_1 + P_2 + P_3 + P_4 \tag{8.76}$$

with

$$P_1 = \sin^2 \theta_{23} \sin^2 2\theta_{13} \left(\frac{\Delta_{13}}{B_\pm}\right)^2 \sin^2 \frac{B_\pm L}{2} \tag{8.77}$$

$$P_2 = \cos^2 \theta_{23} \sin^2 2\theta_{12} \left(\frac{\Delta_{12}}{A}\right)^2 \sin^2 \frac{AL}{2} \tag{8.78}$$

$$P_3 = J \cos \delta \left(\frac{\Delta_{13}}{B_\pm}\right) \left(\frac{\Delta_{12}}{A}\right) \cos \frac{\Delta_{13} L}{2} \sin \frac{AL}{2} \sin \frac{B_\pm L}{2} \tag{8.79}$$

$$P_4 = \mp J \sin \delta \left(\frac{\Delta_{13}}{B_\pm}\right) \left(\frac{\Delta_{12}}{A}\right) \cos \frac{\Delta_{13} L}{2} \sin \frac{AL}{2} \sin \frac{B_\pm L}{2} \tag{8.80}$$

with $\Delta_{ij} = \Delta m_{ij}^2/(2E)$, $J = \cos \theta_{13} \sin 2\theta_{12} \sin 2\theta_{13} \sin 2\theta_{23}$, $A = \sqrt{2} G_F n_e$ (matter effect) and $B_\pm = |A \pm \Delta_{13}|$. The +(-) sign is for neutrinos (antineutrinos). Thus, the three angles and one phase in the PMNS matrix have to be measured with higher accuracies. The final task now is to investigate CP violation in the lepton sector. As in the quark sector, the CP-phase is always part of a product with all mixing angles and the Jarlskog invariant J (see (8.21)). A problem hereby arises in the form of parameter degeneracy. Assuming that all mixing parameters except θ_{13} and δ are known, and a precise measurement of $P(\nu_\mu \rightarrow \nu_e)$ and $P(\bar{\nu}_\mu \rightarrow \bar{\nu}_e)$ has been performed, there is still a situation where four different solutions can be found (two for CP-even, two for CP-odd) [Bur01, Bar02]. The only chance to remove the ambiguities is to perform either an experiment at two different energies or baselines or to combine two different experiments. In matter, the measurement of CP violation can become more complicated because the oscillation probabilities for neutrinos and antineutrinos are, in general, different in matter, even if $\delta = 0$. Indeed, the matter

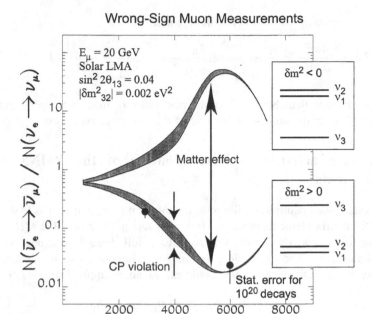

Figure 8.8. Possibilities for observing CP violation and matter effects using beams from a neutrino factory by using wrong sign muons. Matter effects start to significantly split in two bands if a detector is at least 1000 km away from the source. The two bands correspond to normal and inverted mass hierarchies. The width of the band gives the size of the possible CP violation using the parameters stated. With kind permission of Fermilab.

effect can either contaminate or enhance the effect of an intrinsic CP violation effect coming from δ [Ara97, Min98, Min00, Min02]. For the case of T violation, the situation is different. If $\Delta P^T_{\alpha\beta} \neq 0$ for $\alpha \neq \beta$ would be established, then this implies $\delta \neq 0$ even in the presence of matter. The reason is that the oscillation probability is invariant under time reversal even in the presence of matter. Similar to the case of CP violation, T violation effects can either be enhanced or suppressed in matter [Par01]. However, a measurement of T violation is experimentally more difficult to perform because there is a need for a non-muon neutrino beam, like a beta beam. Additionally, matter effects also give a handle on the determination of the sign of Δm^2_{23}. A compilation of expected matter effects and CP violation is shown in Figure 8.8.

To accomplish this physics program, a variety of ideas for new beams with very high intensity has been pushed forward. As already mentioned, nuclear reactors provide a clean measurement of θ_{13} without any degeneracy problems.

8.10 New neutrinos beams

Some principals of creating neutrino beams have already been discussed in Chapter 4. The difference here is now the distance between the target and the detector and a full 3-flavour scenario.

8.10.1 Off-axis superbeams

The first realization is the T2K experiment. The newly built Japanese Hadron Facility (JHF) in Tokai is producing a 0.77 MW beam of protons with 50 GeV on a target. Super-Kamiokande is hereby used as the far detector being about 2.5 degrees off axis [Aok03]. The baseline corresponds to 295 km. The experiment has started data taking and can be updated in a second phase to 4 MW and also a 1 Mt detector (Hyper-Kamiokande). The future detector might be split in two using a site in Korea instead (T2KK). An ongoing experiment is NOνA with a baseline of about 810 km from Fermilab to the detector (see Section 4.2.2). Also the newly planned DUSEL underground facility in the USA is considered for long-baseline experiments. The idea in all experiments is to measure ν_e appearance in a ν_μ beam. The oscillation probability is directly proportional to $\sin^2 \theta_{13}$. In addition, such experiments would also allow $\sin^2 2\theta_{23}$ and Δm_{23}^2 to be measured.

8.10.2 Muon storage rings – neutrino factories

In recent years the idea to use muon storage rings to obtain high-intensity neutrino beams has become very popular, see also Chapter 4 [Gee98, Aut99, Alb00, Als03, Apo02, Ban09]. The two main advantages are the precisely known neutrino beam composition and the high intensity (about 10^{21} muons/yr should be filled in the storage ring). A conceptional design is shown in Figure 8.9. Even if many technical challenges have to be solved, it offers a unique source for future accelerator-based neutrino physics. Muon decay as neutrino source provides $\nu_e(\bar{\nu}_e)$ and $\bar{\nu}_\mu(\nu_\mu)$ as beams simultaneously. Hence, for oscillation searches in the appearance channel for $\nu_e \rightarrow \nu_\mu$ the signal will be a wrong-signed muon and thus charge identification is a crucial part of any detector concept. A compilation of search modes is given in Table 8.3.

To disentangle the 8-fold degeneracy among the parameters, several of the channels have to be used. Furthermore, a measurement at two different locations from the source and thus at two different values of L/E will help. Alternatively, binned energy spectra for a given distance will work in the same spirit. A special distance for an experiment would be the "magic baseline" L of about 7300-7600 km, because of

$$\sqrt{2}G_F n_e L = \frac{\pi}{2} \rightarrow \sin\left(\frac{AL}{2}\right) = 0 \qquad (8.81)$$

As a consequence only the term P_1 in (8.76) is non-vanishing. After discussing

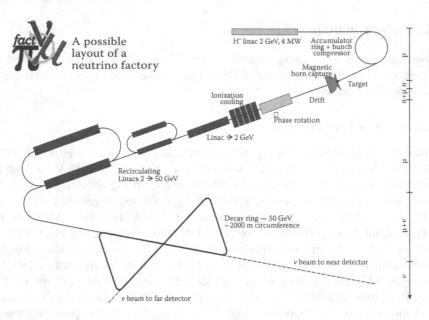

Figure 8.9. Proposed layout for a neutrino factory. The main ingredients are: a high intensity proton linac, a target able to survive the deposited energy and giving a good yield of pions, a cooling device for the decay muons, an accelerator for the muons and a storage ring allowing for muon decay and therefore neutrino beams (from [Taz03]). © IOP Publishing. Reproduced with permission. All rights reserved.

Table 8.3. The various oscillation channels available at a neutrino factory.

$\mu^+ \to e^+ \, \nu_e \bar{\nu}_\mu$	$\mu^- \to e^- \, \bar{\nu}_e \nu_\mu$	
$\bar{\nu}_\mu \to \bar{\nu}_\mu$	$\nu_\mu \to \nu_\mu$	disappearance
$\bar{\nu}_\mu \to \bar{\nu}_e$	$\nu_\mu \to \nu_e$	appearance
$\bar{\nu}_\mu \to \bar{\nu}_\tau$	$\nu_\mu \to \nu_\tau$	appearance (atm. osc.)
$\nu_e \to \nu_\mu$	$\bar{\nu}_e \to \bar{\nu}_\mu$	appearance
$\nu_e \to \nu_e$	$\bar{\nu}_e \to \bar{\nu}_e$	disappearance
$\nu_e \to \nu_\tau$	$\bar{\nu}_e \to \bar{\nu}_\tau$	appearance

neutrino oscillations and first experiments we now move on to atmospheric neutrinos.

Chapter 9

Atmospheric neutrinos

DOI: 10.1201/9781315195612-9

In the last three decades the study of atmospheric neutrinos has been one of the most important fields in neutrino physics. Atmospheric neutrinos are produced in meson and muon decays, created by interactions of cosmic rays within the atmosphere. The study of these neutrinos revealed evidence for neutrino oscillations [Fuk98]. With dominantly energies in the GeV range and baselines from about 10 km to as long as the Earth diameter ($L \approx 10^4$ km) neutrino mass differences in the order of $\Delta m^2 \gtrsim 10^{-4}$ eV^2 or equivalent values in the L/E ratio from 10–10^5 km GeV^{-1} are probed. Most measurements are based on relative quantities because absolute neutrino flux calculations are still affected by large uncertainties. The obtained results depend on four factors: the primary cosmic-ray flux and its modulations, the production cross-sections of secondaries in atmospheric interactions, the neutrino interaction cross-section in the detector and the detector acceptance and efficiency. More quantitatively, the observed number of events is given by [Gai16]

$$\frac{\mathrm{d}N_l(\theta, p_l)}{\mathrm{d}\Omega_\theta \, \mathrm{d}p_l} = t_{\mathrm{obs}} \sum_{\pm} \int N_t \frac{\mathrm{d}\phi_{\nu_l}^{\pm}(E_\nu, \theta)}{\mathrm{d}\Omega_\theta \, \mathrm{d}E_\nu} \frac{\mathrm{d}\sigma^{\pm}(E_\nu, p_l)}{\mathrm{d}p_l} F(q^2) \, \mathrm{d}E_\nu \qquad (9.1)$$

where l stands for $e^{\pm}$ or $\mu^{\pm}$, p_l the lepton momentum, E_ν the neutrino energy, θ the zenith angle, t_{obs} the observation time, N_t the number of target particles, $\phi_{\nu_l}^{\pm}(E_\nu, \theta)$ the neutrino flux and $\sigma(E_\nu, p_l)$ the cross-section. $F(q^2)$ takes into account the dependence of momentum transfer in nuclear effects such as the Fermi momenta of the target nucleons, Pauli blocking of recoil nucleons, etc., see Chapter 4. The summation ($\pm$) is done for ν_l and $\bar{\nu}_l$, since most of the current observations do not distinguish the lepton charge. For further literature see [Ber90a, Lon92, 94, Gri01, Jun01, Lea01, Lip01, Kaj01, Sta04, Kaj14, Gai16, Spu18, Kac19]. The first two steps are now discussed in a little more detail.

9.1 Cosmic rays

The charged component of the primary cosmic rays hitting the atmosphere consist of about 98% hadrons and 2% electrons. The hadronic component itself is dominated by

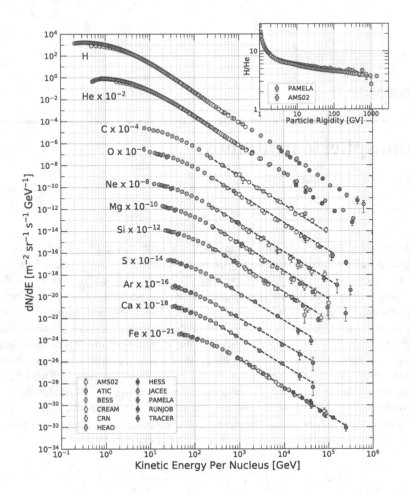

Figure 9.1. Fluxes of nuclei of the primary cosmic radiation in particles per energy and per nucleus are plotted versus energy per nucleus. The inset shows the H/He ratio at constant rigidity. The lower left legend shows the contributing experiments (from [PDG18]). With kind permission from M. Tanabashi et al. (Particle Data Group).

protons ($\approx 87\%$) mixed with α-particles ($\approx 11\%$) and heavier nuclei ($\approx 2\%$) [Gri01]. The chemical composition has been determined by several experiments in an energy range up to 100 TeV (Figure 9.1). For higher energies, indirect methods like air showers are used. As the neutrino flux depends on the number of nucleons rather than on the number of nuclei, a significant fraction of the flux is produced by He, C, N, O and heavier nuclei. The differential energy spectrum follows a power law of the form

$$N(E)\, dE \propto E^{-\gamma}\, dE \tag{9.2}$$

with $\gamma \simeq 2.7$ for energies of a few up to about 10^5 GeV, suggesting that they are produced by non-thermal processes. From this point onward, the spectrum steepens (the 'knee') to $\gamma \simeq 3$. The exact position of the knee depends on the atomic number A as was shown by KASCADE and KASCADE-Grande, with lighter nuclei showing the knee at lower energies [Swo02, Hoe03, Hoe04, Ant05, Ber08, Blü09]. At about 10^9 GeV the spectrum flattens again (the 'ankle') as measured by AUGER. This ultra-high-energy part of cosmic rays will be discussed in more detail in Chapter 12. The part of the cosmic-ray spectrum which is dominantly responsible for the current atmospheric neutrino investigations is in the energy range below 1 TeV. In the GeV energy range several effects can occur. First of all, there is the modulation of the primary cosmic-ray spectrum with solar activity. An indicator of the latter is the sunspot number. The solar wind prohibits low-energy galactic cosmic rays from reaching the Earth, resulting in an 11 yr anti-correlation of cosmic-ray intensity with solar activity. This effect is most prominent for energies below 10 GeV. Such particles have, in contrast, a rather small effect on atmospheric neutrino fluxes, because the geomagnetic field prevents these low-energy particles from entering the atmosphere. The geomagnetic field bends the trajectories of cosmic rays and determines the minimum rigidity called the cutoff rigidity (for an extensive discussion on this quantity see [Hil72]) for particles to arrive at Earth [Lip00a]. The dynamics of any high energy particle in a magnetic field configuration B depends on this rigidity R given by

$$R = \frac{pc}{ze} = r_L \times B \tag{9.3}$$

with p as the relativistic 3-momentum, z as the electric charge and r_L as the gyroradius. Particles with different masses and charge but identical R show the same dynamics in a magnetic field. The cutoff rigidity depends on the position at the Earth's surface and the arrival direction of the cosmic ray. Figure 9.2 shows a contour map of the calculated cutoff rigidity at Kamioka (Japan) [Ric16], where Super-Kamiokande is located. Therefore, the geomagnetic field produces two prominent effects: the latitude (the cosmic-ray flux is larger near the geomagnetic poles) and the east–west (the cosmic-ray flux is larger for east-going particles) effect. The last one is an azimuthal effect not depending on any new physics and can be used to check the air shower simulations [Lip00b]. Such a measurement was performed by Super-Kamiokande [Fut99]. With a statistics of 45 kt $\times$ yr and cuts on the lepton momentum (400 MeV/$c < p_l < $ 3000 MeV/c) and zenith angle ($|\cos\theta| < 0.5$) to gain sensitivity, an east–west effect is clearly visible (Figure 9.3).

For higher energetic neutrinos up to 100 GeV, the primary energy is up to 1 TeV, where the details of the flux are not well measured.

9.2 Interactions within the atmosphere

The atmospheric neutrinos stem from the decay of secondary particles produced in hadronic interactions of primary cosmic rays with the atmosphere, dominantly pions.

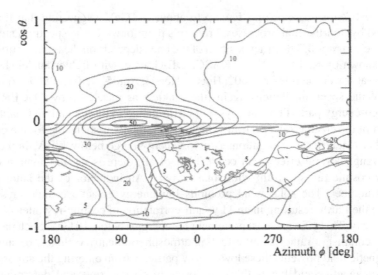

Figure 9.2. Contour map of the cutoff rigidity numbers (in GeV) relevant for Kamioka (from [Ric16]). © 2016 by the American Physical Society.

Their dominant decay chains are

$$\pi^+ \rightarrow \mu^+ \nu_\mu \qquad \mu^+ \rightarrow e^+ \nu_e \bar{\nu}_\mu \tag{9.4}$$

$$\pi^- \rightarrow \mu^- \bar{\nu}_\mu \qquad \mu^- \rightarrow e^- \bar{\nu}_e \nu_\mu. \tag{9.5}$$

Depending on the investigated neutrino energy additional contributions come from kaon decay, especially the decay modes

$$K^\pm \rightarrow \mu^\pm \nu_\mu (\bar{\nu}_\mu) \tag{9.6}$$

$$K_L \rightarrow \pi^\pm e^\pm \nu_e (\bar{\nu}_e). \tag{9.7}$$

The latter, so-called K_{e3} decay, is the dominant source for ν_e above $E_\nu \approx 1\,\text{GeV}$. In the low energy range ($E_\nu \approx 1\,\text{GeV}$) there is the previously mentioned contribution from muon decay. However, for larger energies the Lorentz boost for muons is high enough to reach the Earth surface. For example, most muons are produced in the atmosphere at about 15 km. This length corresponds to the decay length of a 2.4 GeV muon, which is shortened to 8.7 km by energy loss (a vertical muon loses about 2 GeV in the atmosphere by ionization loss according to the Bethe–Bloch formula). Therefore, at E_ν larger than several GeV this component can be neglected. At higher energies the contribution of kaons becomes even more important. To describe this precisely, a more detailed knowledge of pion and kaon production in proton-nucleus collisions is necessary. For very high energies, the prompt neutrinos from charmed meson decays in the atmosphere becomes an important component [Enb08].

Several groups have performed simulations to calculate the atmospheric neutrino flux [Agr96, Gai02, Hon04, Bar06, Gai16]. The general consensus of all these

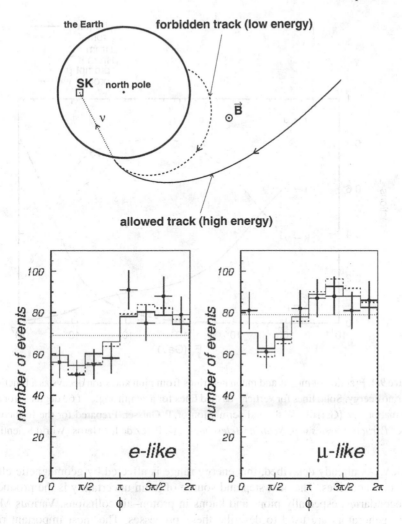

Figure 9.3. Top: Schematic explanation for the occurrence of the east–west effect. Bottom: The east–west effect as observed with Super-Kamiokande (from [Fut99]). © 1999 by the American Physical Society.

studies is that the ratio of fluxes

$$R = \frac{\nu_e + \bar{\nu}_e}{\nu_\mu + \bar{\nu}_\mu} \tag{9.8}$$

can be predicted with an accuracy of about 5 % because several uncertainties cancel. However, in the absolute flux predictions there is some disagreement on the level of 20–30 % in the spectra and overall normalization of the neutrino flux. Let us investigate the differences in more detail. The fluxes for 'contained events' (see Section 9.3) are basically produced by cosmic primaries with energies below about

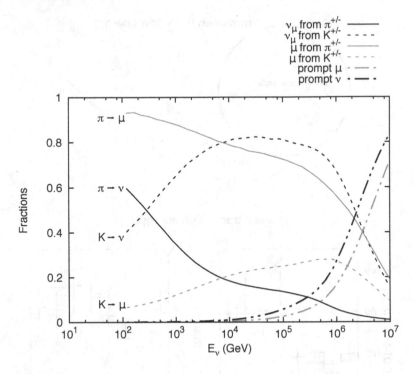

Figure 9.4. Fraction of muons and muon neutrinos from pion and kaon decays as a function of neutrino energy. Solid lines for vertical, dashed lines for a zenith angle of 60 degrees (for more information see [Gai16]). With kind permission of T. Gaisser. Prepared for the forthcoming book *"Particle Physics with Neutrino Telescopes"*, C. Pérez de los Heros (World Scientific).

20 GeV. As already described, this energy range is affected by geomagnetic effects and solar activities. The next step and source of main uncertainty is the production of secondaries, especially pions and kaons in proton–air collisions. Various Monte Carlo generators are used to describe these processes. The most important range of interaction energies for production of neutrinos with energies between 300 MeV and 3 GeV is a primary energy between $5\,\mathrm{GeV} < E_N < 50\,\mathrm{GeV}$ where E_N is the total energy of the incident nucleon in the laboratory system. In general, the primary energy is typically an order of magnitude higher than the corresponding neutrino energy. The production of secondary mesons from proton interactions using the cascade equations is often described by Z-factors (Figure 9.4), given here as an example for a proton and a π^+ (for more details see [Gai16])

$$Z_{p\pi^+} = \int_0^1 x^{\gamma-1} \frac{\mathrm{d}n_{\pi^+}(x, E_N)}{\mathrm{d}x} \mathrm{d}x \qquad (9.9)$$

where $x = E_\pi/E_N$, E_π is the energy of the produced pion and γ as given in (9.2). Analogous factors can be derived for other secondaries like Z_{pK^+}. There is still a lack of experimental data to describe the Z-factors accurately. Furthermore, past accelerator experiments have only measured pion production in pp-collisions and p-Be collisions. They have to be corrected to p-air collisions. The transformation to heavier nuclei with the use of an energy-independent enhancement factor is a further source of severe uncertainty.

Two new experimental approaches might help to improve the situation considerably. First of all, there are better measurements of muons in the atmosphere. Strongly connected with neutrino production from meson decay is the production of muons. Assume the two-body decay of a meson $M \to m_1 + m_2$. The magnitude of the momenta of secondaries in the rest frame of M are then given by

$$p_1^* = p_2^* = p^* = \frac{M^4 - 2M^2(m_1^2 + m_2^2) + (m_1^2 - m_2^2)^2}{2M}. \tag{9.10}$$

In the laboratory frame the energy of the decay product is

$$E_i = \gamma E_i^* + \beta \gamma p^* \cos \theta^* \tag{9.11}$$

where β and γ are the velocity and Lorentz factor of the parent in the laboratory system. Therefore, the limits on the laboratory energy of the secondary i are

$$\gamma(E_i^* - \beta p^*) \leq E_i \leq \gamma(E_i^* + \beta p^*). \tag{9.12}$$

In the absence of polarization there is, in addition,

$$\frac{dn}{d\Omega^*} = \frac{dn}{2\pi \, d\cos\theta^*} \propto \frac{dn}{dE_i} = \text{const.} \tag{9.13}$$

meaning that, in such cases, a flat distribution for a product of a two-body decay between the limits of (9.12) results. For example, for this process this results in (see also Chapter 4)

$$\frac{dn}{dE_\nu} = \frac{dn}{dE_\mu} = \frac{0.635}{1 - (m_\mu^2/m_K^2)p_K} \tag{9.14}$$

with p_K as the laboratory momentum of the kaon and the factor 0.635 stems from the branching ratio of decay (9.6). Often we deal with decays of relativistic particles, resulting in $\beta \to 1$, which would imply for decays like $M \to \mu\nu$ kinematic limits on the laboratory energies of the secondaries of

$$E \frac{m_\mu^2}{m_M^2} \leq E_\mu \leq E \tag{9.15}$$

and

$$0 \leq E_\nu \leq \left(1 - \frac{m_\mu^2}{m_M^2}\right) E, \tag{9.16}$$

with E as the laboratory energy of the decay meson. Average values are:

$$\langle E_\mu \rangle / E_\pi = 0.79 \quad \text{and} \quad \langle E_\nu \rangle / E_\pi = 0.21 \quad \text{for } \pi \to \mu\nu \tag{9.17}$$

$$\langle E_\mu \rangle / E_K = 0.52 \quad \text{and} \quad \langle E_\nu \rangle / E_K = 0.48 \quad \text{for } K \to \mu\nu. \tag{9.18}$$

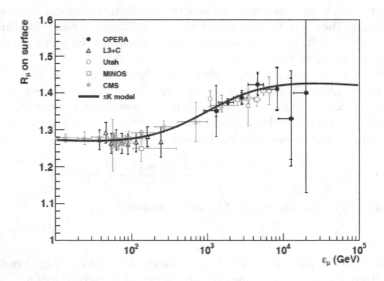

Figure 9.5. Compilation of data for the muon charge ratio at surface from 10 GeV up to 100 TeV. Data are taken from OPERA [Aga14], L3+C [Ach04], Utah [Ash75], MINOS [Ada07, Ada11] and CMS [Kha10]. The solid curve corresponds to Monte Carlo expectation (from [Aga14]). Reproduced with permission of SNCSC.

It is a consequence of the kinematics that if one of the decay products has a mass close to the parent meson, it will carry most of the energy.

There are several measurements of atmospheric muon fluxes, i.e., those by CAPRICE [Boe99], AMS on the ISS [Alc00] and BESS [San00], which are in agreement with each other at a level of ±5%. Other important measurements have been obtained at high altitude (10–30 km) during the ascent of stratospheric balloons by the MASS, CAPRICE, HEAT and BESS detectors. Since low-energy muons are absorbed in the atmosphere and decay with a high probability ($c\tau_\mu \approx 6.3 p_\mu$[GeV] km) only these high altitude measurements allow a precise measurement of muons that are most strictly associated with sub-GeV neutrino events. Also the data set of underground measurements of muon charge ratios has been significantly improved and extended to 100 TeV energies by MINOS and OPERA [Ada07] as shown in Figure 9.5.

A second important step is new measurements on secondary particle yields at accelerators. The first to mention is the HARP experiment at CERN [Har99]. Several targets were used, among them are nitrogen and oxygen targets for beam energies in the range of 1-15 GeV.

Follow-up experiments such as NA49 [Alt07, Abg14] and NA61/SHINE [Adu17] provided a large amount of data for calculations (Figure 9.6). Also proton–oxygen collisions were directly measured with high accuracy [Cat08]. The gap in between low and high energy data was bridged by the MIPP experiment at Fermilab

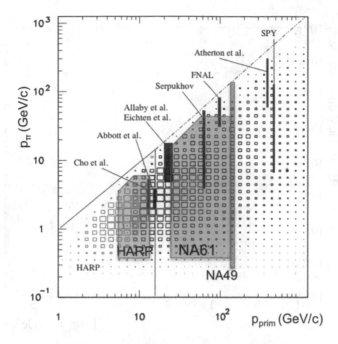

Figure 9.6. Summary of pion momenta as a function of primary and secondary proton momenta (describing the phase space). The vertical bands represent accelerator based experiments. With kind permission of T. Gaisser, adapted from [Bar06]. Prepared for the forthcoming book *"Particle Physics with Neutrino Telescopes"*, C. Pérez de los Heros (World Scientific).

[Raj05] taking data from 5-120 GeV/c.

A compilation of various measured atmospheric neutrino fluxes are shown in Figure 9.7. As can be seen it consists basically of ν_μ and ν_e neutrinos. At very high energies ($E_\nu \gg$ TeV) neutrinos from charm production become an additional source [Thu96]. A possible atmospheric ν_τ flux is orders of magnitude lower than the ν_μ flux. Now with the flux at hand, we discuss the experimental observation.

9.3   Experimental status

Relevant neutrino interaction cross-sections for detection have already been discussed in Chapter 4. The observed ν_μ events can be divided by their experimental separation into contained (fully and partially), stopping, through-going and upward-going events. Basically two types of experiments have been done using either Cherenkov detection or calorimetric tracking devices. Due to its outstanding role in the field, the Super-Kamiokande detector as a Cherenkov detector is described in a little more detail. For a discussion of former experiments see [Fuk94] (Kamiokande), [Bec92] (IMB), [Kaf94] (Soudan2), [Ber90] (Frejus), [Agl89] (Nusex) and MACRO

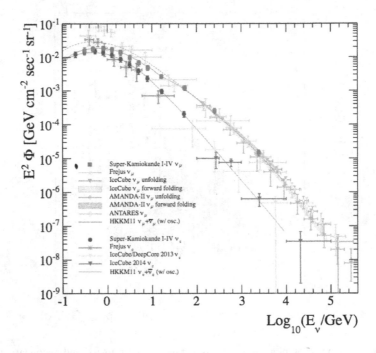

Figure 9.7. Comparison of atmospheric neutrino flux measurements of various experiments (from [Ric16]). © 2016 by the American Physical Society.

[Amb01, Amb02].

9.3.1 Super-Kamiokande

Super-Kamiokande is a water Cherenkov detector containing 50 kt of ultra-pure water in a cylindrical stainless steel tank [Fuk03] (Figure 9.8). The tank is 41.4 m high and 39.3 m in diameter and separated into two regions: a primary inner volume viewed by 11 146 photomultiplier tubes (PMTs) of 50 inch diameter and a veto region, surrounding the inner volume and viewed by 1885 PMTs of 20 inches. For analysis an inner fiducial volume of 22.5 kt is used. Neutrino interactions occurring inside the fiducial volume are called contained events. Fully-contained (FC) events are those which have no existing signature in the outer veto detector and comprise the bulk of the contained event sample. In addition, a partially-contained (PC) sample is identified in which at least one particle (typically an energetic muon) exits the inner detector. The FC sample is further divided into sub-GeV ($E_{\text{vis}} < 1.33$ GeV) and multi-GeV ($E_{\text{vis}} > 1.33$ GeV), where E_{vis} is the total visible energy in the detector (Figure 9.9). The events are characterized as either showering (e-like) or non-showering (μ-like) based on the observed Cherenkov light pattern. Two examples are shown in Figure 9.10. Criteria have been developed to distinguish between both and were confirmed by accelerator beams [Kas96]. Due to its long-running time and

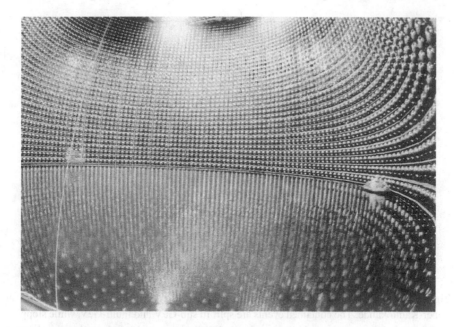

Figure 9.8. Photograph of the Super-Kamiokande detector during filling (from [Sup19]). With permission of the Kamioka Observatory, ICRR (Institute for Cosmic Ray Research) and the University of Tokyo.

associated modifications the data taking in Super-Kamiokande has been split into four phases, lasting from 1996-2001, 2003-2005, 2006-2008 and from 2009 to 2018.

9.3.1.1 The ν_μ/ν_e ratio

Historically important for any hint of neutrino oscillation was the R-ratio defined as observed versus expected ratio of events

$$R = \frac{[N(\mu\text{-like})/N(e\text{-like})]_{\text{obs}}}{[N(\mu\text{-like})/N(e\text{-like})]_{\text{exp}}}. \tag{9.19}$$

Here the absolute flux predictions cancel and if the observed flavour composition agrees with expectation, then $R = 1$. Therefore, any deviation of R from 1 is a hint for possible oscillations, even if it cannot be judged without additional information whether ν_μ or ν_e are responsible. A compilation of R-values is given in Table 9.1. As can be seen, besides Frejus and Nusex all other data sets prefer an R-value different from 1 and centre around $R = 0.6$. More convincing evidence has been found by investigating the zenith-angle dependence of the observed electron and muon events separately.

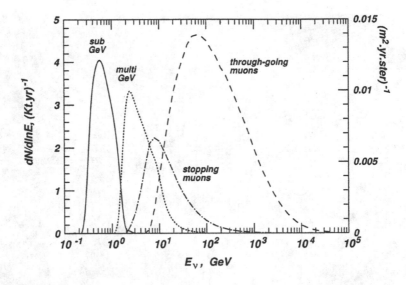

Figure 9.9. Distributions of neutrino energies that give rise to four classes of events at Super-Kamiokande. The contained events are split in sub-GeV and multi-GeV, while stopping and through-going muons refer to neutrino-induced muons produced outside the detector (from [Gai02]). © 2002 Annual Review of Nuclear and Particle Science.

Table 9.1. Compilation of existing R measurements since 1995. The statistics is clearly dominated by Super-Kamiokande. The no-oscillation case corresponds to $R = 1$.

Experiment	R	Stat. significance (kT × y)
Super-Kamiokande (sub-GeV)	$0.638 \pm 0.017 \pm 0.050$	79
Super-Kamiokande (multi-GeV)	$0.675 \pm^{0.034}_{0.032} \pm 0.080$	79
Soudan2	$0.69 \pm 0.10 \pm 0.06$	5.9
IMB	$0.54 \pm 0.05 \pm 0.11$	7.7
Kamiokande (sub-GeV)	$0.60^{+0.06}_{-0.05} \pm 0.05$	7.7
Kamiokande (multi-GeV)	$0.57^{+0.08}_{-0.07} \pm 0.07$	7.7
Frejus	$1.00 \pm 0.15 \pm 0.08$	2.0
Nusex	$0.96^{+0.32}_{-0.28}$	0.74

9.3.1.2 Zenith-angle distributions

Neutrinos are produced everywhere in the atmosphere and can, therefore, reach a detector from all directions. Those produced directly above the detector, characterized by a zenith angle $\cos\theta = 1$, have a typical flight path of about 10 km, while those coming from the other side of the Earth ($\cos\theta = -1$) have to travel more than 12 000 km before entering a detector and might interact. Since

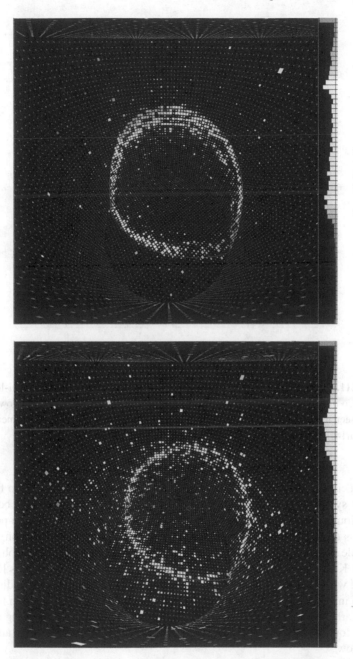

Figure 9.10. Two characteristic events as observed in Super-Kamiokande: top, sharp Cherenkov ring image produced by a muon; bottom, Cherenkov ring image produced by an electron, which is more diffuse due to multiple scattering (from [Sup19]). With permission of the Kamioka Observatory, ICRR (Institute for Cosmic Ray Research) and the University of Tokyo.

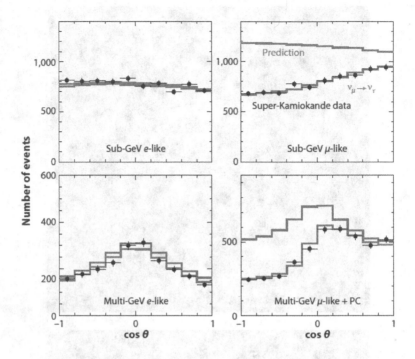

Figure 9.11. Super-Kamiokande zenith-angle distribution for e-like (left) and μ-like events (right), also divided into sub-GeV (upper row) and multi-GeV samples (lower row). A clear deficit is seen in the upward-going muons. The data corresponds to 3903 days of measurement (from [Kaj14]). © Annual Review of Nuclear and Particle Science.

the production in the atmosphere is isotropic we can expect the neutrino flux to be up/down symmetric. Slight modifications at low energies are possible because of the previously mentioned geomagnetic effects. Such an analysis can be performed as long as the created charged lepton (e,μ) follows the neutrino direction, which is reasonable for momenta larger than about 400 MeV. In the combined phases I-IV of Super Kamiokande far more than 40 000 atmospheric neutrino events have been observed with a total statistics of 328 kt×yr [Abe18b]. The zenith angle distribution is shown in Figure 9.11. It is obvious that, in contrast to e-like data which follow the Monte Carlo prediction, there is a clear deficit in the μ-like data becoming more and more profound for zenith angles smaller than horizontal, meaning less ν_μ are coming from below.

An independent check of the results from contained events can be done with upward-going muons. Upward-going events are classified as $\cos\theta < 0$. They are produced by neutrinos interacting in the rock below the detector producing muons which traverse the complete detector from below. The typical neutrino energy is about 100 GeV. Lower energetic neutrinos produce upward going stopping muons

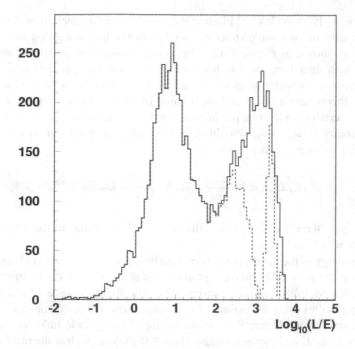

Figure 9.12. L/E double-bump structure. The bump at low values corresponds to downward-going events, the one at high L/E is due to upward-going events (from [Bat99]). With kind permission of G. Battistoni and P. Lipari.

and their energy is comparable to the PC events. This contains two implications. First, the overall expected suppression is larger in this case, since the L/E argument of the oscillation probability is larger. Second, even neutrinos from the horizon will experience significant oscillation. The ratio stopping/through-going events can also be used to remove the normalization uncertainty. Upward-through-going muons have to be compared directly with absolute flux predictions. Now let us take a closer look into the oscillation analysis.

9.3.1.3 Oscillation analysis

All data sets (FC, PC, stopping upward muons and through-going upward muons) are divided into angular bins and their distributions are analysed. Furthermore, the FC events are also binned in energy. In the common fit, the absolute normalization is allowed to vary freely and other systematic uncertainties are taken into account by additional terms, which can vary in the estimated ranges. The best-fit value obtained in a two flavour analysis is $\Delta m^2 = 2.1 \times 10^{-3}$ eV^2 and maximal mixing, having a $\chi_r^2 = 468/420$ degrees of freedom.

A very important check of the oscillation scenario can be done by plotting the L/E ratio. The L/E ratio for atmospheric neutrinos varies over a large range from about 1–10^5 km GeV^{-1}. Plotting the event rate as a function of L/E results in a characteristic two-bump structure, corresponding to down-going and up-going particles as shown in Figure 9.12. The valley between is populated mostly by particles with directions close to horizontal; the event rate per unit L/E is lower here because the neutrino path length L changes rapidly with the zenith angle θ. However, this structure is smeared out because of the imperfect energy measurement and the uncertainty in the real production point of the neutrino. According to (8.24) the probability $P\left(\nu_\mu \rightarrow \nu_\mu\right)$ should show an oscillatory behaviour with minima for L/E ratios and n as an integer number of

$$L/E = n \times \frac{2\pi}{\Delta m^2} = n \times \frac{1236}{\Delta m^2_{-3}} \ \text{km GeV}^{-1} \tag{9.20}$$

with Δm^2_{-3} as the value of Δm^2 in units of 10^{-3} eV^2. Obviously, the first minimum occurs for $n = 1$.

The energy of the neutrino is determined by a correction to the final-state lepton momentum. At $p = 1$ GeV/c the lepton carries about 85% of the neutrino energy, while at 100 MeV/c it typically carries 65%. The flight distance L is determined following [Gai98] using the estimated neutrino energy and the reconstructed lepton direction and flavour. Figure 9.13 shows the data/Monte Carlo ratio for FC data as a function of L/E and momenta larger than 400 MeV/c. A clear decrease in μ-like events can be seen; however, the oscillation pattern cannot be resolved because of the previously mentioned uncertainties in energy measurements. So for large L/E a muon neutrino has undergone numerous oscillations and these averages out to roughly 50% of the initial rate.

There is an additional check on the oscillation scenario by looking at the zenith-angle distribution of upward-going muons and compare it with absolute flux predictions. As can be seen in Figure 9.14, a deficit is also visible here and an oscillation scenario describes the data reasonably well.

Having established a ν_μ disappearance, the question concerning the reason for the deficit arises. Scenarios other than oscillations such as neutrino decay [Bar99], decoherence [Fog03a], flavour-changing neutral currents [Gon99] or violation of the equivalence principle [Hal96] have been proposed. However, they all show a different L/E behaviour and can more or less be ruled out. To prove that the oscillation really is dominantly going into ν_τ was shown by the OPERA experiment (see Section 9.4.3) and was confirmed by Super Kamiokande excluding a non-tau appearance with high significance [Li18].

Last but not least, there could be matter effects because a possible sterile neutrino ν_S does not interact at all, resulting in a different effective potential from that of ν_μ as described in Chapter 7. Density profiles of the Earth, relevant for the prediction, can be calculated using the Earth model. The Earth can be described in a simplified way as a 2-component system: the crust and the core. The crust has an average density of $\rho = 3$ g cm^{-3} and an electron fraction/nucleon of $Y_e = 0.5$ (see Chapter 8). However, for large distances $\rho = \rho(x)$ must be used. For the core, the density increases up to $\rho = 13$ g cm^{-3} and we can use a step function to describe

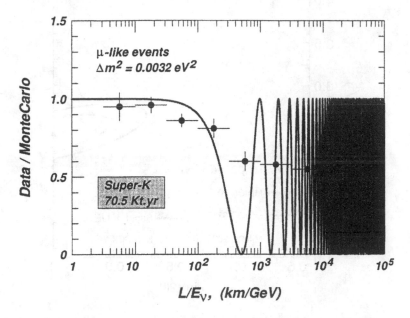

Figure 9.13. Oscillation probability as a function of L/E for the given Δm^2 of the earlier Super-Kamiokande data (from [Gai02]). © Annual Review of Nuclear and Particle Science.

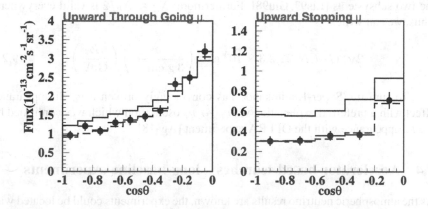

Figure 9.14. Super-Kamiokande stopping and upward-going muons flux as function of the zenith angle. Left: Flux of through-going muons from horizontal ($\cos\theta = 0$) to vertical upward ($\cos\theta = -1$). Right: Upward-going muons which stop in the detector. Also shown are Monte Carlo expectations without oscillations and best-fit values assuming oscillations (from [Tos01]).

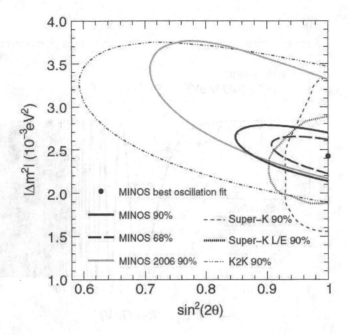

Figure 9.15. Compilation of all observed oscillation evidences from Super-Kamiokande, K2K and MINOS (from [Ada08]). © 2008 by the American Physical Society. A more recent compilation can be found at [Ada13b].

the two subsystems [Lis97, Giu98]. Furthermore, $N_n \approx N_e/2$ is valid everywhere. Thus, we can write

$$2\sqrt{2}G_F E N_e \simeq 2.3 \times 10^4 \text{ eV}^2 \left(\frac{\rho}{3 \text{ g cm}^{-3}}\right)\left(\frac{E}{\text{GeV}}\right). \qquad (9.21)$$

To sum up, Super-Kamiokande has convincingly proven a ν_μ disappearance effect with a preferred explanation via $\nu_\mu \to \nu_\tau$ oscillation which was confirmed by the ν_τ appearance with the OPERA experiment [Aga18].

9.4   Accelerator-based searches – long-baseline experiments

As the atmospheric neutrino results are known, the experiments could be located with a certain distance from the source to cover at least a tiny fraction of the earth diameter and thus a signal. This led to the concept of long-baseline experiments where the source and detectors are more than 100 km away from each other. Two strategies have been followed using accelerator neutrino beams. First of all, experiments should confirm a ν_μ disappearance and, second, perform a ν_e and ν_τ appearance search (see Chapter 4). The latter has to deal with smaller statistics because of the τ - production threshold of 3.5 GeV and, therefore, a reduced cross-section as well as the involved

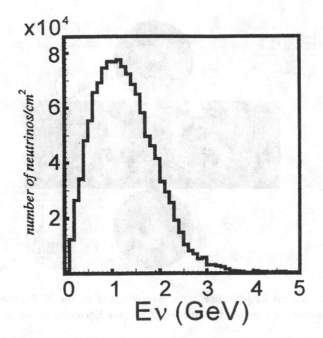

Figure 9.16. Neutrino energy spectrum of the K2K neutrino beam. Because of the relatively low beam energy, no ν_τ appearance searches can be performed.

efficiency for τ-detection. In the following some experiments in chronological order will be described.

9.4.1 K2K

The first of the accelerator-based long-baseline experiments was the KEK-E362 experiment (K2K) [Oya98] in Japan sending a neutrino beam from KEK to Super-Kamiokande. K2K used two detectors: one about 300 m away from the target and Super-Kamiokande at a distance of about 250 km. Super-Kamiokande has already been described in more detail in Section 9.3.1. The neutrino beam was produced by 12 GeV protons from the KEK-PS hitting an Al-target of 2 cm diameter × 65 cm length. Using a decay tunnel of 200 m and a magnetic horn system for focusing π^+ an almost pure ν_μ-beam is produced. The contamination of ν_e from μ and K-decay was of the order 1 %. The protons were extracted in a fast extraction mode allowing spills of a time width of 1.1 μs every 2.2 s. With 6×10^{12} pots (protons on target) per spill about 1×10^{20} pots could be accumulated in three years. The average neutrino beam energy is 1.4 GeV, with a peak at about 1 GeV (Figure 9.16). In this energy range quasi-elastic interactions are dominant. Kinematics allows to reconstruct E_ν

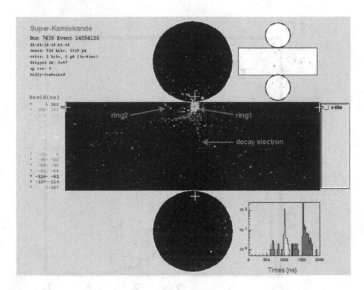

Figure 9.17. The first long-baseline event ever observed by the K2K experiment. With kind permission of the Kamioka Observatory, ICRR, the University of Tokyo and the T2K Collaboration.

even if only the muon is measured via

$$E_\nu = \frac{m_N E_\mu - \frac{m_\mu^2}{2}}{m_N - E_\mu + P_\mu \cos\theta_\mu} \tag{9.22}$$

with m_N as the mass of the nucleon and θ_μ as the angle of the outgoing muon with respect to the beam. The near detector consists of two parts: a 1 kt water-Cherenkov detector and a fine-grained detector. The water detector is implemented with 820 PMTs of 20 inch and its main goal is to allow a direct comparison with Super-Kamiokande events and to study systematic effects of this detection technique. The fine-grained detector basically consists of four parts and should provide information on the neutrino beam profile as well as the energy distribution. The relative energy resolution turned out to be about $8\%/\sqrt{E}$. The muon chambers consisted of 900 drift tubes and 12 iron plates. Muons generated in the water target via CC reactions could be reconstructed with a position resolution of 2.2 mm. The energy resolution was about 8–10%. The detection method within Super-Kamiokande is identical to that of their atmospheric neutrino detection.

Due to the low beam energy, K2K was able to search for $\nu_\mu \rightarrow \nu_e$ appearance and a general ν_μ disappearance. The main background for the search in the electron channel were quasi-elastic π^0 production in NC reactions, which can be significantly reduced by a cut on the electromagnetic energy. Furthermore, the near detector allowed a good measurement of the cross-section of π^0 production in NC.

K2K accumulated 9.2×10^{19} pot (Figure 9.17). They observed 112 events but expected $158^{+9.2}_{-8.6}$ from the near detector measurement, a clear deficit [Ahn03,

Yam06, Ahn06]. The best-fit values are $\sin^2 2\theta = 1$ and $\Delta m^2 = 2.8 \times 10^{-3}$ eV^2. This number is in good agreement with the oscillation parameters deduced from the atmospheric data.

9.4.2 MINOS

Another neutrino program called NuMI (Neutrinos at the Main Injector) has been associated with the new Main Injector at Fermilab. This long-baseline project is sending a neutrino beam to two locations, the first one to the Soudan mine about 735 km away from Fermilab for the MINOS and MINOS+ experiment and the second one at 810 km away to the NOVA experiment (see Chapter 4). At the Soudan mine the MINOS experiment [Mic03] is located. Using a detection principle similar to CDHS (see Chapter 4), it consists of a 980 t near detector located at Fermilab about 900 m away from a graphite target and a far detector at Soudan. The far detector is made of 486 magnetized iron plates, producing an average toroidal magnetic field of 1.3 T. They have a thickness of 2.54 cm and an octagonal shape measuring 8 m across. They are interrupted by about 25 800 m^2 active detector planes in the form of 4.1 cm wide solid scintillator strips with x and y readout to get the necessary tracking information. Muons are identified as tracks transversing at least five steel plates, with a small number of hits per plane. The total mass of the detector is 5.4 kt.

Several strategies are at hand to discriminate among the various oscillation scenarios. The proof of ν_μ–ν_τ oscillations will be the measurement of the NC/CC ratio in the far detector. The oscillated ν_τ will not contribute to the CC reactions but to the NC reactions. In the case of positive evidence, a 10% measurement of the oscillation parameters can be done by comparing the rate and spectrum of CC events in the near and far detector. Three beam options are possible for the low energy are discussed which are shown in Figure 9.18. With an average neutrino energy of 3 GeV for the low energy option, this implies a pure $\nu_\mu$disappearance search. In this channel clear evidence for oscillations has been found [Ada08, Ada08a].

9.4.3 CERN–Gran Sasso

Another program in Europe was a long-baseline experiment using a neutrino beam (CNGS) from CERN to the Gran Sasso Laboratory [Els98]. The distance is 732 km. In contrast to K2K and MINOS, the idea here was an optimised beam to search directly for ν_τ appearance. The beam protons from the SPS at CERN are extracted with energies up to 450 GeV hitting a graphite target at a distance of 830 m from the SPS. After a magnetic horn system for focusing the pions, a decay pipe of 1000 m follows.

Two experiments were located at the Gran Sasso Laboratory to perform an oscillation search. The first one is the T600 test module as a very early prototype for a future liquid argon time projection chamber (LAr TPC) for the DUNE experiment. This 600 ton liquid Ar TPC with a modular design offers excellent energy and position resolution. In addition, very good imaging quality is possible, hence allowing good particle identification (Figure 9.19).

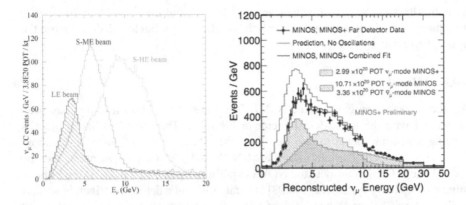

Figure 9.18. Left: Three different options are available for the neutrino beam (NuMI) used by MINOS and MINOS+ experiment at Fermilab, a low Energy (LE), medium (ME) and high energy (HE) (from [Kop04]). © 2004 IEEE. Right: The events/GeV as function of the reconstructed energy for the far detector data spectra. This is shown for the prediction of no oscillations, the best fit spectrum and its uncertainty and exposure (from [Whi16]). © 2016 With permission from Elsevier.

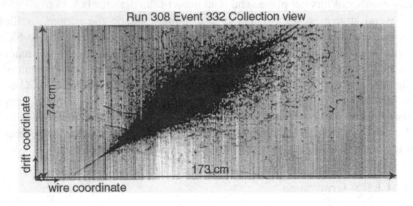

Figure 9.19. A broad electromagnetic shower as observed with the ICARUS T600 test module on the surface. This impressively shows the data quality obtainable with LAr TPCs (from [Ame04]). © 2004 With permission from Elsevier and the ICARUS collaboration.

The second one was a ν_τ-appearance search with a 2 kt lead-emulsion sandwich detector (OPERA) [Gul00], see also Chapter 4. The principle is to use lead as a massive target for neutrino interactions and thin (50 μm) emulsion sheets working conceptually as emulsion cloud chambers (ECC). The detector has a modular design, with a brick as the basic building block, containing 58 emulsion films. Some 3264 bricks together with electronic trackers form a module. Twenty-four modules will form a supermodule of about 652 t mass. Two supermodules interleaved with a muon spectrometer finally form the full detector. In total, about 150 000 bricks were implemented. These encapsulated bricks were built which contained 57 emulsion films with 300 μm thickness interleaved with 56 lead plates of 1mm thickness. These bricks were stapled into walls. Between those walls, orthogonal scintillator strips for positioning and energy measurements were installed. High resolution muon tracking devices in the form of resistive plate chambers and drift tubes were installed for charge and momentum measurements. The τ-lepton, produced by CC reactions in the lead target, decays in the gap region, and the emulsion sheets are used to verify the kink of a τ-decay which is the signal. For τ-decays within the lead, an impact parameter analysis has been done to show that the required track does not come from the primary vertex (see Figure 4.6). For these signals to be found, the scanning of the emulsion sheets has been done using high speed automatic CCD microscopes. The τ, produced by CC reactions in the lead, decays in the gap region, and the emulsion sheets were used to verify the kink in the decay, a principle also used in the CHORUS and DONUT experiments. For decays within the lead, an impact parameter analysis has been performed to show that the required track does not come from the primary vertex. In addition to the $\tau \rightarrow e, \mu, \pi$ decay modes three pion decays could also be examined. The analysis was based on an event by event basis and the experiment was, in general, considered to be background free. Over the years 2008-2012 a statistics of 17.97 10^{19} pot was accumulated. In total, 10 candidates have been identified with an expected background of 2.0$\pm$ 0.4 events proving the appearance of ν_τ in a muon beam. This results in $\Delta m_{23}^2 = 2.7 \times 10^{-3}$ eV^2 under the assumption of $\sin^2 2\theta_{23} = 1$ [Aga18].

Currently running long baseline experiments like MINOS/MINOS+, NOVA and T2K have already been presented in Chapter 4. Here the current results of the experiments NOVA and T2K will be presented.

NOvA: The NOvA experiment is performing a ν_μ- disappearance and also an ν_e- appearance search. They have accumulated 8.85 $\times 10^{20}$ pot [Ace18]. From their measurements they can deduce values for $\mid \Delta m_{23}^2 \mid, \theta_{23}$ and the CP-phase δ_{CP}. The results are giving best fit values of $\Delta m_{23}^2 = 2.44 \times 10^{-3}$ eV^2, $\sin^2\theta_{23}$=0.56 and $\delta_{CP} = 1.21\pi$. In addition, the inverted hierarchy is disfavoured by 95% (see Section 4.2).

T2K: The T2K-experiment has used so far 14.7 $\times 10^{20}$ pot for neutrino beams and 7.6 $\times 10^{20}$ pot on anti-neutrinos [Abe18]. The best fit values obtained are $\Delta m_{23}^2 = 2.463^{+0.071}_{-0.070} \times 10^{-3}$ eV^2 and $\sin^2\theta_{23} = 0.5^{+0.032}_{-0.036}$. The CP-conserving case ($\delta_{CP} = 0, \pi$) is excluded by 2σ. Also here the corresponding significance contours are shown in Fig. 9.20 (see Section 4.2).

9.5 Future experimental plans and ideas

Several new experiments are considered for future long baseline experiments, part of them will be shortly described, with DUNE already described in Chapter 4, see [Diw16].

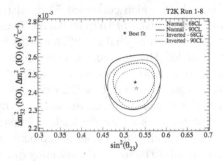 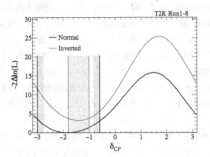

Figure 9.20. Left: Similar contour plots for the extracted parameter region from T2K: Δm^2_{23} versus $\sin^2\theta_{23}$, shown for normal and inverted hierarchy. Right: The logarithmic likelihood as a function of the CP-phase δ_{CP} is shown again for both hierarchies (from [Abe18]). © 2018 by the American Physical Society.

9.5.1   INO-ICAL

It is planned to install a new atmospheric neutrino detector in the Indian Neutrino Observatory (INO). The planned 50 kt magnetized iron tracking calorimeter (ICAL) at INO will follow the idea of the former MONOLITH proposal [Mon00]. It will consist of three modules, each 16 m × 16 m × 12 m, made out of horizontal iron plates which are interleaved by active tracking devices in the form of glass resistive plate chambers (RPCs) [Beh09]. This allows the muon charge to be measured; therefore, discriminating between ν_μ and $\bar{\nu}_\mu$. This will be important in studying matter effects (see Chapter 8). Measuring the hadronic energy and the momentum of the semi-contained muons will allow a reasonably good reconstruction of the neutrino energy. The neutrino angular distribution, which determines the resolution of L, is sufficient to allow a good L/E resolution [Ahm17a].

9.5.2   Hyper-Kamiokande

Hyper-Kamiokande is a 1 Mt device (with 560 kt fiducial volume) water Cherenkov detector made out of eight cubes 50 m × 50 m × 50 m each [Abe14]. It will be installed close to Super-Kamiokande; hence, it will also use the neutrino beam from J-PARC as T2K does.

9.5.3   THEIA

Another large scale project considered is Theia [Fis18]. The innovative concept here is to use a water based liquid scintillator (WbLS). This would allow directional information using the Cherenkov effect combined with more light yield from the scintillation light. Further, progress in detector developments allows to use large area avalanche photo-diodes (LAAPD) with picosecond timing to be used. A 2 kt detector is envisaged with a potential upgrade to 100 kton [Ask19]. .

9.5.4 AQUA-RICH

The basic principle and ideas of AQUA-RICH are summarized in [Ant99]. By using the RICH technique, particle velocities can be measured by the ring radius and direction by the ring centre. An improvement over existing Cherenkov detectors is the measurement of higher ring multiplicities and, therefore, more complicated events can be investigated. However, the key concept is to measure momenta via multiple scattering. Multiple scattering causes a displacement and an angular change as a particle moves through a medium. The projected angular distribution θ of a particle with velocity β, momentum p and charge Z after traversing the path L in a medium of absorption length X_0 is Gaussian with the width

$$\sigma_{\mathrm{ms}} = \theta_{\mathrm{rms}} = \frac{k_{\mathrm{ms}}}{\beta cp} Z \sqrt{\frac{L}{X_0}} \tag{9.23}$$

with $k_{\mathrm{ms}} = 13.6$ MeV as the multiple scattering constant. Momentum resolution better than 10% for 10 GeV muons could be obtained in simulations, sufficient to see the oscillation pattern in atmospheric neutrinos [Gro04]. A 1 Mt detector is proposed.

Chapter 10

Solar neutrinos

DOI: 10.1201/9781315195612-10

Solar neutrinos has been one of the longest standing and most interesting problems in particle astrophysics. From the astrophysical point of view, solar neutrinos are the only objects besides the study of solar oscillations (helioseismology) which allow us a direct view into the solar interior. The study of the fusion processes in the Sun via neutrino spectroscopy offers a unique perspective. From the particle physics point of view, the baseline Sun–Earth with an average of 1.496×10^8 km and neutrino energies of about 1 MeV allows probing of neutrino oscillation parameters down to $\Delta m^2 \approx 10^{-10}$ eV^2, which is not possible by terrestrial means. The Sun is a pure source of ν_e resulting from fusion chains. During recent decades it has been established that significantly fewer solar neutrinos are observed than would be expected from theoretical modeling. It was extremely important to find out to what extent this discrepancy pointed to "new physics" like neutrino oscillations, rather than to an astrophysical problem. This could have been a lack of knowledge of the solar structure or regarding its reactions in the interior. Also "terrestrial" problems due to the limited knowledge of capture cross-sections in neutrino detectors were discussed. Nowadays the amount of data confirmed the neutrino oscillation hypothesis, which is the third piece of evidence for a non-vanishing neutrino mass besides indications from atmospheric and reactor neutrino experiments. In the following chapter the situation is discussed in more detail.

10.1 The standard solar model

If fusion reactions among light elements are responsible for the solar luminosity, then a specific linear combination of solar neutrino fluxes must be equal to the solar constant, which is [Bah02]

$$\frac{L_\odot}{4\pi(A.U.)^2} = \sum_i \alpha_i \Phi_i \tag{10.1}$$

where L⊙ is the solar luminosity measured at the Earth's surface at 1 A.U., which is the average Earth-Sun distance. This equation is called luminosity constraint. First,

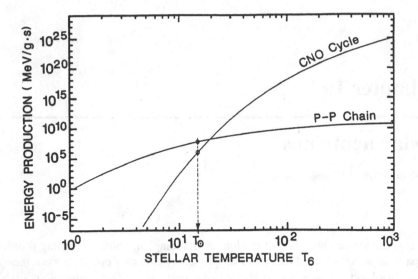

Figure 10.1. Contributions of the pp and CNO cycles for the energy production rate in stars as a function of the central temperature. While the pp cycle is dominant in the Sun, the CNO process becomes dominant above about 20 million degrees (from [Rol88]). With kind permission of University of Chicago Press.

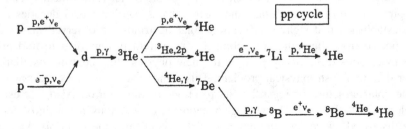

Figure 10.2. The route of proton fusion according to the pp cycle. After the synthesis of ^{3}He the process branches into three different chains. The pp cycle produces 98.4% of the solar energy (from [Kla97]). © Taylor & Francis Group.

we disuss the various nuclear reaction processes, which create the individual neutrino fluxes.

10.1.1   Energy production processes in the Sun

According to our understanding, the Sun, like all stars, creates its energy via nuclear fusion [Gam38, Bet39]. For a general discussion of the structure of stars and stellar energy generation see, e.g., [Cox68, Cla68, Rol88, Sti02, Ili15]. Hydrogen fusion to helium proceeds according to

$$4p \rightarrow {}^4\mathrm{He} + 2e^+ + 2\nu_e \tag{10.2}$$

The two positrons annihilate with two electrons resulting in an energy relevant equation

$$2e^- + 4p \rightarrow {}^4He + 2\nu_e + 26.73 \text{ MeV}. \tag{10.3}$$

Therefore, an energy of $Q = 2m_e + 4m_p - m_{He} = 26.73$ MeV per ^{4}He fusion is released. Using the solar constant $S = 8.5 \times 10^{11}$ MeV cm^{-2} s^{-1} at the Earth a first guess for the total neutrino flux at the Earth can be obtained:

$$\Phi_\nu \approx \frac{S}{13 \text{ MeV per } \nu_e} = 6.5 \times 10^{10} \text{ cm}^{-2} \text{ s}^{-1}. \tag{10.4}$$

However, details of the neutrino flux and, therefore, its creating fusion processes, are more complex. There are two fundamental fusion cycles: one is the pp cycle [Bet38], the other the CNO cycle [Wei37, Bet39]. Figure 10.1 shows the contribution of both processes to energy production as a function of temperature. The pp cycle (see Figure 10.2) is dominant in the Sun and accounts for almost all energy production. Solar neutrinos are labeled according to their production reaction in the fusion network. The first reaction step is the fusion of protons into deuteron (deuterium nucleus):

$$p + p \rightarrow d + e^+ + \nu_e \qquad (E_\nu \leq 0.42 \text{ MeV}). \tag{10.5}$$

This first process is a weak interaction process , that is why stars are long living. The primary pp fusion proceeds this way to 99.6%. In addition, the alternative process occurs with a much lower probability of 0.4%:

$$p + e^- + p \rightarrow d + \nu_e \qquad (E_\nu = 1.44 \text{ MeV}). \tag{10.6}$$

The neutrinos produced in this reaction (pep neutrinos) are mono-energetic. The conversion of the created deuteron to helium is identical in both cases:

$$d + p \rightarrow {}^3He + \gamma + 5.49 \text{ MeV}. \tag{10.7}$$

Neutrinos are not produced in this reaction. From that point onwards the reaction chain divides. With a probability of 85%, the ^{3}He fuses directly into ^{4}He:

$$^3He + {}^3He \rightarrow {}^4He + 2p + 12.86 \text{ MeV}. \tag{10.8}$$

In this step, also known as the pp I-process, no neutrinos are produced. However, two neutrinos are created in total, as the reaction of Equation (10.5) has to occur twice, in order to produce two ^{3}He nuclei which can undergo fusion. Furthermore, ^{4}He can also be created with a probability of 2.4×10^{-5}% by

$$^3He + p \rightarrow {}^4He + \nu_e + e^+ + 18.77 \text{ MeV}. \tag{10.9}$$

The neutrinos produced here are very energetic (up to 18.77 MeV) but they have a very low flux. They are called *hep* neutrinos. The alternative reaction produces ^{7}Be:

$$^3He + {}^4He \rightarrow {}^7Be + \gamma + 1.59 \text{ MeV}. \tag{10.10}$$

Subsequent reactions again proceed via several sub-reactions. The pp II-process leads to the production of ^{7}Li with a probability of 15% via electron capture

$$^7\text{Be} + e^- \rightarrow {}^7\text{Li} + \nu_e \qquad (E_\nu = 0.862\,\text{MeV or } E_\nu = 0.384\,\text{MeV}). \qquad (10.11)$$

This reaction produces ^{7}Li in the ground state 90% of the time and leads to the emission of mono-energetic neutrinos of 862 keV. The remaining 10% are captured into an excited state by emission of neutrinos with an energy of 384 keV. Thus mono-energetic neutrinos are produced in this process. In the next reaction step, helium is created via

$$^7\text{Li} + p \rightarrow 2\,{}^4\text{He} + 17.35\,\text{MeV}. \qquad (10.12)$$

Assuming that ^{7}Be has already been produced, this pp II-branch has a probability of 99.98%. There is also the possibility of proceeding via ^{8}B (the pp III-chain) rather than by ^{7}Li via

$$^7\text{Be} + p \rightarrow {}^8\text{B} + \gamma + 0.14\,\text{MeV}. \qquad (10.13)$$

^{8}B undergoes β^+-decay via

$$^8\text{B} \rightarrow {}^8\text{Be}^* + e^+ + \nu_e \qquad (E_\nu \lesssim 15\,\text{MeV}). \qquad (10.14)$$

The precise endpoint is slightly uncertain because the final state in the daughter nucleus is very broad. The neutrinos produced here are very energetic but also very rare. Nevertheless, they play an important role for experimental detection. ^{8}Be dissociates into two α-particles:

$$^8\text{Be}^* \rightarrow 2\,{}^4\text{He} + 3\,\text{MeV}. \qquad (10.15)$$

The CNO cycle accounts for only about 1.6% of the energy production in the Sun, hence it is mentioned here only briefly. It is a catalytic process relying on the presence of C, N and O in the Sun. The main reaction steps are:

$$\text{CNO I:} \quad {}^{12}\text{C} + p \rightarrow {}^{13}\text{N} + \gamma \qquad\qquad\qquad\qquad\qquad (10.16)$$
$$^{13}\text{N} \rightarrow {}^{13}\text{C} + e^+ + \nu_e \qquad (E_\nu \leq 1.20\,\text{MeV}) \qquad (10.17)$$
$$^{13}\text{C} + p \rightarrow {}^{14}\text{N} + \gamma \qquad\qquad\qquad\qquad\qquad (10.18)$$
$$^{14}\text{N} + p \rightarrow {}^{15}\text{O} + \gamma \qquad\qquad\qquad\qquad\qquad (10.19)$$
$$^{15}\text{O} \rightarrow {}^{15}\text{N} + e^+ + \nu_e \qquad (E_\nu \leq 1.73\,\text{MeV}) \qquad (10.20)$$
$$^{15}\text{N} + p \rightarrow {}^{12}\text{C} + {}^4\text{He} \qquad\qquad\qquad\qquad\qquad (10.21)$$
$$\text{CNO II:} \quad {}^{15}\text{N} + p \rightarrow {}^{16}\text{O} + \gamma \qquad\qquad\qquad\qquad\qquad (10.22)$$
$$^{16}\text{O} + p \rightarrow {}^{17}\text{F} + \gamma \qquad\qquad\qquad\qquad\qquad (10.23)$$
$$^{17}\text{F} \rightarrow {}^{17}\text{O} + e^+ + \nu_e \qquad (E_\nu \leq 1.74\,\text{MeV}) \qquad (10.24)$$
$$^{17}\text{O} + p \rightarrow {}^{14}\text{N} + {}^4\text{He}. \qquad\qquad\qquad\qquad\qquad (10.25)$$

These processes are also illustrated in Figure 10.3.

We have now introduced the processes relevant for neutrino production. To predict the expected neutrino spectrum, we need further information – in particular about the cross-sections of the reactions involved [Par94, Lan94].

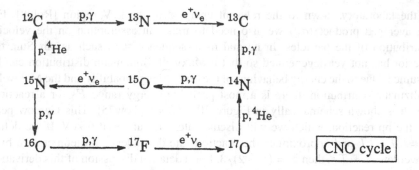

Figure 10.3. Representation of the CNO process. This also burns hydrogen to helium with C, N and O acting as catalysts and is responsible for 1.6% of the solar energy (from [Kla97]). © Taylor & Francis Group.

10.1.2 Reaction rates

Before dealing with details of the Sun, we first state some general comments on the reaction rates [Cla68, Rol88, Bah89, Raf96, Adl11, Tho09, Ili15]. They play an important role in the understanding of energy production in stars. Consider a reaction of two particles T_1 and T_2 of the general form

$$T_1 + T_2 \rightarrow T_3 + T_4. \tag{10.26}$$

Their reaction rate is given by

$$R = \frac{n_1 n_2}{1 + \delta_{12}} \langle \sigma v \rangle_{12} \tag{10.27}$$

where n_i is the particle density, σ the cross-section, v the relative velocity and δ the Kronecker symbol to avoid double counting of identical particles. $\langle \sigma v \rangle$ is the temperature averaged product, assuming a Maxwell-Boltzmann distribution for the particles. At typical thermal energies of several keV inside the stars and Coulomb barriers of several MeV, it can be seen that the dominant process for charged particles is quantum mechanical tunneling, which was used by Gamow to explain α-decay [Gam38]. It is common to write the cross-section in the form

$$\sigma(E) = \frac{S(E)}{E} \exp(-2\pi\eta) \tag{10.28}$$

where the exponential term is Gamow's tunnelling factor, the factor $1/E$ expresses the dependence of the cross-section on the de Broglie wavelength and η is the so-called Sommerfeld parameter, given by $\eta = Z_1 Z_2 e^2 / \hbar v$ with Z being the atomic number and v being the relative velocity. Nuclear physics now enters into calculations only through the so-called S-factor $S(E)$, which, as long as no resonances appear, should have a relatively smooth behaviour. This assumption is critical, since we have to extrapolate from the values at several MeV, measured in the laboratory, down to the relevant energies in the keV region [Rol88]. For the averaged product $\langle \sigma v \rangle$ we also need to make an assumption on the velocity distribution of the particles. In normal main-sequence stars such as our Sun, the interior has not yet degenerated so that a Maxwell–Boltzmann distribution can be assumed. Due to the energy behaviour of the tunnelling probability and the Maxwell–Boltzmann distribution, there is a most probable energy range E_0 for a reaction, which is shown schematically in Figure 10.4 [Bur57, Fow75]. This Gamow peak for the pp reaction, which we will discuss later, lies at about 6 keV. If we define $\tau = 3E_0/kT$ and approximate the reaction rate dependence on temperature by a power law $R \sim T^n$, then $n = (\tau - 2)/3$. For a detailed discussion of this derivation see, e.g., [Rol88, Bah89, Ili15]. Since the energy of the Gamow peak is temperature dependent, $S(E)$ is, for ease of computation, expanded in a Taylor series with respect to energy:

$$S(E) = S(0) + \dot{S}(0)E + \tfrac{1}{2}\ddot{S}(0)E^2 + \cdots \tag{10.29}$$

where $S(0), \dot{S}(0)$, etc. are obtained by a fit to the experimental data.

Due to their comparatively small cross-sections, it is challenging to measure fusion processes directly in the stellar energy region. This was done the first time in the LUNA experiment using accelerators built underground, where the detectors are shielded against cosmic radiation. [Gre94, Fio95, Arp96]. In a first step the LUNA collaboration was operating a 50 kV accelerator at the Gran Sasso Laboratory to investigate the $^3\text{He}(^3\text{He}, 2p)\,^4\text{He}$ reaction as the final step in the pp I chain [Arp98]. An upgrade to a 400 kV accelerator (LUNA II) was performed, which enabled additional measurements of, e.g., the $^3\text{He}(\alpha, \gamma)^7\text{Be}$ [Con08] and $^{14}\text{N}(p, \gamma)\,^{15}\text{O}$ [Lem06] cross-sections. Two underground accelerators in the MV region are the planned LUNA-MV [Bro18] and the Felsenkeller accelerator [Bem18]. As an example of the current status the data points of the $^3\text{He}(\alpha, \gamma)^7\text{Be}$ reaction are shown (Figure 10.5) .

10.1.3   The solar neutrino spectrum

The measurement of neutrinos from the various fusion reactions provides deeper insides of the solar interior. Furthermore, predictions of solar models can be compared in detail with the observed neutrinos fluxes. The actual prediction of the solar neutrino spectrum requires detailed model calculations [Tur88, Bah88, Bah89, Bah92, Tur93a, Tur93b, Bah95, Bah01, Cou02, Bah06, Hax08, Bas09, Ser09, Ser16, Vin17].

10.1.3.1   Standard solar models

Simulations to model the operation of the Sun are using the basic equations of stellar evolution (see [Cla68, Rol88, Bah89, Ser16, Vin17]).

(i)   Hydrodynamic equilibrium, i.e., the gas and radiation pressure, balance the gravitational attraction:

$$\frac{\mathrm{d}p(r)}{\mathrm{d}r} = -\frac{GM(r)\rho(r)}{r^2} \tag{10.30}$$

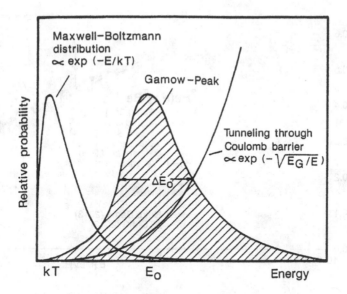

Figure 10.4. The most favourable energy region for nuclear reactions between charged particles at very low energies is determined by the convolution of the opposing effects. The first is the Maxwell–Boltzmann distribution with a maximum at about kT, and an exponentially decreasing number of particles at higher energies. The other effect is that the quantum mechanical tunneling probability E_G rises with growing energy. This results in the Gamow peak (not shown true to scale), at E_0, which can be much larger than kT (from [Rol88]). With kind permission of the University of Chicago Press.

with r as the radial distance in the Sun and mass conservation

$$M(r) = \int_0^r 4\pi r^2 \rho(r)\,\mathrm{d}r.$$

(ii) Energy balance, meaning the observed luminosity L, is generated by an energy generation rate ϵ:

$$\frac{\mathrm{d}L(r)}{\mathrm{d}r} = 4\pi r^2 \rho(r)\epsilon. \tag{10.31}$$

(iii) Energy transport dominantly by radiation and convection which is given in the radiation case by

$$\frac{\mathrm{d}T(r)}{\mathrm{d}r} = -\frac{3}{64\pi\sigma}\frac{\kappa\rho(r)L(r)}{r^2 T^3} \tag{10.32}$$

with σ as the Stefan–Boltzmann constant and κ as the absorption coefficient. The inner 70% of the solar radius energy transport is radiation dominated, while the outer 30% forms a convection zone. These equations are governed by three additional equations of state for the pressure p, the absorption coefficient κ and the energy generation rate ϵ:

$$p = p(\rho, T, X) \qquad \kappa = \kappa(\rho, T, X) \qquad \epsilon = \epsilon(\rho, T, X) \tag{10.33}$$

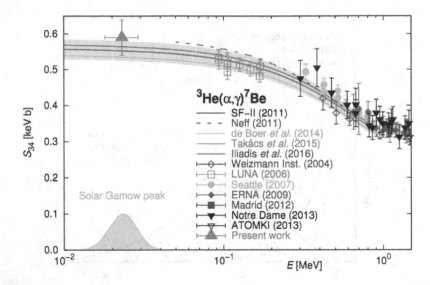

Figure 10.5. Compilation of data points for the $^3\text{He}(\alpha, \gamma)^7\text{Be}$ reaction as function of energy. The lowest point in the solar region is calculated from the precise neutrino flux measurements (from [Tak18]). © 2018 With permission from Elsevier.

where X denotes the chemical composition. The Russell–Vogt theorem then ensures, that for a given M and X a unique equilibrium configuration will evolve, resulting in certain radial pressure, temperature and density profiles of the Sun [Car14]. Under these assumptions, solar models can be calculated as an evolutionary sequence from an initial chemical composition. The boundary conditions are that the model has to reproduce the age, luminosity, surface temperature and mass of the present Sun. The two typical adjustable parameters are the ^4He abundance and the relation of the convective mixing length to the pressure scale height. Input parameters include the age of the Sun and its luminosity, as well as the equation of state, nuclear parameters, chemical abundances and opacities.

10.1.3.2 *Diffusion*

Evidence from several experiments strongly suggests a significant mixing and gravitational settling of He and the heavier elements in the Sun. The longstanding problem of ^7Li depletion in the solar photosphere can be explained if ^7Li is destroyed by nuclear burning processes which, however, require temperatures of about 2.6×10^6 K. Such temperatures do not exist at the base of the convection zone; therefore, ^7Li has to be brought to the inner regions. This so-called ^7Li problem is still not solved. Also the measured sound speed profiles in the solar interior obtained by helioseismological data can be better reproduced by including diffusion processes. Therefore, these effects were included in newer solar models.

10.1.3.3 Initial composition

The chemical abundance of the heavier elements (beyond helium) forms an important ingredient for solar modeling. Their abundance influences the radiative opacity and, therefore, the temperature profile within the Sun. Under the assumption of a homogeneous Sun, the elemental abundance in the solar photosphere still corresponds to the initial values. This has been questioned by 3-dimensional studies of photospheric line shapes [Asp09], which lead to a lower agreement for the sound speed and abundances. The relative abundances of the heavy elements are best determined in a certain type of meteorite, the type I carbonaceous chondrite, which can be linked and found to be in good agreement with the photospheric abundances [Gre93, Gre93a, Asp09]. The abundance of C, N and O is taken from photospheric values, whereas the ^{4}He abundance cannot be measured and is used as an adjustable parameter.

10.1.3.4 Opacity and equation of state

The opacity is a measure of the photon absorption capacity. It depends on the chemical composition and complex atomic processes. The influence of the chemical composition on the opacity can be seen, for example, in different temperature and density profiles of the Sun. The ratio of the "metals" Z (in astrophysics all elements heavier than helium Y are known as metals) to hydrogen X is seen to be particularly sensitive. The experimentally observable composition of the photosphere is used as the initial composition of elements heavier than carbon. In the solar core ($T > 10^7$ K) the metals do not play the central role for the opacity, which is more dependent on inverse bremsstrahlung and photon scattering on free electrons (Thomson-scattering). The opacity or Rosseland mean absorption coefficient κ is defined as a harmonic mean integrated over all frequencies ν:

$$\frac{1}{\kappa} = \frac{\int_0^\infty \frac{1}{\kappa_\nu} \frac{dB_\nu}{dT} \, d\nu}{\int_0^\infty \frac{dB_\nu}{dT} \, d\nu} \tag{10.34}$$

where B_ν denotes a blackbody Planck spectrum. The implication is that more energy is transported at frequencies where the material is more transparent and at which the radiation field is more temperature dependent. The calculation of the Rosseland mean requires a knowledge of all the involved absorption and scattering cross-sections of photons on atoms, ions and electrons. The calculation includes bound–bound (absorption), bound–free (photoionization), free–free (inverse bremsstrahlung) transitions and Thomson scattering. Corrections for electrostatic interactions between the ions and electrons and for stimulated emissions have to be taken into acount. The number densities n_i of the absorbers can be extracted from the Boltzmann and Saha equations [Sah20]. The radiative opacity per unit mass can then be expressed as (with the substitution $u \equiv h\nu/kT$)

$$\frac{1}{\kappa} = \rho \int_0^\infty \frac{15u^4 e^u / 4\pi^4 (e^u - 1)^2}{(1 - e^u) \sum_i \sigma_i n_i + \sigma_s n_e} \, du \tag{10.35}$$

Table 10.1. Exemplaric evolution of the Sun by modeling, at the beginning of fusion (t=0) and today according to the standard solar model (SSM) of [Bah89]. However, the values for X, Y and Z should be replaced with the ones given by [Vin17]).

	$t = 4.6 \times 10^9$ yr (today)	$t = 0$
Luminosity $L_\odot$	$\equiv 1$	0.71
Radius $R_\odot$	696 000 km	605 500 km
Surface temperature T_S	5773 K	5 665 K
Core temperature T_c	15.6×10^6 K	—
Core density	148 g cm^{-3}	—
X (H)	34.1%	71%
Y (He)	63.9%	27.1%
Z	1.96%	1.96%

where σ_s denotes the Thomson scattering cross-section. Comprehensive compilation of opacities is given by the Livermore group (OPAL) [Ale94, Igl96] and STAR [Kri16].

A further ingredient for solar model calculations is the equation of state, meaning the density as a function of p and T or, as widely used in the calculations, the pressure expressed as a function of density and temperature. Except for the solar atmosphere, the gas pressure exceeds the radiation pressure anywhere in the Sun. The gas pressure is given by the ideal gas law, where the mean molecular weight μ must be determined by the corresponding element abundances. The different degrees of ionization can be determined using the Saha equations. An equation of state in the solar interior has to consider plasma effects and the partial electron degeneracy deep in the solar core. The latest equation of state is given by [Rog02,Cas03]. It is assumed here that the Sun has been a homogeneous star since joining the main sequence.

10.1.3.5   Predicted neutrino fluxes

With all these inputs it is then possible to calculate a sequence of models of the Sun that finally predict values of $T(r), \rho(r)$ and the chemical composition of its current state (see Table 10.1). These models are called *standard solar models* (SSM) [Tur88, Bah89, Bah92, Tur93a, Bah95, Bah01, Cou02, Bah06, Ser09, Vin17]. They predict the location and rate of the nuclear reactions that produce neutrinos (see Figure 10.6). Finally, these models give predictions for the expected neutrino spectrum and the observable fluxes on Earth (see Figure 10.7 and Table 12.2). It can clearly be seen that the largest part of the flux comes from the pp neutrinos. In addition to the flux, in order to predict the signal to be expected in the various detectors, it is necessary to know the capture or reaction cross-sections for neutrinos.

Although the total neutrino flux on Earth has a value of the order of

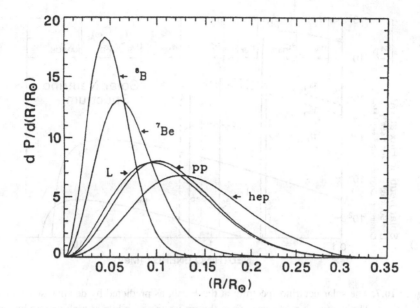

Figure 10.6. Production of neutrinos from different nuclear reactions as a function of the distance from the Sun's centre, according to the standard solar model. The luminosity produced in the optical region (denoted by L) as a function of radius is shown as a comparison. This is coupled very strongly to the primary pp fusion (from [Bah89]). © Cambridge University Press.

Table 10.2. Three examples of SSM predictions for the flux Φ_ν of solar neutrinos on the Earth (from [Bah01, Cou02] and [Ber16]).

Source	Φ_ν (10^{10} cm^{-2} s^{-1})		
	[Bah01]	[Cou02]	[Ber16]
pp	5.95	5.92	5.98
pep	1.40×10^{-2}	1.43×10^{-2}	1.44×10^{-2}
^{7}Be	4.77×10^{-1}	4.85×10^{-1}	4.93×10^{-1}
^{8}B	5.05×10^{-4}	4.98×10^{-4}	5.46×10^{-4}
^{13}N	5.48×10^{-2}	5.77×10^{-2}	2.78×10^{-2}
^{15}O	4.80×10^{-2}	4.97×10^{-2}	2.05×10^{-2}
^{17}F	5.63×10^{-4}	3.01×10^{-4}	5.29×10^{-4}

10^{10} cm^{-2} s^{-1}, their detection is extremely difficult because of the small cross-sections. We now turn to the experiments, results and interpretations.

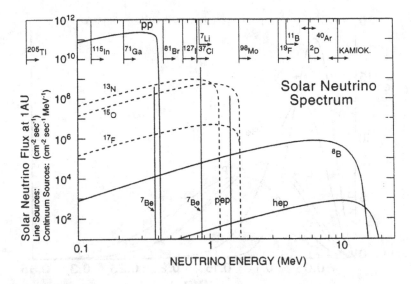

Figure 10.7. The solar neutrino spectrum at the Earth, as predicted by detailed solar model calculations. The dominant part comes from the pp neutrinos, while at high energy hep and ^{8}B neutrinos dominate. The threshold energies for solar neutrinos for different radiochemical (used and potential ones) detections and real time detector thresholds are shown on the upper axis (from [Win00]). © Cambridge University Press.

10.2   Solar neutrino experiments

In principle there are two kinds of solar neutrino experiments: radiochemical and real-time experiments. The principle of the *radiochemical experiments* is the reaction

$$^A_N Z + \nu_e \rightarrow\ ^A_{N-1}(Z+1) + e^- \tag{10.36}$$

where the daughter nucleus is unstable and decays back with a "reasonable" half-life since it is this radioactive decay of the daughter nucleus which is used for detection. The production rate of the daughter nucleus is given by

$$R = N \int \Phi(E)\sigma(E)\, \mathrm{d}E \tag{10.37}$$

where Φ is the solar neutrino flux above a certain threshold (see Figure 10.7), N the number of target atoms and σ the cross-section for the reaction of Equation (10.36). Dealing with discrete nuclear states, this implies knowledge of the involved Gamow–Teller strengths of the transition (see Chapter 7) . Given an incident neutrino flux of about 10^{10} cm^{-2} s^{-1} and a cross-section of about 10^{-45} cm^2, about 10^{30} target atoms are required to produce one event per day. Therefore, very large detectors are required of the order of several tons to convert one atom per day. Thus, the detection is not trivial. It is convenient to define a new unit more suitable for such low event rates, the SNU (solar neutrino unit) where

$$1 \text{ SNU} = 10^{-36} \text{ captures per target atom per second.}$$

Any information about the time of the event, the direction and energy (with the exception of the lower limit, which is determined by the energy threshold of the detector) of the incident neutrino is lost in these experiments since only the average production rate of the unstable daughter nuclei over a certain time period can be measured.

The situation in *real-time experiments* is different. The main detection method here is neutrino–electron scattering and neutrino reactions on deuterium, in which either Cherenkov or scintillation light is created by electrons, which can then be detected. In the case of scattering, the electron direction is closely correlated with the direction of the incoming neutrino. However, so far detectors achieved an energy threshold of about 3.5 MeV and are, therefore, sensitive only to ^{8}B and hep-neutrinos. The ^{8}B flux is about four orders of magnitude lower than the pp-flux and, therefore, the target mass here has to be in the kiloton range. In discussing the existing experimental data, we will follow the historic sequence.

10.2.1 The chlorine experiment

The first solar neutrino experiment, and the birth of neutrino astrophysics in general, is the chlorine experiment of Davis [Dav64, Dav68, Dav94, Dav94a, Cle98], which has been running since 1968. The reaction used to detect the neutrinos is

$$^{37}\text{Cl} + \nu_e \rightarrow {}^{37}\text{Ar} + e^- \tag{10.38}$$

which has an energy threshold of 814 keV. The detection method utilizes the decay

$$^{37}\text{Ar} + e \rightarrow {}^{37}\text{Cl} + \nu_e \tag{10.39}$$

which has a half-life of 35 days and results in 2.82 keV X-rays or Auger electrons from K-capture (90%). With the given threshold this experiment is not able to measure the pp neutrino flux. The contributions of the various production reactions for neutrinos to the total flux are illustrated in Table 10.2 according to one of the current solar models. The solar model calculations predict values of (7.5 ± 1.0) SNU [Bah01, Cou02], where the major part comes from the ^{8}B neutrinos. All, except the ^{8}B neutrinos, only lead to the ground state of ^{37}Ar whereas ^{8}B is also populating excited states including the isobaric analogue state. The cross-section for the reaction (10.33) averaged over the ^{8}B spectrum has been measured to be $\sigma = 1.14 \pm 0.11 \times 10^{-42}$ cm^2 [Auf94, Bah95]. The experiment (Figure 10.8) operated in the Homestake gold mine in South Dakota (USA), where a tank with 615 t perchloro-ethylene (C_2Cl_4), which served as the target, was situated at a depth corresponding to 4100 m.w.e. (metre water equivalent). The natural abundance of ^{37}Cl is about 24%, so that the number of target atoms is 2.2×10^{30}. The argon atoms that are produced are volatile in the solution and are extracted about once every 60–70 days. The extraction efficiency is controlled by adding a small amount of isotopical pure

Figure 10.8. The chlorine detector of R. Davis for the detection of solar neutrinos in the approximately 1400 m deep Homestake Mine in Lead, South Dakota (USA) in about 1967. The 380 000 l tank full of perchloro-ethylene is shown. Dr. Davis is standing above (Courtesy Brookhaven National Laboratory.)

inert ^{36}Ar or ^{38}Ar. To do this, helium is flushed through the tank taking the volatile argon out of the solution and allowing the collection of the argon atoms in a cooled charcoal trap. The trapped argon is then purified, concentrated in several steps and finally filled into special miniaturized proportional counters. These are then placed in a very low activity lead shielding and then the corresponding ^{37}Ar-decay can be observed. In order to further reduce the background, both the energy information of the decay and the pulse shape are used. A production rate of one argon atom per day corresponds to 5.35 SNU. The results from more than 20 years of measuring are shown in Figure 10.9. The average counting rate of 108 runs is [Cle98]

$$2.56 \pm 0.16(\text{stat.}) \pm 0.15(\text{sys.}) \text{ SNU}. \tag{10.40}$$

This is less than the value predicted by the standard solar models. This discrepancy is the primary source of the so-called *solar neutrino problem*.

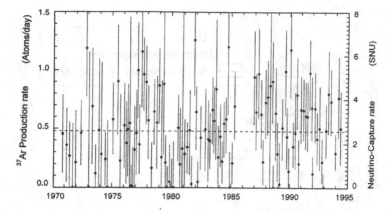

Figure 10.9. The neutrino flux measured from the Homestake ^{37}Cl detector since 1970. The average measured value (broken line) is significantly smaller than the predicted one. This discrepancy is the origin of the so-called solar neutrino problem (from [Dav96]). © 1996 With permission from Elsevier.

10.2.2 Super-Kamiokande

A real-time experiment for solar neutrinos is being carried out with the Super-Kamiokande detector [Fuk03], an enlarged follow-up version of the former Kamioka detector. This experiment is situated in the Kamioka mine in Japan and has a shielding depth of 2700 m.w.e. Super-Kamiokande started operation on 1 April 1996 and has already been described in detail in Chapter 9. The fiducial mass used for solar neutrino searches is 22 kt. The detection principle is the Cherenkov light produced in neutrino-electron scattering within the water. Energy and directional information are reconstructed from the corresponding number and timing of the hit photomultipliers. The cross-section for neutrino–electron scattering is given in Chapter 3. From that, the differential cross-section with respect to the scattering angle can be deduced as

$$\frac{d\sigma}{d\cos\theta} = 4\frac{m_e}{E_\nu}\frac{(1+m_e/E_\nu)^2\cos\theta}{[(1+m_e/E_\nu)^2-\cos^2\theta]^2}\frac{d\sigma}{dy} \tag{10.41}$$

with

$$y = \frac{2(m_e/E_\nu)\cos^2\theta}{(1+m_e/E_\nu)^2-\cos^2\theta}. \tag{10.42}$$

Therefore, for $E_\nu \gg m_e$ the electron keeps the neutrino direction with $\theta \lesssim (2m_e/E_\nu)^{1/2}$. This directional information is clearly visible as shown in Figure 10.10.

Given the threshold of about 4 MeV, the detector can measure only the ^{8}B and hep neutrino flux. The experimental observation up to the end of March 2017 is shown in Figure 10.10, resulting in 89, 285 neutrino events. From the measurements

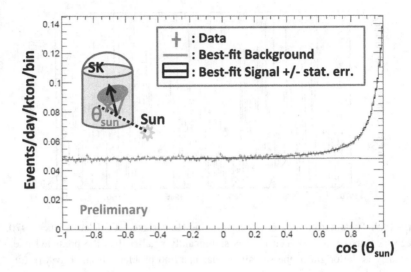

Figure 10.10. Angular distribution of the events in the Super-Kamiokande detector, relative to the direction of the Sun, after a measuring time of several years. More than 84000 solar neutrinos have been detected (from [Yan17]).

a time-averaged flux of ^{8}B neutrinos of [Yan17]

$$\Phi(^8B) = 2.80 \pm 0.19(\text{stat.}) \pm 0.33(\text{sys.}) \cdot 10^6 \, \text{cm}^{-2} \, \text{s}^{-1} \, \text{Kamiokande} \quad (10.43)$$

$$\Phi(^8B) = 2.38 \pm 0.02(\text{stat.}) \pm 0.08(\text{sys.}) \cdot 10^6 \, \text{cm}^{-2} \, \text{s}^{-1} \, \text{Super-K I} \quad (10.44)$$

$$\Phi(^8B) = 2.41 \pm 0.05(\text{stat.})^{+0.16}_{-0.15}(\text{sys.}) \quad \cdot 10^6 \, \text{cm}^{-2} \, \text{s}^{-1} \, \text{Super-K II} \quad (10.45)$$

$$\Phi(^8B) = 2.40 \pm 0.04(\text{stat.}) \pm 0.05(\text{sys.}) \cdot 10^6 \, \text{cm}^{-2} \, \text{s}^{-1} \, \text{Super-K III} \quad (10.46)$$

$$\Phi(^8B) = 2.36 \pm 0.02(\text{stat.}) \pm 0.04(\text{sys.}) \cdot 10^6 \, \text{cm}^{-2} \, \text{s}^{-1} \, \text{Super-K IV} \quad (10.47)$$

has been measured, leading to an overall averaged flux of $\Phi(^8B) = 2.37 \pm 0.02(\text{stat.}) \pm 0.04(\text{sys.}) \times 10^6 \, \text{cm}^{-2} \, \text{s}^{-1}$, being little bit less than 50 % of the SSM prediction of [Bah01]. It should be mentioned that the phases SK-I and SK-III were running with a threshold of 4.5 MeV, while SK-II was running with 6.5 MeV and SK-IV with 3.5 MeV threshold. In total about 100000 solar ^{8}B neutrinos have been detected. A possible hep flux is constrained to be

$$\Phi(\text{hep}) < 7.3 \times 10^4 \, \text{cm}^{-2} \, \text{s}^{-1} \quad (10.48)$$

corresponding to less than 7.9 times the SSM prediction.

It should be noted that the observed flux at Super-Kamiokande for neutrino oscillations is a superposition of ν_e and ν_μ, ν_τ. The dominant number of events is produced by ν_e scattering because of the higher cross-section (see Chapter 4). The high statistics of Super-Kamiokande not only allows the total flux to be measured; they also yield more detailed information which is very important for neutrino oscillation discussions. In particular, the spectral shape of the ^{8}B spectrum is measured, annual variations in the flux and day/night effects and these issues will be discussed later in this chapter.

10.2.3 The gallium experiments

Both experiments described so far were unable to measure directly the pp flux, which is the reaction coupled to the Sun's luminosity (Figure 10.6). A suitable material to detect these neutrinos is gallium [Kuz66]. At that time there have been two experiments built that were sensitive to the pp neutrino flux: GALLEX/GNO and SAGE. The detection relies on the reaction

$$^{71}Ga + \nu_e \rightarrow {}^{71}Ge + e^- \tag{10.49}$$

with a threshold energy of 233.5 ± 1.2 keV [Fre15]. The natural abundance of ^{71}Ga is 39.9%. The detection reaction is via electron capture

$$^{71}Ge + e^- \rightarrow {}^{71}Ga + \nu_e \tag{10.50}$$

resulting in Auger electrons and X-rays from K and L capture from the ^{71}Ge decay producing two lines at 10.37 keV and 1.2 keV. The detection of the ^{71}Ge decay (half-life 11.4 days, 100% via electron capture) is achieved using miniaturized proportional counters similar to the chlorine experiment. Both energy and pulse shape information are also used for the analysis.

10.2.3.1 GALLEX

The GALLEX collaboration (gallium experiment) [Ans92a, Ans95b] used 30.3 t of gallium in the form of 101 t of $GaCl_3$ solution. Their experiment was carried out in the Gran Sasso underground laboratory from 1991 to 1997. The produced $GeCl_4$ is volatile in $GaCl_3$ and was extracted every three weeks by flushing nitrogen through the tank. Inactive carriers of ^{72}Ge, ^{74}Ge and ^{76}Ge were added to control the extraction efficiency, which was at 99%. The germanium was concentrated in several stages and subsequently transformed into germane (GeH_4), which has similar characteristics as methane (CH_4), which when mixed with argon, is a standard gas mixture (P10) in proportional counters. The germane is, therefore, also mixed with a noble gas (Xe) to act as a counter gas, the mixture being optimized for detection efficiency, drift velocity and energy resolution.

In addition, there was a first attempt to demonstrate the total functionality of a solar neutrino experiment using an artificial 2 MCi (7.4×10^{16} Bq) ^{51}Cr source. This yielded mono-energetic neutrinos, of which 81% had $E_\nu = 746$ keV. This test was performed twice. The results of the two calibrations of GALLEX with the artificial chromium neutrino source resulted in the following ratios between the observed number of ^{71}Ge decays and the expectation from the source strength [Ans95a, Ham96]

$$R = 1.04 \pm 0.12 \quad \text{and} \quad 0.83 \pm 0.08. \tag{10.51}$$

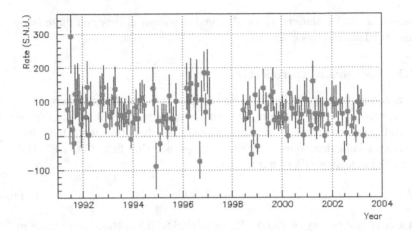

Figure 10.11. Results from 65 GALLEX runs and 58 GNO runs (from [Alt05]). © 2005 With permission from Elsevier.

This confirms the full functionality and sensitivity of the GALLEX experiment to solar neutrinos. At the end of the experiment ^{71}As, which also decays into ^{71}Ge, was added to study extraction with *in situ* produced ^{71}Ge. The final result of GALLEX is $77.5 \pm 6.2(\text{stat.})^{+4.3}_{-4.7}(\text{sys.})$ SNU(1σ) [Ham99], with theoretical predictions of 128 ± 8 SNU [Bah01, Cou02]. The data were re-evaluated later and gave an event rate of [Kae10]

$$73.4^{+7.1}_{-7.3} \quad \text{SNU} \tag{10.52}$$

Clearly the experiment is far off from expectation.

10.2.3.2   GNO

After some maintenance and upgrades of the GALLEX equipment, the experiment was renewed in the form of a new collaboration—GNO. After 58 runs, GNO reported a value of [Alt05]

$$62.9 \pm 5.4(\text{stat.}) \pm 2.5(\text{sys.}) \text{ SNU} \tag{10.53}$$

which combined with the 65 GALLEX runs averages to

$$69.3 \pm 5.5 \text{ SNU.} \tag{10.54}$$

The single run signal is shown in Figure 10.11.

10.2.3.3   SAGE

The Soviet–American collaboration, SAGE [Gav03], uses 57 t of gallium in metallic form as the detector and has operated the experiment in the Baksan underground

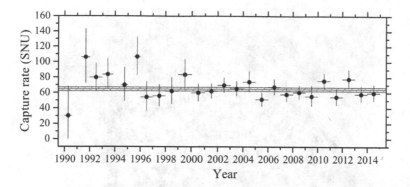

Figure 10.12. Results of the SAGE solar neutrino measurements combined by years. The shaded region corresponds to the combined SAGE result of 64.6 ± 2.4(stat.) SNU. The vertical bars at each point correspond to a statistical error of 68%, and the horizontal bars correspond to the time interval of the combined analysis of measurements (from [Mey19]). With kind permission of V. Gavrin. © 2019 World Scientific

laboratory since 1990. The main difference with respect to GALLEX lies in the extraction of ^{71}Ge from metallic gallium.

The SAGE experiment was calibrated in a similar way to GALLEX [Abd99] and also with ^{37}Ar [Abd06]. The result of SAGE is [Mey19]

$$64.6 \pm 2.4(\text{stat.}) \, \text{SNU.} \tag{10.55}$$

The data are shown in Figure 10.12. Both gallium experiments are in good agreement and show fewer events than expected from the standard solar models. The combined results of both experiments is

$$66.1 \pm 3.1 \, \text{SNU.} \tag{10.56}$$

Both GALLEX and SAGE provide the first observation of pp neutrinos and an experimental confirmation that the Sun's energy really does come from hydrogen fusion.

10.2.4 The Sudbury Neutrino Observatory (SNO)

All experiments so far indicate a deficit of solar neutrinos compared to the theoretically predicted flux (see Table 10.2). If we accept that there is a real discrepancy between experiment and theory, there are two main solutions to the problem. One is that the model of the Sun's structure may not be correct or our knowledge of the neutrino capture cross-sections may be insufficient; the other is the possibility that the neutrino has as yet unknown properties. The aim of the Sudbury Neutrino Observatory (SNO) was to make a measurement of the solar neutrino flux independent of solar models.

Figure 10.13. Construction of the Sudbury Neutrino Observatory (SNO) in a depth of 2070 m in the Craighton mine near Sudbury (Ontario). This Cherenkov detector used heavy water rather than normal water. The heavy water tank was shielded by an additional 7300 t of normal water. The support structure for the photomultipliers is also shown. With kind permission of the SNO collaboration and Lawrence Berkeley National Laboratory.

Being a real-time Cherenkov detector like Super-Kamiokande, this experiment used 1000 t of heavy water (D_2O) instead of H_2O. Placed in a transparent acrylic vessel, it was surrounded by 9700 photomultipliers and several kilotons of H_2O as shielding [Bog00] (Figure 10.13). The threshold of SNO was about 5 MeV which in later analyses could be reduced down to 3.5 MeV. Heavy water allows several reaction channels to be studied. The first one is charged weak currents sensitive only to ν_e:

$$\nu_e + d \rightarrow e^- + p + p \qquad \text{(CC)} \qquad (10.57)$$

with a threshold of 1.442 MeV. In addition to this charge current reaction on deuterium (10.57), a second process for detecting all types of neutrinos is neutrino-electron scattering

$$\nu + e^- \rightarrow \nu + e^- \quad \text{(ES)} \tag{10.58}$$

with a dominant contribution from ν_e scattering as in Super-Kamiokande. The remarkable new aspect, however, was the additional determination of the *total* neutrino flux, independent of any oscillations, due to the flavour-independent reaction via neutral weak currents

$$\nu + d \rightarrow \nu + p + n \quad \text{(NC)} \tag{10.59}$$

which has a threshold of 2.225 MeV (the binding energy of D). Cross-section calculations can be found in [Nak02]. The released neutrons were detected in phase I of the experiment via 6.3 MeV gamma-rays produced in the reaction

$$n + d \rightarrow {}^3\text{H} + \gamma. \tag{10.60}$$

To enhance the detection efficiency of neutrons and therefore improve on the NC flux measurement, two further phases were performed: To enhance the NC sensitivity, 2 t of NaCl were added to the heavy water (phase II) to use the gamma-rays up to 8.6 MeV produced in the ${}^{35}\text{Cl}(n, \gamma) {}^{36}\text{Cl}$ process, which also has a higher cross-section for neutron capture than deuterium alone. A discrimination of NC and CC reactions on an event-by-event basis (phase III) was possible by deploying a set of ${}^3\text{He}$-filled proportional counters (neutral current detectors, NCDs). Here neutron detection occurred via

$$^3\text{He} + n \rightarrow {}^3\text{H} + p. \tag{10.61}$$

A measurement of the ratio of the two processes, (10.57) and (10.59), provides a direct test of the oscillation hypothesis. However, a comparison of the CC absorption with the scattering process also provides important information. As mentioned, ν_μ and ν_τ also contribute to elastic scattering but with a lower cross-section (see Chapter 4) and the ratio is given by

$$\frac{\text{CC}}{\text{ES}} = \frac{\nu_e}{\nu_e + 0.14(\nu_\mu + \nu_\tau)}. \tag{10.62}$$

The first measurement by SNO on pure D$_2$O of the CC reaction resulted in [Ahm01, Aha07]

$$\Phi(^8\text{B}) = 1.76 \pm 0.06(\text{stat.}) \pm 0.09(\text{sys.}) \times 10^6 \text{ cm}^{-2} \text{ s}^{-1} \tag{10.63}$$

significantly less than the value of Super-Kamiokande. This was already a hint that additional active neutrino flavours are coming from the Sun as they participate in the scattering process. The real breakthrough came with the measurement of the first NC data [Ahm02]. This flavour-blind reaction indeed measured a total solar neutrino flux of

$$\Phi_{\text{NC}} = 5.09^{+0.44}_{-0.43}(\text{stat.})^{+0.46}_{-0.43}(\text{sys.}) \times 10^6 \text{ cm}^{-2} \text{ s}^{-1} \tag{10.64}$$

in excellent agreement with the standard model (Figure 10.14). The flux in non-ν_e neutrinos is

$$\Phi(\nu_\mu, \nu_\tau) = 3.41 \pm 0.45 \pm 0.43 \times 10^6 \text{ cm}^{-2} \text{ s}^{-1} \tag{10.65}$$

on a 5.3σ level different from zero. This result is in good agreement with the measurement using the data set including salt and therefore having an enhanced NC sensitivity. Here, it is assumed a total NC flux with no constraint on the spectral shape of [Ahm04, Aha05]

$$\Phi_{NC}^{Salt} = 5.21 \pm 0.27(\text{stat.}) \pm 0.38(\text{sys.}) \times 10^6 \text{ cm}^{-2} \text{ s}^{-1}$$

and assuming just a ^{8}B spectral shape results in

$$\Phi_{NC}^{Salt} = 4.90 \pm 0.24_{-0.27}^{+0.29} \times 10^6 \text{ cm}^{-2} \text{ s}^{-1}.$$

The result of the neutral current flux from the NCD phase is [Aha08]

$$\Phi_{NC}^{NCD} = 5.54_{-0.31}^{+0.33}(\text{stat.})_{-0.34}^{+0.36}(\text{sys.}) \times 10^6 \text{ cm}^{-2} \text{ s}^{-1}.$$

The CC/NC ratio is given by 0.301 ± 0.033. After several decades the problem of missing solar neutrinos is finally solved. The measured total solar neutrino flux is in accordance with the Standard Solar Model predictions but only about 30% of them arrive at Earth as electron neutrinos. It is no longer a problem of missing neutrinos but a fact that the bulk of solar neutrinos arrive at the Earth in a different flavour. A further reduction of the threshold down towards lower energy has been achieved (about 3.5 MeV, called low energy threshold analysis, LETA) for a combined analysis on phase I and II resulting in a total ^{8}B neutrino flux of [Aha10]

$$\Phi_{NC}^{LETA} = 5.046_{-0.152}^{+0.159}(\text{stat.})_{-0.123}^{+0.107}(\text{sys.}) \times 10^6 \text{ cm}^{-2} \text{ s}^{-1}.$$

Combined with the results from the reactor experiment KamLAND (see Chapter 5) this restricts the oscillation parameters to an extremely small region as discussed later in Section 10.3.3. SNO also searched for potential $\bar{\nu}_e$ via the reaction

$$\bar{\nu}_e + d \rightarrow e^+ + n + n \tag{10.66}$$

without seeing a signal [Aha04]. In addition, an upper limit on the hep neutrino flux of $2.3 \times 10^4 \text{ cm}^{-2} \text{ s}^{-1}$ could be derived [Aha06].

10.2.5   The Borexino experiment

Over about the last decade, the Borexino experiment at the Gran Sasso underground laboratory (LNGS) has been the dominant source of information for solar neutrino measurements. Borexino has been designed - among others - to measure the important (862 keV) ^{7}Be line in real time [Bor91]. The Borexino experiment uses 300 t of a liquid scintillator, of which 100 t can be used as a fiducial volume (i.e., in order to reduce background only this reduced volume is used for data taking). The detection reaction is

$$\nu_e + e^- \rightarrow \nu_e + e^- \tag{10.67}$$

but, in contrast to existing real time experiments, scintillation instead of the Cherenkov effect is used for detection, allowing a much lower threshold.

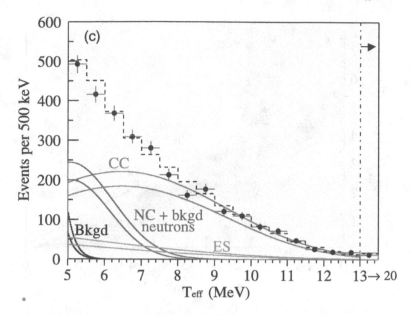

Figure 10.14. Kinetic energy T_{eff} of events with a vertex within a fiducial volume of 550 cm radius and $T_{eff} > 5$ MeV. Also shown are the Monte Carlo predictions for neutral currents (NC) and background neutrons, charged currents (CC) and elastic scattering (ES), scaled to the fit results. The broken lines represent the summed components and the bands show $\pm 1\sigma$ uncertainties (from [Aha07]). © 2007 by the American Physical Society.

The pp-neutrino detection is massively prohibited by the ^{14}C content of the organic scintillator, even though Borexino has several orders of magnitude lower contamination than normal ^{14}C. The mono-energetic ^{7}Be line produces a recoil spectrum of electrons which has a maximum energy of 665 keV ('Compton edge'). The signal is visible as a plateau in a region between 250–650 keV. This was indeed observed and thus is the first real time solar neutrino detection below 1 MeV [Arp08, Arp08a]. In the meantime with Borexino phase II the experiment has improved again with background reduction and it was finally even able to measure the pp-neutrinos in real time [Bel14]. Borexino measured also the pep-neutrinos and ^{7}Be neutrinos with high precision [Ago17a]. This even allowed a common fit in the analysis of all these components [Ago18].

10.3 Theoretical solutions–matter effects

The problem of missing solar neutrinos is basically solved by the SNO results in combination with KamLAND because there is no way to produce ν_μ or ν_τ in the fusion processes of the Sun. The total observed flux agrees well with the SSM predictions (Figure 10.15) and the corresponding fit values are reproduced on Earth completely independently with nuclear reactor experiments. For historical reasons

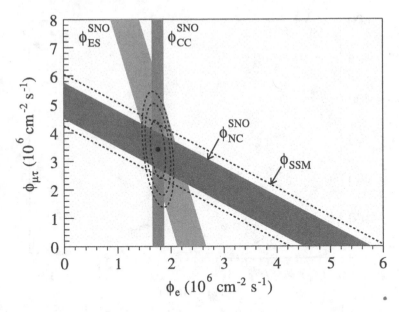

Figure 10.15. Flux of ^{8}B solar neutrinos which are μ or τ flavour *versus* flux of ν_e deduced from the three neutrino reactions in the SNO phase II using salt. The diagonal bands show the total ^{8}B flux as predicted by the SSM (dashed lines) and the ones measured with the NC reaction at SNO (solid line). The intercepts of these bands with the axes represent the $\pm 1\sigma$ errors (from [Aha05]). © 2005 by the American Physical Society.

and because of the physics involved, which might be important in supernovae as well, we discuss two possible sources for the flavour conversion: neutrino oscillations and a neutrino magnetic moment.

10.3.1 Neutrino oscillations as a solution to the solar neutrino problem

Two kinds of solutions are provided by oscillations. Either the ν_e oscillates in vacuum on its way to Earth or it has already been converted within the Sun by matter effects (see Chapter 8). In the case of vacuum solutions, the baseline is about 1.5×10^8 km and E_ν about 10 MeV, resulting in Δm^2 regions of around 10^{-10} eV^2. However, this is in disagreement with the observation of KamLAND and can be ruled out. The other attractive solution is a conversion in matter via the Mikheyev, Smirnov and Wolfenstein (MSW) effect [Wol78, Mik86].

10.3.2 Neutrino oscillations in matter and the MSW effect

Matter influences the propagation of neutrinos by elastic, coherent forward scattering. The basic idea of this effect is the differing interactions of different neutrino flavours within matter. While interactions with the electrons of matter via neutral weak currents are possible for all kinds of neutrinos, only the ν_e can interact

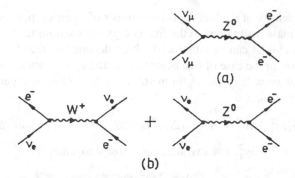

Figure 10.16. Origin of the Mikheyev, Smirnov and Wolfenstein effect. Whereas weak NC interactions are possible for all neutrino flavours, only the ν_e also has the possibility of interacting via charged weak currents (from [Kla95]). © Taylor & Francis Group.

via charged weak currents (see Figure 10.16). The CC for the interaction with the electrons of matter leads to a contribution to the interaction Hamiltonian of

$$H_{WW} = \frac{G_F}{\sqrt{2}} [\bar{e}\gamma^\mu(1-\gamma_5)\nu_e][\bar{\nu}_e\gamma_\mu(1-\gamma_5)e]. \tag{10.68}$$

By a Fierz transformation (see Chapter 3) this term can be brought to the form

$$H_{WW} = \frac{G_F}{\sqrt{2}} [\bar{\nu}_e\gamma^\mu(1-\gamma_5)\nu_e][\bar{e}\gamma^\mu(1-\gamma_5)e]. \tag{10.69}$$

Calculating the four-current density of the electrons in the rest frame of the Sun, we obtain

$$\langle e|\bar{e}\gamma^i(1-\gamma_5)e|e\rangle = 0 \tag{10.70}$$

$$\langle e|\bar{e}\gamma^0(1-\gamma_5)e|e\rangle = N_e. \tag{10.71}$$

The spatial components of the current must disappear (no permanent current density throughout the Sun) and the zeroth component can be interpreted as the electron density of the Sun. For left-handed neutrinos we can replace $(1-\gamma_5)$ by a factor of 2, so that Equation (10.68) can be written as

$$H_{WW} = \sqrt{2}G_F N_e \bar{\nu}_e\gamma_0\nu_e. \tag{10.72}$$

Thus the electrons contribute to an additional potential V for the electron neutrino $V = \sqrt{2}G_F N_e$, with N_e as the electron density in the Sun. With this additional term, the free energy–momentum relation becomes (see Chapter 7)

$$p^2 + m^2 = (E-V)^2 \simeq E^2 - 2EV \qquad \text{(for } V \ll E\text{)}. \tag{10.73}$$

In more practical units this can be written as [Bet86]

$$2EV = 2\sqrt{2}\left(\frac{G_F Y_e}{m_N}\right)\rho E = A \tag{10.74}$$

where ρ is the density of the Sun, Y_e is the number of electrons per nucleon and m_N is the nucleon mass. In analogy to the free energy–momentum relation, an effective mass $m_{\text{eff}}^2 = m^2 + A$ can be introduced which depends on the electron density of the solar interior. In the case of two neutrinos ν_e and ν_μ in matter, the matrix of the squares of the masses of ν_e and ν_μ in matter have the following eigenvalues for the two neutrinos $m_{1m,2m}$:

$$m_{1m,2m}^2 = \tfrac{1}{2}(m_1^2 + m_2^2 + A) \pm [(\Delta m^2 \cos 2\theta - A)^2 + \Delta m^2 \sin^2 2\theta]^{1/2} \quad (10.75)$$

where $\Delta m^2 = m_2^2 - m_1^2$. The two states are closest together for

$$A = \Delta m^2 \cos 2\theta \quad (10.76)$$

which corresponds to an electron density of

$$N_e = \frac{\Delta m^2 \cos 2\theta}{2\sqrt{2}G_F E} \quad (10.77)$$

(see also [Bet86, Gre86, Sch97]). At this point (the resonance region) the oscillation amplitude is maximal, meaning ν_e and ν_μ oscillation between the extremes 0 and 1 can occur, independent of the vacuum mixing angle. Also the oscillation length has a maximum of

$$L_{mR} = \frac{L_0}{\sin 2\theta} = \frac{1.64 \times 10^7 \text{m}}{\tan 2\theta \times Y_e \rho/\text{g cm}^{-3}} \quad (10.78)$$

with $L_0 = L_C \cos 2\theta$ at the resonance. L_C, the scattering length of coherent forward scattering, is given as $(p \approx E)$

$$L_C = \frac{4\pi E}{A} = \frac{\sqrt{2}\pi}{G_F N_e} = \frac{\sqrt{2}\pi m_N}{G_F Y_e \rho} \quad (10.79)$$

which can be written numerically as

$$L_C = 1.64 \times 10^7 \frac{N_A}{N_e/\text{cm}^{-3}} \text{ m} = \frac{1.64 \times 10^7}{Y_e \rho/\text{g cm}^{-3}} \text{ m}. \quad (10.80)$$

The oscillation length is stretched by a factor $\sin^{-1} 2\theta$ with respect to vacuum. The width (FWHM) of the resonance

$$\Gamma = 2\Delta m^2 \sin 2\theta \quad (10.81)$$

becomes broader for larger mixing angles θ. It is worthwile mentioning that such a resonance cannot occur for antineutrinos $(A \to -A)$ because now in the denominator of (8.67), the term $A/\Delta m^2 + \cos 2\theta$ cannot vanish $(0 < \theta < \pi/4$ for $\Delta m^2 > 0)$.

10.3.2.1   Constant density of electrons

The energy difference of the two neutrino eigenstates in matter is modified compared to that in the vacuum by the effect discussed in the previous section to become

$$(E_1 - E_2)_m = C \cdot (E_1 - E_2)_V \quad (10.82)$$

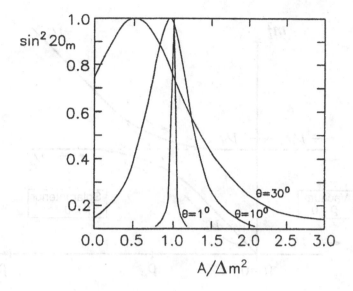

Figure 10.17. Dependence of the oscillation amplitude as a function of $A/\Delta m^2$ for different values of the vacuum mixing angle $\theta (\theta = 1°, 10°, 30°)$. In principle, this oscillation can be maximal, but this is not observed (from [Sch97]). Reproduced with permission of SNCSC.

where C is given by

$$C = \left[1 - 2 \left(\frac{L_V}{L_e} \right) \cos 2\theta_V + \left(\frac{L_V}{L_e} \right)^2 \right]^{\frac{1}{2}} \tag{10.83}$$

and the neutrino–electron interaction length L_e is given by

$$L_e = \frac{\sqrt{2}\pi\hbar c}{G_F N_e} = 1.64 \times 10^5 \left(\frac{100 \text{ g cm}^{-3}}{\mu_e \rho} \right) \quad [\text{m}]. \tag{10.84}$$

The equations describing the chance of finding another flavour eigenstate after a time t correspond exactly to Equation (8.24) with the additional replacements

$$L_m = \frac{L_V}{C} \tag{10.85}$$

$$\sin 2\theta_m = \frac{\sin 2\theta_V}{C}. \tag{10.86}$$

In order to illustrate this, we consider the case of two flavours in three limiting cases. Using Equations (8.24) and (10.83), the oscillation of ν_e into a flavour ν_x is given by

$$|\langle \nu_x | \nu_e \rangle|^2 = \begin{cases} \sin^2 2\theta_V \sin^2(\pi R/L_V) & \text{for } L_V/L_e \ll 1 \\ (L_e/L_V)^2 \cdot \sin^2 2\theta_V \sin^2(\pi R/L_e) & \text{for } L_V/L_e \gg 1 \\ \sin^2(\pi R \sin 2\theta_V/L_V) & \text{for } L_V/L_e = \cos 2\theta_V. \end{cases}$$

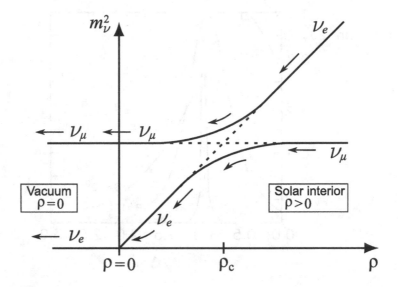

Figure 10.18. The MSW effect. The heavy mass eigenstate is almost identical to ν_e inside the Sun: in the vacuum, however, it is almost identical to ν_μ. If a significant jump at the resonance density ρ_c can be avoided, the produced electron neutrino remains on the upper curve and therefore escapes detection in radiochemical experiments. The transformation close to the crossing point can be described by the Landau-Zener theory. The experimental results can be interpreted in adiabatic and diabatic conversions. The quantum mechanical perturbation theory requires a gap between the states. (For additional information see [Bet86]). With kind permission of S. Turkat.

The last case corresponds exactly to the resonance condition mentioned earlier. In the first case, corresponding to very small electron densities, the matter oscillations reduce themselves to vacuum oscillations. In the case of very high electron densities, the mixture is suppressed by a factor $(L_e/L_V)^2$. The third case, the resonance case, contains an energy-dependent oscillatory function, whose energy average results typically in a value of 0.5. This corresponds to maximal mixing. In a medium with constant electron density N_e, the quantity A is constant for a fixed E and, in general, does not fulfil the resonance condition. Therefore, the effect described here does not show up. However, in the Sun we have varying density which implies that there are certain resonance regions where this flavour conversion can happen.

10.3.2.2   *Variable electron density*

A variable density causes a dependence of the mass eigenstates $m_{1m,2m}$ on A (N_e) which is shown in Figure 10.18. Assume the case $m_1^2 \approx 0$ and $m_2^2 > 0$ which implies

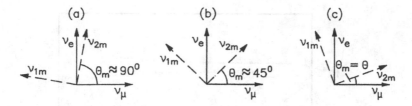

Figure 10.19. Representation of the three phases of the MSW effect in the (ν_e, ν_μ) plane: (a) $\theta_m \approx 90°$ in the solar interior; (b) $\theta_m \approx 45°$ in the resonance layer; and (c) $\theta_m \approx 0°$ at the surface of the Sun (from [Sch97]). Reproduced with permission of SNCSC.

$\Delta m^2 \approx m_2^2$. For $\theta = 0$, resulting in $\theta_m = 0$ as well for all A, this results in

$$\nu_{1m} = \nu_1 = \nu_e \qquad \text{with} \qquad m_{1m}^2 = A$$
$$\nu_{2m} = \nu_2 = \nu_\mu \qquad \text{with} \qquad m_{2m}^2 = m_2^2. \qquad (10.87)$$

The picture changes for small $\theta > 0$. Now for $A = 0$ the angle $\theta_m = \theta$ which is small and implies

$$\nu_{1m} = \nu_1 \approx \nu_e \qquad \text{with} \qquad m_{1m}^2 = 0$$
$$\nu_{2m} = \nu_2 \approx \nu_\mu \qquad \text{with} \qquad m_{2m}^2 = m_2^2. \qquad (10.88)$$

For large A there is $\theta_m \approx 90°$ and the states are given as

$$\nu_{1m} \approx -\nu_\mu \qquad \text{with} \qquad m_{1m}^2 \approx m_2^2$$
$$\nu_{2m} \approx \nu_e \qquad \text{with} \qquad m_{2m}^2 \approx A \qquad (10.89)$$

opposite to the $\theta = 0°$ case (Figure 10.19). This implies an inversion of the neutrino flavour. While ν_{1m} in vacuum is more or less ν_e, at high electron density it corresponds to ν_μ; the opposite is valid for ν_{2m}. This flavour flip is produced by the resonance where maximal mixing is possible.

Solar neutrinos are produced in the interior of the Sun, where the density is $\rho \approx 150$ g cm^{-3}. Therefore, assuming $A/\Delta m^2 \gg 1$ equivalent to $\theta_m \approx 90°$, the produced ν_e are basically identical to ν_{2m}, the heavier mass eigenstate. A ν_e produced in the interior of the Sun, therefore, moves along the upper curve and passes a layer of matter where the resonance condition is fulfilled. Here maximal mixing occurs, $\theta_m \approx 45°$, and

$$\nu_{2m} = \frac{1}{\sqrt{2}}(\nu_e + \nu_\mu). \qquad (10.90)$$

Passing the resonance from right to left and remaining on the upper curve (adiabatic case), the state ν_{2m} at the edge of the Sun is now associated with ν_μ. The average probability that a ν_e produced in the solar interior passes the resonance and leaves the Sun still as ν_e is given by

$$P(\nu_e \to \nu_e) = \tfrac{1}{2}(1 + \cos 2\theta_m \cos 2\theta) \qquad (10.91)$$

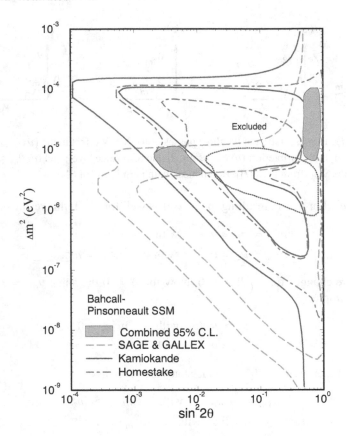

Figure 10.20. Contour ('Iso-SNU') plot in the Δm^2 *versus* $\sin^2 2\theta$ plane. For each experiment the total rate defines a triangular-shaped region as an explanation of the experimental results. The different energy thresholds cause a shift of the curves and only the overlap regions describe all data. Additional information like day–night effects constrain the regions further (from [Hat94]). © 1994 by the American Physical Society.

with θ_m as the mixing angle at the place of neutrino production. The conversion is, therefore,

$$P(\nu_e \to \nu_\mu) = \tfrac{1}{2}(1 - \cos 2\theta_m \cos 2\theta) \approx \cos^2 \theta. \qquad (10.92)$$

The smaller the vacuum mixing angle is, the larger the flavour transition probability becomes.

10.3.3 Experimental signatures and results

Having discussed matter effects, we now want to examine which parameter space is consistent with the experimental results. Every experiment measures the probability $P(\nu_e \to \nu_e)$ which manifests itself in a triangular-shaped Iso-SNU band in the Δm^2–$\sin^2 2\theta$ plot (Figure 10.20). Because of the different energy intervals investigated by the experiments, the bands are shifted against each other, but further information is available. An energy-dependent suppression could be visible in two observables: a distortion in the ^{8}B β-spectrum and a day–night effect. The MSW effect could occur on Earth during the night, if the neutrinos have to travel through matter, just as it does on the Sun. The density in the mantle is 3–5.5 g cm^{-3} and that of the core 10–13 g cm^{-3}. This density change is not big enough to allow for the full MSW mechanism but for fixed E_ν there is a region in Δm^2 where the resonance condition is fulfilled. Taking $E_\nu = 10$ MeV, $\rho \approx 5$ g cm^{-3} and $Y_e \approx 0.5$, a value of $\Delta m^2 \approx 4 \times 10^{-6}$ eV^2 for $\cos 2\theta \approx 1$ results. Because of the now strong oscillations, a reconversion of ν_μ or ν_τ to ν_e can result (ν_e -regeneration). Therefore, the measured ν_e flux could be higher at night than during the day (day–night effect). For the same reason there should be an annual modulation between summer and winter because neutrinos have to travel shorter distances through the Earth in summer than in winter time for an individual experiment. However, this effect is smaller than the day–night effect. One additional requirement exists for the day–night effect to occur, namely the resonance oscillation length L_{mR} should be smaller than the Earth's diameter; therefore, $\sin^2 2\theta$ should not be too small. Taking these values for ρ and Y_e (see Chapter 11), it follows that, at resonance, half of the oscillation length is smaller than the Earth's diameter results if $\sin^2 2\theta \gtrsim 0.07$. Last but not least, there should be annual modulations due to the eccentricity of the orbit of the Earth around the Sun.

While SNO did not observe any of these effects, the longer running Super-Kamiokande and its larger mass provided first evidence of such effects. The day–night effect has been measured to [Abe16]

$$A = 2\frac{D - N}{D + N} = -0.033 \pm 0.010(\text{stat.}) \pm -0.005(\text{sys.}) \tag{10.93}$$

Taking all the solar data together and performing combined fits, only the large mixing angle (LMA) solution with $\Delta m^2 \approx 7 \times 10^{-5}$ eV^2 (Figure 10.22) with a non-maximal mixing survives. In addition, using the KamLAND result (see Chapter 8) as well, confining Δm^2 very strongly, the current best fit parameters are $\theta_{12} = 34.06°^{+1.16}_{-0.84}$ and $\Delta m^2 = 7.59^{+0.20}_{-0.21} \times 10^{-5}$ eV^2. How sensitive the fit parameters are on potential CC/NC ratios and day-night asymmetries is shown for SNO in Figure 10.24. The nice agreement between these two completely independent measurements is shown in Figure 10.23. It also shows that the vacuum mixing angle θ_{12} is already very large; hence no real resonance behaviour is needed.

With the given experimental results, the three-flavour electron neutrino survival probability is related to an effective two-flavour oscillation probability

$$P(\nu_e \to \nu_e) = \cos^4 \theta_{13} \times P^{2\nu}(\Delta m^2, \theta_{12}, \cos^2 \theta_{13} N_e) + \sin^4 \theta_{13} \tag{10.94}$$

with a rescaled electron density $\cos^2 \theta_{13} N_e$. The effect of Δm^2 from atmospheric oscillations averages out on the energies and distances of relevance for solar neutrinos. The effect of a non-vanishing θ_{13} is the introduction of the factor $\cos^4 \theta_{13}$. It is convenient to describe the importance of matter effect by a parameter β as the

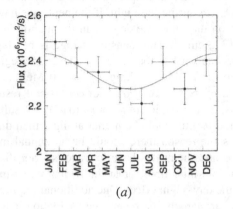

(*a*)

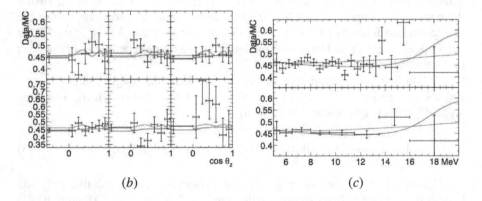

(*b*)   (*c*)

Figure 10.21. The high statistics of Super-Kamiokande allows a search for various effects on the ⁸B spectrum. A similar measurement was also done by Borexino [Ago17b]. (*a*) Seasonal effects: Only the annual modulation of the flux due to the eccentricity of the Earth orbit could be observed (solid line). Vacuum oscillations would have produced an additional effect. (*b*) Day–night effect: The solar zenith angle (θ_e) dependence of the neutrino flux normalized to the SSM prediction. The width of the night-time bins was chosen to separate solar neutrinos that pass through the Earth core ($\cos \theta_e > 0.84$) from those that pass through the mantle. (*c*) Spectral distortions: The measured ⁸B and hep spectrum relative to the SSM predictions using the spectral shape from [Ort00]. The data from 14–20 MeV are combined in a single bin. The horizontal solid line shows the measured total flux, while the dotted band around this line indicates the energy-correlated uncertainty (from [Hos06]). © 2006 by the American Physical Society.

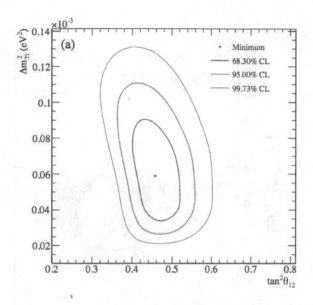

Figure 10.22. Regions in the Δm^2–$\tan^2 2\theta$ describing all obtained solar neutrino observations. As can be seen the LMA solution is the only one remaining. Contour plots are made for two different abundances (AGSS09 and GS98). For GS98 also contours with or without Super-Kamiokande day/night results. The use of $\tan^2 2\theta$ instead of $\sin^2 2\theta$ stems from the following fact: The transition probability for vacuum oscillations is symmetric under $\Delta m^2 \rightarrow -\Delta m^2$ or $\theta \rightarrow \theta + \pi/4$. However, the MSW transition is symmetric only under simultaneous transformations $(\Delta m^2, \theta) \rightarrow (-\Delta m^2, \theta \pm \pi/4)$ (see [Fog99, deG00]). For $\Delta m^2 > 0$ resonance is possible only for $\theta < \pi/4$ and thus traditionally MSW solutions were plotted in $(\Delta m^2, \sin^2 2\theta)$. In principle, solutions are possible for $\theta > \pi/4$. To account for that, $\tan^2 2\theta_{12}$ is used now. Similar results can be found in [Aha05]. For updated information see also [Est19]. Reproduced with permission of SNCSC.

ratio of oscillation length in matter and in vacuum

$$\beta = \frac{2\sqrt{2}G_F \cos^2_{13} N_e E_\nu}{\Delta m^2} \tag{10.95}$$

or in more convenient units

$$\beta = 0.22 \cos^2 \theta_{13} \left(\frac{E_\nu}{1\,\mathrm{MeV}} \right) \left(\frac{\mu_e \rho}{100\,\mathrm{g\,cm^{-3}}} \right) \left(\frac{7 \times 10^{-5}\,\mathrm{eV^2}}{\Delta m^2} \right) \tag{10.96}$$

with μ_e as the electron mean molecular weight ($\mu_e \approx 0.5 \times (1 + X)$, with X as mass fraction of hydrogen). The daytime survival probability can be expressed in a good approximation as

$$P(\nu_e \rightarrow \nu_e) = \cos^4 \theta_{13} (\tfrac{1}{2} + \tfrac{1}{2} \cos 2\theta_{12}^m \cos 2\theta_{12}) \tag{10.97}$$

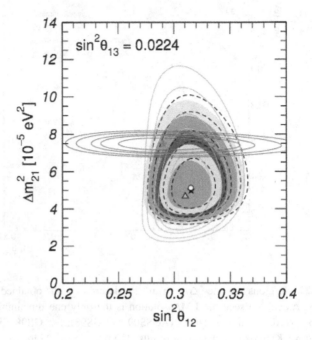

Figure 10.23. Regions in the Δm^2–$sin\theta_{12}^2$ plot from global fits describing all solar neutrino results. This time reactor data by KamLAND were included (horizontal, narrow ellipses). As can be seen, there is good overlap of allowed parameters with the LMA solution of solar neutrinos (from [Est19]). With kind permission from T. Schwetz-Mangold.

with (see also Chapter 8)

$$\cos 2\theta_{12}^m = \frac{\cos 2\theta_{12} - \beta}{\sqrt{(\cos 2\theta_{12} - \beta)^2 + \sin^2 2\theta_{12}}} \tag{10.98}$$

where β is calculated at the production point of the neutrino. The evolution is adiabatic; i.e., the mass eigenstate will always remain an eigenstate. Thus, only the initial and final densities are of importance, but not the details on the density profile. If $\beta < \cos 2\theta_{12} \approx 0.4$ the survival probability corresponds to vacuum averaged oscillations

$$P(\nu_e \rightarrow \nu_e) = \cos^4 \theta_{13}(1 - \tfrac{1}{2}\sin^2 2\theta_{12}) \tag{10.99}$$

while for $\beta > 1$ it corresponds to matter dominated oscillations

$$P(\nu_e \rightarrow \nu_e) = \cos^4 \theta_{13} \sin^2 \theta_{12}. \tag{10.100}$$

The actual transition range depends on the neutrino flavours as they are created at different solar radii; see Figure 10.6. The critical energy where $\beta = \cos 2\theta_{12}$ is for a given value of $\tan^2 \theta_{12}= 0.41$ about 3.3 MeV (pp), 2.2 MeV (^{7}Be) and 1.8 MeV (^{8}B), respectively. Thus, ^{8}B neutrinos are always in the matter-region, while pp and ^{7}Be neutrinos are in the vacuum averaged oscillation region. The pep-neutrinos with 1.44 MeV are in the transition range and the measurement of them by Borexino is constraining the exact shape of the survival probability curve.

A fit to all experimental data shows that the LMA solution is realized in nature and the MSW effect is at work [Est19] as can be seen in Figure 10.26. The full information will be available only if the whole solar neutrino spectrum is measured in real time, but most of the flux components have now been measured. Furthermore, Borexino also has seen seasonal variation in the neutrino flux [Ago17b].

10.4 Future potential experiments

Despite the fact that the solar neutrino problem is solved, still great efforts are being made to improve measurements of the solar neutrino spectrum due to its importance to both astrophysics and elementary particle physics.

For example, the measurement of the ratio

$$R = \frac{\langle ^3\text{He} + {}^4\text{He}\rangle}{\langle ^3\text{He} + {}^3\text{He}\rangle} = \frac{2\phi(^7\text{Be})}{\phi(\text{pp}) - \phi(^7\text{Be} + {}^8\text{B})} = 0.174 \quad \text{(SSM)} \quad (10.101)$$

is a very important test of SSM predictions and stellar astrophysics. Its value reflects the competition between the two primary ways of terminating the pp chain [Bah03]. The Borexino experiment measured R = 0.18 ± 0.03 [Ago18] in good agreement with the calculated values of R=0.180 $\pm$ 0.011 (HZ) and R=0.161 $\pm$ 0.010 (LZ) [Vin17]. Motivated by this, the use of detectors with various threshold energies in radiochemical experiments and through direct measurements of the energy spectrum of solar neutrinos in real time and at low energies are explored. As reaction processes to detect them, scattering and inverse β-decay are usually considered.

Another new feature is a newly arising debate on the composition of the Sun [Asp09]. Using different compositions of the Sun changes the solar neutrino flux and challenges the assumption of the amount and distribution of heavy elements in the Sun. Thus, a detection of the CNO neutrinos will be a crucial test for this. In addition, the existence of this fundamental process for stellar burning has not been experimentally verified yet. This is one measurement which could be done by Borexino or future experiments like the Jinping-experiment [Bea17] and potentially JUNO [An16, An16a].

10.4.1 Real-time measurement of pp neutrinos using coincidence techniques

The final goal of solar neutrino spectroscopy will be a real-time measurement of pp neutrinos. A proposal for doing this with nuclear coincidence techniques was made

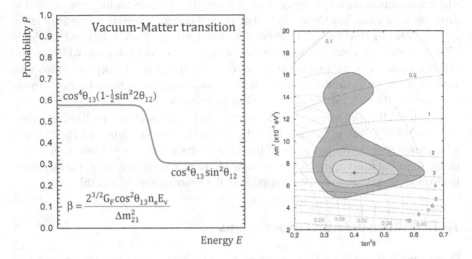

Figure 10.24. Left: Schematic picture of the (daytime) electron survival probability of solar neutrinos as function of energy for the given LMA parameters. As can be seen, there is a transition range around 1-2 MeV, above which matter effects dominate; below that is the region of vacuum oscillations (for more information see [Bah03b]). With kind permission of S. Turkat. Right: Predictions for the CC/NC ratio and day-night asymmetry for SNO. The dashed lines belong to constant CC/NC ratios and the dotted lines show constant day-night asymmetries, with the numbers given in %. Although the plot is based on early data, it shows how oscillations parameters manifest themselves in the experimental observables (from [deH04]). © 2004 With permission from Elsevier.

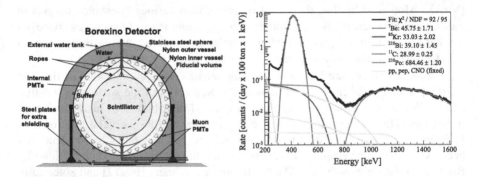

Figure 10.25. Left: Schematic picture of the Borexino detector. Right: An exemplaric plot of the number of events as a function of the energy within the Borexino experiment. A common fit to all neutrino components as well as major background components are shown. Monoenergetic neutrinos like ^{7}Be produce a Compton edge-like feature in the data (from [Bel14a]). © 2014 by the American Physical Society.

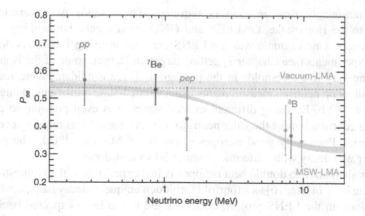

Figure 10.26. Survival probability of electron neutrinos coming from the Sun as a function of energy. The Borexino experiment was able to measure all dominant chains for neutrinos in a single experiment (from [Ago18]). Reproduced with permission of SNCSC.

Table 10.3. Real-time experiments under study for sub-MeV neutrinos using electron scattering (ES) or charged current reactions (CC) for detection.

Experiment	Idea	Principle
DEAP/CLEAN	Liquid Ar	ES
XMASS, XENON, DARWIN	Liquid Xe	ES
KamLAND	Liquid Sci.	ES
SNO+	Liquid Sci.	ES
JUNO	Liquid Sci.	ES
Jingping	Liquid Sci.	ES
MOON	^{100}Mo	CC
COBRA	^{116}Cd	CC
Li-Exp.	^{7}Li	CC

recently [Rag97]. The detection principle using coincidences relies on the following two reactions:

$$\nu_e + (A, Z) \rightarrow (A, Z + 1)_{\text{GS}} + e^- \qquad (10.102)$$
$$\hookrightarrow (A, Z + 2) + e^- + \bar{\nu}_e \qquad (10.103)$$
$$\nu_e + (A, Z) \rightarrow (A, Z + 1)^* + e^- \qquad (10.104)$$
$$\hookrightarrow (A, Z + 1)_{\text{GS}} + \gamma. \qquad (10.105)$$

Therefore, either coincidences between two electrons for the ground-state transitions or the coincidence of an electron with the corresponding de-excitation photon(s) and a reasonably time relation is required. These kinds of experiments are similar to the Homestake, GALLEX and GNO experiments, but looking for short coincidences. One example was the LENS experiment using 115In. As double β-decay experiments (see Chapter 7) getting larger and larger, some of the isotopes are providing very low thresholds in the pp-region. Three candidates were found that would allow pp neutrino measurements using excited-state transitions, namely ^{82}Se, ^{160}Gd or ^{176}Yb. By using different excited states, it is even possible to compare different contributions of the solar neutrino flux. All these ideas require several tons of material. Potentially good isotopes might be ^{100}Mo and ^{116}Cd. The produced daughter will decay with a lifetime of about 30 s via β-decay.

An alternative to double beta isotopes is just experiments like Homestake. One candidate is ^{115}In [Rag76] as a fourfold forbidden unique β-decay isotope which was under study in the LENS project, but discarded to the large expected background [Bac04]. Due to the low threshold of $E_\nu = 128$ keV it has been considered for some time, but only recently scintillator technology has made it feasible. Additional isotopes were proposed [Zub03]. Here also a possible antineutrino tag with a threshold of 713 keV is proposed by using the $\beta^+\beta^+$ emitter ^{106}Cd. Solar pp antineutrinos are, in principle, unobservable by the nuclear coincidence method, because one always has to account for at least the positron mass.

Besides this coincidence technique, there always remains the possibility of using neutrino–electron scattering as a real-time pp-neutrino reaction. As most liquid scintillators suffer from the existence of ^{14}C to perform such a measurement, new large scale noble gas detectors might be suitable for that [Bar14a]. These include XMASS, XENON and DARWIN (Liquid Xe) and DEAP/CLEAN (LAr,LNe).

In summary, after several decades of missing solar neutrinos, the problem has finally been solved by SNO. Their result shows clearly that the full solar neutrino flux is arriving on Earth, but the dominant part is not ν_e. From the various solutions discussed, a single one (the LMA solution) was shown to be right with the help of new data from all solar neutrino experiments especially Borexino and from reactors. Matter effects are responsible for the flavour conversion and the LMA solution is the correct one with $\Delta m^2 \approx 7.59^{-5}$ eV^2 and $\theta \approx 34°$ implying non-maximal mixing.

Having dealt in great detail with solar neutrinos, we now discuss another astrophysical source of neutrinos which has caused a lot of excitement and discussion over the past few years.

Chapter 11

Neutrinos from supernovae

DOI: 10.1201/9781315195612-11

Among the most spectacular events in astrophysics are phenomena from the late phase of stellar evolution, namely the explosion of massive stars. Such events are called supernovae and some of them are extremely luminous neutrino sources. Neutrinos are emitted in a period of about 10 s and roughly equal in number to those emitted by the Sun during its life. However, various effects have an impact on the released flavour composition. The physics of supernova explosions is rather complex and still far from being completely understood; hence, additional information can be found in [Sha83, Woo86, Arn89, Pet90, Whe90, Bet90, Woo92, Arn96, Raf96, Raf99, Ful01, Fry04, Mez05, Woo05, Kot06, Jan07, Kot11, Jan12, Bur13, Jan16].

11.1 Supernovae

Supernovae arise from two different final stages of stars. Either they are caused by thermonuclear explosions of a white dwarf within a binary system or they are explosions caused by the core collapse of massive stars ($M \geq 8M_\odot$). In the first case a compact star accretes matter from its main sequence companion until it is above a critical mass called the Chandrasekhar mass. The second mechanism is due to the fact that no further energy can be produced by nuclear fusion of iron-group nuclei like ^{56}Fe created in the interiors of massive stars, a result of reaching the maximal binding energy per nucleon. The most stable nucleus, ^{62}Ni, cannot be synthesized by nuclear reactions in stars starting from ^{56}Fe by a one- or two-step fusion process. Therefore, such stars become unstable with respect to gravity collapse. Hence, these are called core collapse supernovae.

Supernovae are classified spectroscopically by their optical properties. A major discrimination is the appearance of H-lines in the spectrum. Those with no H-lines are called type I supernovae and those with H-lines correspond to supernovae of type II. Supernovae are further subdivided due to other spectral features like appearance or absence of He-lines and those of heavier elements, plateaus in the light curve, etc. (see [Whe90, Sma09] for details). In addition to type II, type Ib (signatures of He) and Ic (neither H nor He) are nowadays considered as core collapse supernovae as well. The whole supernova phenomenon is more complex

Table 11.1. Hydrodynamic burning phases during stellar evolution (from [Gro90]).

Fuel	$T\ (10^9\ \mathrm{K})$	Main product	Burning time for $25M_\odot$	Main cooling process
^{1}H	0.02	^{4}He, ^{14}N	7×10^6 a	Photons, neutrinos
^{4}He	0.2	^{12}C, ^{16}O, ^{22}Ne	5×10^5 a	Photons
^{12}C	0.8	^{20}Ne, ^{23}Na, ^{24}Mg	600 a	Neutrinos
^{20}Ne	1.5	^{16}O, ^{24}Mg, ^{28}Si	1 a	Neutrinos
^{16}O	2.0	^{28}Si, ^{32}S	180 d	Neutrinos
^{28}Si	3.5	^{54}Fe, ^{56}Ni, ^{52}Cr	1 d	Neutrinos

than the simple classification system suggests, which is supported by the fact that during the evolution of the light curve occasionally the classification changes. Furthermore, a link between supernovae and gamma-ray bursters (see Chapter 12) has been established [Woo06] adding more features for classifications. Since no neutrinos are produced in association with type Ia supernovae, only core collapse supernovae will be considered here.

11.1.1 The evolution of massive stars

Stars generate their energy via nuclear fusion. Various burning cycles exist and the final stage of a star depends on its initial mass and the mass loss during its life, for details on stellar evolution see [Cox68, Cla68, Rol88, Kip90, Ibe13, Ili15].

After ignition of hydrogen fusion (see Chapter 10) and achieving hydrostatic equilibrium, stars appear on the zero age main sequence (ZAMS) in the Hertzsprung-Russell diagram. This starts the longest phase in the life of a star of static hydrogen burning. After the exhaustion of hydrogen in the core, which occurs over about the inner 13% of its mass, a He-core with degenerate electrons forms and shell burning of hydrogen will start. The inner part of the star does not compensate for this change by a reduced luminosity but rather by contraction. According to the virial theorem, only half of the energy released in this process produces an internal rise in pressure, while the other half is released. From the equation of state for a non-degenerate ideal gas ($p \sim \rho \cdot T$) it follows that a pressure increase is connected with a corresponding temperature increase. If a sufficiently high temperature has been reached in the He-core, typically about 0.5 M$_\odot$, the burning of helium via the triple α-process to ^{12}C ignites (helium flash). This causes the outer shells to inflate and a red giant develops. After a considerably shorter burning time than the hydrogen burning phase, the helium in the core has been fused, mainly into ^{12}C. In the following stage, ^{16}O is produced via the ^{12}C$(\alpha, \gamma)^{16}$O reaction. This then continues further to ^{20}Ne, performing another (α, γ) reaction. Hence, the configuration of the star at this stage is a core composed of C and O (for initial masses of roughly

2.25 - 8.5 $M_{\odot}$) or O and Ne (if the initial mass is in the range of about 8.5-10.5 $M_{\odot}$) with degenerated electrons. Detailed descriptions of the produced isotopes and abundances depend on the properties of the stars (temperature, density) and all involved potential nuclear reactions (captures, resonances, decays). Hence, typically stellar model simulations are combined with complex nuclear reaction network codes for more realistic predictions. Nevertheless, the same cycle, i.e., contraction with an associated temperature increase, now leads to a successive burning of these elements into heavier ones. The lower temperature burning phases (He, H) move towards the stellar surface. Starting from C burning, neutrino emission becomes the dominant energy loss mechanism of the star, which results in a further reduction of the burning time scales. Additional burning phases follow as shown in Table 11.1. The two major reactions of neon burning are either the photo disintegration of ^{20}Ne which dominates at higher T and ρ with respect to the inverse reaction ^{16}O$(\alpha, \gamma)^{20}$Ne. The released α-particles are then captured and produce (α, γ) reactions to form ^{24}Mg and ^{28}Si, respectively. Furthermore, carbon burning and oxygen burning reactions like ^{12}C + ^{12}C and ^{16}O + ^{16}O will lead to heavier elements like Ne, Si, P and S as well. The last possible reaction is the burning of silicon to nickel and iron elements involving a large amount of different photodissociation and α-capture processes, finally resulting in isotopes like ^{54,56}Fe and ^{58}Ni. More details can be found in [Cla68, Rol88, Arn96, Ibe13, Ili15]. This final burning process only lasts in the order of a days. This ends the hydrostatic burning phases of the star. Further energy gain from fusion of elements is no longer possible, as the maximum binding energy per nucleon, of about 8 MeV/nucleon, has been reached in the region of iron. Before discussing the fate of the created iron core, a discussion of the produced neutrinos is in place.

11.1.2   Energy loss of massive stars due to neutrino emission

As seen in Table 11.1 energy losses of higher burning phases are dominated by neutrinos. However, these are not fusion neutrinos but rather find their origin in electromagnetic processes. Individual contributions strongly depend on the involved temperatures and densities. In principle four processes can be considered (for details see [Raf96, Ibe13]). First of all there is pair production of $(\gamma \to \bar{\nu}_e + \nu_e)$ due to high energy photons in the nuclear matter which can decay into neutrino pairs. This increases with temperature because more energetic photons will be present. However, high densities, forcing the electrons to be degenerate, will reduce this process because of the energy needed in the Fermi-Dirac distribution of electrons. Furthermore there are photo neutrinos $(\gamma e^- \to e^- \nu \bar{\nu})$, bremsstrahlung neutrinos $(e^-(Ze) \to e^-(Ze)\nu\bar{\nu})$ and plasma neutrinos $(\gamma \to \nu\bar{\nu})$.

11.1.3   The actual collapse phase

Stars with more than about $12 M_{\odot}$ carry out burning up to iron group elements. Such stars then have a small, dense, iron core and an extended envelope. The burning regions form shells on top of each other and give the interior of stars an onion-like

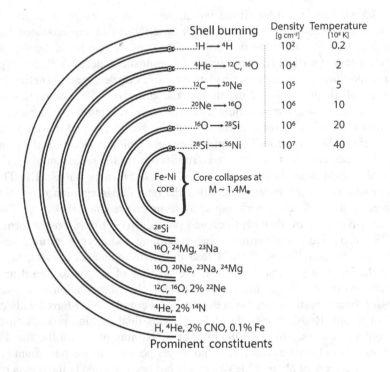

Shell burning	Density [g cm⁻³]	Temperature [10⁸ K]

Figure 11.1. Schematic representation of the structure, composition and development of a heavy star. In the hydrostatic burning phases of the shells, elements of higher atomic number, up to a maximum of Fe and Ni, are built up from the initial composition (the major components of which are labelled). The gravitational collapse of the core leads to the formation of a neutron star or a black hole and the ejection of $\approx 95\%$ of the mass of the star (supernova explosion). The ejected outer layers are traversed by the detonation shock wave which initiates explosive burning. With kind permission of A. Jansen.

structure (see Figure 11.1). The stability of the iron core is mainly guaranteed by the pressure of the degenerate electrons. The origin of the degenerate electron gas can be understood from Figure 11.3. It shows a phase diagram which characterizes the state of the matter inside stars. For very large densities the Pauli principle has to be taken into account. This implies that each cell in phase space of size h^3 can contain a maximum of two electrons. The entire phase space volume V_{Ph} is then given by

$$V_{Ph} = \frac{4}{3}\pi R^3 \frac{4}{3}\pi p^3. \tag{11.1}$$

Higher densities at a constant radius produce an increased degeneracy. In the degenerate case the electron pressure p_e is no longer determined by the kinetic energy but rather by the Fermi momentum p_F. Hence, in this case the relations $p_e \sim p_F^5$ (non-relativistic) and $p_e \sim p_F^4$ (relativistic) apply. In addition, the pressure no longer depends on the temperature but exclusively on the electron density n_e, according

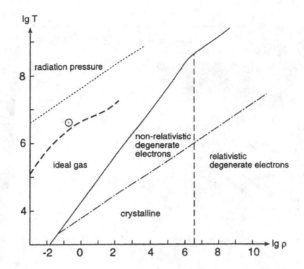

Figure 11.2. Schematic ρ–T phase diagram for the characterization of matter inside stars. The areas shown are those in which the equation of state is dominated either by radiation pressure (above the dotted line) or by a degenerate electron gas (below the solid line). The latter can be relativistic or non-relativistic (to the right or left of the vertical dashed line). The dot-dashed line characterizes a temperature below which the ions prefer a crystalline state. The heavy dotted line shows the standard solar model evolution path and the present Sun (from [Kip90]). Reproduced with permission of SNCSC.

to [Sha83]

$$p = \frac{1}{m_e}\frac{1}{5}(3\pi^2)^{\frac{2}{3}}n_e^{\frac{5}{3}} \qquad \text{non-relativistic} \qquad (11.2)$$

$$p = \frac{1}{4}(3\pi^2)^{\frac{1}{3}}n_e^{\frac{4}{3}} \qquad \text{relativistic.} \qquad (11.3)$$

This leads to the fatal consequence of gravitational collapse. The previous cycle of pressure increase $\rightarrow$ temperature increase $\rightarrow$ ignition $\rightarrow$ expansion $\rightarrow$ temperature drop now no longer functions. Released energy leads only to a temperature increase and thus to unstable processes but no longer to pressure increase.

Three types of supernova explosions can be considered according to their phase diagram [Jan12]. The first one would be those triggered by electron capture (see Chapter 6) and hence are called EC supernovae. It is an ONeMg core with a steep density rise at the edge of the O,Ne core. Degeneracy of electrons occurs before Ne burning can start. Due to that, the Fermi energy increases and, supported by the low reaction thresholds of Mg and Ne, this leads to strong EC. For a solar metallicity star this would happen in a mass range of 9 - 9.25 M$_\odot$, but this region can broaden and shift for other compositions.

The second type of SN is the collapse of iron cores, which is of major interest here. Stellar rotation is not expected to play a crucial role, as mass loss during

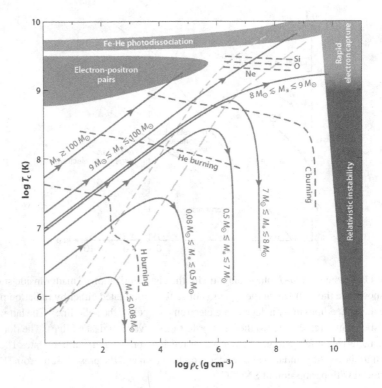

Figure 11.3. A phase diagram of stellar evolution shown in a log-log form of central density versus central temperature (ρ_c and T_c). The arrowed lines suggest some rough birth mass of the considered stars. Also the various burning stages are shown as dashed lines. The parallel lines from the lower left to the upper right mark the onset of degeneracy and also strong degeneracy of the electron plasma. The regions at high values which are named cause core collapses. The stellar evolution tracks have been very simplified (from [Jan12]). © Annual Review of Nuclear and Particle Science.

evolution, especially in the pre-red giant phase, it carries away angular momentum.

Third, there are supernovae associated with very powerful events in the sky as gamma ray bursters (see Chapter 12).

The stability condition of a star with mass M and radius R in hydrostatic equilibrium is given by [Lan75, Sha83]

$$E = \frac{3\gamma - 4}{5\gamma - 6} \frac{GM^2}{R} \qquad (11.4)$$

$\gamma = \partial \ln p / \partial \ln \rho$ is the adiabatic index, p the pressure and ρ the density. The adiabatic index for a non-relativistic degenerate electron gas is $5/3$, while it is $4/3$ in the relativistic case (see Figure 11.2). The appearance of a critical mass, the *Chandrasekhar limit*, reflects the violation of the stability condition ($\gamma > 4/3$) for hydrostatic equilibrium, because of the pressure dependence $p \propto \rho^{4/3} \propto n_e^{4/3}$

of a non-relativistic degenerate electron gas. The Chandrasekhar mass is given by
([Cha39, 67], also see [Hil88])

$$M_{Ch} = 5.72\,Y_e^2 M_\odot \qquad (11.5)$$

where Y_e is the number of electrons per nucleon. Such an instability is causing core
collapse in massive stars as a pre-condition for a supernova explosion.

The typical parameter values for a $15 M_\odot$ star with a core mass of $M_{Ch} \approx 1.5 M_\odot$ are a central temperature of approximately 8×10^9 K, a central density of 3.7×10^9 g cm^{-3} and a Y_e of 0.42. The Fermi energy of the degenerate electrons is roughly 4 MeV to 8 MeV and stabilises the core. These are typical values at the start of the collapse of a star. The main processes causing the core collapse are the photo-disintegration of ^{56}Fe into α-particles via the reaction

$$\gamma +^{56} \text{Fe} \rightarrow 13\alpha + 4n - 124.4\,\text{MeV} \quad \text{(endothermic)} \qquad (11.6)$$

and electron capture on free protons and heavy nuclei

$$e^- + p \rightarrow n + \nu_e \qquad e^- + (A, Z) \rightarrow (A, Z - 1) + \nu_e. \qquad (11.7)$$

The latter process becomes possible because of the high Fermi energy of the
electrons. The number of electrons is strongly reduced by these processes (11.7)
and mainly neutron-rich, unstable nuclei are produced. Since it was the pressure
of degenerate electrons which balanced the gravitational force, the core collapses
quickly. The lowering of the electron concentration can also be expressed by an
adiabatic index $\gamma < 4/3$. The inner part ($\approx 0.6 M_\odot$) keeps $\gamma = 4/3$ and collapses
homologously ($v/r \approx 400$–700 s^{-1}), while the outer part collapses at supersonic
speed. Homologous means that the density profile is kept during the collapse of
the neutron star. The behaviour of the infall velocities within the core is shown in
Figure 11.4 at a time of 2 ms before the total collapse. The matter outside the sonic
point defined by $v_{coll} = v_{sound}$ collapses with a velocity characteristic for free fall.
The outer layers of the star do not notice the collapse of the iron core, due to the
low speed of sound. More and more neutron-rich nuclei are produced in the core,
which is reflected in a further decrease of Y_e taken away by the produced neutrinos.
Initially the emitted neutrinos can leave the collapsing core zone unhindered, but they
get trapped starting at density regions around 10^{12} g cm^{-3} in which typical neutron
rich nuclei have masses between 80 and 100, with about 50 neutrons. Here, matter
becomes opaque for neutrinos as the outward diffusion speed becomes considerably
smaller than the inward collapse speed of a few milliseconds. The dominant process
for the neutrino opacity is coherent neutrino–nucleus scattering via neutral weak
currents with a cross-section of (see, e.g., [Fre74, Hil88])

$$\sigma \simeq 10^{-44} \text{ cm}^2 N^2 \left(\frac{E_\nu}{\text{MeV}}\right)^2 \qquad (11.8)$$

where N is the neutron number in the nucleus. The reason is that 10 MeV neutrinos
interact with the nucleus coherently and the NC cross-section on protons is reduced

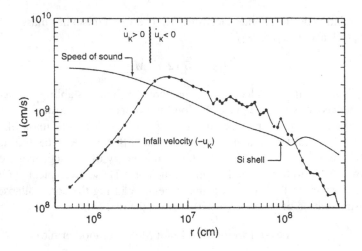

Figure 11.4. Infall velocity of the material in the core of a supernova about 2 ms before the complete collapse of the star. Within the homologous inner core ($r < 40$ km) the velocity is smaller than the local velocity of sound. In the region $r > 40$ km (outer core), the material collapses with supersonic speed (from [Arn77]). Reproduced by permission of the AAS.

by $1-4\sin^2\theta_W$ (see Chapter 4). Hence, neutrinos are trapped and move with the collapsing material (*neutrino trapping*). The transition between the "neutrino optical" opaque and the free-streaming region defines a *neutrino sphere* (see Equation 11.10). The increase in neutrino capture by neutrons acts as an inverse process to that of (11.7), consequently stabilizing electron loss and leads to an equilibrium with respect to weak interactions. Therefore, no further neutronization occurs and the lepton number per baryon $Y_L = Y_e + Y_\nu$ is conserved at the value of Y_e at the beginning of neutrino-trapping $Y_L \approx 0.35$ [Bet86a].

Henceforth, the collapse progresses adiabatically. This is equivalent to a constant entropy, as now neither significant energy transport nor an essential change in composition takes place. Figure 11.5 shows the mean mass numbers and nuclear charge number of the nuclei formed during neutronization, together with the mass fractions of neutrons, protons, as well as the number of electrons per nucleon. We, thus, have a gas of electrons, neutrons, neutrinos and nuclei whose pressure is still determined by the relativistic degenerate electrons. Thus the "neutron star" begins as a hot lepton-rich quasi-static object, which develops into its final state via neutrino emission; i.e., it starts off as a *quasi-neutrino star* [Arn77].

The collapsing core finally reaches densities that normally appear in atomic nuclei ($\rho > 3 \times 10^{14}$ g cm^{-3}). For higher densities, however, the short range force between nucleons becomes strongly repulsive and matter becomes incompressible and bounces back (equivalent to $\gamma > 4/3$) (see, e.g., [Lan75, Bet79, Kah86]). This creates an outward going shock wave. Exactly how much energy this shock wave contains depends, among other factors, on the equation of state of the very strongly compressed nuclear matter [Kah86, Bet88, Lat12, Lat16, Ann18, Tew18] and

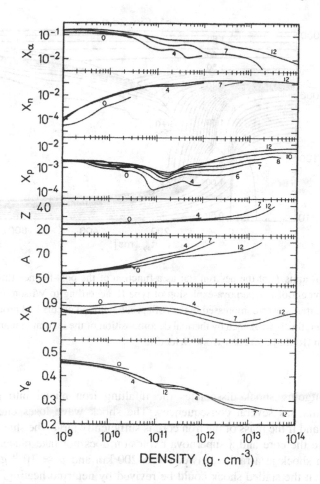

Figure 11.5. Change in the core composition during the gravitational collapse (the numbers correspond to various stages of the collapse). X_n, X_p, X_α, X_A denote the *mass* fraction (not the *number* densities) of the neutrons, protons, α-particles and nuclei. Y_e denotes the electrons per nucleon (from [Bru85]). © IOP Publishing. Reproduced with permission. All rights reserved.

whether the bounce-back of the core is hard or soft. A soft bounce-back provides the shock with less initial energy. Unfortunately, the equation of state is not very well known since extrapolation into areas of supernuclear density is required, but the new observations of neutron star mergers producing gravitational waves will lead to deeper insights in the future. At bounce a strong sound wave is produced, which propagates outwards into a shock wave near $M \approx 0.6 M_\odot$ after about 1 ms after bounce. This is illustrated in Figure 11.6. This shock formation corresponds to a radius of about 100 km, where the core density is about the nuclear density $\rho \approx 10^{14}$ g cm^{-3}.

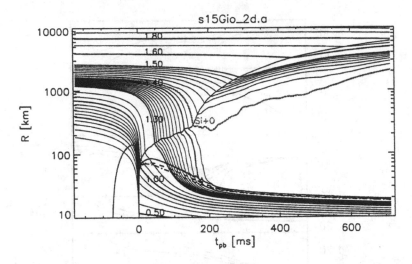

Figure 11.6. Radius R of the neutron star as a function of the post-bounce time t_{pb}. Shown is the development of a supernova explosion of type II, according to Wilson and Mayle. As the nuclear matter is over-compressed in the collapse, a rebound occurs and produces a shock wave. However, this is weakened by thermal decomposition of the incoming matter during the collapse (from [Raf03]). © 2003 World Scientific.

The outgoing shock dissociates the infalling iron nuclei into protons and neutrons. This has several consequences. The shock wave loses energy by this mechanism and if the mass of the iron core is sufficiently large, the shock wave does not penetrate the core and a supernova explosion does not take place. It becomes an accretion shock at a radius between 100–200 km and $\rho \approx 10^{10}$ g cm^{-3}. It is suggested that the stalled shock could be revived by neutrino heating. This should finally result in an explosion and has been called *delayed explosion* mechanism [Bet85]. However, new simulations seem to indicate that still no explosion is happening. The nuclear binding energy of $0.1 M_\odot$ iron is about 1.7×10^{51} erg and thus comparable to the explosion energy. However, the dissociation into nucleons leads to an enormous pressure increase, which results in a reversal of the direction of motion of the incoming matter in the shock region. This transforms a collapse into an explosion. As the shock moves outwards, it still dissociates heavy nuclei, which are mainly responsible for neutrino trapping. Moreover, the produced free protons allow quick neutronization via $e^- + p \rightarrow n + \nu_e$ if passing the neutrino sphere in which the density is below 10^{11} g cm^{-3}. The produced neutrinos are released immediately within a few milliseconds and are often called 'prompt ν_e burst' or 'deleptonization burst'. If the shock wave leaves the iron core without getting stalled, the outer layers represent practically no obstacle and are blown away, which results in an optical supernova in the sky. Also in the outer shells the MSW effect occurs. Such a mechanism is known as a *prompt explosion* [Coo84, Bar85].

The structure of a star at the end of its hydrostatic burning phases can only be understood with the help of complex numerical computer simulations. One-dimensional spherical computer simulations suggest that core bounce and shock formations are not sufficient to cause a supernova explosion [Jan95, Jan96, Mez98, Bur00, Ram00, Mez01, Bur02, Jan12, Jan16]; further, decisive boost is given to the shock by the formation of neutrino-heated hot bubbles [Bet85, Col90, Col90a], which furthermore produce a considerable mixing of the emitted material. This strong mixing has been observed in supernova 1987A (see Section 11.4.1.2) and has been confirmed by two-dimensional computer simulations, which allow these effects to be described. The simulation also showed the importance of large-scale convection in the process. The problem during the last 20 years has been that the shock has been stalled about 100–150 km from the center of the star and, in general, the inclusion of neutrino absorption and the implied energy deposition permitted only a moderate explosion. The newest supercomputers allowed the simulation of dying stars in *two* or *three* dimensions, following both the radial and lateral directions. Convection brings hot material from near the neutrino sphere quickly up in regions behind the shock and cooler material down to the neutrino sphere where it helps absorbing energy from the neutrino flow. This helps in revitalizing the stalled (for about 100 ms) shock. The simulations also led to the general belief that newly born neutron stars are convective [Her94, Jan95, Jan96, Mez98, Mez05, Mar09, Jan16]. A further aspect of convection is that typical explosions show asymmetrically ejected matter, resulting in a recoil of the remaining core which gives it a speed of hundreds of kilometres per second (rocket effect) [Bur95a, Jan95a]. Such asymmetries might also point to the importance of rotation and magnetic fields in supernova explosions [Kho99]. A new phenomenon was the description of collective neutrino interactions [Dua06, Dua06a] which also have a large effect on the expected neutrino spectrum [Dua10].

Meanwhile the object below the shock has become a protoneutron star. It consists basically of two parts: an inner settled core within the radius where the shock wave was first formed, consisting of neutrons, protons, electrons and neutrinos ($Y_{\text{lepton}} \approx 0.35$); the second part is the bloated outer part, which lost most of its lepton number during the ν_e burst at shock breakout. This part settles within 0.5–1 s of core bounce, emitting most of its energy in neutrinos. After about 1 s the protoneutron star is basically an object by itself. It has a radius of about 30 km which slowly contracts further and cools by emission of (anti)neutrinos of all flavours and deleptonizes by the loss of ν_e. After 5–10 s it has lost most of its lepton number and energy; a period called Kelvin–Helmholtz cooling.
The energy released in a supernova corresponds to the binding energy of the neutron star produced:

$$E_B \approx \frac{3}{5} \frac{GM_{\text{neutron star}}}{R} = 5.2 \times 10^{53} \text{ erg} \left(\frac{10 \text{ km}}{R_{\text{neutron star}}} \right) \left(\frac{M_{\text{neutron star}}}{1.4 M_\odot} \right)^2 .$$

(11.9)

This is the basic picture of an exploding, massive star. For a detailed account see [Arn77, Sha83, Woo86, Pet90, Col90, Bur95, Mez05, Woo05, Jan07, Mar09, Jan16].

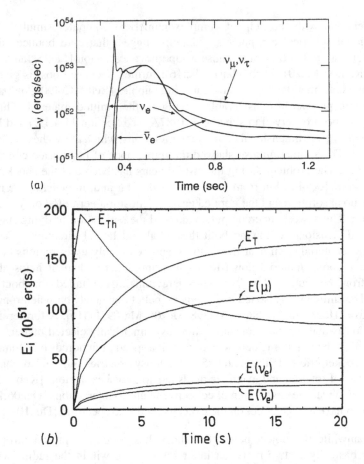

Figure 11.7. (*a*) Calculated neutrino luminosity of a $2M_\odot$ iron core of an $\approx 25M_\odot$ main sequence star as a function of the time from the start of the collapse for the various neutrino flavours (from [Bru87]). © 1987 by the American Physical Society. (*b*) Cooling of a hot proto-neutron star of $1.4M_\odot$ in the first 20 s after gravitational collapse. E_{Th} denotes the integrated internal energy, E_T is the total energy released and E_{ν_e} and $E_{\bar{\nu}_e}$ are the total energies emitted as ν_e and $\bar{\nu}_e$, respectively. E_μ is the energy emitted as ν_μ, $\bar{\nu}_\mu$, ν_τ and $\bar{\nu}_\tau$. All energies are in units of 10^{51} erg (from [Bur86]). With kind permission of A. Burrows. © IOP Publishing. Reproduced with permission. All rights reserved.

11.2   Neutrino emission in supernova explosions

We now discuss the neutrinos which could be observed from supernova explosions. This picture has been dramatically changed in recent years, as more sophisticated supernova models have been developed but also by including neutrino oscillations and new collective phenomena of neutrinos, which alter the classical prediction significantly. Nevertheless, the classic prediction will firstly be discussed before including phenomena linked to neutrinos.

11.2.1 The classical prediction

The observable spectrum originates from two processes. First the deleptonization burst as the outgoing shock passes the neutrino sphere, resulting in the emission of ν_e with a duration of a few milliseconds. The radius R_ν of this sphere can be defined via the optical depth τ_ν and be approximated by

$$\tau_\nu(R_\nu, E_\nu) = \int_{R_\nu}^{\infty} \kappa_\nu(E_\nu, r)\rho(r)\,\mathrm{d}r = \tfrac{2}{3} \tag{11.10}$$

where $\kappa_\nu(E_\nu, r)$ is the opacity and $\tau_\nu < 2/3$ characterizes the free streaming of neutrinos. The second part comes from the Kelvin–Helmholtz cooling phase of the protoneutron star resulting in an emission of all flavours (ν_e, $\bar{\nu}_e$, ν_μ, $\bar{\nu}_\mu$, ν_τ, $\bar{\nu}_\tau$; in the following ν_μ is used for the last four because their spectra are quite similar). The emission lasts typically for about 10 s (Figure 11.7). It is reasonable to assume that the energy is equipartitioned among the different flavours.

Neutrinos carry away about 99% of the energy released in a supernova explosion. This global property of emitting 0.5×10^{53} erg in each neutrino flavour with typical energies of 10 MeV during about 10 s remains valid; however, it is worthwhile taking a closer look at the details of the individual spectra. Their differences offer the opportunity to observe neutrino flavour oscillations.

From considerations concerning the relevant opacity sources for the different neutrino flavours, a certain energy hierarchy might be expected. For ν_e the dominant source is $\nu_e n \to ep$ and for $\bar{\nu}_e$ it is $\bar{\nu}_e p \to e^+ n$. The spectrum of the $\bar{\nu}_e$ corresponds initially to that of the ν_e, but, as more and more protons vanish, the $\bar{\nu}_e$s react basically only via neutral currents also resulting in a lower opacity and, therefore, higher average energy (about 14–17 MeV). The typical average energy for ν_e is in the region of 10–12 MeV. As $\nu_\mu(\bar{\nu}_\mu)$ and $\nu_\tau(\bar{\nu}_\tau)$ can only interact via neutral currents (energies are not high enough to produce muons and tau-leptons), they have smaller opacities and their neutrino spheres are further inside, resulting in a higher average energy (about 24–27 MeV). The expected post-bounce energy hierarchy is $\langle E_{\nu_e} \rangle < \langle E_{\bar{\nu}_e} \rangle < \langle E_{\nu_\mu} \rangle$. Obviously this implies for the radii of the neutrino spheres that $\langle R_{\nu_\mu} \rangle < \langle R_{\bar{\nu}_e} \rangle < \langle R_{\nu_e} \rangle$. However, recent more sophisticated simulations including additional neutrino processes, such as $NN \to NN\nu_\mu\bar{\nu}_\mu$ (nucleon-nucleon bremsstrahlung), $e^+e^- \to \nu_\mu\bar{\nu}_\mu$ and $\nu_e\bar{\nu}_e \to \nu_\mu\bar{\nu}_\mu$ (annihilation) and scattering on electrons $\nu_\mu e \to \nu_\mu e$, have revealed that the difference in $\langle E_{\nu_\mu} \rangle$ and $\langle E_{\bar{\nu}_e} \rangle$ is much smaller than previously assumed [Han98, Raf01, Kei03]. It is only on the level of 0–20% even if the fluxes themselves may differ by a factor of two. For a recent simulation see [Fis10, Jan16].

The spectra from the cooling phase are not exactly thermal. Since neutrino interactions increase with energy, the effective neutrino sphere (last energy exchange with medium) increases with energy. Thus, even if neutrinos in their respective neutrino sphere are in thermal equilibrium with matter, the spectrum becomes 'pinched', i.e., depleted at higher and lower energies in comparison with a thermal spectrum. A typical parametrisation is given as a pinched Fermi-Dirac distribution

$$F(E,t) = \frac{E^2}{e^{E/T(t)-\eta(t)} + 1} \tag{11.11}$$

with $\eta(t)$ as degeneracy parameter. This is a consequence of the facts that the temperature decreases with increasing radius and the density decreases faster than $1/r$. Furthermore, the neutrino emission and thus the signal might abruptly be stopped by the formation of a black hole [Bea01]. Last, but not least, there is interest to detect the much smaller prae-supernova neutrinos, which would also add important information [Pat17].

11.2.2 Neutrino oscillations and supernova signals

After describing this basic picture now effects have to be discussed which are linked to neutrino oscillations and the impact they might have on supernova explosions and *vice versa*. A supernova as a complex object has a variety of densities and density gradients to enable the matter effects to work (see Chapter 10), and some of these effects will be discussed now. There are 3 major differences with respect to matter effects discussed within the context of the Sun:

- The matter density close to the proto neutron star is so high that two resonance regions can be expected and linked to the two Δm^2 observed [Lun04, Dua09, Das09, Dig10].
- The density profile is highly dynamic, i.e., it changes quickly as a function of time due to the shock propagation and thus also matter effects.
- The neutrino density around the proto neutronstar is so high that new phenomena like forward neutrino-neutrino scattering (this is including antineutrinos) have to be considered, which lead to collective phenomena as well and severely influence the final neutrino spectrum.

11.2.2.1 *Effects on the prompt ν_e burst*

One consequence which could be envisaged is that the deleptonization burst of ν_e could be much harder to detect because of ν_e–$\nu_{\mu,\tau}$ oscillations and the correspondingly smaller interaction cross-section. The solar ν_e flux is depleted by the MSW mechanism so the same should occur in supernovae (see also Section 10.3.2). The main difference is that instead of the possible exponential electron density profile used for solar neutrinos, the electron density is now better approximated by a power law

$$n_e \approx 10^{34} \ \text{cm}^{-3} \left(\frac{10^7 \ \text{cm}}{r} \right)^3 \tag{11.12}$$

where 10^7 cm is the approximate radius where the ν_e are created. [Noe87] found a conversion probability of more than 50% if

$$\Delta m^2 \sin^3 2\theta > 4 \times 10^{-9} \ \text{eV}^2 \frac{E_\nu}{10 \ \text{MeV}} \tag{11.13}$$

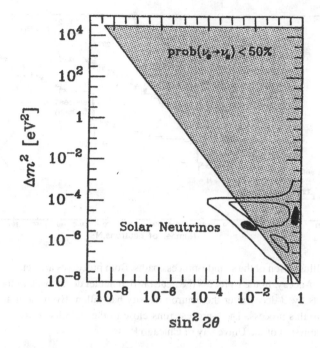

Figure 11.8. MSW triangle for the prompt ν_e burst from a stellar collapse. In the shaded area the conversion probability exceeds 50% for $E_\nu = 20$ MeV. The solar LMA and SMA (ruled out by observations) solutions as well as the old Kamiokande allowed range are also shown (from [Raf96]). With kind permission of the University of Chicago Press.

assuming that $\Delta m^2 \lesssim 3 \times 10^4$ eV$^2 r^{-3} E_\nu / 10$ MeV to ensure that the resonance is outside the neutrino sphere [Raf96]. The region with larger than 50% conversion probability for $E_\nu = 20$ MeV is shown in Figure 11.8. As can be seen, the solar LMA solution implies a significant conversion, making a direct observation of this component more difficult.

11.2.2.2 Cooling phase neutrinos

Neutrino oscillations could cause a partial swap $\nu_e \leftrightarrow \nu_{\mu,\nu_\tau}$ and $\bar\nu_e \leftrightarrow \bar\nu_{\mu,\nu_\tau}$, so that the measured flux at Earth could be a mixture of original $\bar\nu_e$, $\bar\nu_\mu$ and $\bar\nu_\tau$ source spectra. However, note that in a normal mass hierarchy no level crossing occurs for antineutrinos. The LMA mixing angles are large and imply significant spectral swapping. Applying this swapping to SN 1987A data leads to contradictory results [Smi94, Lun01, Min01, Kac01]. Cooling phase neutrinos might face another signature, namely the abrupt ending of the neutrinos due to black hole formation [Bea01].

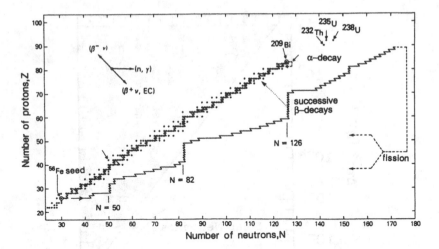

Figure 11.9. Illustration of the s- and r-process paths. Both processes are determined by (n, γ) reactions and β-decay. In the r-process the neutron-rich nuclei decay back to the peninsula of stable elements by β-decay after the neutron density has fallen. Even the uranium island is still reached in this process. The s-process runs close to the stability valley (from [Rol88]). With kind permission of the University of Chicago Press.

11.2.2.3   *Production of r-process isotopes*

Supernovae are cauldrons for the production of heavy elements. As already discussed, lighter elements are converted by fusion up to iron-group elements, where no further energy can be obtained by fusion and thus no heavier elements can be created. These heavier elements could also not have been created in sufficient amounts via charged particle reactions, due to the increased Coulomb barriers—other mechanisms must have been at work. As proposed by Burbidge, Burbidge, Fowler and Hoyle (B^{2}FH) [Bur57] heavier isotopes are produced by the competing reactions of neutron capture (n, γ-reactions) and nuclear β-decay. Depending now on the β-decay lifetimes and the lifetime against neutron capture, two principal ways can be followed in the nuclide chart (Figure 11.9). For rather low neutron fluxes and, therefore, a *slow* production of isotopes via neutron capture, at one point β-decay dominates and the process is called the s-process (slow process). Element production follows the 'valley of stability' up to ^{209}Bi, where strong α-decay stops this branch. However, in an environment with very high neutron densities ($n_n \approx 10^{20}$ cm^{-3}) there could be several neutron captures on one isotope before β-decay lifetimes get short enough to compete. This process is called the r-process (rapid process) and pushes elements in the poorly known region of nuclei far beyond stability. In this way elements up to U and Th can be produced. These are the dominant processes for element production. There might be other processes at work, i.e., the νp-process (γ, α) and (γ, n) [Frö06] for producing neutron-depleted isotopes, but they are of minor importance here.

Naturally, the r-process can occur only in a neutron-rich medium ($Y_e < 0.5$). The p/n ratio in this region is governed by neutrino spectra and fluxes because of the much higher number density with respect to the ambient e^+e^- population. The system is driven to a neutron-rich phase because normally the $\bar{\nu}_e$ are more energetic than ν_e, therefore preferring β-reactions of the type $\bar{\nu}_e p \leftrightarrow n e^+$ with respect to $\nu_e n \leftrightarrow p e^-$. The production of the r-process nuclides was considered to be supernovae as there are enough iron nuclei available as seeds and sufficient neutron densities can be achieved. However, first observations of gravitational waves from black-hole and neutron star merging seem to be a more efficient source for nucleosynthesis [Sie19].

If $\nu_e \leftrightarrow \nu_{\mu,\nu_\tau}$ oscillations happen outside the neutrino sphere, a subsequent flux of ν_e can be produced which is more energetic than $\bar{\nu}_e$ because the original ν_{μ,ν_τ} are more energetic (see Section 11.2). In addition, the energetic neutrinos might produce new nuclei by the ν-process [Dom78, Woo88, Woo90, Heg05]. Their abundance might serve as a 'thermometer' and allows information on the neutrino spectrum and oscillations. Also the yield of produced isotopes like ^{7}Li and ^{11}B can depend on neutrino parameters like the mass hierarchy and the value of θ_{13} [Yos06].

11.2.2.4 Neutrino mass hierarchies from supernova signals

Studies have been performed to disentangle information about the neutrino mass hierarchies from a high statistics supernovae observation [Dig00, Tak02, Lun03]. Assume only the solar and atmospheric evidence and that $\Delta m_{32}^2 = \Delta m_{atm}^2$ and $\Delta m_{21}^2 = \Delta m_\odot^2$. The two key features for the discussion are the neutrino mass hierarchy $|\Delta m_{32}^2| \approx |\Delta m_{31}^2| \gg |\Delta m_{21}^2|$ and the upper bound $|\Delta m_{ij}^2| < 10^{-2}$ eV^2. It should be noted that $\Delta m_{32}^2 > 0$ implies $m_3 > m_2, m_1$ and $\Delta m_{32}^2 < 0$ means $m_3 < m_2, m_1$ (see Chapter 5). The major difference is that, in the first case, the small U_{e3} admixture to ν_e is the heaviest state while, in the inverted case, it is the lightest one. Since the ν_μ and ν_τ spectra are indistinguishable only the U_{ei} elements are accessible. Unitarity ($\sum U_{ei} = 1$) further implies that a discussion of two elements is sufficient, for example U_{e2} and U_{e3}, where the latter is known to be small. In that case U_{e2} can be obtained from the solar evidence:

$$4|U_{e2}|^2|U_{e1}|^2 \approx 4|U_{e2}|^2(1 - |U_{e2}|^2) = \sin^2 2\theta_\odot. \tag{11.14}$$

The system is then determined by two pairs of parameters

$$(\Delta m_L^2, \sin^2 2\theta_L) \simeq (\Delta m_\odot^2, \sin^2 2\theta_\odot)$$
$$(\Delta m_H^2, \sin^2 2\theta_H) \simeq (\Delta m_{atm}^2, 4|U_{e3}|^2). \tag{11.15}$$

The resonance density for the MSW effect is given by

$$\rho_{res} \approx \frac{1}{2\sqrt{2}G_F} \frac{\Delta m^2}{E} \frac{m_N}{Y_e} \cos 2\theta$$

$$\approx 1.4 \times 10^6 \text{ g cm}^{-3} \frac{\Delta m^2}{1 \text{ eV}^2} \frac{10 \text{ MeV}}{E} \frac{0.5}{Y_e} \cos 2\theta. \tag{11.16}$$

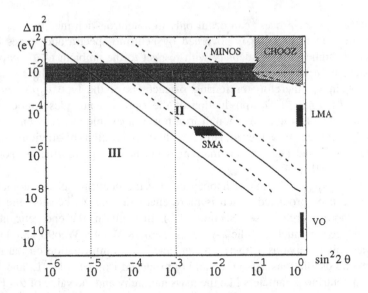

Figure 11.10. The contours of equal flip probability. The solid lines denote the contours of flip probability for a 5 MeV neutrino, with the left-hand one representing a 90% flip (highly diabatic) and the right-hand one for 10% (adiabatic). The dashed lines shows the corresponding curves for 50 MeV. The two vertical lines indicate the values of 4 $|U_{e3}|^2 = \sin^2 2\theta$ lying on the borders of adiabatic (I), diabatic (III) and transition (II) regions for Δm^2 corresponding to the best-fit value of the atmospheric solution. The regions LMA, SMA and VO are indicating the solutions for a large mixing angle, a small mixing angle and vacuum oscillations respectively (from [Dig00]) © 2000 by the American Physical Society.

This implies, for the two-parameter sets, a resonance at $\rho = 10^3$–10^4 g cm^{-3} and $\rho = 10$–30 g cm^{-3}, the latter if the L parameters lie within the LMA region. Both resonance regions are outside the supernova core, more in the outer layers of the mantle. This has some immediate consequences: the resonances do not influence the dynamics of the collapse and the cooling. In addition the possible r-process nucleosynthesis does not occur. The produced shock wave has no influence on the MSW conversion. The density profile assumed for resonant conversion can be almost static and is identical to that of the progenitor star. Furthermore, in regions with $\rho > 1$ g cm^{-3}, Y_e is almost constant and

$$\rho Y_e \approx 2 \times 10^4 \left(\frac{r}{10^9 \text{ cm}} \right)^{-3} \tag{11.17}$$

is a good approximation. However, the exact shape depends on details of the composition of the star.

The transition regions can be divided into three parts: a fully adiabatic part, a transition region and a section with strong violation of adiabaticity. This is shown in Figure 11.10. As can be seen, the H-resonance is adiabatic if $\sin^2 2\theta_{e3} = 4|U_{e3}|^2 > 10^{-3}$ and the transition region corresponds to $\sin^2 2\theta_{e3} \approx 10^{-5}$–$10^{-3}$. The features

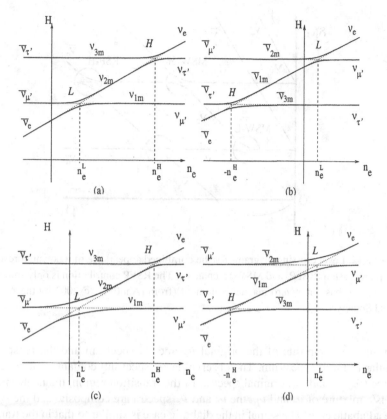

Figure 11.11. Level crossing diagrams for small solar mixing (upper line) and large mixing (lower line) as is realized by LMA. The left-hand column corresponds to normal and the right-hand to inverted mass hierarchies. The part of the plot with $N_e < 0$ is the antineutrino channel, where H refers to High and L to Low (from [Dig00]). © 2000 by the American Physical Society.

of the final neutrino spectra strongly depend on the position of the resonance. The smallness of $|U_{e3}|^2$ allows the two resonances to be discussed independently. The locations of the resonances in the different mass schemes are shown in Figure 11.11. Note that for antineutrinos the potential is $V = -\sqrt{2}G_F N_e$ (see Chapter 10). They can be drawn in the same diagram and can be envisaged as neutrinos moving through matter with an effective $-N_e$. It is important that the starting points are at the extremes and that the neutrinos move towards $N_e \to 0$. As can be seen in the plot, for the LMA solution the L-resonance is always in the neutrino sector independent of the mass hierarchy, while the H-resonance is in the neutrino (antineutrino) sector for the normal (inverted) hierarchy.

As mentioned, the value of $|U_{e3}|^2$ and, therefore, the position of the H-resonance is important. Consider the normal hierarchy first. In the adiabatic region, the neutronization burst almost disappears from ν_e and appears as ν_μ. The $\bar{\nu}_e$

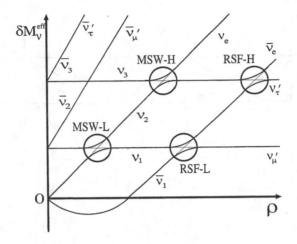

Figure 11.12. Level crossing diagrams for the combined appearance of a potential resonance spin flip precession (RSFP) and MSW resonances. The RSFP contribution is only showing up if the neutrino has a non-zero magnetic moment (from [And03]). © 2003 by the American Physical Society.

spectrum is a composite of the original $\bar{\nu}_e$ and ν_μ spectrum and the ν_e spectrum is similar to the ν_μ spectrum. The observable ν_μ spectrum contains components of all three ($\bar{\nu}_e$, ν_e and ν_μ) original spectra. In the transition region, the neutronization burst is a mixture of ν_e and ν_μ, the ν_e and $\bar{\nu}_e$ spectra are composite and the ν_μ is as in the adiabatic case. The signal in the diabatic case is similar to that in the transition region but now there can be significant Earth matter effects for ν_e and $\bar{\nu}_e$.

In the inverted mass scheme, the adiabatic region shows a composite neutronization burst, the ν_e and ν_μ spectra are composite and $\bar{\nu}_e$ is practically all ν_μ. A strong Earth matter effect can be expected for ν_e. In the transition region, all three spectra are a composite of the original ones, the neutronization burst is a composite and Earth matter effects show up for ν_e and $\bar{\nu}_e$.

Therefore, for a future nearby supernova, an investigation of the neutronization burst and its possible disappearance, the composition and hardness of the various spectra and the observation of the Earth matter effect might allow conclusions on the mass hierarchy to be drawn.

11.3   Detection methods for supernova neutrinos

The expected supernova spectrum contains all neutrino flavours, however ν_μ and ν_τ due to their low energy can only be detected via neutral currents (see Chapter 4). As is clear from the discussion in the last section, basically all solar and reactor neutrino detectors can be used for supernova detection [Bur92,Sch12]. Ideally a measurement would consist of energy, flavour, detection time and direction of each neutrino; however, this is hard to realise. The expected rate in a detector is a convolution of three quantities: The supernova neutrino spectrum discussed in the last section, the reaction cross-section and the detector response. The number of events will scale with $1/d^2$ if d is the distance to the supernova and the rate is proportional to the detector mass, assuming a detector response independent from its mass. The major cross-section is inverse β-decay [Vog99, Mar03, Str03]

$$\bar{\nu}_e + p \rightarrow n + e^+ \tag{11.18}$$

making most experiments mainly sensitive to the $\bar{\nu}_e$ component. The only directional information is available in scattering experiments like neutrino-electron scattering (see Chapter 4). Recently, it was proposed to measure NC interactions by using elastic scattering on free protons $\nu + p \rightarrow \nu + p$ where the proton recoil produces a signal-like scintillation light [Bea02]. Most running experiments are relying on these reactions. Additionally neutrino capture on nuclei or neutral current excitation of a nucleus can occur

$$\nu_x + (A, Z) \rightarrow (A - 1, Z)^* \rightarrow (A - 1, Z) + \gamma, n + \nu_x \quad \text{(NC)} \tag{11.19}$$
$$\nu_e + (A, Z) \rightarrow (A - 1, Z + 1) + e^- \quad \text{(CC)} \tag{11.20}$$
$$\bar{\nu}_e + (A, Z) \rightarrow (A - 1, Z - 1) + e^+ \quad \text{(CC)} \tag{11.21}$$

with potential emission of gamma rays or neutrons if excited states are involved. The neutrino threshold energy E_{Th} for the CC interaction is given by

$$E_{\text{Th}} = \frac{m_f^2 + m_e^2 + 2m_f m_e - m_i^2}{2m_i} \simeq m_f - m_i + m_e \tag{11.22}$$

where m_f, m_i are the masses of the final and the initial nucleus, which might be modified if capture occurs in excited states. Among the interesting ones are

$$\nu_e + {}^{16}\text{O} \rightarrow {}^{16}\text{F} + e^- \quad \text{(CC)} \quad E_{\text{Th}} = 15.4 \, \text{MeV} \tag{11.23}$$
$$\bar{\nu}_e + {}^{16}\text{O} \rightarrow {}^{16}\text{N} + e^+ \quad \text{(CC)} \quad E_{\text{Th}} = 11.4 \, \text{MeV} \tag{11.24}$$
$$\bar{\nu} + {}^{16}\text{O} \rightarrow {}^{16}\text{O}^* + \bar{\nu} \quad \text{(NC)} \tag{11.25}$$
$$\nu_i(\bar{\nu}_i) + e^- \rightarrow \nu_i(\bar{\nu}_i) + e^- \quad (i \equiv e, \mu, \tau) \tag{11.26}$$
$$\nu_e + {}^{12}\text{C} \rightarrow {}^{12}\text{C}^* + \nu_e \quad \text{(NC)} \tag{11.27}$$
$$\nu_e + {}^{12}\text{C} \rightarrow {}^{12}\text{N} + e^- \quad \text{(CC)} \tag{11.28}$$
$$\nu_i + {}^{208}\text{Pb} \rightarrow {}^{208}\text{Pb}^* + \nu_i \quad \text{(NC)} \tag{11.29}$$
$$\nu_e + {}^{208}\text{Pb} \rightarrow {}^{208}\text{Bi} + e^- \quad \text{(CC)} \tag{11.30}$$
$$\nu_e + {}^{40}\text{Ar} \rightarrow {}^{40}\text{K}^* + e^- \quad \text{(CC)} \tag{11.31}$$

with a threshold of 5.9 MeV and a 4.3 MeV gamma ray from the ${}^{40}\text{K}^*$ de-excitation. Additional processes with smaller cross-sections are: The CC reaction cross-section on ${}^{16}\text{O}$ which has a threshold of 13 MeV and rises very fast with respect to energy, making it the dominant ν_e detection mode at higher energies. Liquid organic scintillators lack the reactions on oxygen but can rely on CC and NC reactions on ^{12}C: The superallowed NC reaction on ^{12}C might be detected by the associated 15 MeV de-excitation gamma (see Chapter 4).

As the neutron excess in stable nuclei is increasing for heavier masses, the $\bar{\nu}_e$ capture gets suppressed (Pauli-blocking), making any detector based on this principle more sensitive to ν_e and NC interactions. Hence, reactions on ^{208}Pb via neutron emission have been discussed [Har96, Smi97, Ful98, Smi01, Kol01, Zac02, Eng03, Vaa11]. The CC reaction has a threshold of $E_\nu \gtrsim 30$ MeV. Unfortunately neutrino cross-sections on nuclei in the energy range of 10-30 MeV are experimentally not known, activities at neutron spallation sources have started to improve the situation [Bol12]. In principle, the discovered coherent neutrino-nucleus scattering should occur as well. The behaviour of the relevant cross-sections as function of E_ν is shown in Figure 11.13.

Running real-time experiments can be divided into water Cherenkov detectors such as Super-Kamiokande (now also doped with Gd for better neutron capture) and IceCube as well as liquid scintillator detectors like KamLAND, Borexino, Baksan Scintillation Telescope, SNO+, MiniBooNE and the Large Volume Detector (LVD), a 1.8 kt detector which has been running in the Gran Sasso Laboratory since 1992 [Ful99]. In the future the Jingping experiment and JUNO will also add to this. Furthermore, the HALO experiment at SNOLAB is online using 79 tons of lead and the SNO neutral current detectors (see Chapter 10), with a potential for an update to 1 kt. Nevertheless, at the moment any directional information is based on water Cherenkov detectors, but also water-based liquid scintillators are explored.

The detection method of neutrino telescopes (see Chapter 12) like ANTARES and Icecube is different from the typical underground water Cherenkov detectors. The photomultipliers used here are too sparse to allow an event-by-event reconstruction but a large burst of supernova neutrinos producing enough Cherenkov photons could cause a coincident increase in single count rates for many phototubes, i.e., a sudden deviation from the summed photomultiplier noise rate in the detector [Pry88, Hal94, Hal96a]. The Deep Core extension of Icecube might allow a mixed mode of both techniques [Sal12]. Sea-based telescopes like ANTARES (see Chapter 12) might need an external trigger signal to investigate their data due to the higher background. The advantage of neutrino telescopes is their large effective volume, for example it is 3 Mt for Icecube and thus larger than any underground experiment and allows for higher statistics [Abb11]. According to the cross-sections, it can be considered that mainly $\bar{\nu}_e$ were detected from the supernova 1987A which is discussed in the next section.

11.4 Supernova 1987A

One of the most important astronomical events of the last century was the supernova (SN) 1987A [Arn89, Che92, Kos92, Woo97, Imm07, McC16, IAU17] (see Figure 11.14) (the numbering scheme for supernovae contains the year of their discovery and another letter following the alphabet which indicates the order of

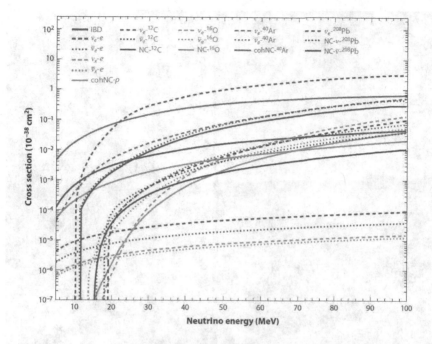

Figure 11.13. Some relevant total neutrino cross-sections for supernova detection using various types of detector materials. The curves refer to the total cross-section per water molecule so that a factor of two for protons and 10 for electrons is already included (from [Sch12]). © Annual Review of Nuclear and Particle Science.

occurrence). This was the brightest supernova since Kepler's supernova in 1604 and provided astrophysicists with an overwhelming amount of new data and insights, as it was for the first time possible to observe a supernova at all wavelengths and to follow the evolution up to now. Moreover, for the first time neutrinos could be observed from this spectacular event. This first detection of neutrinos which do not originate from the Sun for many scientists marked the birth of neutrino astrophysics. Further details can be found, e.g., in [Arn89, Bah89, Che92, Kos92, McG93].

11.4.1   Characteristics of supernova 1987A

11.4.1.1   *Properties of the progenitor star and the event*

Supernova 1987A was discovered on 23 February 1987 at a distance of 150 000 light years (corresponding to 50 kpc) in the Large Magellanic Cloud (LMC), a companion galaxy of our own Milky Way [McN87]. The evidence that the supernova was of type II, an exploding star, was confirmed by the detection of hydrogen lines in the spectrum. However, the identification of the progenitor star Sanduleak $-69°$ 202 was a surprise since it was a blue B3I supergiant with a mass of about $20 M_\odot$. Until then it was assumed that only red giants could go supernova. The explosion of a massive

(a) (b)

Figure 11.14. The supernova 1987A. (a) The Large Magellanic Cloud on 26 February 1987 at 1^{h} 25^{min} where the 4.4^{m} brightness supernova 1987A can be seen. The length of the horizontal scale is 1 arcmin. (b) The same field of view as (a) before the supernova on 9 December 1987. The precursor star Sanduleak $-69°$ 202 is shown (from [Aao19]). With kind permission of David Malin (Australian Astronomical Observatory).

blue supergiant could be explained by the smaller 'metal' abundance in the Large Magellanic Cloud, which is only one-third of that found in the Sun, together with a greater mass loss, which leads to a change from a red giant to a blue giant. The oxygen abundance plays a particularly important role. On one hand, the oxygen is relevant for the opacity of a star and, on the other hand, less oxygen results in less efficient catalysis of the CNO process, which causes a lower energy production rate in this cycle. Indeed it is possible to show with computer simulations that blue stars can also explode in this way [Arn91, Lan91].

A comparison of the bolometric brightness L of the supernova in February 1992 of $L = 1 \times 10^{37}$ erg s^{-1} shows that this star really did explode. This is more than one order of magnitude less than the value of $L \approx 4 \times 10^{38}$ erg s^{-1}, which was measured *before* 1987, i.e., the original star has vanished. SN 1987A went through a red giant phase but developed back into a blue giant about 20 000 years ago. The large mass ejection in this process was discovered by the Hubble Space Telescope,

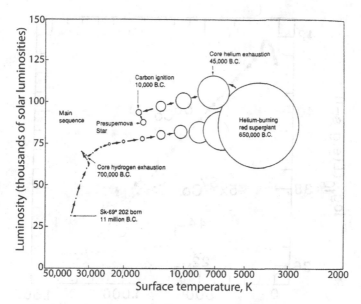

Figure 11.15. According to the theoretical model of S. Woosley et al. (from [Woo88]), Sanduleak −62° 202 was probably born some 11 million years ago, with a mass about 18 times that of the Sun. Its initial size predetermined its future life, which is mapped in this diagram showing the luminosity against surface temperature at various stages, until the moment immediately before the supernova explosion. Once the star had burned all the hydrogen at its centre, its outer layers expanded and cooled until it became a red supergiant, on the right-hand side of the diagram. At that stage when helium started burning in the core to form carbon, and by the time the supply of helium at the centre was exhausted, the envelope contracted and the star became smaller and hotter, turning into a blue supergiant. With kind permission of T. Weaver and S. Woosley.

as a ring around the supernova. Moreover, an asymmetric explosion seems now to be established [Wan02]. The course of evolution is shown in Figure 11.15. The total explosive energy amounted to $(1.4 \pm 0.6) \times 10^{51}$ erg [Che92].

11.4.1.2 γ-radiation

γ-line emission could also be observed for the first time. It seems that the double magic nucleus ^{56}Ni is mostly produced in the explosion. It has the following decay chain:

$$^{56}\text{Ni} \xrightarrow{\beta^+} {}^{56}\text{Co} \xrightarrow{\beta^+} {}^{56}\text{Fe}^* \xrightarrow{\gamma} {}^{56}\text{Fe}. \tag{11.32}$$

^{56}Ni decays with a half-life of 6.1 days. ^{56}Co decays with a half-life of 77.1 days, which is very compatible with the decrease in the light curve as shown in Figure 11.16 [Che92]. Two gamma lines at 847 and 1238 keV, which are characteristic lines of the ^{56}Co decay, were detected by the Solar Maximum Mission satellite (SMM)

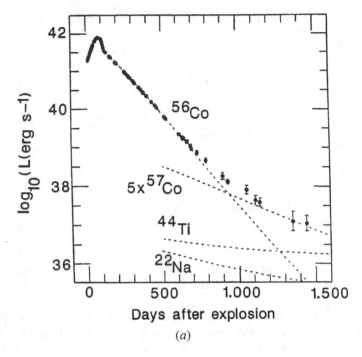

Figure 11.16. Behaviour of the light curve of supernova 1987A. (*a*) The early phase (from [Sun92]). With kind permission of N. Suntzeff, CTIO/NOAO. © IOP Publishing. Reproduced with permission. All rights reserved. (*b*) V-band light curve (named after the visual region at 555 nm using narrow light band classifications). The early phase of an expanding photosphere is driven by the shock breakout, resulting in an early peak lasting from a few hours to a couple of days. After a rapid, initial cooling the supernova enters a phase when its temperature and luminosity remain nearly constant. In this "plateau phase" it is powered by the recombination of the previously ionized atoms in the supernova shock. Once the photosphere has receded deep enough, additional heating due to radioactive decays of ^{56}Ni and ^{56}Co dominates. Afterwards the light curve is powered solely by radioactive decay in the remaining nebula; the light curve enters the "radioactive tail", typically after about 100 days. SN 1987A suffered from dust forming within the ejecta (after $\approx$450 days), which resulted in an increase in the decline rate in the optical as light was shifted to the infrared. After about 800 days the light curve started to flatten again due to energy released of ionized matter ('freeze out'). Later, the flattening is caused by long-lived isotopes, especially ^{57}Co and ^{44}Ti. At very late times, the emission is dominated by the circumstellar inner ring, which was ionized by the shock breakout. After about 2000 days the emission of the ring is stronger than from the supernova ejecta itself (from [Lei03]). Reproduced with permission of SNCSC.

at the end of August 1987 [Mat88]. From the intensity of the lines the amount of ^{56}Fe produced in the explosion can be estimated to be 0.075 $M_\odot$. Recently, also the production of ^{44}Ti could be confirmed, its decay is supposed to dominate the optical light curve after several years (see Figure 11.16) [Gre12].

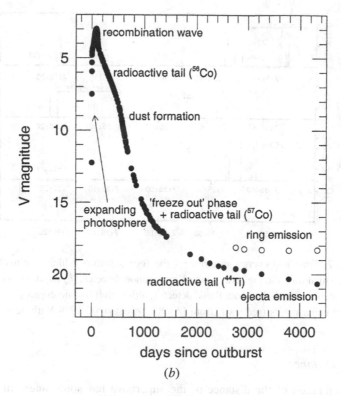

Figure 11.16. (Continued.)

In general, photometric measurements of light curves provide important information about supernovae [Lei03]. They depend mainly on the size and mass of the progenitor star and the strength of the explosion. However, various additional energy inputs exist which results in modulations of the emerging radiation. By far the longest observed light curve is SN 1987A [Sun92]. The early development of the bolometric light curve together with the V-band ($\lambda = 540$ nm) light curve cover more than 30 years; the first 10 years are shown in Figure 11.16.

A direct search for a pulsar at the center of SN 1987A with the Hubble Space Telescope has still been unsuccessful [Per95]. The evidence for a pulsar in SN 1987A by powering the light curve would be very interesting insofar as it has never been possible to observe a pulsar and supernova directly from the same event and no hint of a pulsar contribution to a supernova light curve has been established. The implication of a non-observation of the pulsar associated with SN 1987A is discussed in [Man07].

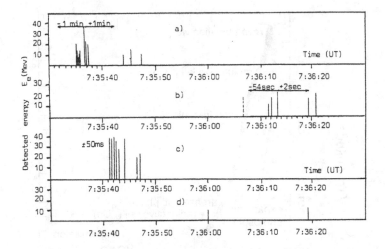

Figure 11.17. Time and energy spectrum of the four detectors which saw neutrinos from SN1987A as mentioned in the text: (*a*) the Kamiokande detector, (*b*) the Baksan detector, (*c*) the IMB detector and (*d*) the Mont Blanc detector, although it had no events at the time seen by the other experiments (see text) (from [Ale87a, Ale88]). © 1988 With permission from Elsevier.

11.4.1.3   Distance

The determination of the distance of the supernova has some interesting aspects. The Hubble telescope discovered a ring of diameter (1.66 ± 0.03) arcsec around SN 1987A in the UV region in a forbidden line of doubly ionized oxygen [Pan91]. Using the permanent observations of UV lines from the International Ultraviolet Explorer (IUE), in which these lines also appeared, the ring could be established as the origin of these UV lines. The diameter was determined to be $(1.27 \pm 0.07) \times 10^{18}$ cm, from which the distance to SN 1987A can be established as $d = (51.2 \pm 3.1)$ kpc [Pan91]. Correcting to the centre of mass of the LMC leads to a value of $d = (50.1 \pm 3.1)$ kpc, another estimate based on the ring resulted in $d = (47.2 \pm 0.9)$ kpc [Gou98]. These values are not only in good agreement with those of other methods but have a relatively small error, which makes the use of this method for distance measurements at similar events in the future very attractive.

11.4.1.4   Summary

In the previous sections we have discussed only a small part of the observations and details of SN 1987A. Many more have been obtained, mainly related to the increasing interaction of the ejected layers with the interstellar medium. The detection of the expected neutron star (pulsar) created in the supernova, or even a black hole, are eagerly awaited. Even though SN 1987A has provided a huge amount of new information and observations in all regions of the spectrum, the most exciting event was, however, the first detection of neutrinos from the star's collapse.

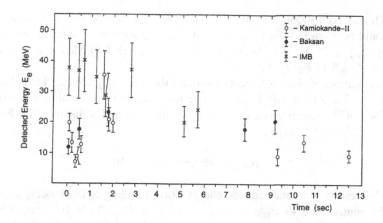

Figure 11.18. Energies of all neutrino events detected at 7.35 UT on 23 February 1987 *versus* time: The time of the first event in each detector has been set to $t = 0$ s (from [Ale88,Ale88a]). © 1988 With permission from Elsevier.

11.4.2   Neutrinos from SN 1987A

A total of four detectors claim to have seen neutrinos from SN 1987A [Agl87, Ale87, Ale88, Bio87, Hir87, Hir88, Bra88]. Two of these are water Cherenkov detectors (KamiokandeII and Irvine–Michigan–Brookhaven (IMB) detector) and two are liquid scintillator detectors (Baksan Scintillator Telescope (BST) and Mont Blanc). The Cherenkov detectors had a far larger amount of target material, the fiducial volumes used are 2140 t (Kamiokande), 6800 t (IMB) compared to the liquid scintillation detector of 200 t (BST). Important for detection is the trigger efficiency for $e^{\pm}$ reaching about 90% (80%) for KII (BST) at 10–20 MeV and being much smaller for IMB at these energies. In addition, IMB reports a dead time of 13% during the neutrino burst [Bra88]. The observed events are listed in Table 11.2 and were obtained by the reactions described in Section 11.3. Within a certain timing uncertainty, three of the experiments agree on the arrival time of the neutrino pulse, while the Mont Blanc experiment detected them about 4.5 hr before the other detectors. Since all five events are lying very close to the trigger threshold of 5 MeV and as the larger Cherenkov detectors saw nothing at that time, it is generally assumed that these events are a statistical fluctuation and are not related to the supernova signal. The other three experiments also detected the neutrinos before the optical signal arrived, as expected. The relatively short time of a few hours between neutrino detection and optical discovery points to a compact progenitor star. The time structure and energy distribution of the neutrinos is shown in Figure 11.18. If *assuming* the first neutrino events have been seen by each detector at the same time, i.e., setting the arrival times of the first event to $t = 0$ for each detector, then within 12 s, 24 events were observed (KamiokandeII + IMB + Baksan). The overall important results tested experimentally with these observations can be summarized as follows:

(i) All observed events are due to $\bar{\nu}_e$ interactions (maybe the first event of KII could be from ν_e). Fitting a Fermi–Dirac distribution an average temperature of $\langle T_\nu \rangle = (4.0 \pm 1.0)$ MeV and $\langle E_\nu \rangle = (12.5 \pm 3.0)$ MeV can be obtained ($\langle E_\nu \rangle = 3.15\, k_{\rm B} \langle T_\nu \rangle$).

(ii) The number of observed events estimates the time integrated $\bar{\nu}_e$ flux to be about $\Phi = (5 \pm 2.5) \times 10^9$ $\bar{\nu}_e$ cm^{-2}. The total number of neutrinos emitted from SN 1987A is then given (assuming six flavours and a distance SN–Earth of $L = 1.5 \times 10^{18}$ km) by

$$N_{\rm tot} = 6\Phi\, 4\pi L^2 \approx 8 \times 10^{57}. \tag{11.33}$$

This results in a total radiated energy corresponding to the binding energy of the neutron star of $E_{\rm tot} = N_{\rm tot} \times \langle E_\nu \rangle \approx (2 \pm 1) \times 10^{53}$ erg, which is in good agreement with expectation. This observation also for the first time experimentally verified that indeed more than 90% of the total energy released is carried away by neutrinos and that the visible signal corresponds to only a minute fraction of the released energy.

(iii) The duration of the neutrino pulse was of the order 10 s.

The number of more detailed and specific analyses, however, far exceeds the number of observed events. A systematic and comprehensive study based on a solid statistical treatment is given in [Lor02]. Several SN models were explored, dominantly from [Woo86], half being prompt explosions and half delayed explosions. Their conclusion is in strong favor of the data of the delayed explosion mechanism and a typical radius of the resulting neutron star in the order of 10 km; see also [Pag09]. For more extensive discussion of neutrinos from SN 1987A, we refer to [Kos92, Raf96, Vis15].

11.4.2.1 Possible anomalies

Unexplained facts remain from the observation; however, in the derived conclusions the small statistics has to be considered. The least worrisome is a discrepancy in $\langle E_\nu \rangle$ between KII and IMB, implying a harder sloped spectrum for IMB. However, IMB has a high energy threshold and relies on the high energy tail of the neutrino spectrum, which depends strongly on the assumed parameters. Therefore, both might still be in good agreement. A second point is the 7.3 s gap between the first nine and the following three events in Kamiokande. But given the small number of events this is possible within statistical fluctuations and additionally the gap is filled with events from IMB and BST. Probably the most disturbing is the deviation from the expected isotropy for the $\bar{\nu}_e + p \rightarrow e^+ + n$ reaction (even a small backward bias is expected), especially showing up at higher energies. Various explanations have been proposed, among them ν_e forward scattering. However, the cross-section for this reaction is too small to account for the number of observed events. Another anomaly would be the preferred forward direction of the lepton. Unless some new imaginative idea is born, the most common explanation relies again on a possible statistical fluctuation.

Table 11.2. Table of the neutrino events registered by the four neutrino detectors KamiokandeII [Hir87], IMB [Bio87], Mont Blanc [Agl87] and Baksan [Ale87a, Ale88]. T gives the time of the event, E gives the visible energy of the electron (positron). The absolute uncertainties in the given times are: for Kamiokande ± 1 min, for IMB ± 50 ms and for Baksan -54 s, $+2$ s.

Detector	Event number	T (UT)	E (MeV)
Kamioka	1	7 : 35 : 35.000	20 ± 2.9
	2	7 : 35 : 35.107	13.5 ± 3.2
	3	7 : 35 : 35.303	7.5 ± 2.0
	4	7 : 35 : 35.324	9.2 ± 2.7
	5	7 : 35 : 35.507	12.8 ± 2.9
	(6)	7 : 35 : 35.686	6.3 ± 1.7
	7	7 : 35 : 36.541	35.4 ± 8.0
	8	7 : 35 : 36.728	21.0 ± 4.2
	9	7 : 35 : 36.915	19.8 ± 3.2
	10	7 : 35 : 44.219	8.6 ± 2.7
	11	7 : 35 : 45.433	13.0 ± 2.6
	12	7 : 35 : 47.439	8.9 ± 1.9
IMB	1	7 : 35 : 41.37	38 ± 9.5
	2	7 : 35 : 41.79	37 ± 9.3
	3	7 : 35 : 42.02	40 ± 10
	4	7 : 35 : 42.52	35 ± 8.8
	5	7 : 35 : 42.94	29 ± 7.3
	6	7 : 35 : 44.06	37 ± 9.3
	7	7 : 35 : 46.38	20 ± 5.0
	8	7 : 35 : 46.96	24 ± 6.0
Baksan	1	7 : 36 : 11.818	12 ± 2.4
	2	7 : 36 : 12.253	18 ± 3.6
	3	7 : 36 : 13.528	23.3 ± 4.7
	4	7 : 36 : 19.505	17 ± 3.4
	5	7 : 36 : 20.917	20.1 ± 4.0
Mt Blanc	1	2 : 52 : 36.79	7 ± 1.4
	2	2 : 52 : 40.65	8 ± 1.6
	3	2 : 52 : 41.01	11 ± 2.2
	4	2 : 52 : 42.70	7 ± 1.4
	5	2 : 52 : 43.80	9 ± 1.8

11.4.3   Neutrino properties from SN 1987A

Several interesting results on neutrino properties can be drawn from the fact that practically all the expected neutrinos were detected within about 12 s and the observed flux is in agreement with expectations. For additional bounds on exotic particles see [Raf96, Raf99, Pay15].

11.4.3.1   Lifetime of the neutrino

As the expected flux of antineutrinos has been measured on Earth, no significant number could have decayed in transit, which leads to a lower limit on the lifetime for $\bar{\nu}_e$ of [Moh91]

$$\left(\frac{E_\nu}{m_\nu}\right) \tau_{\bar{\nu}_e} \geq 5 \times 10^{12} \text{ s} \approx 1.6 \times 10^5 \text{ yr}. \tag{11.34}$$

In particular, the radiative decay channel for a heavy neutrino ν_H

$$\nu_H \to \nu_L + \gamma \tag{11.35}$$

can be limited independently. No enhancement in γ-rays coming from the direction of SN 1987A was observed in the Gamma Ray Spectrometer (GRS) on the Solar Maximum Satellite (SMM) [Chu89]. The photons emanating from neutrino decay would arrive with a certain delay, as the parent heavy neutrinos do not travel at the speed of light. The delay with respect to the first observed neutrino is given by

$$\Delta t \simeq \frac{1}{2} d \frac{m_\nu^2}{E_\nu^2}. \tag{11.36}$$

with d being the distance to the LMC. For neutrinos with a mean energy of 12 MeV and a mass smaller than 20 eV the delay is about 10 s, which should be reflected in the arrival time of any photons from the decay. The study by [Blu92] using SMM data resulted in

$$\frac{\tau_H}{B_\gamma} \geq 2.8 \times 10^{15} \frac{m_H}{\text{eV}} \text{ s} \qquad m_H < 50 \text{ eV} \tag{11.37}$$

$$\frac{\tau_H}{B_\gamma} \geq 1.4 \times 10^{17} \text{ s} \qquad 50 \text{ eV} < m_H < 250 \text{ eV} \tag{11.38}$$

$$\frac{\tau_H}{B_\gamma} \geq 6.0 \times 10^{18} \frac{\text{eV}}{m_H} \text{ s} \qquad m_H > 250 \text{ eV}. \tag{11.39}$$

Here B_γ is the branching ratio of a heavy neutrino into the radiative decay channel. Thus, there is no hint of a neutrino decay.

11.4.3.2   Mass of the neutrino

Direct information about the mass is obtained from the observed spread in arrival time. The time of flight t_F of a neutrino with mass m_ν and energy E_ν ($m_\nu \ll E_\nu$)

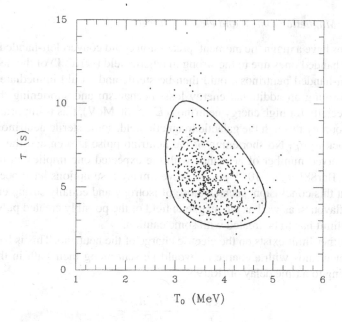

Figure 11.19. Loredo plot (time τ versus energy T_0) of parameter estimation of neutrino mass (from [Lor02]). © 2002 by the American Physical Society.

from the source (emission time t_0) to the detector (arrival time t) in a distance L is given by

$$t_F = t - t_0 = \frac{L}{v} = \frac{L}{c}\frac{E_\nu}{p_\nu c} = \frac{L}{c}\frac{E_\nu}{\sqrt{E_\nu^2 - m_\nu^2 c^4}} \approx \frac{L}{c}\left(1 + \frac{m_\nu^2 c^4}{2E^2}\right). \quad (11.40)$$

If $m_\nu > 0$, the time of flight is getting shorter if E_ν increases. For two neutrinos with E_1 and E_2 ($E_2 > E_1$) emitted at times t_{01} and t_{02} ($\Delta t_0 = t_{02} - t_{01}$) the time difference on Earth is

$$\Delta t = t_2 - t_1 = \Delta t_0 + \frac{L m_\nu^2}{2c}\left(\frac{1}{E_2^2} - \frac{1}{E_1^2}\right). \quad (11.41)$$

Here $\Delta t, L, E_1, E_2$ are known; Δt_0 and m_ν are unknown. Depending on which events from Table 11.2 are combined and assuming simultaneously a reasonable emission interval Δt_0, mass limits lower than 30 eV (or even smaller at the price of model dependence) are obtained [Arn89, Kol87]. The analysis by [Lor02] sharpens this bound even more and concludes that (see Figure 11.19)

$$m_{\bar{\nu}_e} < 5.7\,\text{eV} \quad (95\%\,\text{CL}) \quad (11.42)$$

which is comparable with current β-decay results (see Chapter 6).

11.4.3.3   *Magnetic moment and electric charge*

If neutrinos have a magnetic moment, precession could convert left-handed neutrinos into right-handed ones due to the strong magnetic field ($\sim 10^8$ T) of the neutron star. Such right-handed neutrinos would then be sterile and would immediately escape, thereby forming an additional energy loss mechanism and shortening the cooling time. Especially for high energy neutrinos ($E_\nu > 30$ MeV), this is important. During the long journey through the galactic magnetic field, some sterile neutrinos might be rotated back to ν_L. No shortening of the neutrino pulse is seen and the agreement of the observed number of neutrinos with the expected one implies an upper limit of [Lat88, Bar88] $\mu_\nu < 10^{-12} \mu_B$. However, many assumptions have been made in arriving at this conclusion, such as those of isotropy and equally strong emission of different flavours, as well as the magnetic field of the possibly created pulsar, so that the exact limit has to be treated with some caution.

A further limit exists on the electric charge of the neutrino. This is based on the fact that neutrinos with a charge e_ν would be sent along their path in the galactic field leading to a time delay of [Raf96]

$$\frac{\Delta t}{t} = \frac{e_\nu^2 (B_T d_B)^2}{6 E_\nu^2} \tag{11.43}$$

with B_T as the transverse magnetic field and d_B the path length in the field. The fact that all neutrinos arrived within about 10 s results in [Bar87]

$$\frac{e_\nu}{e} < 3 \times 10^{-17} \left(\frac{1\,\mu\text{G}}{B_T} \right) \left(\frac{1\,\text{kpc}}{d_B} \right). \tag{11.44}$$

More bounds, e.g., a test of relativity can be found in [Raf96].

11.4.3.4   *Conclusion*

Our knowledge of supernova explosions has grown enormously in recent years because of SN 1987A, not least from the first confirmation that supernovae type II really are phenomena from the late phase in the evolution of massive stars and that the energy released corresponds to the expectations. Also, the first detection of SN neutrinos has to be rated as a particularly remarkable event. As to how far the observed data are specific to SN 1987A, and to what extent they have general validity, only further supernovae of this kind can show. SN 1987A initiated a great deal of experimental activity in the field of detectors for supernova neutrinos. We now turn to the prospects for future experiments. The experiments themselves as well as the likely occurrence of supernovae are important.

11.5   Supernova rates and future experiments

How often do type II SN occur in our galaxy and allow us additional observations of neutrinos? Two ways of determining are generally considered [Raf02]. Historical records of supernovae in our galaxy can be explored, by counting supernova remnants or historic observations. Another way is to study supernova rates in other nearby galaxies [And05], which depends on the morphological structure of the galaxy. This can then be converted into a proposed rate for our Milky Way. Because of the small statistics involved and further systematic uncertainties, a rate of between one and six supernovae per century in our Milky Way seems to be realistic [Ada13]. The rate could be increased to about 1 per year if larger detectors are used like Icecube (see Chapter 13) with the interior Deep Core detector. To achieve this observed rate, a minimal fiducial mass around 5 Mt must be considered, which should allow the detection of supernovae up to 10 Mpc.

The future prospects for supernova detection are based on the reactions described in Section 11.3. Several running experiments are upgrading their detectors or plan larger follow-up projects: ICECUBE with the installation of Deep Core and Pingu , ANTARES with Orca. SuperKamiokande has added Gd into water to increase the neutron capture efficiency (GdZooks). Furthermore, Hyper-Kamiokande as a Mt detector is under construction. HALO, a 1 kt lead based ^{3}He counter is considered for the second phase. New scintillator experiments like SNO+ (1kt) and NOvA and larger versions of running scintillator experiments like Daya Bay 2 and RENO-50 might add to it as well as the Jingping experiment and JUNO both of about 20-50 kton scale. A new technology, pioneered by ICARUS, is LAr time projection chambers, intensively studied worldwide for the DUNE neutrino beam, but a 50 kton underground detector would also be very sensitive to supernova neutrinos. This would offer directional information as well. A possible coherent scattering measurement could be realized with future large-scale dark matter experiments, as the signal is a recoiling nucleus in the range of several keV and hence the same as for direct dark matter searches, however confined to a 10 s time interval [And05].

The total number of available and planned experiments is impressive and will, at the time of the next nearby supernova, provide a lot of new information.

It is apparent from the discussion of supernova explosions and confirmed by SN 1987A that neutrinos are arriving earlier than optical signals from the explosion. Hence a global network of available detectors called SuperNova Early Warning System (SNEWS) has been established [Sch01, Ant04] to alert telescopes. With two detectors producing directional information, a ring on the sky for the potential source can be defined. This might be improved by three (2 intersecting rings forming 2 spots) or four detectors, which would pin down the source to a spot. It has been considered to use triangulation among the different experiments. Given the fact that the time of flight of neutrinos through the earth is in the region of milliseconds and the supernova pulse duration in the order of 10 s, this needs large statistics to be applied.

A completely different way of detecting supernovae utilizes the vibrations of spacetime due to the powerful explosion. This appears in the form of gravitational waves [Sau94, Mag08, Sat09, Hug09, Cre11, Ril13]. The indirect evidence of their existence was done by studying the slowdown in the period of the pulsar PSR1913+16 in a binary system with an unseen neutron star (NS) but can be described exactly by assuming the emission of gravitational waves [Tay94]. In 2017 finally the direct observation of gravitational waves has been done by LIGO

[Abb16]. In the meantime more than 10 events including black holes (BH) have been observed. Preferential sources are BH-BH, BH-NS and NS-NS mergers. Several detectors in the form of massive resonance masses in the form of bars (e.g., Nautilus, EXPLORER, Auriga, ALLEGRO) and spheres (MiniGrail) have been used. The breakthrough however was the laser interferometers (e.g., advanced VIRGO, advanced LIGO, GEO-HF and KAGRA). Some frequency regions cannot be studied on earth due to the seismic noise, hence a laser interferometer in space is considered. Such a project (eLISA) is proposed as a cornerstone project of the European Space Agency; other proposals exist. Additional searches are performed in radioastronomy on pulsar timing arrays. The search for the small vibrations of spacetime of which supernovae are a good source has started and exciting results can be expected in the near future.

11.5.1   Diffuse supernova neutrino background

Besides the supernova rate in our galaxy, for direct detection it might be worthwhile asking whether a diffuse supernova neutrino background (DSNB) accumulated over cosmological times can be detected. However, an estimate of the expected flux is not an easy task to perform [Tot95, Tot96, Mal97, Kap00, Bea04, Hor09, Lun09, Bea10,Lun16,Hor18] depending on various uncertainties like the supernova rate as a function of redshift, the assumed neutrino spectrum of a supernova and all parameters linked to the detection. The DSNB flux depends approximately quadratic on the Hubble constant and weakly on the density parameter (Ω_0) and the cosmological constant (Λ) (see Chapter 13).

A calculated flux [Tot95] is shown in Figure 11.20. As can be seen, solar and terrestrial neutrino sources are overwhelming below 15 MeV; starting at about 50 MeV atmospheric neutrinos dominate. Therefore, only a slight window between 15–40 MeV exists for possible detection. Converting this flux into an event rate prediction for Super-Kamiokande results in 1.2 events per year with an uncertainty of a factor three [Tot95]. Super-Kamiokande measured an upper bound of 1.2 $\bar{\nu}_e$ cm^{-2} s^{-1} for the DSNB flux in the energy range $E_\nu > 19$ MeV [Mal03], recently updated to less than $2.8 - 3.0$ $\bar{\nu}_e$ cm^{-2} s^{-1} for $E_\nu > 17.3$ MeV [Bay12], which is in the order of some theoretical predictions. In addition, SNO is giving a limit on the ν_e flux to be less than 70 ν_e cm^{-2} s^{-1} [Aha06]. Additional measurements from Borexino [Bel11] and KamLAND [Gan12] exist. The Super-K-Gd phase running will have a high sensitivity for $\bar{\nu}_e$ from the DSNB.

This has been only a short glimpse of the strong interplay between neutrino and supernova physics. Many uncertainties and caveats exist, so the dominant part of the presented numbers should be handled with some caution and there is hope that future observations will help to sharpen our view. Having discussed solar and supernova neutrinos, where experimental observations exist, we now come to other astrophysical neutrino sources where the first neutrino discoveries occurred just recently.

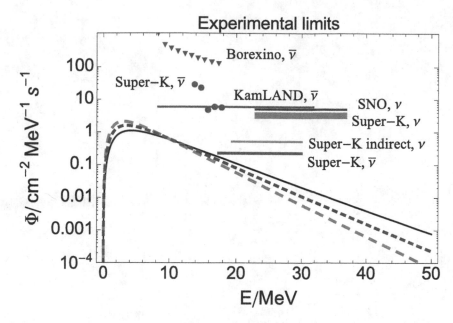

Figure 11.20. Supernova relic neutrino background compared with upper limits from experiments (from [Lun16]). © 2016 With permission from Elsevier.

Chapter 12

Ultra-high energetic cosmic neutrinos

DOI: 10.1201/9781315195612-12

Having discussed neutrinos from stars with energies of $E_\nu \leq 100$ MeV we now want to explore additional astrophysical neutrinos produced with much higher energy. The observation of cosmic rays with energies up to 10^{20} eV, by the Fly's Eye, AGASA and the Auger air shower array, supports the possibility of observing neutrinos up to this energy range. Neutrinos mostly originate from the decay of secondaries like pions resulting from "beam dump" (see Figure 12.1) interactions of protons with other protons (or nuclei) or photons as in accelerator experiments. Like photons, neutrinos are not affected by the presence of a magnetic field and thus could point to sources. Even absorption which might prohibit photon detection is not an issue and this also allows a search for hidden sources, which cannot be seen otherwise. Therefore, neutrinos are an excellent candidate for finding point sources in the sky and might help to identify the sources of cosmic rays as will be shown later. In addition, our view of the universe in photons is limited for energies beyond 1 TeV. The reason is the interaction of such photons with background photons $\gamma + \gamma_{BG} \to e^+ e^-$. This reaction has a threshold of $4E_\gamma E_{\gamma_{BG}} \approx (2m_e)^2$. In this way TeV photons are attenuated due to reactions on the infrared background and PeV photons by the cosmic microwave background (see Chapter 13). For additional literature see [Sok89, Ber91, Lon92, 94, Gai95, Lea00, Sch00, Gri01, Hal02, Aha03, Sta04, Bec08, Blü09, Hin09, Anc10, Kot11, Gai16, Fed18, Spu18].

12.1 Sources of high-energy cosmic neutrinos

The search for high-energy neutrinos can be split into two categories. One is the obvious search for point sources, in the hope that a signal will shed light on the question of what the sources of cosmic rays are. The second one is a diffuse neutrino flux like the one observed in gamma rays. This flux is created by pion decays, produced in cosmic-ray interactions within the galactic disc. Instead of this observational-motivated division, the production mechanism itself can be separated roughly in two categories. Annihilation in combination with the decay of heavy particles and acceleration processes. The acceleration process can be subdivided further into those of galactic and extragalactic origin.

12.1.1   Neutrinos produced in acceleration processes

The observation of TeV γ-sources together with the detection of a high-energy ($E_\gamma >$ 100 MeV) diffuse galactic photon flux by the EGRET experiment on the Compton Gamma Ray Observatory opened a new window into high-energy astrophysics. Such highly energetic photons might be produced by electron acceleration due to synchrotron radiation and inverse Compton scattering. In addition, it is known that cosmic rays with energies up to 10^{20} eV exist, implying the acceleration of protons in some astrophysical sources ("cosmic accelerators"). Observations of ultra-high energy neutrinos would prove proton acceleration because of charged pion (meson) production

$$p + p, p + \gamma \rightarrow \pi^0, \pi^\pm, K^\pm + \mathrm{X} \tag{12.1}$$

with X as further hadronic final states particles. This process is similar to the production of artificial neutrino beams as described in Chapter 4. Associated with charged pion-production is π^0-production creating highly-energetic photons. It offers another source for TeV photons via π^0-decay. Two types of sources exist in that way: diffusive production within the galaxy by interactions of protons with the interstellar medium; and point-like sources, where the accelerated protons interact directly in the surrounding of the source. The latter production mechanism corresponds to an astrophysical beam-dump experiment also creating neutrinos (Figure 12.1). The dump must be partially transparent for protons; otherwise, it cannot be the source of cosmic rays. These are guaranteed sources of neutrinos because it is known that the beam and the target both exist (Table 12.1). High energy photons are affected by interactions with the cosmic microwave (see Section 13.3) and infrared backgrounds, mostly through $\gamma\gamma \rightarrow e^+e^-$ reactions, as they traverse intergalactic distances. This limits the range for the search of cosmic sources, a boundary not existing for neutrinos. The current TeV (the units are GeV/TeV/PeV/EeV/ZeV in ascending factors of 10^3) γ-observations could not prove the existence of proton accelerators convincingly, but the field has made major progress forward with the advent of Cherenkov telescopes like HESS, MAGIC, VERITAS, CANGAROO [Hin09] and in the future the Cherenkov Telescope Array (CTA) [Dor13]. A positive observation of neutrinos from point sources would be strong evidence for proton acceleration. In addition, in proton acceleration, the neutrino and photon fluxes are related [Hal97,Wax99,Bah01a,Man01a]. Some of the possible galactic point-source candidates where acceleration can happen are:

- Young supernova remnants. Two mechanisms for neutrino generation by accelerated protons are considered. First, the inner acceleration, where protons in the expanding supernova shell are accelerated, e.g., by the strongly rotating magnetic field of the neutron star or black hole. The external acceleration is done by two shock fronts running towards each other.
- Binary systems. Here matter is transformed from an expanded star like a red giant towards a compact object like a neutron star or black hole. This matter forms an accretion disc which acts as a dynamo in the strong magnetic field of the compact object and also as a target for beam-dump scenarios.

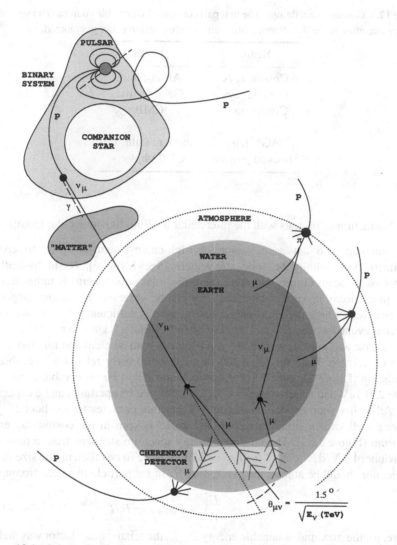

Figure 12.1. Schematic drawing of a "cosmic beam dump" experiment at a cosmic "accelerator". Protons are accelerated in a binary system and hitting matter in the accretion disc. The result is hadroproduction, especially pions. While neutral pions decay into photons, charged pions decay into $\nu_\mu(\bar{\nu}_\mu)$. This shows a strong correlation between high-energy gamma rays and neutrinos. While photons might get absorbed or downscattered to lower energy, neutrinos will find their way to the Earth undisturbed, allowing a search for "hidden" sources. On Earth neutrinos might be detected by CC interactions, resulting in upward-going muons in a detector, because otherwise the atmospheric muon background would be too large (from [Ant97]). With kind permission of the ANTARES collaboration.

Table 12.1. Cosmic beam dumps. The first part consists of calculable sources. The second part has uncertainties in the flux determination and an observation is not guaranteed.

Beam	Target
Cosmic rays	Atmosphere
Cosmic rays	Galactic disc
Cosmic rays	CMB
AGN jets	Ambient light, UV
Shocked protons	GRB photons

- Interaction of protons with the interstellar medium like molecular clouds.

The Sun can also act as a source of high-energy neutrinos due to cosmic-ray interactions within the solar atmosphere [Sec91, Ing96]. As in basically all astrophysical beam dumps the target (here the solar atmosphere) is rather thin and most pions decay instead of interact. They can be observed at energies larger than about 10 TeV, where the atmospheric background is sufficiently low. However, the expected event rate for $E_\nu > 100$ GeV is only about 17 km^{-3} yr^{-1} [Gri01]. The galactic disc and galactic centre are good sources with a calculable flux most likely to be observable. Here the predicted flux can be directly related to the observed γ-emission [Dom93, Ing96a]. The galactic centre might be observable in neutrinos above 250 TeV and about 160 events km^{-2} yr^{-1} in a 5° aperture can be expected.

All stellar sources are considered to accelerate particles up to about 10^{15} eV, where a well-known structure called "the knee" is seen in the cosmic-ray energy spectrum (Figure 12.2). The observed energy spectrum steepens from a power law behaviour of dN/d$E \propto E^{-\gamma}$ from $\gamma = 2.7$ to $\gamma = 3$. To be efficient, the size R of an accelerator should be at least the gyroradius R_g of the particle in an electromagnetic field:

$$R > R_g = \frac{E}{B} \rightarrow E_{\max} = \gamma BR \qquad (12.2)$$

where, for the maximal obtainable energy $E_{\max}$, the relativistic γ factor was included because we may not be at rest in the frame of the cosmic accelerator. Using reasonable numbers for supernova remnants $E_{\max}$, values are obtained close to the knee position. For higher energies, stronger and probably extragalactic sources have to be considered; possible sources are shown in Figure 12.3. The best candidates are active galactic nuclei (AGN), which in their most extreme form are also called quasars. A schematic picture of the quasar phenomenon is shown in Figure 12.4. They are among the brightest sources in the universe and measurable to high redshifts. Moreover, they must be extremely compact because variations in the luminosity are observed on the time scales of days. The appearance of two jets perpendicular to the accretion disc of these objects is probably an efficient place for particle acceleration. AGNs where the jet is in the line of sight to the Earth

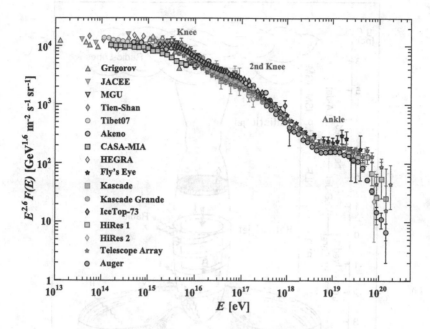

Figure 12.2. The all particle energy spectrum of cosmic rays as a function of E (energy per nucleus) from many air shower measurements. Some features like the first and second knee as well as the ankle are shown (from [PDG18]). With kind permission from M. Tanabashi et al. (Particle Data Group).

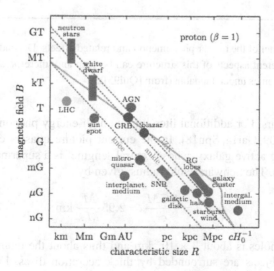

Figure 12.3. Hillas plot of sources. Shown are the magnetic field strength as a function of "accelerator" size (from [Ahl10a]). With kind permission of Fermilab and M. Ahlers.

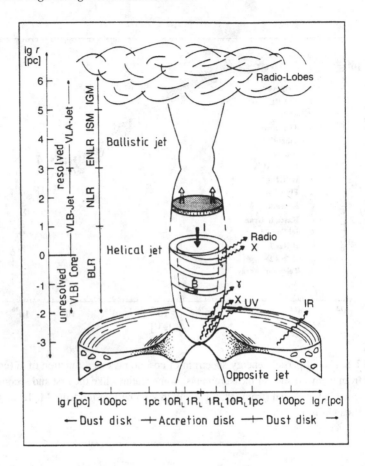

Figure 12.4. Model of the quasar phenomenon and related effects. Depending on the angle of observation, different aspects of this structure can be seen and, therefore, the richness of the phenomenon becomes understandable (from [Qui93]).

are called blazars. For additional literature on high-energy phenomena see [Lon92, 94, Sch00, Hin09, Gai14, Spu18]. In the current picture quasars correspond to the core of a young active galaxy, whose central 'engine' is a supermassive black hole ($M \approx 10^8 M_\odot$). The Schwarzschild radius, given by

$$R_S = \frac{2GM}{c^2} = 2.95 \frac{M}{M_\odot} \text{ km,} \tag{12.3}$$

of such black holes is about 3×10^8 km and thus about the diameter of the Earth orbit. These objects are surrounded by thick accretion discs. From there matter spirals towards the black hole and will be strongly accelerated and transformed into a hot, electrically conducting plasma producing strong magnetic fields. One part of this infalling matter is absorbed into the black hole; the other part is redirected by the magnetic field, which then forms two plasma jets leaving to opposite sides and perpendicular to the disc. In such jets or their substructures (blobs) protons can be accelerated to very high energies.

Another extragalactic neutrino source becoming more prominent during the last years are gamma-ray bursters (GRBs), especially now with its association to gravitational waves [Wax97, Vie98, Boe98, Mes02, Wax03, Geh09, Abb17]. The phenomenon has been discovered only in the gamma-ray region with bursts lasting from 6 ms up to 2000 s. After being a mystery for more than 20 years, major progress in the last years has been achieved with an improved rate of several events per day and it is known that they are of cosmological origin. The most distant GRB observed has a redshift of larger than z>8. In addition, also a link between GRBs and supernova explosions has been established [Hjo03, Sta03, Del03, Woo06]. A new boost in the field has been the launch of the SWIFT and Fermi LAT satellites as well as the HAWC telescope. Even though there is still not a detailed understanding of the internal mechanism of GRBs, the relativistic fireball model is phenomenologically successful [Zha16]. Expected neutrino energies cluster around 1–100 TeV for GRBs and 100 PeV for AGN jets assuming no beaming effects.

12.1.2   Neutrinos produced in annihilation or decay of heavy particles

Three kinds of such sources are typically considered:

- evaporating black holes,
- topological defects from phase transitions in the early Universe and
- annihilation or decay of (super-)heavy particles.

We concentrate here qualitatively only on the last possibility, namely neutralinos as heavy relics ($m > $ GeV) of the Big Bang and candidates for cold dark matter. The neutralino χ (discussed in Chapter 5) is one of the preferred candidates for weakly interacting massive particles (WIMPs) to act as dark matter in the universe (see Chapter 13). They can be accumulated in the centre of objects like the Sun or the Earth [Ber98]. The reason is that by coherent scattering on nuclei they lose energy and if they fall below the escape velocity they get trapped and, finally, by additional scattering processes they accumulate in the core. The annihilation can proceed via

$$\chi + \bar{\chi} \to b + \bar{b} \qquad \text{(for } m_\chi < m_W) \tag{12.4}$$

or

$$\chi + \bar{\chi} \to W^+ + W^- \qquad \text{(for } m_\chi > m_W). \tag{12.5}$$

Detailed predictions depend on the assumed nature of the neutralino. The ν_μ component might be observed by the detectors already described, by looking for the CC reaction. As the created muon follows the incoming neutrino direction at these energies, it should point towards the Sun or the core of the Earth. Within that context, the galactic centre has become more interesting because simulations show that the dark matter halos of galaxies may be sharply "cusped" toward a galaxy's centre [Gon00, Gon00a, Nes13, Wan15, Cha18]; however, bounds exist from TeV-gamma observations [Aha06a].

12.1.3 Event rates

For experimental detection of any of these sources, three main parameters have to be known: the predicted neutrino flux from the source including oscillation effects, the interaction cross-section of neutrinos and the detection efficiency.

For various sources like AGNs and GRBs, the flux still depends on the model and, hence, the predictions have some uncertainties. The rate of neutrinos produced by $p\gamma$ interactions in GRBs and AGNs is essentially dictated by the observed energetics of the source. In astrophysical beam dumps, like AGNs and GRBs, typically one neutrino and one photon is produced per accelerated proton [Gai95, Gan96]. The accelerated protons and photons are, however, more likely to suffer attenuation in the source before they can escape. So, a hierarchy of particle fluxes emerges with protons < photons < neutrinos. Using these associations, one can constrain the energy and luminosity of the accelerator from the gamma- and cosmic-ray observations and subsequently anticipate the neutrino fluxes. These calculations represent the basis for the construction of kilometre-scale detectors as the goal of neutrino astronomy.

12.1.4 Neutrinos from active galactic nuclei

Active galactic nuclei (AGNs) are the brightest sources in the universe. It is anticipated that the beams accelerated near a central black hole are dumped on the ambient matter in the active galaxy. Typically two jets emerge in opposite directions, perpendicular to the disk of the AGN. An AGN viewed from a position illuminated by the cone of a relativistic jet is called a blazar. Particles are accelerated by Fermi shocks in blobs of matter, travelling along the jet with a bulk Lorentz factor of $\gamma \approx 10$ or higher.

In the estimate (following [Hal98]) of the neutrino flux from a proton blazar, primes will refer to a frame attached to the blob moving with a Lorentz factor γ relative to the observer. In general, the transformation between blob and observer frame is $R' = \gamma R$ and $E' = \frac{1}{\gamma}E$ for distances and energies, respectively. For a burst of 15 min duration, the strongest variability observed in TeV emission, the size of the accelerator, is only

$$R' = \gamma c \Delta t \sim 10^{-4}\text{--}10^{-3}\text{pc} \tag{12.6}$$

for $\gamma = 10\text{--}10^2$. So the jet consists of relatively small structures with short lifetimes. High-energy emission is associated with the periodic formation of these blobs.

Shocked protons in the blob will photoproduce pions on the photons whose properties are known from the observed multi-wavelength emission. From the observed photon luminosity L_γ, the energy density of photons in the shocked region can be deduced:

$$U'_\gamma = \frac{L'_\gamma \Delta t}{\frac{4}{3}\pi R'^3} = \frac{L_\gamma \Delta t}{\gamma} \frac{1}{\frac{4}{3}\pi(\gamma c \Delta t)^3} = \frac{3}{4\pi c^3} \frac{L_\gamma}{\gamma^4 \Delta t^2}. \tag{12.7}$$

(Geometrical factors of order unity will be ignored throughout.) The dominant photon density is at UV wavelengths, the UV bump. Assume that a luminosity $\mathcal{L}_\gamma$ of

10^{45} erg s^{-1} is emitted in photons with energy $E_\gamma = 10$ eV. Luminosities larger by one order of magnitude have actually been observed. The number density of photons in the shocked region is

$$N'_\gamma = \frac{U'_\gamma}{E'_\gamma} = \gamma \frac{U'_\gamma}{E_\gamma} = \frac{3}{4\pi c^3} \frac{L_\gamma}{E_\gamma} \frac{1}{\gamma^3 \Delta t^2} \sim 6.8 \times 10^{14}\text{--}6.8 \times 10^{11} \text{ cm}^{-3}. \quad (12.8)$$

From now on, the range of numerical values will refer to $\gamma = 10$–10^2, in that order. With such a high density the blob is totally opaque to photons with 10 TeV energy and above. As photons with such energies have indeed been observed, one must essentially require that the 10 TeV γ are below the $\gamma\gamma \rightarrow e^+e^-$ threshold in the blob, i.e.,

$$E_{\text{thr}} = \gamma E'_{\gamma \text{ thr}} \geq 10 \text{ TeV} \quad (12.9)$$

or

$$E_{\text{thr}} > \frac{m_e^2}{E_\gamma} \gamma^2 > 10 \text{ TeV} \quad (12.10)$$

or

$$\gamma > 10. \quad (12.11)$$

To be more conservative, the assumption $10 < \gamma < 10^2$ is used.

The accelerated protons in the blob will produce pions, predominantly at the Δ-resonance, in interactions with the UV photons. The proton energy for resonant pion production is

$$E'_p = \frac{m_\Delta^2 - m_p^2}{4} \frac{1}{E'_\gamma} \quad (12.12)$$

or

$$E_p = \frac{m_\Delta^2 - m_p^2}{4 E_\gamma} \gamma^2 \quad (12.13)$$

$$E_p = \frac{1.6 \times 10^{17} \text{ eV}}{E_\gamma} \gamma^2 = 1.6 \times 10^{18}\text{--}1.6 \times 10^{20} \text{ eV}. \quad (12.14)$$

The secondary ν_μ has energy

$$E_\nu = \tfrac{1}{4}\langle x_{p\rightarrow\pi}\rangle E_p = 7.9 \times 10^{16}\text{--}7.9 \times 10^{18} \text{ eV} \quad (12.15)$$

for $\langle x_{p\rightarrow\pi}\rangle \simeq 0.2$, the fraction of energy transferred, on average, from the proton to the secondary pion produced via the Δ-resonance. The $\tfrac{1}{4}$ is because each lepton in the decay $\pi \rightarrow \mu\nu_\mu \rightarrow e\nu_e\nu_\mu\bar{\nu}_\mu$ carries roughly equal energy. The fraction of energy f_π lost by protons to pion production when travelling a distance R' through a photon field of density N'_γ is

$$f_\pi = \frac{R'}{\lambda_{p\gamma}} = R' N'_\gamma \sigma_{p\gamma\rightarrow\Delta} \langle x_{p\rightarrow\pi}\rangle \quad (12.16)$$

where $\lambda_{p\gamma}$ is the proton interaction length, with $\sigma_{p\gamma \to \Delta \to n\pi^+} \simeq 10^{-28}$ cm^2 resulting in

$$f_\pi = 3.8\text{--}0.038 \qquad \text{for } \gamma = 10\text{--}10^2. \tag{12.17}$$

For a total injection rate in high-energy protons $\dot{E}$, the total energy in ν is $\frac{1}{2}f_\pi t_H \dot{E}$, where $t_H \approx 10$ Gyr is the Hubble time. The factor $\frac{1}{2}$ accounts for the fact that half of the energy in charged pions is transferred to $\nu_\mu + \bar{\nu}_\mu$, (see earlier). The neutrino flux is

$$\Phi_\nu = \frac{c}{4\pi} \frac{(\frac{1}{2}f_\pi t_H \dot{E})}{E_\nu} e^{f_\pi}. \tag{12.18}$$

The last factor corrects for the absorption of the protons in the source, i.e., the observed proton flux is a fraction e^{-f_π} of the source flux which photoproduces pions. We can write this as

$$\Phi_\nu = \frac{1}{E_\nu} \frac{1}{2} f_\pi e^{f_\pi} (E_p \Phi_p). \tag{12.19}$$

For $E_p \Phi_p = 2 \times 10^{-10}$ TeV (cm^2 s sr)$^{-1}$, we obtain

$$\Phi_\nu = 8 \times 10^5 \text{ to } 2 \text{ (km}^2 \text{ yr)}^{-1} \tag{12.20}$$

over 4π sr. (Neutrino telescopes are background free for such high-energy events and should be able to identify neutrinos at all zenith angles.) The detection probability is computed from the requirement that the neutrino has to interact within a distance of the detector which is shorter than the range of the muon it produces. Therefore,

$$P_{\nu \to \mu} \simeq \frac{R_\mu}{\lambda_{\text{int}}} \simeq A E_\nu^n \tag{12.21}$$

where R_μ is the muon range and λ_{int} the neutrino interaction length. For energies below 1 TeV, where both the range and cross-section depend linearly on energy, $n = 2$. At TeV and PeV energies $n = 0.8$ and $A = 10^{-6}$, with E in TeV units. For EeV energies $n = 0.47$, $A = 10^{-2}$ with E in EeV [Gai95, Gan96]. The observed neutrino event rate in a detector is

$$N_{\text{events}} = \Phi_\nu P_{\nu \to \mu} \tag{12.22}$$

with

$$P_{\nu \to \mu} \cong 10^{-2} E_{\nu,\text{EeV}}^{0.4} \tag{12.23}$$

where E_ν is expressed in EeV. Therefore,

$$N_{\text{events}} = (3 \times 10^3 \text{ to } 5 \times 10^{-2}) \text{ km}^{-2} \text{ yr}^{-1} = 10^{1\pm2} \text{ km}^{-2} \text{ yr}^{-1} \tag{12.24}$$

for $\gamma = 10\text{--}10^2$. This estimate brackets the range of γ factors considered. Notice, however, that the relevant luminosities for protons (scaled to the high-energy cosmic rays) and the luminosity of the UV target photons are themselves uncertain. The large uncertainty in the calculation of the neutrino flux from AGN is predominantly associated with the boost factor γ.

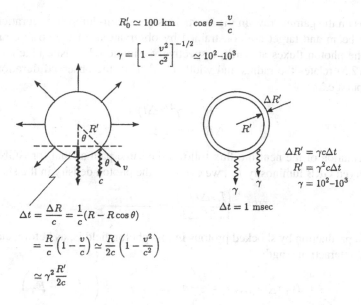

Figure 12.5. Simplified kinematics of Gamma Ray Bursters (from [Hal98]). © 1998 World Scientific

12.1.5   Neutrinos from gamma ray bursters

Recently, Gamma Ray Bursters (GRBs) may have become the best motivated source for high-energy neutrinos. Their neutrino flux can be calculated in a relatively model-independent way. Although neutrino emission may be less copious and less energetic than that from AGNs, the predicted fluxes can probably be bracketed with more confidence.

In GRBs, a fraction of a solar mass of energy ($\sim 10^{53}$ erg) is released over a time scale of order 1 s into photons with a very hard spectrum. It has been suggested that, although their ultimate origin is a matter of speculation, the same cataclysmic events also produce the highest energy cosmic rays. This association is reinforced by more than the phenomenal energy and luminosity. The phenomenon consists of three parts. First of all there must be a central engine, whose origin is still under debate. Hypernovae and merging neutron stars are among the candidates. The second part is the relativistic expansion of the fireball. Here, an Earth-sized mass is accelerated to 99.99% of the speed of light. The fireball has a radius of 10–100 km and releases an energy of 10^{51-54} erg. Such a state is opaque to light. The observed gamma rays are the result of a relativistic shock with $\gamma \approx 10^2$–10^3 which expands the original fireball by a factor of 10^6 over 1 s. For this to be observable, there must be a third condition, namely an efficient conversion of kinetic energy into non-thermal gamma rays.

The production of high-energy neutrinos is a feature of the fireball model because, as in the AGN case, the protons will photoproduce pions and, therefore, neutrinos on the gamma rays in the burst. This is a beam-dump configuration where both the beam and target are constrained by observation: of the cosmic-ray beam and of the photon fluxes at Earth, respectively. Simple relativistic kinematics (see Figure 12.5) relates the radius and width $R', \Delta R'$ to the observed duration of the photon burst $c\Delta t$:

$$R' = \gamma^2(c\Delta t) \tag{12.25}$$
$$\Delta R' = \gamma c\Delta t. \tag{12.26}$$

The calculation of the neutrino flux follows the same path as that for AGNs. From the observed GRB luminosity L_γ, we compute the photon density in the shell:

$$U'_\gamma = \frac{(L_\gamma \Delta t/\gamma)}{4\pi R'^2 \Delta R'} = \frac{L_\gamma}{4\pi R'^2 c\gamma^2}. \tag{12.27}$$

The pion production by shocked protons in this photon field is, as before, calculated from the interaction length

$$\frac{1}{\lambda_{p\gamma}} = N_\gamma \sigma_\Delta \langle x_{p\to\pi} \rangle = \frac{U'_\gamma}{E'_\gamma} \sigma_\Delta \langle x_{p\to\pi} \rangle \qquad \left(E'_\gamma = \frac{1}{\gamma} E_\gamma \right) \tag{12.28}$$

σ_Δ is the cross-section for $p\gamma \to \Delta \to n\pi^+$ and $\langle x_{p\to\pi} \rangle \simeq 0.2$. The fraction of energy going into π production is

$$f_\pi \cong \frac{\Delta R'}{\lambda_{p\gamma}} \tag{12.29}$$

$$f_\pi \simeq \frac{L_\gamma}{E_\gamma} \frac{1}{\gamma^4 \Delta t} \frac{\sigma_\Delta \langle x_{p\to\pi} \rangle}{4\pi c^2} \tag{12.30}$$

$$f_\pi \simeq 0.14 \left\{ \frac{L_\gamma}{10^{51} \text{ ergs}^{-1}} \right\} \left\{ \frac{1 \text{ MeV}}{E_\gamma} \right\} \left\{ \frac{300}{\gamma} \right\}^4 \left\{ \frac{1 \text{ ms}}{\Delta t} \right\}$$
$$\times \left\{ \frac{\sigma_\Delta}{10^{-28} \text{ cm}^2} \right\} \left\{ \frac{\langle x_{p\to\pi} \rangle}{0.2} \right\}. \tag{12.31}$$

The relevant photon energy within the problem is 1 MeV, the energy where the typical GRB spectrum exhibits a break. The number of higher energy photons is suppressed by the spectrum and lower energy photons are less efficient at producing pions. Given the large uncertainties associated with the astrophysics, it is an adequate approximation to neglect the explicit integration over the GRB photon spectrum. The proton energy for production of pions via the Δ-resonance is

$$E'_p = \frac{m_\Delta^2 - m_p^2}{4E'_\gamma}. \tag{12.32}$$

Therefore,

$$E_p = 1.4 \times 10^{16} \text{ eV} \left(\frac{\gamma}{300} \right)^2 \left(\frac{1 \text{ MeV}}{E_\gamma} \right) \tag{12.33}$$

$$E_\nu = \tfrac{1}{4} \langle x_{p\to\pi} \rangle E_p \simeq 7 \times 10^{14} \text{ eV}. \tag{12.34}$$

We are now ready to calculate the neutrino flux:

$$\phi_\nu = \frac{c}{4\pi} \frac{U'_\nu}{E'_\nu} = \frac{c}{4\pi} \frac{U_\nu}{E_\nu} = \frac{c}{4\pi} \frac{1}{E_\nu} \left\{ \frac{1}{2} f_\pi t_H \dot{E} \right\} \tag{12.35}$$

where the factor 1/2 accounts for the fact that only half of the energy in charged pions is transferred to $\nu_\mu + \bar{\nu}_\mu$. As before, $\dot{E}$ is the injection rate in cosmic rays beyond the 'ankle', a flattening of the cosmic-ray energy spectrum around 10^{18} eV, ($\sim 4 \times 10^{44}$ erg Mpc^{-3} yr^{-1}) and t_H is the Hubble time of $\sim 10^{10}$ Gyr. Numerically,

$$\phi_\nu = 2 \times 10^{-14} \text{ cm}^{-2} \text{ s}^{-1} \text{ sr}^{-1} \left\{ \frac{7 \times 10^{14} \text{ eV}}{E_\nu} \right\} \left\{ \frac{f_\pi}{0.125} \right\} \left\{ \frac{t_H}{10 \text{ Gyr}} \right\}$$

$$\times \left\{ \frac{\dot{E}}{4 \times 10^{44} \text{ erg Mpc}^{-3} \text{ yr}^{-1}} \right\}. \tag{12.36}$$

The observed muon rate is

$$N_{\text{events}} = \int_{E_{\text{thr}}}^{E_\nu^{\max}} \Phi_\nu P_{\nu \to \mu} \frac{dE_\nu}{E_\nu} \tag{12.37}$$

where $P_{\nu \to \mu} \simeq 1.7 \times 10^{-6} E_\nu^{0.8}$ (TeV) for TeV energy. Therefore,

$$N_{\text{events}} \cong 26 \text{ km}^{-2} \text{ yr}^{-1} \left\{ \frac{E_\nu}{7 \times 10^{14} \text{ eV}} \right\}^{-0.2} \left\{ \frac{\Delta\theta}{4\pi} \right\}. \tag{12.38}$$

This number might be reduced by a factor of five due to absorption in the Earth. The result is insensitive to beaming. Beaming yields more energy per burst but fewer bursts are actually observed. The predicted rate is also insensitive to the neutrino energy E_ν because higher average energy yields less ν but more are detected. Both effects are approximately linear. As can be seen, the fluxes at high energies are very low, requiring large detectors. Below about 10 TeV the background due to atmospheric neutrinos will dominate. The rate is expected to be approximately the same in the case of GRBs coming from supernova explosions. The burst lasts for only about 10 s, where the burst is formed by the shock created in the transition of a supernova into a black hole.

12.1.6 Cross-sections

Neutrinos can be divided into VHE (very high energy) neutrinos from pp reactions with $E_\nu > 50$ GeV and UHE (ultra-high energy) neutrinos from $p\gamma$ reactions with $E_\nu > 10^6$ GeV [Ber91]. The reason for this UHE comes from the high threshold for pion production in photoproduction of nuclei

$$N + \gamma \to N' + \pi. \tag{12.39}$$

In a collinear collision the threshold is given by $s = (m_N + m_\pi)^2$. Using

$$s = (p_N + p_\gamma)^2 \approx m_N^2 + 2p_N p_\gamma \approx m_N^2 + 4E_N E_\gamma \tag{12.40}$$

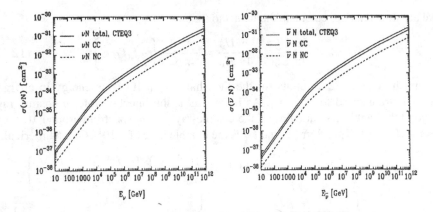

Figure 12.6. Total as well as NC and CC cross-sections for high-energy neutrinos (left) and antineutrinos (right) (from [Gan96]). © 1996 With permission from Elsevier.

it follows ($N \equiv p$)

$$E_P^S = \frac{(2m_p + m_\pi)}{4E_\gamma} = 7 \times 10^{16} \frac{\text{eV}}{E_\gamma} \, \text{eV}. \tag{12.41}$$

Taking the cosmic microwave background as the photon source (see Chapter 13) with $\langle E_\gamma \rangle \approx 7 \times 10^{-4}$ eV, a threshold of $E_P^S = 10^{20}$ eV $= 10^{11}$ GeV follows. This is the well-known Greisen–Zatsepin–Kuzmin (GZK) cutoff [Gre66, Zat66].

The cross-section for CC νN interactions has already been given in Chapter 4. However, we are dealing now with neutrino energies far beyond the ones accessible in accelerators, implying a few modifications. At energies of 10^4 GeV, a deviation of the linear rise with E_ν has to be expected because of the W-propagator, leading to a damping, an effect already observed by H1 and ZEUS at DESY (see Chapter 4). Considering parton distribution functions in the UHE regime the heavier quarks, e.g., charm, bottom and top, have to be included in the sea (see Chapter 4). In good approximation the top sea contribution can be neglected and the charm and bottom quarks can be considered as massless. In addition, perturbative QCD corrections are insignificant at these energies. The dominant contribution to the cross-section comes from the region $Q^2 \approx m_W^2$ implying that the involved partons have x-values of around $m_W^2 / 2ME_\nu$. This requires extrapolations towards small x-values, not constrained by experiments. Data obtained at HERA [Abt16] and LHC give important constraints up to energies of 10^8 GeV. Beyond that, one has to rely on the various extrapolations available, causing the main uncertainty in the cross-section for higher energetic neutrino interactions [Gan96, Gan98, Gan01]. Moreover, at $E_\nu > 10^6$ GeV νN and $\bar{\nu}$N cross-sections become equal because the $(1-y)^2$ term from valence quark scattering (4.74) is now of minor importance and the sea-quarks dominate the cross-section. A reasonable parametrization of the cross-sections in the

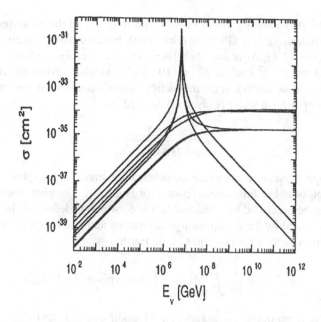

Figure 12.7. Glashow resonance in the $\bar{\nu}_e e$ cross-section. The curves correspond, in the low-energy region from highest to lowest, to (i) $\bar{\nu}_e e \rightarrow$ hadrons, (ii) $\nu_\mu e \rightarrow \mu \nu_e$, (iii) $\nu_e e \rightarrow \nu_e e$, (iv) $\bar{\nu}_e e \rightarrow \bar{\nu}_\mu \mu$, (v) $\bar{\nu}_e e \rightarrow \bar{\nu}_e e$, (vi) $\nu_\mu e \rightarrow \nu_\mu e$ and (vii) $\bar{\nu}_\mu e \rightarrow \bar{\nu}_\mu e$ (from [Gan96]). © 1996 With permission from Elsevier.

region 10^{16} eV $< E_\nu < 10^{21}$ eV is given within 10% by [Gan98] (see also [Alb15])

$$\sigma_{CC}(\nu N) = 5.53 \times 10^{-36} \text{ cm}^2 \left(\frac{E_\nu}{1 \text{ GeV}} \right)^{0.363} \tag{12.42}$$

$$\sigma_{NC}(\nu N) = 2.31 \times 10^{-36} \text{ cm}^2 \left(\frac{E_\nu}{1 \text{ GeV}} \right)^{0.363} \tag{12.43}$$

$$\sigma_{CC}(\bar{\nu} N) = 5.52 \times 10^{-36} \text{ cm}^2 \left(\frac{E_\nu}{1 \text{ GeV}} \right)^{0.363} \tag{12.44}$$

$$\sigma_{NC}(\bar{\nu} N) = 2.29 \times 10^{-36} \text{ cm}^2 \left(\frac{E_\nu}{1 \text{ GeV}} \right)^{0.363} \tag{12.45}$$

Below 10^{16} eV all the different PDF parametrizations agree, and at energies around 10^{21} eV a factor of two uncertainty is reasonable. The total cross-sections on nucleons are shown in Figure 12.6. It should be noted that new physics beyond the Standard Model might affect these cross-sections. The cross-section on electrons is, in general, much smaller than on nucleons except in a certain energy range between 2×10^{15} eV to 2×10^{17} eV for $\bar{\nu}_e$. Here the cross-section for

$$\bar{\nu}_e + e^- \rightarrow W^- \rightarrow \text{hadrons} \tag{12.46}$$

can dominate. At an energy of 6.3 PeV ($= 6.3 \times 10^{15}$ eV) the cross-section shows a resonance behaviour (the Glashow resonance), because here $s = 2m_e E_\nu = m_W^2$ [Gla60] (Figure 12.7). At resonance $\sigma(\bar{\nu}_e e) = (3\pi/\sqrt{2})G_F = 3.0 \times 10^{-32}$ cm^2 while $\sigma(\nu N) \approx 10^{-33}$ cm^2 at $E_\nu \approx 10^7$ GeV. Another severe effect associated with the rising cross-section, is the interaction rate of neutrinos within the Earth. The interaction length L (in water equivalent) defined as

$$L = \frac{1}{\sigma_{\nu N}(E_\nu) N_A} \qquad (12.47)$$

in rock is approximately equal to the diameter of the Earth for energies of 40 TeV. At higher energies the Earth becomes opaque for neutrinos. The phenomenon of Earth shielding can be described by a shadow factor S, which is defined to be an effective solid angle divided by 2π for upward-going muons and is a function of the energy-dependent cross-section for neutrinos in the Earth:

$$S(E_\nu) = \frac{1}{2\pi} \int_{-1}^{0} d\cos\theta \int d\phi \, \exp[-z(\theta)/L(E_\nu)] \qquad (12.48)$$

with z the column-depth, as a function of nadir angle θ and $N_A = 6.022 \times 10^{23}$ mol$^{-1} = 6.022 \times 10^{23}$ cm^{-3} (water equivalent). The shadowing increases from almost no attenuation to a reduction of the flux by about 93% for the highest energies observed in cosmic rays. For energies above 10^6 GeV the interaction length is about the same for neutrinos and antineutrinos as well as for ν_e- and ν_μ-type neutrinos. The damping for ν_τ is more or less absent, because with the CC production of a τ-lepton, its decay produces another ν_τ [Hal98a]. Below 100 TeV their interaction is not observable. However, a special situation holds for $\bar{\nu}_e$ because of the previously mentioned resonance. A similar length scale for interactions with electrons can be defined:

$$L = \frac{1}{\sigma_{\nu e}(E_\nu)(10/18)N_A} \qquad (12.49)$$

where the factor $(10/18)N_A$ is the number of electrons in a mole of water. This interaction length is very small in the resonance region; hence, damping out $\bar{\nu}_e$ in this energy range is very efficient. The high energy of the neutrinos results in a strong correlation of the muon direction from ν_μ CC interactions with the original neutrino direction resulting in a typical angle of $\theta_{\mu\nu} \approx 1.5°/\sqrt{E(\text{TeV})}$. This allows point sources to be sought and identified. VHE neutrinos can be detected only as point sources because of the overwhelming atmospheric neutrino background.

Combining all numbers the expected event rates for AGNs and GRBs are likely of the order 1–100 events/(km^3 yr) requiring detectors of km^3 sizes.

12.2 Detection

The most promising way of detection is by looking for upward-going muons, produced by ν_μ CC interactions. Such upward-going muons can barely be

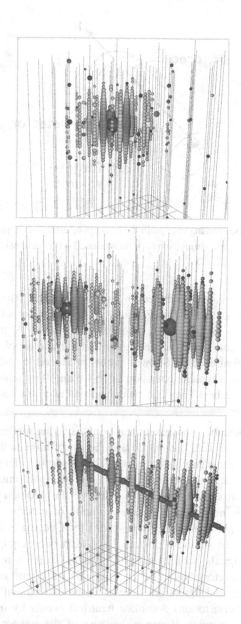

Figure 12.8. The three kinds of signals in water Cherenkov detectors shown here exemplaric for IceCube. (Top) Charged current muon neutrino event. The produced muon produces a long track. The size of the circles is the detected light on the phototube. (Middle) Neutral current interactions and electron events. The calorimetric energy deposition has no long tracks like a muon signature. (Bottom) Double bang signature, i.e., two separate energy depositions. This signature is expected in case of a potential ν_τ-interaction (from [Ice17]). With kind permission of M. Ahlers and the IceCube collaboration.

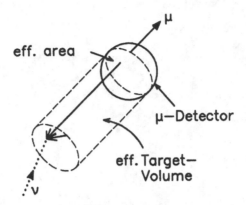

Figure 12.9. Schematic drawing to explain the concept of effective target volume and effective area in a neutrino telescope (from [Sch97]). Reproduced with permission of SNCSC.

misidentified from muons produced in the atmosphere. The obvious detection strategy relies on optical identification with the help of Cherenkov light, producing signals in an array of photomultiplier tubes. A muon can be found by track reconstruction using the timing, amplitude and topology of the triggered photomultipliers (Figure 12.8). Shower events produced by NC interactions or ν_e CC interactions have a typical extension of less than 10 m, smaller than the typical spacing of the phototubes. They can be considered as point sources of light within the detector.

Water is a reasonably transparent and non-scattering medium available in large quantities. The two crucial quantities are the absorption and scattering lengths, both of which are wavelength dependent. The absorption length should be large because this determines the required spacing of the photomultipliers. Moreover, the scattering length should be long to preserve the geometry of the Cherenkov pattern. The idea is now to equip large amounts of natural water resources like oceans and ice with photomultipliers to measure the Cherenkov light of the produced muons. By now, most of these devices have also been equipped with detectors sensitive to other signals, such as acoustic signals. To get a reasonable event rate, the size of such detectors has to be on the scale of 1 km^3 and hence cannot be installed in underground laboratories. Even if the experiment is installed deep in the ocean, atmospheric muons dominate neutrino events by orders of magnitude, especially at lower energies. However, because of the steeper energy dependence ($\propto E^{-3}$) they fall below predicted astrophysical fluxes starting from around 10–100 TeV because their energy dependence is typically assumed to be more like $\propto E^{-2}$ due to Fermi acceleration [Lon92, 94]. The effective size of a detector is actually enhanced because the range of high-energy muons also allows interactions between the surrounding ice and muons flying into the detector. Hence, it is not the volume of the detector that is important but the area pointing towards the neutrino flux ('effective area A_μ'). The larger this effective area is, the larger the effective

Figure 12.10. Installation of one of the rods of the NT-200 experiment in Lake Baikal. As the lake freezes over in winter, this season is ideal for installation. With kind permission of C. Spiering.

volume $V(E_\mu) \approx A_\mu \times R(E_\mu)$ outside the detector (Figure 12.9) will be. If the interesting neutrinos arrive from various directions as in the diffuse case, the detector area in all directions should be large, finally resulting in a sufficiently large detector volume. The mean energy loss rate of muons with energy E_μ due to ionization, bremsstrahlung, pair-production, hadroproduction and catastrophic losses is given by [Gai16]

$$\left\langle \frac{dE_\mu}{dX} \right\rangle = -\alpha - \frac{E_\mu}{\xi}. \tag{12.50}$$

The constant α describing ionization loss (Bethe–Bloch formula) is about 2 MeV/g cm^{-2} in rock. The constant ξ describing the catastrophic losses is $\xi \approx 2.5 \times 10^5$ g cm^{-2} in rock. Above a critical energy, $\epsilon = \alpha\xi$, they dominate with respect to ionization. For muons in rock, $\epsilon \approx 500$ GeV. This leads to a change in energy dependence from linear to logarithmic. If α and ξ are energy independent, the range of the average loss for a muon of initial energy E_μ and final energy $E_\mu^{\min}$ is given by

$$R(E_\mu, E_\mu^{\min}) = \int_{E_\mu^{\min}}^{E_\mu} \frac{dE_\mu}{\langle dE_\mu/dX \rangle} \simeq \frac{1}{\xi} \ln\left(\frac{\alpha + \xi E_\mu}{\alpha + \xi E_\mu^{\min}} \right). \tag{12.51}$$

For $E_\mu \ll \epsilon$, the range of muons is correctly reproduced by $R \propto E_\mu$; for higher energies, detailed Monte Carlo studies are neccessary to propagate them. For a muon with initial energy larger than 500 GeV, the range exceeds 1 km.

The rate at which upward-going muons can be observed in a detector with effective area A is

$$A = \int_{E_\mu^{\min}}^{E_\mu^{\max}} dE_\nu P_\mu(E_\nu, E_\mu^{\min}) S(E_\nu) \frac{dN}{dE_\nu} \tag{12.52}$$

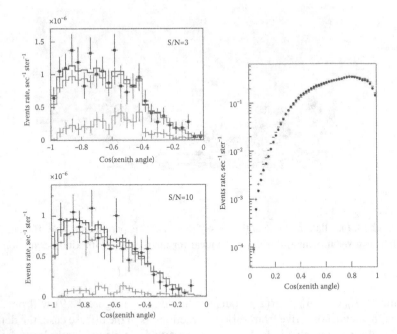

Figure 12.11. Angular distribution of muon tracks in the Baikal NT-200. Left: Upward going muon events for two different signal/noise ratios. Shown are data (points) as well as simulations of signal and background. Additionally the angular distribution is also shown with and without oscillations. Right: Downward going atmospheric muons, data and simulation (from [Ayn09]). © 2009 with permission from Elsevier.

Figure 12.12. Schematic view of ANTARES. With kind permission of F. Montanet.

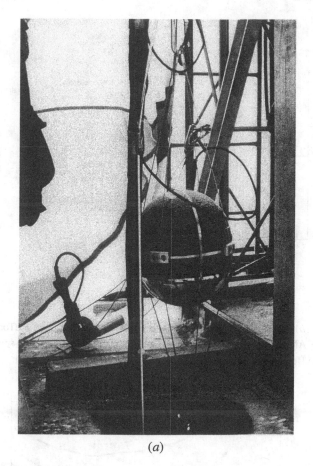

(*a*)

Figure 12.13. (a) Deployment of a string of photomultipliers. With kind permission of the AMANDA collaboration.

with S defined in (12.48) and the probability P for a muon arriving in the detector with an energy threshold of $E_\mu^{\min}$ is given by

$$P\left(\mu(E_\nu, E_\mu^{\min})\right) = N_A \sigma_{CC}(E_\nu) R(E_\mu, E_\mu^{\min}) \tag{12.53}$$

with R given in (12.51). The actual threshold is a compromise between large detector volume (large spacing of the optical modules) and low-energy threshold for physics reasons (requiring small spacing). In this way various experiments might be complementary as well as by the fact that some are sensitive to the Northern and some to the Southern Sky.

In addition to such long tracks, cascades might also be detected, e.g., from ν_e CC interactions where the electrons produce an electromagnetic shower. Therefore, the effective volume is close to the real geometrical volume of the Cherenkov telescope. For the ν_τ interaction a signature has been proposed in the form of a

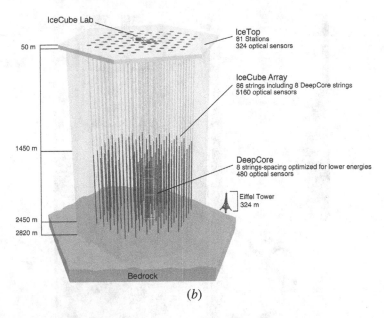

(*b*)

Figure 12.13. (*b*) The IceCUBE experiment and its various subdetectors: IceTop as muon veto on the surface, AMANDA which is part of IceCube now and DeepCore to be sensitive to lower energies (from [Ice19]) . With kind permission of the IceCube collaboration.

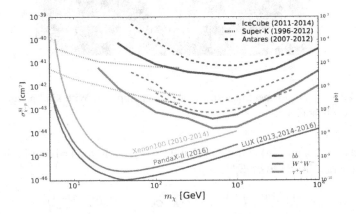

Figure 12.14. Upper limits on spin independent neutralino-proton cross-section σ as a function of neutralino mass from neutralino annihilation in the Sun for the three Cherenkov detectors Super-Kamiokande, IceCube and Antares as function of neutralino mass m_χ. Also shown are the limits from underground dark matter experiments (from [Aar17]). Reproduced with permission of SNCSC.

double bang [Lea95]. The pathlength of a τ-lepton in CC interactions is $c\tau E/m_\tau = 86.93\ \mu\text{m}E/1.777\ \text{GeV}$. At energies of 2 PeV this corresponds to a distance of about

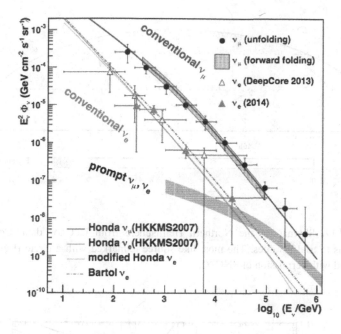

Figure 12.15. Neutrino fluxes as function of energy measured by IceCube. A factor E^2 has been singled out to expose the spectrum better (from [Aar18b]). © 2018 With permission from Elsevier.

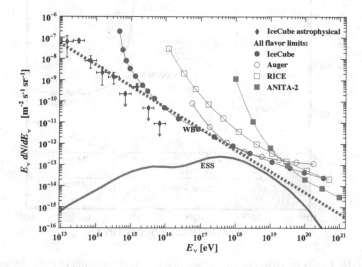

Figure 12.16. An all neutrino spectrum as measured by different detectors [Aar15]. A factor E has been singled out to expose the spectrum better (from [PDG18]). With kind permission from M. Tanabashi et al. (Particle Data Group).

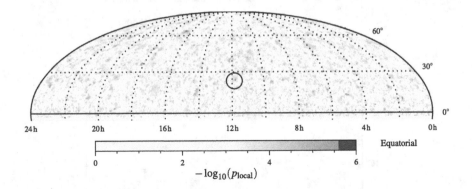

Figure 12.17. Skymap of the Northern Hemisphere from IceCube data. Color coding corresponds to trial p-values. The most like source is marked with a circle [Rei18, Aar19]. Reproduced with permission of SNCSC.

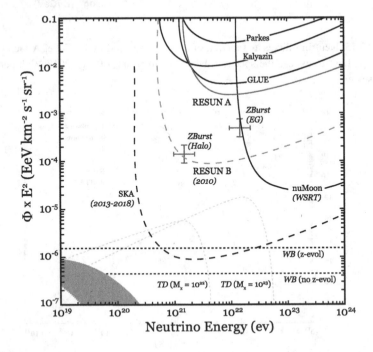

Figure 12.18. Limits on UHE neutrino fluxes due to non-observation of radio emission from the Moon. Also several theoretical model predictions are included (from [Jae10]). © 2010 With permission from Elsevier.

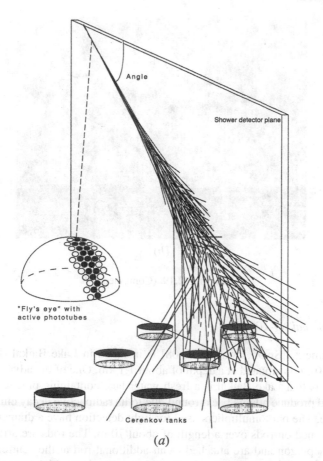

Figure 12.19. (*a*) Principle of the Auger experiment, combining two techniques: the detection of Cherenkov light with a huge array of water tanks and the detection of nitrogen fluorescence with telescopes. (*b*) One of the Auger Cherenkov tanks in Argentina. With kind permission from Enrique Zas and the Pierre Auger Collaboration.

100 m before its decay. This results in two light-emitting processes, the production of a τ-lepton and its decay, where the initial burst shows about half of the energy of the τ-decay burst.

We will now discuss the various Cherenkov detectors—other possible detection methods such as acoustic and radio detection will be briefly described later.

12.2.1 Water Cherenkov detectors

The pioneering effort to build a large-scale neutrino telescope in the ocean was started by DUMAND in the 1970s. However, the project was stopped in the 1990s. The first one to run was Baikal NT-200 in Lake Baikal in Russia, which started data taking in 1995 and fully completed in 1998.

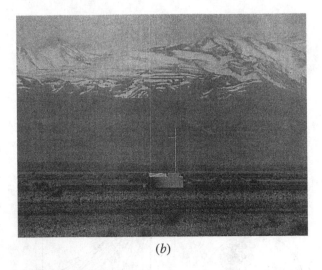

(*b*)

Figure 12.19. (Continued.)

12.2.1.1   *Baikal NT-200*

This experiment [Spi96, Dom02, Wis08] is installed in Lake Baikal (Russia) (see Figure 12.10). It is situated at a depth of about 1.1 km. One of the advantages of this experiment is that Lake Baikal is a fresh water lake containing practically no ^{40}K, which could produce a large background due to its radioactive decay emitting a 1.46 MeV γ-line. The photomultipliers used for light detection have a diameter of 37 cm and are fastened on rods over a length of about 70 m. The rods are arranged in the form of a heptagon and are attached to an additional rod at the centre. The whole arrangement is supported by an umbrella-like construction, which keeps the rods at a distance of 21.5 m from the centre. The photomultipliers are arranged in pairs, with one facing upwards and the other downwards. The distance between two phototubes with the same orientation is about 7.5 m and between two with opposite orientation about 5 m. The full array with 192 optical modules (OMs) has been operational since April 1998. Figure 12.11 shows the reconstructed upward-going muons in a dataset of 234 days. From a smaller prototype NT-96 and only 70 days of data-taking, an upper limit on a diffuse neutrino flux assuming a $\propto E^{-2}$ shape for the neutrino spectrum of [Avr09]

$$\frac{\mathrm{d}\Phi_\nu}{\mathrm{d}E}E^2 < 2.9 \times 10^{-7}\ \mathrm{cm}^{-2}\,\mathrm{s}^{-1}\,\mathrm{sr}^{-1}\,\mathrm{GeV} \tag{12.54}$$

in the energy range 20–20000 TeV has been obtained. With a rather small spacing for the OMs the detector is well suited for the detection of WIMPs. An upgraded version of the detector (NT200+) by adding three more strings in a larger distance was done in 2005 as a first step towards an 1 km^3 detector which is almost completed as the Baikal-GVD experiment. Data taking is about starting soon [Avr17].

12.2.1.2 ANTARES

The ANTARES project is located in the Mediterranean about 40 km from the coast near Toulon (France) at a depth of 2400 m [Ant97, Tho01, Mon02, Age07]. The optical parameters measured are an absorption length of about 40 m at 467 nm and 20 m at 375 nm and an effective scattering length defined as

$$\Lambda_{eff} = \frac{\Lambda}{1 - \langle \cos \theta \rangle} \tag{12.55}$$

of about 300 m. The array consists of 12 lines of 450 m electro-mechanical cables, carrying 25 storeys each with a separation of 14.5 m. A storey itself consists of a triplet of 10" photomultipliers (Figure 12.12). The tubes look sideways and downwards to avoid background from biofouling. The detector is fully operational since May 2008. Results and measurements on the atmospheric muon flux can be found in [Agu10, Pre09]. All of these projects will merge into a single Mediterranean 1 km^3 neutrino telescope called km3NET [Dis09]. This experiment will be made out of two parts at two places: The ORCA experiment at the ANTARES side of the Megaton size to study neutrinos in the tens of GeV range. Besides astrophysical results also the neutrino hierarchy will be explored. The second experiment ARCA is planned as a Gigaton size at Capo Passero (Italy). This is designed to measure the UHE region in the TeV-PeV range.

12.2.2 Ice Cherenkov detectors—IceCube

Another way of using water is in its frozen form, i.e., building an experiment in the ice of Antarctica. This is exactly the idea of AMANDA [Wis99, And00] which later became IceCube to be the first 1 km^3 neutrino telescope. Photomultipliers of 8-inch diameter are used as OMs and plugged into holes in the ice, obtained by hot-water drilling, along long strings. After an exploratory phase in which AMANDA-A was installed at a depth of 800–1000 m, AMANDA B-10 was deployed between 1995 and 1997 (Figure 12.13). It consists of 302 modules at a depth of 1500–2000 m below the surface. The instrumented volume forms a cylinder with an outer diameter of 120 m. In January 2000, AMANDA-II, consisting of 19 strings with 677 OMs, where the ten strings from AMANDA-B10 form the central core, was completed. The measured absorption length is about 110 m at 440 nm, while the effective scattering length is about 20 m. The final goal is IceCube [Hal01, Spi01, Kar02], a real 1 km^3 detector, consisting of 80 strings spaced by 125 m, each with 60 OMs with a spacing of 17 m, resulting in a total of 4800 photomultipliers started data taking in 2009. Now even an upgrade of a 10 km^3 detector called IceCube Gen2 is considered. A large amount of physics results based on the various stages of AMANDA and IceCube have already been obtained. The detection of atmospheric neutrinos in the form of upward going muons is discussed first [Ach07]. An obvious search is the one for neutrino point sources. So far no signal could be found and the flux limits on a muon neutrino source showing a E_ν^{-2} behaviour as well as a sky map are shown in Figure 12.16 and Figure 12.17 [Abb09, Abb09a]. No obvious excess at any specific point in the sky is seen. For any source with a differential energy spectrum and a declination larger than $+40°$ a limit on the flux of $E^2(\Phi)_{\nu_\mu} \leq 1.410^{-11}$ TeV cm^{-2} s^{-1} in the range 3-3000 TeV is obtained [Abb09].

12.2.3   Multi-messenger approaches

Over the last years it became more and more important, and is also convenient, to study phenomena over a wide range of the electromagnetic spectrum, but also adding neutrinos and gravitational waves. In this way much more information can be gained, a wonderful example is [Lig17]. As much as it is assumed that ultra-high energy (UHE) cosmic rays and neutrinos are not from galactic sources, only recently first identifiable UHE neutrinos have been detected [Aar18a]. The understanding is that UHE neutrinos are produced close to the acceleration site but these are unknown. A striking event has been observed on 22. Sept. 2017 at the IceCube experiment which gives first indication of such a site. At this date a well-reconstructed muon neutrino with about 290 TeV [Aab18] has been observed, while the Large Area Telescope (LAT) on the Fermi Gamma Ray Space Telescope has detected gamma rays at this time within 0.06 degrees of the IceCube event [Atw09]. The analysis led to the source being the blazar TXS 0506+056 [Aar18a]. Blazars are a special group of active galactic nuclei (AGN) with powerful relativistic jets pointing close to our line of sight. Their electromagnetic emission is highly variable on time scales of minutes to years. At the time of observation of TXS 0506+056 it was in an enhanced state. This could be confirmed by follow-up observations with the MAGIC [Ale12] and VERITAS gamma ray telescopes [Abe18a]. Triggered by this event, IceCube analysed their whole data period of 9.5 years and could find more events especially in a 158-day period from September 2014-March 2015 [Aar18], where the significance became larger than 3σ (see Figure 12.21). Finally, a first very good candidate for the long searched UHE cosmic-ray accelerator could be found.

12.2.4   Gravitational waves

Another milestone in physics has been reached with the observation of gravitational waves by the LIGO experiment [Abb16]. These "ripples" of space time are associated with supernova explosions (see Chapter 12) and with the merging of neutron stars and black holes. Far more than a dozen events have been observed. After some improvements now Advanced LIGO is online and also the Advanced VIRGO and KAGRA experiment, so the number of observations will rise quickly and will provide extremely valuable data also in the context of following the multi-messenger approach. A highlight of the early gravitational waves was one of the events (GW170817) which has been seen also at the same time in the electromagnetic part with an associated GRB 170817A and thus allowed to follow up the light curve by many different telescopes and satellites. Finally, like neutrinos, gravitational waves are the only messengers at very high energies being unaffected.

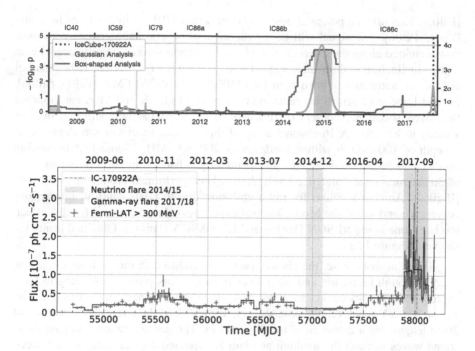

Figure 12.20. Top: Neutrino events of IceCube as a function of time. After the observation in 2017 (vertical dashed line) the 9.5 years of data backwards are analysed by two different methods (box-shaped and Gaussian analysis). A highly active period in the time window from September 2014-March 2015 could be identified with good statistical significance (from [Aar18]). © American Society for the adancement of Science. Bottom: The Fermi-LAT data as a function of time. The observed event coincident with the 2017 IceCube event is marked (dashed vertical line). At the period of the IceCube major event around 2015 (upper shaded grey bar) very little activity has been seen in gamma rays (from [Gar19]). With kind permission of S. Garrappa.

12.2.5   Alternative techniques—acoustic and radio detection

Associated with almost all of the previously mentioned projects are alternative detectors using different techniques, namely acoustic and radio detection. An electromagnetic shower in matter develops a net charge excess due to the photon and electron scattering pulling additional electrons from the surrounding material in the shower and by positron annihilation (Askaryan effect) [Ash62]. This can result in a 20–30% net charge excess, which has been observed. The effect leads to a strong coherent radio Cherenkov emission which has been verified experimentally in the laboratory [Sal01] and later in ice [Gor07]. In this way large area antenna arrays can be built for the detection of UHE neutrinos. Coherent geosynchrotron radiation has also been proposed as a source for radio emission, due to the deflection of electrons and positrons in the geomagnetic field resulting in a dipole radiation

[Fal03]. This offers a potential detection method for UHE ν_e interactions in matter [Zas92, Pet06]. The signal will be a radiopulse of several nanoseconds with most power emitted along the Cherenkov angle and neutrino showers require events with a large inclination to discrimate against normal cosmic rays. A clear radiosignal of cosmic-ray showers has been seen by LOPES and CODALEMA [Fal05, Ard06]. Further large scale arrays for radio astronomy like LOFAR and SKA can improve the current situation. Another experiment is RICE [Seu01] at the South Pole in close vicinity to IceCube. A 16-channel array of dipole radio receivers was deployed at a depth of 100–300 m with a bandpass of 200–500 MHz. Some first interesting limits could be obtained [Kra06]. Independently, a balloon mission flying an array of 36 antennas over Antarctica (ANITA) has been performed producing new limits [Bar06a]. Another medium for radio emission due to UHE neutrino interactions besides ice and salt is the Moon. Various experiments have searched for this effect, the latest one being RESUN [Jae10] and LUNASKA [Jam09]. Obtained limits are shown in Figure 12.18.

In the acoustic case, the shower particles produced in the ν_e interaction lose energy through ionization and other known energy loss processes leading to local heating and a density change localized along the shower. A neutrino interaction with $E_\nu = 10^{20}$ eV creates a hadronic shower, with 90% of the energy in a cylinder of 20 m length and a diameter of roughly 20 cm. The density change propagates as sound waves through the medium and can be detected with an array of detectors, e.g., hydrophones. A reconstruction of the event can be performed by measuring the arrival times and amplitudes. The speed of sound in water is about 1.5 km s^{-1}, so the frequency range of interest is between 10–100 kHz. This interesting option of using hydrophone arrays is explored [Leh02, Van06], also ANTARES has equipped its experiments with acoustic detectors (AMADEUS) [Sim09].

12.2.6 Horizontal air showers—the AUGER experiment

An alternative method to water Cherenkov detection is use of extended air showers (EAS) in the atmosphere, the largest experiment being the AUGER experiment [All08]. This is a well-established technique to measure the cosmic-ray spectrum by the cosmic rays' interactions with air and, hence, to determine their chemical composition and energy by measuring various shower parameters. In this way, a suppression of the cosmic-ray flux above 4×10^{19} eV has been found in accordance with the GZK cutoff [Abb08, Abr08]. The possible origin of very high UHE cosmic rays is discussed in [Bha00, Nag00]. As mentioned, "beyond GZK" events have to come from our cosmological neighborhood, basically a sphere of 50–100 Mpc radius. UHE neutrinos of the order 10^{21} eV could come from cosmological distances. One interesting explanation combining highest and lowest neutrino energies in the universe is given by [Wei82] (the 'Z-burst' model). Hadrons could be produced from Z-decays created by interactions of UHE neutrinos with low-energy antineutrinos from the relic neutrino background (see Chapter 13). The cross-section $\sigma(\nu\bar{\nu} \to Z^0)$

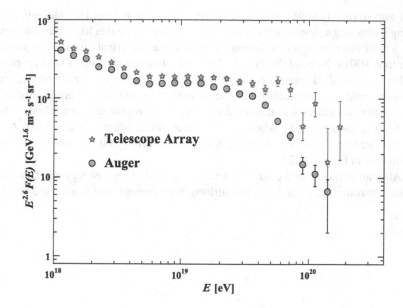

Figure 12.21. The very highest cosmic ray flux as measured by Auger [Fen17] and the Telescope Array [Iva15]. The thresholds for these events are 10^{18} eV (from [PDG18]). With kind permission from M. Tanabashi et al. (Particle Data Group). For additional information see [Aab19].

shows a resonance at energy

$$E_\nu = \frac{m_Z^2}{2m_\nu} = 4 \left(\frac{\text{eV}}{m_\nu} \right) \times 10^{21} \text{ eV}. \qquad (12.56)$$

Here the cross-section is $\sigma(\nu\bar{\nu}) = (4\pi/\sqrt{2})G_F = 4.2 \times 10^{-32}$ cm^2.

The experimental statistics in the region beyond 10^{18} eV is still limited but the situation is improving fast due to the Auger experiment in Argentina [Aug96]. Their detection system combines two major techniques (Figure 12.19): a fluorescence detector system to measure the longitudinal profile of the EAS; and a surface array of detectors to sample its lateral distribution on the ground. The Auger site has a detection acceptance of more than 16 000 km^2 sr. It is composed of 1600 Cherenkov stations (their surface detector units) and four fluorescence eyes located at the periphery of the array Each eye is composed of six 30° × 30° mirror and camera units looking inwards over the surface station network. First major results have been obtained such as the observations of anisotropies in UHE cosmic rays and evidence for the GZK-cutoff [Abr07, Abr08]. A counterpart in the Northern Hemisphere is under consideration in the USA. Also, projects to observe such air showers from space (JEM-EUSO at the International Space Station) are under consideration.

A striking feature for neutrino detection is their deeper interaction in the atmosphere which allows them to be discriminated from hadrons, interacting high in the atmosphere [Cap98]. Horizontal EAS produced by neutrino interactions are "young", meaning a shower at its beginning, showing properties like a curved shower front, a large electromagnetic component and a spread in arrival times of the particles larger than 100 ns. None of this is valid for well-advanced showers. These properties could be measured adequately if the interactions happen in the air above the array. τ-leptons could also be produced in the mountains or the ground around the array and thus produce a clear signal if the decay occurs above the detector. Most of them stem from upward-going ν_τ where the CC interactions occur in the ground. A first limit obtained by Auger has been published in [Abr08] and is shown with other experiments in Figure 12.21.

After an overview of the rapidly developing field of high-energy neutrino astrophysics, the role of very low energy neutrinos in cosmology will finally be discussed.

Chapter 13

Neutrinos in cosmology

DOI: 10.1201/9781315195612-13

It is a reasonable assumption that, on the scales that are relevant for a description of the development of the present universe, from all the interactions only gravity plays a role. All other interactions are neutralized by the existence of opposite charges in the neighbourhood and have an influence only on the detailed course of the initial phase of the evolution of the universe. Currently, the accepted theory of gravitation is Einstein's *general theory of relativity*. This is *not* a gauge theory: gravitation is interpreted purely geometrically as the curvature of four-dimensional spacetime. For a detailed introduction to general relativity see [Wei72, Mis73, Sex87, Lid15]. While general relativity was being developed (1917), the accepted model was that of a stationary universe. In 1922 Friedman examined *non-stationary* solutions of Einstein's field equations. Almost all models based on expansion contain an initial singularity of infinitely high density. From this the universe developed via an explosion (the Big Bang). Hubble discovered galactic redshifts in 1929 [Hub29], something which has already been seen by Lemaitre in 1916. This velocity of recession is interpreted as a consequence of this explosion. With the discovery of the cosmic microwave background in 1964 [Pen65], which is interpreted as the echo of the Big Bang, the Big Bang model was finally established in preference to competing models, such as the steady-state model. The abundance of the light elements could also be predicted correctly over 10 orders of magnitude within this model (see Section 13.8). All this has resulted in the Big Bang model being today known as the *standard model of cosmology*. For further literature see [Boe03, Gut89, Kol90, Kol93, Nar93, Pee93, Kla97, Bot98a, Pea98, Ber99, Ric01, Dol02, Dod03, Sch06a, Wei08, Dol10, Les13, Lid15]. Standard cosmology predicts a 1.95K relic neutrino background as an analogue to the cosmic 2.73 K microwave background. Neutrino properties can be deduced from cosmology as they affect, for example, the cosmic microwave background and large scale structures of the Universe as well as Big Bang nucleosynthesis. For more detailed information see [Les06, Han06, Les13, Ili15].

13.1   Cosmological models

Our present conception of the universe is that of a homogeneous, isotropic and expanding universe. Even though the observable spatial distribution of galaxies seems decidedly lumpy, it is generally assumed that, at distances large enough, these inhomogeneities will average out and an even distribution will exist. At least, this seems to be a reasonable approximation today. The high isotropy of the microwave background radiation (Section 13.3) also testifies to the very high isotropy of the universe. These observations are embodied in the so-called *cosmological principle*, which states that there is no preferred observer, which means that the universe looks the same from any point in the cosmos. The spacetime structure is described with the help of the underlying metric. In three-dimensional space the distance is given by the line element $\mathrm{d}s^2$ with

$$\mathrm{d}s^2 = \mathrm{d}x_1^2 + \mathrm{d}x_2^2 + \mathrm{d}x_3^2 \tag{13.1}$$

whereas in the four-dimensional spacetime of the *theory of special relativity*, a line element is given by

$$\mathrm{d}s^2 = \mathrm{d}t^2 - (\mathrm{d}x_1^2 + \mathrm{d}x_2^2 + \mathrm{d}x_3^2) \tag{13.2}$$

which in the general case of non-inertial systems can also be written as

$$\mathrm{d}s^2 = \sum_{\mu\nu=1}^{4} g_{\mu\nu} \, \mathrm{d}x^\mu \, \mathrm{d}x^\nu. \tag{13.3}$$

Here $g_{\mu\nu}$ is the metric tensor which, in the case of the special theory of relativity, takes on the simple diagonal form of

$$g_{\mu\nu} = \mathrm{diag}(1, -1, -1, -1). \tag{13.4}$$

The simplest metric with which to describe a homogeneous isotropic universe in the form of space of constant curvature is the *Robertson–Walker metric* [Wei72], in which a line element can be described by

$$\mathrm{d}s^2 = \mathrm{d}t^2 - a^2(t) \left[\frac{\mathrm{d}r^2}{1 - kr^2} + r^2 \, \mathrm{d}\theta^2 + r^2 \sin^2 \theta \, \mathrm{d}\phi^2 \right]. \tag{13.5}$$

Here r, θ and ϕ are the three co-moving spatial coordinates, $a(t)$ is the scale-factor and k characterizes the curvature. A closed universe has $k = +1$, a flat Euclidean universe has $k = 0$ and an open hyperbolic one has $k = -1$. In the case of a closed universe, a can be interpreted as the 'radius' of the universe. The complete dynamics is embodied in this time-dependent scale-factor $a(t)$,[1] which is described by Einstein's field equations

$$R_{\mu\nu} - \tfrac{1}{2} a g_{\mu\nu} = 8\pi G T_{\mu\nu} + \Lambda g_{\mu\nu}. \tag{13.6}$$

---

[1] This name implies that the spatial separation of two adjacent 'fixed' space points (with constant r, ϕ, θ coordinates) is scaled in time by $a(t)$.

In this equation $R_{\mu\nu}$ is the Ricci tensor, $T_{\mu\nu}$ corresponds to the energy–momentum tensor and Λ is the cosmological constant [Wei72, Mis73, Sex87]. When space is locally observed, at first approximation is can be assumed to be flat, which means the metric is given by the Minkowski metric of the special theory of relativity (13.4). As $g_{\mu\nu}$ is diagonal, the energy-momentum tensor also has to be diagonal. Its spatial components are equal due to isotropy. The dynamics can be described in analogy to the model of a perfect liquid with density $\rho(t)$ and pressure $p(t)$ and has the form

$$T_{\mu\nu} = \mathrm{diag}(\rho, -p, -p, -p). \tag{13.7}$$

The cosmological constant acts as a contribution to the energy momentum tensor in the form

$$T_{\mu\nu}^{\Lambda} = \mathrm{diag}(\rho_{\Lambda}, -\rho_{\Lambda}, -\rho_{\Lambda}, -\rho_{\Lambda}) \tag{13.8}$$

with $\rho_{\Lambda} = 3\Lambda/(8\pi G)$. Thus, vacuum energy has a very unusual property, that in the case of a positive ρ_{Λ} it has a negative pressure. In an expanding universe this even accelerates the expansion. From the zeroth component of Einstein's equations, it follows that

$$\frac{\dot{a}^2}{a^2} + \frac{k}{a^2} = \frac{8\pi G}{3}(\rho + \rho_V) \tag{13.9}$$

while the spatial components give

$$2\frac{\ddot{a}}{a} + \frac{\dot{a}^2}{a^2} + \frac{k}{a^2} = -8\pi Gp. \tag{13.10}$$

These equations (13.9) and (13.10) are called the *Einstein–Friedmann–Lemaitre equations*. From these equations, it is easy to show that

$$\frac{\ddot{a}}{a} = -\frac{4\pi G}{3}(\rho + 3p - 2\rho_V). \tag{13.11}$$

Since currently $\dot{a} \geq 0$ (i.e., the universe is expanding), and on the assumption that the expression in brackets has always been positive, i.e., $\ddot{a} \leq 0$, it inevitably follows that a was once 0. This singularity at $a = 0$ can be seen as the 'beginning' of the development of the universe. Evidence for such an expanding universe came from the redshift of far away galaxies by Lemaitre and Hubble [Hub29]. The farther away galaxies are from us, the more redshifted are their spectral lines, which can be interpreted as a consequence of the velocity of recession v. This can be demonstrated by expanding $a(t)$ as a Taylor series around the value it has today, giving

$$\frac{a(t)}{a(t_0)} = 1 + H_0(t - t_0) - \frac{1}{2}q_0 H_0^2(t - t_0)^2 + \cdots. \tag{13.12}$$

The index 0 represents the current value both here and in what follows. The Hubble constant H_0 is, therefore,

$$H_0 = \frac{\dot{a}(t_0)}{a(t_0)} \tag{13.13}$$

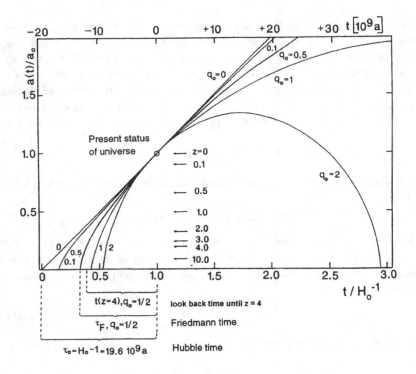

Figure 13.1. Expected behaviour of the scale factor $a(t)$ for different models of the universe. For all models $\Lambda = 0$ was assumed. Also shown are the various redshifts, as well as the influence of various deceleration parameters q_0. A Hubble constant of 50 km s^{-1} Mpc^{-1} has been used. However, the Universe provided us even with an accelerating model (from [Uns92]). Reproduced with permission of SNCSC.

corresponding to the current expansion rate of the universe and the deceleration parameter q_0 is given by

$$q_0 = \frac{-\ddot{a}(t_0)}{\dot{a}^2(t_0)} a(t_0).$$ (13.14)

These measurements result for low redshifts in the Hubble relation in

$$v = cz = H_0 r$$ (13.15)

using the redshift z. In general, the Hubble parameter defined in Equation 13.13 describes the expansion rate at a given time. The behaviour of the scale factor for various cosmological models is shown in Figure 13.1.

13.1.1   The cosmological constant Λ

A $\Lambda \neq 0$ would also be necessary if the Hubble time H^{-1} (for $\Lambda = 0$) and astrophysically determined data led to different ages for the universe. Λ has experienced a revival through modern quantum field theories. In these the vacuum is not necessarily a state of zero energy, but the latter can have a finite expectation value. The vacuum is defined only as the state of lowest energy. Due to the Lorentz invariance of the ground state it follows that the energy–momentum tensor in every local inertial system has to be proportional to the Minkowski metric $g_{\mu\nu}$. This is the only 4×4 matrix which in special relativity theory is invariant under Lorentz 'boosts' (transformations along a spatial direction). According to this, the cosmological constant can be associated with the energy density ϵ_V of the vacuum to give

$$\epsilon_V = \frac{c^4}{8\pi G}\Lambda = \rho_V c^2. \tag{13.16}$$

All terms contributing in some form to the vacuum energy density also provide a contribution to the cosmological constant. There exists, in principle, three different contributions:

- The static cosmological constant Λ_{geo} impinged by the underlying spacetime geometry. It is identical to the free parameter introduced by Einstein [Ein17].
- Quantum fluctuations Λ_{fluc}. According to Heisenberg's uncertainty principle, virtual particle–antiparticle pairs can be produced at any time even in a vacuum. That these quantum fluctuations really exist was proved clearly via the Casimir effect [Cas48, Lam97].
- Additional contributions of the same type as the previous one due to invisible, currently unknown, particles and interactions Λ_{inv}.

The sum of all these terms is what can be experimentally explored

$$\Lambda_{tot} = \Lambda_{geo} + \Lambda_{fluc} + \Lambda_{inv}. \tag{13.17}$$

Consider first the *static solutions* ($\dot{a} = \ddot{a} = 0$). The equations are then written (for $p = 0$) as

$$\frac{8\pi G}{3}(\rho + \rho_V) = \frac{k}{a^2} \tag{13.18}$$

$$\rho = 2\rho_V. \tag{13.19}$$

From Equation (13.19) it follows that $\rho_V > 0$ and, therefore, Equation (13.18) has a solution only for $k = 1$:

$$a^2 = \frac{1}{4\pi G\rho}. \tag{13.20}$$

Equation (13.18) represents the equilibrium condition for the universe. The attractive force due to ρ has to exactly compensate for the repulsive effect of a positive cosmological constant in order to produce a static universe. This closed static universe is, however, unstable, since if we increase a by a small amount, ρ decreases, while Λ remains constant. The repulsion then dominates and leads to a further increase in a, so that the solution moves away from the static case. We now consider *non-static* solutions. As can easily be seen, a positive Λ always leads to

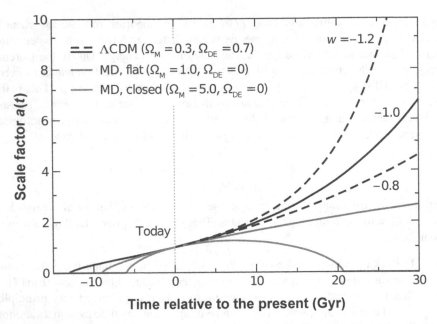

Figure 13.2. The behaviour of the scale factor in two matter-dominated models without a cosmological constant and three with a non-vanishing cosmological constant for different equations of state characterized by w (from [Fri08]). © 2008 by Annual Reviews Inc. Reproduced with permission of Annual Reviews Inc.

an acceleration of the expansion, while a negative Λ acts as a brake. Λ always dominates for large a, since ρ_V is constant. A negative Λ, therefore, always implies a contracting universe and the curvature parameter k does not play an important role. For positive Λ and $k = -1$ or 0 the solutions are always positive which, therefore, results in a continuously expanding universe. For $k = 1$, there exists a critical value

$$\Lambda_c = 4 \left(\frac{8\pi G}{c^2} M \right)^{-2} \tag{13.21}$$

exactly the value of Einstein's static universe, producing two regimes. For $\Lambda > \Lambda_c$ static, expanding and contracting solutions all exist. A very interesting case is that with $\Lambda = \Lambda_c(1 + \epsilon)$ with $\epsilon \ll 1$ (Lemaitre universe). It contains a phase in which the universe is almost stationary, before continuing to expand again (Figure 13.2).

Striking evidence for a non-vanishing cosmological constant has arisen by investigating high redshift supernovae of type Ia [Per97,Rie98,Sch98,Per99,Ton03]. They are believed to behave as standard candles, because the explosion mechanism is assumed to be the same. Therefore, the luminosity as a function of distance scales with a simple quadratic behaviour. By investigating the luminosity distance *versus* redshift relation, equivalent to a Hubble diagram, at high redshift the expected behaviour is sensitive to cosmological parameters. As it turned out [Sch98, Per99, Lei01, Fri08], the best fit describing the data is a universe with a density Ω (see

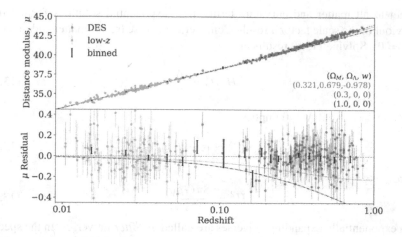

Figure 13.3. Hubble diagram in the form of a magnitude–redshift diagram of supernovae type Ia. Shown are the low redshift supernovae of the Dark Energy Survey project (DES-SN). In the lower part, the residuals are shown. The horizontal line is the best fit (Ω_{Model}), while the two bended curves show a model with no dark energy and one with ($\Omega_M = 0.3$ and 1.0) respectively (from [Abb19]). With kind permission of the Dark Energy Survey Collaboration. © IOP Publishing. Reproduced with permission. All rights reserved. For additional information see [Sco18].

(13.27)) $\Omega_M \approx 0.24$ and $\Omega_\Lambda = 0.76 \pm 0.02$ [Fri08]. A vast amount of new data of supernova Ia has been collected to enhance the statistics and to study potential systematic effects as a function of redshift. Here, especially ESSENCE and SNLS have provided a large amount of data (Figure 13.3). Also other probes such as galaxy clusters, baryon acoustic oscillations (BAO) and weak gravitational lensing support the existence of a non-vanishing Λ.

It is a striking puzzle that any estimated contribution to ρ_V is 50–100 orders of magnitude larger than the cosmological value [Wei89, Kla97, Dol97, Cal09]. In addition, there are many phenomenological models with a variable cosmological 'constant' [Sah00]. A special class of them with a generalized equation of state of Equation 13.3.

$$p = w\rho \qquad \text{with } 1- < w < 0 \qquad (13.22)$$

has been named 'quintessence' [Cal98]. The classical cosmological constant would correspond to $w = -1$. Nowadays due to its unknown character this contribution to the energy density of the Universe is called dark energy. As the identification of dark energy is one of the outstanding problems of cosmology and particle astrophysics, a large number of projects are planned; for a compilation see [Alb06, Fri08].

13.1.2 The inflationary phase

As mentioned, a positive vacuum energy corresponds to a negative pressure $p_V = -\rho_V$. Should this vacuum energy at some time be the dominant contribution with respect to all matter and curvature terms, new exponential solutions for the time behaviour of the scale factor a result. Consider a universe free of matter and radiation ($T_{\mu\nu} = 0$). Solving (13.9) results in

$$H^2(t) = \frac{\Lambda}{3} \tag{13.23}$$

which for $k = 0$ and $\Lambda > 0$ implies

$$a(t) \propto \exp(Ht) \tag{13.24}$$

where

$$H^2 = \frac{8\pi G \rho_V}{3}. \tag{13.25}$$

Such exponentially expanding universes are called *de Sitter* universes. In the specific case in which the negative pressure of the vacuum is responsible for this, it is referred to as *inflationary* universes. Such an inflationary phase, where the exponential increase is valid for only a limited time in the early universe, helps to solve several problems within standard cosmology. Inflation is generally generated by scalar fields ϕ, sometimes called inflaton fields, which couple only weakly to other fields. As the period for the limited inflationary phase, in general the GUT phase transition is considered. Here a new vacuum ground state emerged due to spontaneous symmetry breaking (see Chapter 3). For more detailed reviews on inflation see [Gut81, Alb82, Lin82, Lin84, Kol90, Lin02, Tur02, Boy06, Mar16, Cli18, Akr18]. The extension of the Big Bang hypothesis through an inflationary phase 10^{-35} s after the Big Bang has been proven to be very promising and successful. This is the reason why today the combination of the Big Bang model with inflation is often called the standard model of cosmology.

13.1.3 The density in the universe

From Equation (13.9) it is clear that a flat universe ($k = 0$) is reached for only a certain density, the so-called *critical density*. This is given as [Kol90]

$$\rho_{c0} = \frac{3H_0^2}{8\pi G} \approx 18.8 h^2 \times 10^{-27} \text{ kg m}^{-3} \approx 11 h^2 \text{ H-atoms m}^{-3} \tag{13.26}$$

where $h = H_0/100 \text{ km s}^{-1}\text{Mpc}^{-1}$. Its value is about $H_0 = 70 \text{ km s}^{-1}\text{Mpc}^{-1}$. It is convenient to normalize to this density and, therefore, a *density parameter* Ω is introduced, given by

$$\Omega = \frac{\rho}{\rho_c}. \tag{13.27}$$

$\Omega = 1$, therefore, means a Euclidean universe. This is predicted by inflationary models. An $\Omega > 1$ implies a closed universe, which means that at some time the gravitational attraction will stop the expansion and the universe will collapse again (the 'Big Crunch'), which current data strongly disfavour. An $\Omega < 1$, however, means

Table 13.1. Current experimental values of the most important cosmological parameters with the 5-year data of the WMAP satellite alone and also taking into account baryon acoustic oscillations (BAO) and supernova data (after [Hin09a]).

Quantity	WMAP	WMAP + BAO + SN
Ω_0	$1.099^{+0.100}_{-0.085}$	1.0050 ± 0.0060
Ω_{DM}	0.214 ± 0.027	0.228 ± 0.013
Ω_Λ	0.742 ± 0.030	0.726 ± 0.015
H_0	71.9 ± 2.7 km s^{-1} Mpc^{-1}	70.5 ± 1.3 km s^{-1} Mpc^{-1}
t_0	$13.69 \pm 0.13 \times 10^9$ yr	$13.72 \pm 0.12 \times 10^9$ yr

a universe which expands forever. If the Friedmann Equation (13.9) is solved for the $\mu = 0$ component, the first law of thermodynamics results:

$$d(\rho a^3) = -p d(a^3). \tag{13.28}$$

This means simply that the change in energy in a co-moving volume element is given by the negative product of the pressure and the change in volume. Assuming a simple equation of state $p = k\rho$, where k is a time-independent constant, it immediately follows that

$$\rho \sim a^{-3(1+k)} \tag{13.29}$$
$$a \sim t^{\frac{2}{3}(1+k)}. \tag{13.30}$$

The dependence of the density on a can, hence, be derived for different energy densities using the known thermodynamic equations of state. For the two limiting cases—relativistic gas (the early radiation-dominated phase of the cosmos, particle masses are also negligible) and cold, pressure-free matter (the later, matter-dominated phase)—we have:

$$\text{Radiation} \rightarrow p = 1/3\rho \rightarrow \rho \sim a^{-4} \tag{13.31}$$
$$\text{Matter} \rightarrow p = 0 \rightarrow \rho \sim a^{-3}. \tag{13.32}$$

Hence, in the considered Euclidean case, a simple time dependence for the scale parameter (see Figure 13.1) follows:

$$a \sim t^{\frac{1}{2}} \quad \text{radiation dominated} \tag{13.33}$$
$$a \sim t^{\frac{2}{3}} \quad \text{matter dominated.} \tag{13.34}$$

For the vacuum energy which is associated with the cosmological constant Λ, one has

$$\text{vacuum energy} \rightarrow p = -\rho \rightarrow \rho \sim \text{constant.} \tag{13.35}$$

The current experimental numbers of cosmological parameters is shown in Table 13.1.

13.2   The evolution of the universe

13.2.1   The standard model of cosmology

In this section we consider how the universe evolved from the Big Bang to what we see today. We start from the assumption of thermodynamic equilibrium for the early universe, which is a good approximation because the particle number densities n were so large, that the rates of reactions $\Gamma \propto n\sigma$ (σ being the cross-section of the relevant reactions) were much higher than the expansion rate $H = \dot{a}/a$. A particle gas with g internal degrees of freedom, number density n, energy density ρ and pressure p obeys the following thermodynamic relations [Kol90]:

$$n = \frac{g}{(2\pi)^3} \int f(\boldsymbol{p})\, \mathrm{d}^3 p \qquad (13.36)$$

$$\rho = \frac{g}{(2\pi)^3} \int E(\boldsymbol{p}) f(\boldsymbol{p})\, \mathrm{d}^3 p \qquad (13.37)$$

$$p = \frac{g}{(2\pi)^3} \int \frac{|\boldsymbol{p}|^2}{3E} f(\boldsymbol{p})\, \mathrm{d}^3 p \qquad (13.38)$$

where $E^2 = |\boldsymbol{p}|^2 + m^2$. The phase space partition function $f(\boldsymbol{p})$ is given, depending on the particle type, by the Fermi–Dirac (+ sign in Equation (13.39)) or Bose–Einstein (− sign in Equation (13.39)) distribution

$$f(\boldsymbol{p}) = [\exp((E - \mu)/kT) \pm 1]^{-1} \qquad (13.39)$$

where μ is the chemical potential of the corresponding type of particle. In the case of equilibrium, the sum of the chemical potentials of the initial particles equals that of the end products and particles and antiparticles have equal magnitude in μ but opposite sign. Consider a gas at temperature T. Since non-relativistic particles ($m \gg T$) give an exponentially smaller contribution to the energy density than relativistic ($m \ll T$) particles, the former can be neglected and, thus, for the radiation-dominated phase, we obtain:

$$\rho_R = \frac{\pi^2}{30} g_{\mathrm{eff}} T^4 \qquad (13.40)$$

$$p_R = \frac{\rho_R}{3} = \frac{\pi^2}{90} g_{\mathrm{eff}} T^4 \qquad (13.41)$$

where g_{eff} represents the sum of all effectively contributing massless degrees of freedom and is given by [Kol90]

$$g_{\mathrm{eff}} = \sum_{i=\mathrm{bosons}} g_i \left(\frac{T_i}{T}\right)^4 + \frac{7}{8} \sum_{i=\mathrm{fermions}} g_i \left(\frac{T_i}{T}\right)^4, \qquad (13.42)$$

In this relation the equilibrium temperature T_i of the particles i is allowed to differ from the photon temperature T. The statistical weights are $g_\gamma = 2$ for photons,

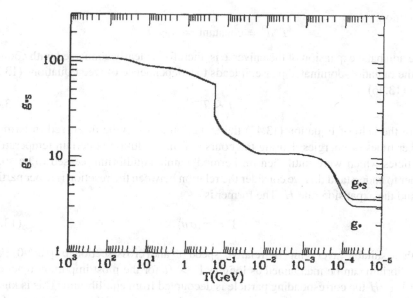

Figure 13.4. The cosmological standard model: behaviour of the summed effective degrees of freedom g_{eff} and g_S as a function of decreasing temperature. Only the particles of the standard model have been taken into consideration. One can see that both g_{eff} and g_S are identical over a wide range (from [Kol90]). © Westview Press.

$g_e = 4$ for e^+, e^- and $g_\nu = 6$ for ν_α with $\alpha \equiv e, \mu, \tau$. This is valid for Dirac neutrinos contributing with four helicity or Majorana neutrinos contributing with two helicity states, resulting in a weight of $g_\nu = 12$. Figure 13.4 illustrates the behaviour of g_{eff}. Starting at 106.75 at high energies where all particles of the standard model contribute, it decreases down to 3.36 if only neutrinos are participating.

In addition to the temperature, the entropy also plays an important role. The entropy is given by

$$S = \frac{R^3(\rho + p)}{T} \tag{13.43}$$

or, in the specific case of relativistic particles, by [Kol90]

$$S = \frac{2\pi^2}{45} g_s T^3 a^3 \tag{13.44}$$

where

$$g_s = \sum_{i=\text{bosons}} g_i \left(\frac{T_i}{T}\right)^3 + \frac{7}{8} \sum_{i=\text{fermions}} g_i \left(\frac{T_i}{T}\right)^3. \tag{13.45}$$

For the major part of the evolution of the universe, the two quantities g_{eff} and g_s were identical [Kol90]. The entropy per co-moving volume element is a conserved quantity in thermodynamic equilibrium, which together with constant g_s leads to the condition

$$T^3 a^3 = \text{constant} \Rightarrow a \sim T^{-1}. \tag{13.46}$$

The adiabatic expansion of the universe is, therefore, clearly connected with cooling. In the radiation-dominated phase, it leads to a dependence of (see Equations (13.33) and (13.46)

$$t \sim T^{-2}. \tag{13.47}$$

With the help of Equation (13.47) the evolution can now be discussed in terms of either times or energies. During the course of the evolution at certain temperatures particles which were until then in thermodynamic equilibrium ceased to be so. In order to understand this we consider the relation between the reaction rate per particle Γ and the expansion rate H. The former is

$$\Gamma = n\langle \sigma v \rangle \tag{13.48}$$

with a suitable averaging of relative speed v and cross-section σ [Kol90]. The equilibrium can be maintained as long as $\Gamma > H$ for the most important reactions. For $\Gamma < H$ the corresponding particle is decoupled from equilibrium. This is known as *freezing out*. Let us assume a temperature dependence of the reaction rate of the form $\Gamma \sim T^n$. Consider the interaction of two particles mediated either by massless bosons such as the photon or by massive bosons with a mass m_M as the Z^0. In the first case for the scattering of two particles a cross-section of

$$\sigma \sim \frac{\alpha^2}{T^2} \quad \text{with} \quad g = \sqrt{4\pi\alpha} = \text{gauge coupling strength} \tag{13.49}$$

results. In the second case, the same behaviour can be expected for $T \gg m_M$. For $T \leq m_M$,

$$\sigma \sim G_M^2 T^2 \quad \text{with } G_M = \frac{\alpha}{m_M^2} \tag{13.50}$$

holds. With a thermal number density, i.e., $n \sim T^3$, for the case of massless exchange particles it follows that

$$\Gamma \sim \alpha^2 T. \tag{13.51}$$

For reactions involving the exchange of massive particles the corresponding relation is

$$\Gamma \sim G_M^2 T^5. \tag{13.52}$$

During the early radiation-dominated phase, the Hubble parameter can be written as [Kol90]

$$H = 1.66 g_{\text{eff}}^{1/2} \frac{T^2}{m_{Pl}}. \tag{13.53}$$

The Planck mass m_{Pl} is given by

$$m_{Pl} = \left(\frac{\hbar c}{G}\right)^{\frac{1}{2}} = 1.221 \times 10^{19} \text{GeV}/c^2. \tag{13.54}$$

For massless particles, it then follows that

$$\frac{\Gamma}{H} \propto \frac{\alpha^2 m_{Pl}}{T}.$$

(13.55)

As long as $T > \alpha^2 m_{Pl} \approx 10^{16}$ GeV, the reactions occur rapidly: in the opposite case they 'freeze out'. For massive particles, the analogous relation is

$$\frac{\Gamma}{H} \propto G_M^2 m_{Pl} T^3.$$

(13.56)

This means that as long as

$$m_M \geq T \geq G_M^{-\frac{2}{3}} m_{Pl}^{-\frac{1}{3}} \approx \left(\frac{m_M}{100 \text{ GeV}} \right)^{\frac{4}{3}} \text{MeV}$$

(13.57)

holds, such processes remain in equilibrium. If a particle freezes out, its evolution is decoupled from the general thermal evolution of the universe.

We will now discuss the evolution of the universe step-by-step (see Figure 13.5). The earliest moment to which our present description can be applied is the Planck time. Planck time t_{Pl} and Planck length l_{Pl} are given by

$$l_{Pl} = \left(\frac{\hbar G}{c^3} \right)^{\frac{1}{2}} = 1.6 \times 10^{-33} \text{ cm}$$

(13.58)

$$t_{Pl} = \left(\frac{\hbar G}{c^5} \right)^{\frac{1}{2}} = 5.4 \times 10^{-44} \text{ s}.$$

(13.59)

Here, the Schwarzschild radius and Compton wavelength of the electron are of the same order. Before this point, a quantum mechanical description of gravity is necessary which does not exist currently. All particles are highly relativistic and the universe is radiation dominated. At the moment at which energies drop to around 10^{16} GeV GUT symmetry breaking takes place, where the potential existing heavy gauge bosons X and Y (see Chapter 5) freeze out. At about 300 GeV a second symmetry breaking occurs, which leads to the interactions that can be observed in today's particle accelerators. At about 10^{-6} s, the quarks and antiquarks annihilate and the surplus of quarks represents the whole of today's observable baryonic matter. The slight surplus of quarks is reflected in a baryon–photon ratio of about 10^{-10}. After about 10^{-5} s equivalent to 100–300 MeV, characterized by Λ_{QCD}, a further phase transition takes place. This is connected with the breaking of the chiral symmetry of the strong interaction and the transition from free quarks in the form of a quark–gluon plasma to quarks confined in baryons and mesons. At temperatures of about 1 MeV several things happen simultaneously. During the period $1–10^2$ s, the process of primordial nucleosynthesis takes place. Therefore, the observation of the lighter elements provides the furthest look back into the history of the universe. Around the same time or, more precisely, a little before, the neutrinos decouple and develop further independently. As a result, a cosmic neutrino background is produced, which has, however, not yet been observed. The almost total annihilation

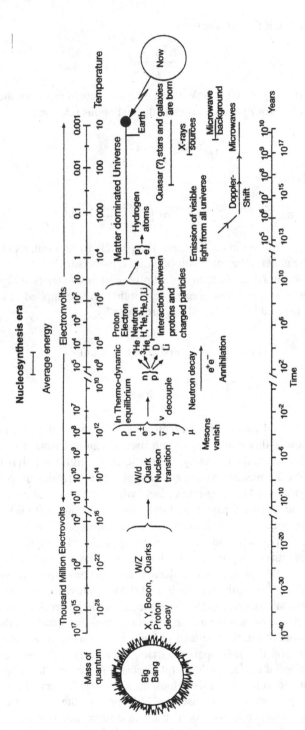

Figure 13.5. The chronological evolution of the universe since the Big Bang (from [Wil93]).

Table 13.2. GUT cosmology (from [Gro90]).

	Time	Energy	Temperature	'Diameter' of the universe
	t	$E = kT$	T	a
	(s)	(GeV)	(K)	(cm)
Planck time t_{Pl}	10^{-44}	10^{19}	10^{32}	10^{-3}
GUT SU(5) breaking M_X	10^{-36}	10^{15}	10^{28}	10
SU(2)$_L \otimes$ U(1) breaking M_W	10^{-10}	10^2	10^{15}	10^{14}
Quark confinement $p\bar{p}$ annihilation	10^{-6}	1	10^{13}	10^{16}
ν decoupling, e^+e^- annihilation	1	10^{-3}	10^{10}	10^{19}
light nuclei form	10^2	10^{-4}	10^9	10^{20}
γ decoupling, hline transition from radiation-dominated to matter-dominated universe, atomic nuclei form, stars and galaxies form	10^{12} ($\approx 10^5$ yr)	10^{-9}	10^4	10^{25}
Today, t_0	$\approx 5 \times 10^{17}$ ($\approx 2 \times 10^{10}$ yr)	3×10^{-13}	3	10^{28}

of electrons and positrons happens at this time as well. The resulting annihilation photons make up part of the cosmic microwave background. The next crucial stage takes place only about 300 000 years later. By then the temperature has sunk so far that nuclei can recombine with the electrons. As Thomson scattering (scattering of photons on free electrons) is strongly reduced, the universe suddenly becomes transparent and the radiation decouples from matter. This can still be detected today as 3 K background radiation. Starting at this time density fluctuations can now increase and, therefore, the creation of large-scale structures which will finally result in galaxies can begin. At that time the universe also passes from a radiation-dominated to a matter-dominated state. This scenario, together with the discussed characteristics, is called the *standard model of cosmology* (see Table 13.2).

13.3   The cosmic microwave background

The cosmic microwave background (CMB) is one of the most important supports for the Big Bang theory. Gamov, Alpher and Herman already predicted in the 1940s that if the Big Bang model was correct, a remnant noise at a temperature of about 5 K should still be present [Gam46,Alp48]. Furthermore at this time McKellar studied the rotational spectra of CN-molecules and could have found the CMB temperature, but it was not recognised. By now, the ultimate mission to study the CMB is the Planck-satellite. For extensive literature concerning this cosmic microwave background we refer to [Par95, Ber02, Sil02, Hu02, Hu03, Sam07, Dur08, Ada16a, Ade16, Agh19].

13.3.1   Spectrum and temperature

During the radiation-dominated era, radiation and matter were in a state of thermodynamic equilibrium. Thompson scattering on free electrons resulted in an opaque universe. As the temperature continued to fall, it became possible for more and more of the nucleons and electrons to recombine to form hydrogen. As most of the electrons were now bound, the mean free path of photons became much larger (of the order c/H) and they decoupled from matter. As the photons were in a state of thermodynamic equilibrium at the time of decoupling, their intensity distribution $I(\nu)\,d\nu$ corresponds to a black-body spectrum:

$$I(\nu)\,d\nu = \frac{2h\nu^3}{c^2} \frac{1}{\exp\left(\frac{h\nu}{kT}\right) - 1}\,d\nu. \tag{13.60}$$

The black-body shape in a homogeneous Friedmann universe remains unchanged despite expansion. The maximum of this distribution lies, according to Wien's law, at a wavelength of

$$\lambda_{\max}T = 2.897 \times 10^{-3} \text{ K m} \tag{13.61}$$

which for 5 K radiation corresponds to about 1.5 mm. Indeed, in 1964 Penzias and Wilson of the Bell Laboratories discovered an isotropic radiation at 7.35 cm, with a temperature of (3.5 ± 1) K [Pen65]. The energy density of the radiation is found by integrating over the spectrum (Stefan–Boltzmann law):

$$\rho_\gamma = \frac{\pi^2 k^4}{15 h^3 c^3} T_\gamma^4 = a T_\gamma^4. \tag{13.62}$$

From Equation (13.36) we obtain the following relationship:

$$n_\gamma = \frac{30\zeta(3)a}{\pi^4 k} T_\gamma^3 \approx 20.3 T_\gamma^3 \ \text{cm}^{-3} \tag{13.63}$$

for the number density of photons. Here $\zeta(3)$ is the Riemann ζ function of 3, which is approximately $1.202\,06$.

The probability distribution of events at the time of last scattering, the so-called last scattering surface, is approximately Gaussian with a mean at a redshift of $z = 1070$ and a standard deviation of 80. This means that roughly half of the last scattering events took place at redshifts between 990 and 1150. This redshift interval today corresponds to a length scale of $\lambda \simeq 7(\Omega h^2)^{1/2}$ Mpc, and an angle of $\theta \simeq 4\Omega^{1/2}$ [arcmin]. Structures on smaller angular scales are smeared out.

13.3.2 Measurement of the spectral form and temperature of the CMB

The satellite COBE (cosmic background explorer) brought a breakthrough in the field [Smo90]. It surveyed the entire sky in different wavelengths. In previous measurements only a few wavelengths had been measured and these were different in every experiment. The measured spectrum shows a perfect black-body form at a temperature of (2.728 ± 0.004) K [Wri94, Fix96] (Figure 13.6). No deviations whatsoever are seen in the spectral form. From that the number density of photons can be determined as $n_\gamma = (412 \pm 2)$ cm^{-3}. The number density is particularly interesting for the photon–baryon ratio η.

In addition to the spectral form and its distortions, the homogeneity and isotropy are also of extraordinary interest, as they allow conclusions as to the expansion of the universe and are an extremely important boundary condition for all models of structure formation.

13.3.3 Anisotropies in the 3 K radiation

Anisotropies in the cosmic background radiation are of extraordinary interest, on one hand for our ideas about the formation of large-scale structures and galaxies in the universe and, on the other hand, for our picture of the early universe. The former reveals itself through anisotropies on small angular scales (arc minutes up to a few degrees), while the latter is noticeable on larger scales (up to 180 degrees). We consider only the small angular scales in more detail [Whi94] because this part is important for neutrino physics. For an overview see [Rea92, Hu95, Ber02, Sil02, Hu02, Hu03, Wri03, Zal03, Sam07, Dur08].

13.3.3.1 *Measurement of the anisotropy*

The temperature field of the CMB can be expanded into its spherical harmonics Y^{lm}

$$\frac{\Delta T}{T}(n) = \sum_{l=2}^{\infty} \sum_{m=-l}^{l} a_{lm} Y^{lm}(n). \tag{13.64}$$

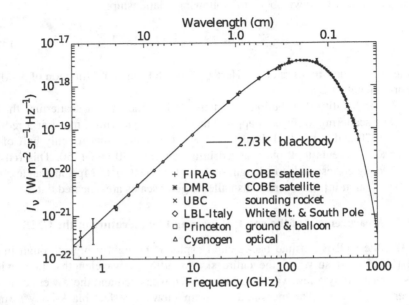

Figure 13.6. Spectrum of the cosmic background radiation, measured with the FIRAS and DMR detectors on the COBE satellite. It shows a perfect black-body behaviour. The smooth curve is the best-fit black-body spectrum with a temperature of 2.728 K. Also shown are the original data point of Penzias and Wilson as well as further terrestrial measurements from NASA (from [PDG02]).

By definition the mean value of a_{lm} is zero. The correlation function $C(\theta)$ of the temperature field is the average across all pairs of points in the sky separated by an angle θ:

$$C(\theta) = \left\langle \frac{\Delta T}{T}(n_1)\frac{\Delta T}{T}(n_2) \right\rangle = \frac{1}{4\pi}\sum_l (2l+1)C_l P_l(\cos\theta) \qquad (13.65)$$

with the Legendre polynomials $P_l(\cos\theta)$ and C_l as a cosmological ensemble average

$$C_l = \langle |a_{lm}^2| \rangle. \qquad (13.66)$$

In the case of random phases the power spectrum is related to a temperature difference ΔT via

$$\Delta T_l = \sqrt{C_l \frac{l(l+1)}{2\pi}}. \qquad (13.67)$$

The harmonic index ℓ is associated with an angular scale θ via $\ell \approx 180°/\theta$. The main anisotropy observed is the dipole component due to the Earth movement with respect to the CMB. It was measured by COBE to be 3.353 ± 0.024 mK [Ben96]

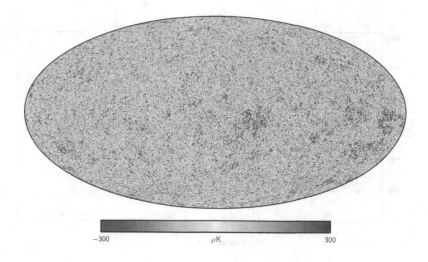

Figure 13.7. The ultimate all sky view of the CMB as measured with the Planck-satellite (from [Bou15]). © 2015 With permission from Elsevier.

and, more recently, by WMAP as 3.346 ± 0.017 mK [Ben03]. After subtracting the dipole component, anisotropies are observed by various experiments on the level of 10^{-5}. These result mainly from thermo-acoustic oscillations of baryons and photons [Hu95, Smo95, Teg95, Ber02, Sil02, Wri03]. It produces a series of peaks in the power spectrum whose positions, heights and numbers depend critically on various cosmological parameters, which is of major importance. The position of the first acoustic peak depends on the total density in the universe as $l \approx 200\sqrt{\Omega_0}$. In addition, Ω_b will increase the odd peaks with respect to even ones. Both h and Ω_Λ will change the height and location of the various peaks. The existence of such acoustic peaks has been shown by various experiments [Tor99, Mau00, Mel00, Net02, Hal02a, Lee01, Pea03, Sco03, Ben03a, Ruh03, Gra03, Kuo04, Ben03, Raj05a, Jon06, Hin09, Nol09, Kom09, Rei09, Agh18, Agh19] and is shown for some experiments in Figure 13.8. The radiation power spectrum shows two characteristic angular scales. A prominent peak occurs at $l \approx 220$ or about 30 arcmin. This is the angular scale that corresponds to the horizon at the moment of last scattering of radiation. The corresponding co-moving scale is about 100 Mpc. The second scale is the damping scale of about 6 arcmin, equivalent to the thickness of the last scattering surface of about 10 Mpc.

The latest and probably final measurement was done with the PLANCK mission to determine these cosmological quantities even more precisely and also aimed to measure polarization. How both high-z supernova observations and CMB peak position measurements restrict cosmological parameters in a complementary way is shown in Figure 13.9.

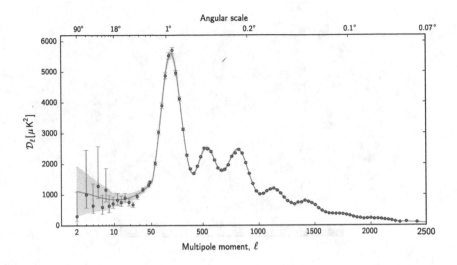

Figure 13.8. The anisotropy power spectrum as a function of the multipole order l as observed by the Planck-mission. The first acoustic peak is clearly visible at $l \approx 200$; also higher order peaks up to 1500 are visible (from [Ade14]). Reproduced with permission © ESO

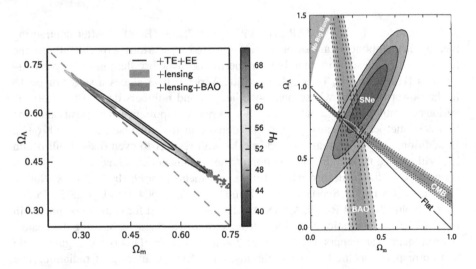

Figure 13.9. Left: Ω_Λ *versus* Ω_m plot. The determination of cosmological parameters using the Planck data, gravitational lensing and baryon acoustic oscillations leads to a small parameter space. Many large-scale galaxy surveys are contributing too (from [Bou15]). © 2015 With permission from Elsevier. Right: A similar plot showing Ω_m and the dark energy contribution. From all data used it is quite striking that about 70% are coming from this contribution. The combined parameter values show with high significance a non-vanishing cosmological constant and an $\Omega_\Lambda \approx 0.7$ and $\Omega_m \approx 0.3$ (from [Kow08]). © IOP Publishing. Reproduced with permission. All rights reserved.

13.3.3.2 *Anisotropies on small scales*

Anisotropies have to be divided into two types, depending on the horizon size at the time of decoupling. Fluctuations outside the event horizon are independent of the microphysics present during decoupling and so reflect the primordial perturbation spectrum, while the sub-horizontal fluctuations depend on the details of the physical conditions at the time of decoupling. The event horizon at the time of decoupling today corresponds to an angular size of [Kol90]

$$\Theta_{\text{dec}}[\text{deg}] = 0.87\Omega_0^{1/2}\left(\frac{z_{\text{dec}}}{1100}\right)^{-1/2}.$$ (13.68)

Below about 1°, therefore, the fluctuations mirror those that show up in structure formation (see Section 13.6). There is a correlation between the mass scale and the corresponding characteristic angular size of the anisotropies (see, e.g., [Nar93]):

$$(\delta\theta)[\text{arcsec}] \simeq 23\left(\frac{M}{10^{11}M_{\odot}}\right)(h_0 q_0^2)^{1/3}.$$ (13.69)

Typical density fluctuations that led to the formation of galaxies, therefore, correspond today to anisotropies on scales of 20 arcsec. Assuming that density fluctuations $\delta\rho/\rho$ develop adiabatically, the temperature contrast in the background radiation should be given by

$$\left(\frac{\delta T}{T}\right)_R = \frac{1}{3}\left(\frac{\delta\rho}{\rho}\right)_R$$ (13.70)

where the subscript R stands for 'at the recombination time'. In order to produce the density currently observed in galaxies, observable temperature anisotropies of $\delta T/T \approx 10^{-3}$–$10^{-4}$ would be expected. However, this has not been observed. The density fluctuations in the baryon sector are damped and additional terms are needed in the density fluctuations. The damping results from the period of recombination. Due to the suddenly increasing mean free path for photons, these can also effectively flow away from areas of high density. Their spreading, due to frequent collisions, does, however, correspond to diffusion rather than to a free flow. This kind of damping is called collisional damping or *Silk-damping* [Sil67, Sil68, Efs83]. Here smaller scales are effectively smeared out as, due to the frequent interaction of the photons, inhomogeneities in the photon–baryon plasma are damped. A significant amount of dark matter in the form of weakly interacting massive particles (WIMPs) could, for example, produce a similar effect. They already form a gravitational potential before recombination which the baryons experience later. Anyhow, the imprint of acoustic peaks as in the CMB should be also visible in baryons, called baryon acoustic oscillations (BAOs). They have been discovered in the Sloan Digital Sky Survey (SDSS) and 2dF galaxy redshift surveys [Col05, Eis05, Rei10]. A lot of experimental activities are going on to study the baryon acoustic peak as a function of redshift in more detail.

After having briefly discussed the basic picture of cosmology, we now want to discuss some special topics that are influenced by neutrinos in more detail.

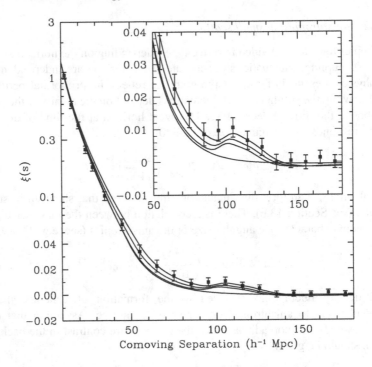

Figure 13.10. The redshift space galaxy correlation function as observed by the Sloan Digital Sky Survey (SDSS). The baryon acoustic peak is visible at around 100 h^{-1} Mpc, the different curves correspond to different cosmological models (from [Eis05]). © IOP Publishing. Reproduced with permission. All rights reserved. For additional information see [Rei10].

13.4 Neutrinos as dark matter

Among the most interesting problems of modern particle astrophysics is dark matter. In general the problem is that there seems to be much more gravitationally interacting matter in the universe than is luminous. It shows up on various scales:

- Rotational curves $v(r)$ of spiral galaxies: They show a flat behaviour even in regions far out of the optical detection region where, according to Newton's law, a Keplerian $v(r) \sim r^{-1/2}$ should be expected.
- Dark matter in galaxy clusters: It has been established by the kinematics of clusters (virial theorem), x-ray emission (gravitational binding of a hot electron gas) and weak gravitational lensing that a large fraction of the mass of clusters is dark. Estimates of cluster masses result in $\Omega \approx 0.3$.
- Cosmology: Big Bang nucleosynthesis with galaxy clusters estimates on matter densities and weak gravitational lensing in the universe are consistent only if assuming non-baryonic dark matter. Also the discrepancy between the observed value and the theoretical prediction of $\Omega = 1$ from inflation requires a large

amount of unknown matter. As already stated, CMB and other cosmological probes have shown that about 75% are made of dark energy typically linked to the vacuum energy density and about 24% are assumed to be dark non-baryonic matter.

For more detailed information, see [Jun96,Kla97,Ber99,Ber05,Gai04,Asz06,Han06, Ahl10,Bam15,Kna17,Pro17]. In summary, it is clear that the observed visible matter is insufficient to close the universe. An explanation of the rotation curves of galaxies and the behaviour of galaxy clusters also does not seem possible.

13.5 Candidates for dark matter

Having shortly presented the evidence for the existence of dark matter the discussion now turns to some particle physics candidates; among them are neutrinos.

13.5.1 Non-baryonic dark matter

The possible candidates for this are limited not so much by physical boundary conditions as by the human imagination and the resulting theories of physics. Consider first the abundance of relics (such as massive neutrinos) from the early period of the universe, which was then in thermodynamic equilibrium. For temperatures T very much higher than the particle mass m, their abundance is similar to that of photons, while at low temperatures ($m > T$) the abundance is exponentially suppressed (Boltzmann factor). How long a particle remains in equilibrium depends on the ratio between the relevant reaction rates and the Hubble expansion discussed before. Pair production and annihilation determine the abundance of long-lived or stable particles. The particle density n is then determined by the Boltzmann equation [Kol90]

$$\frac{dn}{dt} + 3Hn = -\langle\sigma v\rangle_{\text{ann}}(n^2 - n_{\text{eq}}^2) \qquad (13.71)$$

where H is the Hubble constant, $\langle\sigma v\rangle_{\text{ann}}$ the thermally averaged product of the annihilation cross-section and velocity and n_{eq} is the equilibrium abundance. The annihilation cross-section of a particle results from a consideration of all of its decay channels. It is useful to parametrize the temperature dependence of the reaction cross-section σ as follows: $\langle\sigma v\rangle_{\text{ann}} \sim v^p$, where in this partial wave analysis $p = 0$ corresponds to an s-wave annihilation, $p = 2$ to a p-wave annihilation, etc. As furthermore, $\langle v\rangle \sim T^{1/2}$, it follows that $\langle\sigma v\rangle_{\text{ann}} \sim T^n$, with $n = 0$ s-wave, $n = 1$ p-wave, etc. This parametrization is useful in the calculation of abundances for Dirac and Majorana particles. While the annihilation of Dirac particles occurs only via s-waves, i.e., independent of velocity, Majorana particles also have a contribution from p-wave annihilation, leading to different abundances.

There is a lower mass limit on any dark matter particle candidate, which relies on the fact of conservation of phase space (Liouville theorem). The evolution of dark matter distributions is collisionless; therefore, they accumulate in the centre of astronomical objects. Consider hot dark matter and an initial particle density distribution given by Fermi–Dirac statistics:

$$dN = g\frac{V}{(2\pi\hbar)^3} \exp[(E/kT) \pm 1]^{-1} d^3p. \tag{13.72}$$

Having an average occupation number of $\bar{n} = 1/2$, this results in a phase space density ρ of

$$\rho_i < (2\pi\hbar)^{-3}g/2. \tag{13.73}$$

Assuming that the velocity dispersion σ_V has relaxed to a Maxwellian

$$dp = (2\pi\sigma_V^2)^{3/2} \exp(-v^2/2\sigma_V^2) d^3v \tag{13.74}$$

resulting in

$$\rho_f < \left(\frac{\rho}{m}\right)(2\pi\sigma_V^2)^{-3/2} \, m^{-3}. \tag{13.75}$$

Then the condition that the maximum phase space density has not increased results in

$$m^4 > \frac{\rho}{Ng}\left(\frac{\sqrt{2\sigma_V}\hbar}{\sigma_V}\right)^3 \tag{13.76}$$

assuming a number N of neutrinos with mass m. This bound is known as the Tremaine–Gunn limit [Tre79]. For example, a simple isothermal sphere with radius r, having a density according to $\rho = \sigma_V^2/2\pi G r^2$ and the simple case $Ng = 1$ can be assumed. For galaxy clusters with $\sigma_V = 1000$ km s^{-1}, $r = 1$ Mpc it follows that $m_\nu > 1.5$ eV and there is no problem in fitting in neutrinos there. However, for a galactic halo with $\sigma_V = 150$ km s^{-1}, $r = 5$ kpc, it follows that $m_\nu > 33$ eV already close to a value necessary for the critical density and are ruled out anyhow. Therefore, neutrinos cannot be the dark matter in the latter case like regular galaxies—this is further supported by the observation that dwarf galaxies also contain a significant amount of dark matter.

13.5.1.1   Hot dark matter, light neutrinos

Light neutrinos remain relativistic and freeze out at about 1 MeV (see (13.11)), so that their density is given by

$$\rho_\nu = \sum_i m_{\nu i} n_{\nu i} = \Omega_\nu \rho_c. \tag{13.77}$$

From this it follows that for $\Omega \approx 1$ a mass limit for light neutrinos (masses smaller than about 1 MeV) is given by [Cow72]

$$\sum_i m_{\nu i}\left(\frac{g_\nu}{2}\right) = 94 \text{ eV } \Omega_\nu h^2. \tag{13.78}$$

Given the experimentally determined mass limits mentioned in Chapter 6, and the knowledge that there are only three light neutrinos, ν_e is already eliminated as a dominant contribution. Data coming from CMB alone point to a bound on the sum of neutrino masses [PDG18]

$$\sum m_\nu < 0.7 \, \text{eV} \, (90\% \, \text{CL}) \tag{13.79}$$

valid for $w = -1$. In the case of degenerated neutrinos this would imply $m_\nu < 0.23$ eV. However, adding background evolution via BAO and also large-scale structure (LSS) data, most of the estimated upper limits are around

$$\sum m_\nu < 0.3 \, \text{eV} \, (90\% \, \text{CL}) \tag{13.80}$$

However, there should be some caution because a bound on the neutrino mass is strongly correlated with a lot of other cosmological quantities mentioned before (see Section 13.6) and are always calculated in a certain cosmological model. Instead, one should use laboratory limits from β-decay. However, combining cosmological, beta and double decay results leads to quite stringent values.

13.5.1.2 Cold dark matter, heavy particles, WIMPs

The freezing-out of non-relativistic particles with masses of GeV and higher has the interesting characteristic that their abundance is inversely proportional to the annihilation cross-section. This follows directly from the Boltzmann equation and implies that the weaker particles interact, the more abundant they are today. Such "weakly interacting massive particles" are generally known as WIMPs. If we assume a WIMP with mass m_{WIMP} smaller than the Z^0 mass, the cross-section is roughly equal to $\langle \sigma v \rangle_{\text{ann}} \approx G_F^2 m_{\text{WIMP}}^2$ [Kol90], i.e.,

$$\Omega_{\text{WIMP}} h^2 \approx 3 \left(\frac{m_{\text{WIMP}}}{\text{GeV}} \right)^{-2}. \tag{13.81}$$

Above the Z^0 mass the annihilation cross-section decreases as m_{WIMP}^{-2}, due to the momentum dependence of the Z^0 propagator and, hence, a correspondingly higher abundance results. Figure 13.11 shows, as an example, the contribution of massive neutrinos to the mass density in the universe. Neutrinos between 100 eV and about 2 GeV as well as beyond the TeV region should, if they exist, be unstable according to these cosmological arguments [Lee77a].

In order to be cosmologically interesting ($\Omega \approx 1$), stable neutrinos must either be lighter than 100 eV as mentioned before or heavier than about 2 GeV (Dirac neutrinos) or 5 GeV (Majorana neutrinos). The latter case is excluded experimentally.

With heavy neutrinos a little bit out of fashion, currently the most preferred class of possible candidates for dark matter are supersymmetric particles, especially the neutralino as a possible lightest supersymmetric particle. As this has not been found at LHC (see Chapter 3), other perhaps lighter particles like the axion are becoming more attractive for searches.

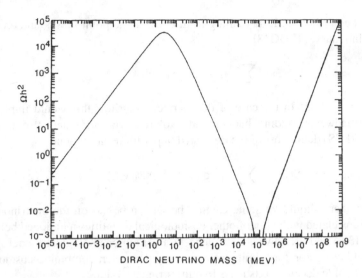

Figure 13.11. The contribution of stable neutrinos of mass m to the matter density of the universe. Only neutrino masses smaller than 100 eV and heavier than several GeV but lighter than TeV are cosmologically acceptable. Otherwise, neutrinos have to be unstable (from [Tur92a]). © 1992 World Scientific

13.5.2 Direct and indirect experiments

The experimental search for dark matter is currently one of the most active fields in particle astrophysics. Two basic strategies are followed: either the direct detection of dark matter interactions mostly via elastic scattering of WIMPs on nuclei or indirect detection by looking mostly for their annihilation products. Recent reviews of the direct detection efforts can be found in [Smi90, Jun96, Kla97, Ahl10]. Indirect experiments do not detect the interaction of dark matter in the laboratory but the products of dark matter particle reactions taking place extra-terrestrially or inside the Earth. For dark matter this is mainly particle–antiparticle annihilation. Two main types of annihilation are considered:

- annihilation inside the Sun or Earth and
- annihilation within the galactic halo.

13.5.2.1 Annihilation inside the Sun or Earth

It is possible that dark matter may accumulate gravitationally in stars and annihilate there with anti-dark matter. One signal of such an indirect detection would be high-energy solar neutrinos in the GeV–TeV range. These would be produced through the capture and annihilation of dark matter particles within the Sun [Pre85, Gou92, Car06]. An estimate of the expected signal due to photino (neutralino) annihilation results in about two events per kiloton detector material and year. These high-energy neutrinos would show up in the large water detectors via both charged and neutral weak currents. The charged weak interactions are about three times as frequent as the neutral ones. So far no signal has been found in detectors like Super-Kamiokande and IceCube (see Chapter 12). The capture of particles of dark matter in the Earth and their annihilation have also been discussed. Neutrinos from neutralino–antineutralino annihilation in the Earth are being sought by looking for vertical upward-going muons (see chapter 12). Again no signal has been observed yet.

13.6 Neutrinos and large-scale structure

One assumption in describing our universe is homogeneity. However, even though this seems to be justified on very large scales, observations have revealed a lot of structure on scales going beyond 100 Mpc. Galaxies group themselves into clusters and the clusters into superclusters, separated by enormous regions with low galaxy density, the so-called voids. The existence of large-scale structure (LSS) depends on the initial conditions of the Big Bang and on how physics processes have operated subsequently. The general picture of structure formation, as the ones described, is gravitational instability, which amplifies the growth of density fluctuations, produced in the early universe. The most likely source for producing density perturbations is quantum zero-point fluctuations during the inflationary era. Initial regions of higher density, which after the recombination era can concentrate through gravity, thereby form the starting points for the formation of structure. Hence, defining an initial spectrum of perturbation, the growth of the various scales has to be explored. The most viable framework of describing structure formation is the self-gravitating fluid which experiences a critical instability, as already worked out in classical mechanics by Jeans. We refer to further literature with respect to a more detailed treatment of structure formation [Pee80,Pad93,Pee93,Bah97,Pea98,Les06,Spr06,Mar08,Tsa08].

For the theoretical description of the development of the fluctuations, it is convenient to introduce the dimensionless density perturbation field or density contrast

$$\delta(x) = \frac{\rho(x) - \langle \rho \rangle}{\langle \rho \rangle}. \tag{13.82}$$

The correlation function of the density field is given by

$$\xi(r) = \langle \delta(x)\delta(x+r) \rangle \tag{13.83}$$

with brackets indicating an averaging over the normalization volume V. The density contrast can be decomposed into its Fourier coefficients:

$$\xi(r) = \frac{V}{(2\pi)^3} \int \delta_k e^{-ik \cdot r} \, d^3k. \tag{13.84}$$

It can be shown that

$$\xi(r) = \frac{1}{V(2\pi)^3} \int |\delta_k|^2 e^{-ik \cdot r} \, d^3k. \tag{13.85}$$

The quantity $|\delta_k|^2$ is known as the *power spectrum*. The correlation function is, therefore, the Fourier transform of the power spectrum. If we assume an isotropic correlation function, integration over the angle coordinates gives

$$\xi(r) = \frac{V}{(2\pi)^3} \int |\delta_k|^2 \frac{\sin kr}{kr} 4\pi k^2 \, dk. \tag{13.86}$$

The aim is to predict this power spectrum theoretically, in order to describe the experimentally determined correlation function. Theory suggests a spectrum with no preferred scale, called the Harrison–Zeldovich spectrum, equivalent to a power law

$$|\delta_k|^2 \sim k^n. \tag{13.87}$$

For Gaussian-like and, therefore, uncorrelated fluctuations, there is a connection between the mean square mass fluctuation and the power spectrum $|\delta_k|^2$, which contains all the information about the fluctuation [Kol90]:

$$\langle \delta^2 \rangle_\lambda \simeq V^{-1} (k^3 |\delta_k|^2 / 2\pi^2)_{k \approx 2\pi/\lambda}. \tag{13.88}$$

with

$$\lambda = \frac{2\pi}{k} a(t). \tag{13.89}$$

Weakly interacting particles, such as light neutrinos, can escape without interaction from areas of high density to areas of low density, which can erase small scale perturbations entirely. This process of free streaming, or collision-less damping, is important before the Jeans instability becomes effective. Light, relativistic neutrinos travel approximately with the speed of light, so any perturbation that has entered the horizon will be damped. The relevant scale is the redshift of matter radiation equality [Bon80, Bon84]

$$\lambda_{\mathrm{FS}} \simeq 1230 \left(\frac{m_\nu}{\mathrm{eV}} \right)^{-1} \mathrm{Mpc} \tag{13.90}$$

corresponding to a mass scale of

$$M_{\mathrm{FS}} \simeq 1.5 \times 10^{17} \left(\frac{m_\nu}{\mathrm{eV}} \right)^{-2} M_\odot. \tag{13.91}$$

Such masses are the size of galaxy superclusters and such objects are the first to form in a neutrino-dominated universe. Below this scale, perturbations are completely erased or at least strongly damped by neutrinos, resulting in a suppression of small scales in the matter power spectrum by roughly [Hu98]

$$\frac{\Delta P_M}{P_M} \sim -\frac{8\Omega_\nu}{\Omega_M}. \tag{13.92}$$

The larger m_ν and Ω_ν, the stronger is the suppression of density fluctuations at small scales. While massive neutrinos have little impact on the CMB power spectrum, it is still necessary to include the CMB data to determine the other cosmological

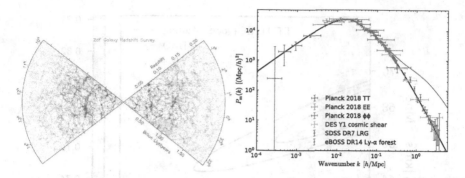

Figure 13.12. Left: The distribution of galaxies (wedge diagram) as part of the 2dFGRS with a slice thickness of 4 degrees. In total 213 703 galaxies are drawn. The filament-like structures, i.e., areas of very high density (superclusters), as well as voids, can clearly be seen (from [Pea02]). With kind permission of the 2dF Galaxy Redshift Survey team [2dfgrs]. Right: Power spectrum as a function of wavenumber as it is obtained from the different experiments (from [Cha19]). With kind permission of M. Millea.

parameters and to normalize the matter power spectrum. Ω_M has been restricted by WMAP and Planck, so the measurements of the power spectrum obtained by large-scale galaxy surveys can now be normalized to the CMB data. Two recent large-scale surveys are the 2dF galaxy redshift survey (2dFGRS) [Elg02] and the Sloan Digital Sky Survey (SDSS) [Dor04]. The galaxy distribution as observed in the 2dFGRS is shown in Figure 13.13. To obtain a bound on neutrino masses, the correlation with other cosmological parameters has to be taken into account accordingly, especially the Hubble parameter H, the matter density Ω_M and the bias parameter b [Han03a]. The bias parameter relates the matter spectrum to the observable galaxy–galaxy correlation function ξ_{gg} in large-scale structure surveys via

$$b^2(k) = \frac{P_g(k)}{P_M(k)} \tag{13.93}$$

with the power spectrum P(k) of mass fluctuations as function of wavenumber k for matter $P_M(k)$ and galaxies $P_g(k)$. These parameters can be connected to a bias parameter σ_8 which somehow defines a threshold for how light traces matter [Kol90]. As mentioned before, the sum of all neutrino masses can be obtained from cosmological studies. Cosmology is also important for the neutrino mass ordering [Han10, Han16]. All current data sets seem to suggest (according to the assumptions made) an upper limit of [PDG18]

$$\sum m_\nu < 1\,\text{eV}. \tag{13.94}$$

However adding more and more data, stronger upper limits up to about $m_\nu < 0.3\,\text{eV}$ can be obtained. Furthermore, this could change if there would be more effective degrees of freedom N_{eff}, which could be for example a fourth neutrino. Hence, there

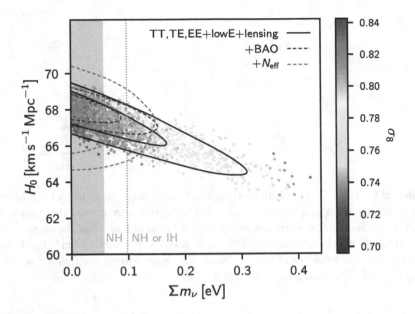

Figure 13.13. Correlation plot of the Hubble parameter H$_0$ as a function of the sum of the neutrino masses $\sum m_\nu$ as obtained by Planck data and further cosmological quantities. On the right side, the scale of σ_8 between 0.7-0.84 is given. The dashed vertical line is marking a value of all neutrino masses of 0.1 eV. The grey area on the left side of the graph is excluded by neutrino oscillation data for the IH. The plot indicates that the inverted hierarchy (IH) scenario is under tension (from [Agh18]).

are a lot of correlations among the various parameters involved like σ_8, H_0, N_{eff} and more.

13.7   The cosmic neutrino background

Analogous to the photon background, a cosmic neutrino background should also . exist. At temperatures above 1 MeV, neutrinos, electrons and photons are in thermal equilibrium with each other via reactions such as $e^+e^- \leftrightarrow \gamma\gamma$ or $e^+e^- \leftrightarrow \nu\bar{\nu}$. As the temperature drops further to less than the rest mass of the electron, all energy is transferred to the photons via pair annihilation, thereby increasing their temperature. Next consider the entropy of relativistic particles which is given by (see Equations (13.44) and (13.45))

$$S = \frac{4}{3}k_B \frac{R^3}{T}\rho \tag{13.95}$$

where ρ represents the energy density. As $\rho \sim T^4$, according to the Stefan–Boltzmann law, it follows that, for constant entropy,

$$S = (Ta)^3 = \text{constant.} \tag{13.96}$$

From the relations mentioned in Section 13.2.1, it follows that

$$\rho_{\nu_i} = \tfrac{7}{16}\rho_\gamma \tag{13.97}$$

and, for $kT \gg m_e c^2$,

$$\rho_{e^\pm} = \tfrac{7}{8}\rho_\gamma. \tag{13.98}$$

Using the appropriate degrees of freedom and entropy conservation, we obtain

$$(T_\gamma a)^3_B(1 + 2\tfrac{7}{8}) + (T_\nu a)^3_B \sum_{i=1}^{6} \rho_{\nu_i} = (T_\gamma a)^3_A + (T_\nu a)^3_A \sum_{i=1}^{6} \rho_{\nu_i} \tag{13.99}$$

where B ("before") represents times $kT > m_e c^2$ and A ("after") times $kT < m_e c^2$. Since the neutrinos had already decoupled, their temperature developed proportional to a^{-1} and, therefore, the last terms on both sides cancel. Hence,

$$(T_\gamma a)^3_B \frac{11}{4} = (T_\gamma a)^3_A. \tag{13.100}$$

However, since prior to the annihilation phase of the e^+e^--pairs $T_\gamma = T_\nu$, this means that

$$\left(\frac{T_\gamma}{T_\nu}\right)_B = \left(\frac{11}{4}\right)^{1/3} \simeq 1.4. \tag{13.101}$$

On the assumption that no subsequent significant changes to these quantities have taken place, the following relation between the two temperatures exists today:

$$T_{\nu,0} = \left(\frac{4}{11}\right)^{1/3} T_{\gamma,0}. \tag{13.102}$$

A temperature of $T_{\gamma,0} = 2.728$ K corresponds then to a neutrino temperature of 1.95 K. If the photon background consists of a number density of $n_{\gamma,0} = 412$ cm^{-3}, the particle density of the neutrino background is

$$n_{\nu_0} = \frac{3}{4}\frac{g_\nu}{g_\gamma}\frac{4}{11}n_{\gamma_0} = \frac{3g_\nu}{22}n_{\gamma_0} = 336 \text{ cm}^{-3} \tag{13.103}$$

with $g_\nu = 6$ ($g_\gamma = 2$). The energy density and average energy are:

$$\rho_{\nu_0} = \frac{7}{8}\frac{g_\nu}{g_\gamma}\left(\frac{4}{11}\right)^{\frac{4}{3}}\rho_{\gamma_0} = 0.178 \text{ eV cm}^{-3} \qquad \langle E_\nu \rangle_0 = 5.28 \times 10^{-4} \text{ eV}. \tag{13.104}$$

These relations remain valid even if neutrinos have a small mass. It is also valid for Majorana neutrinos and light left-handed Dirac neutrinos because, in both cases, $g_\nu = 2$. For heavy Dirac neutrinos ($m > 300$ keV) with a certain probability, the "wrong" helicity states can also occur and $g_\nu = 4$ [Kol90].

The very small cross-section of relic neutrinos has so far thwarted, but an experimental attempt in the form of the PTOLEMY project is ongoing [Bet13, Bar18]. However, cosmic neutrino background plays an important role in the Z-burst model mentioned in Chapter 12 to explain UHE cosmic-ray events.

13.8   Primordial nucleosynthesis

In this chapter we turn our attention to another very important support of the Big Bang model, namely the synthesis of the light elements in the early universe. These are basically H, D, ^{3}He, ^{4}He and ^{7}Li. Together with the synthesis of elements in stars and the production of heavy elements in supernova explosions (see Chapter 11) and neutron star mergers, this is the third important process in the formation of the elements. The fact that their relative abundances are predicted correctly over more than ten orders of magnitude can be seen as one of the outstanding successes of the standard Big Bang model. Studying the abundance of ^{4}He allows statements to be made about the number of possible neutrino flavours in addition to the precise measurements made at LEP. For a more detailed discussion, we refer the reader to [Yan84, Boe03, Den90, Mal93, Pag97, Sch98a, Tyt00, Ste07, Cyb08, Ioc09, Ili15, Cyb16, Pit18].

13.8.1   The process of nucleosynthesis

The synthesis of the light elements took place in the first three minutes after the Big Bang, which means at temperatures of about 0.1–10 MeV. The first step begins at about 10 MeV, equivalent to $t = 10^{-2}$ s. Protons and neutrons are in thermal equilibrium through the weak interaction via the reactions

$$p + e^- \longleftrightarrow n + \nu_e \tag{13.105}$$

$$p + \bar{\nu}_e \longleftrightarrow n + e^+ \tag{13.106}$$

$$n \longleftrightarrow p + e^- + \bar{\nu}_e \tag{13.107}$$

and the relative abundance is given in terms of their mass difference $\Delta m = m_n - m_p$ (neglecting chemical potentials) as

$$\frac{n}{p} = \exp\left(-\frac{\Delta m c^2}{kT}\right). \tag{13.108}$$

The weak interaction rates are (see (13.52) and (13.54))

$$\Gamma \propto G_F^2 T^5 \qquad T \gg Q, m_e. \tag{13.109}$$

If this is compared with the expansion rate H, it follows that

$$\frac{\Gamma}{H} = \frac{T^5 G_F^2}{T^2/m_{Pl}} \approx \left(\frac{T}{0.8 \text{ MeV}}\right)^3 \tag{13.110}$$

implying that for temperatures below about 0.8 MeV the weak reaction rate becomes less than the expansion rate and freezes out. The neutron–proton ratio begins to deviate from the equilibrium value. One would expect a significant production of light nuclei here, as the typical binding energies per nucleon lie in the region of 1–8 MeV. However, the large entropy, which manifests itself in the very small baryon–photon ratio η, prevents such production as far down as 0.1 MeV.

The second step begins at a temperature of about 1 MeV or, equivalently, at 0.02 s. The neutrinos have just decoupled from matter and, at about 0.5 MeV, the electrons and positrons annihilate. This is also the temperature region in which these interaction rates become less than the expansion rate, which implies that the weak interaction freezes out, which leads to a ratio of

$$\frac{n}{p} = \exp\left(-\frac{\Delta mc^2}{kT_f}\right) \simeq \frac{1}{6}. \tag{13.111}$$

The third step begins at 0.3–0.1 MeV, corresponding to about 1–3 min after the Big Bang. Here practically all neutrons are converted into ^{4}He via the reactions shown in Figure 13.16 beginning with

$$n + p \leftrightarrow D + \gamma. \tag{13.112}$$

This is a certain bottleneck, because high energetic photons (above 2.2 MeV) dissociate the deuteron. Once it builds up, the reaction chain continues via

$$D + D \leftrightarrow He + \gamma \tag{13.113}$$
$$D + p \leftrightarrow {}^3He + \gamma \tag{13.114}$$
$$D + n \leftrightarrow {}^3H + \gamma. \tag{13.115}$$

The amount of primordial helium Y is then

$$Y = \frac{2n_n}{n_n + n_p}. \tag{13.116}$$

Meanwhile the initial n/p fraction has fallen to about $1/7$, due to the decay of the free neutrons. The equilibrium ratio, which follows from an evolution according to Equation (13.108), would be $n/p = 1/74$ at 0.3 MeV. The non-existence of stable nuclei of mass 5 and 8, as well as the now essential Coulomb barriers, very strongly inhibit the creation of ^{7}Li, and practically completely forbid that for even heavier isotopes (see Figure 13.14). Because of the small nucleon density, it is also not possible to get over this bottleneck via 3α reactions, as stars do. Therefore, BBN comes to an end if the temperature drops below about 30 keV, when the Universe was about 20 min old.

Current experimental numbers of the elemental abundances are [Pit18]:

$$Y = 0.2449 \pm 0.0040 \tag{13.117}$$
$$Li/H = 1.58 \pm 0.3 \times 10^{-10} \tag{13.118}$$
$$D/H = 2.527 \pm 0.0305 \times 10^{-5}. \tag{13.119}$$

Using the D/H value this corresponds to

$$\eta = 6.1^{+0.7}_{-0.3} \times 10^{-10}. \tag{13.120}$$

with η being the baryon to photon ratio. It can be converted into a baryonic density $(\eta \times 10^{10} = 274\,\Omega_b h^2)$

$$\Omega_b h^2 = 0.022^{+0.003}_{-0.002}. \tag{13.121}$$

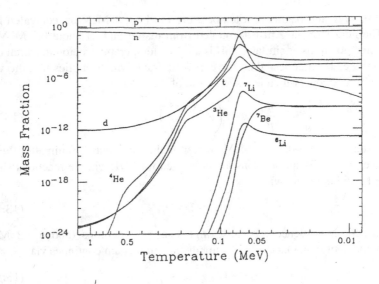

Figure 13.14. Development of the abundances of the light elements during primordial nucleosynthesis (from [Ree94]). © 1994 by the American Physical Society.

WMAP and Planck results combined with SN and BAO data imply a $\Omega_b h^2 = 0.02214 \pm 0.00024$ which results in the value of $\eta = 6.2 \pm 0.16 \times 10^{-10}$ [Hin09a]. While both numbers are in good agreement, there is a certain tension in the ^{4}He measurement and a disagreement in the ^{7}Li abundance. The given Ω_b implies $Y = 0.248 \pm 0.001$, higher than the value of (13.117). Further studies will show whether there is reasonable agreement. Therefore, according to primordial nucleosynthesis, it is not possible to produce a closed universe from baryons alone. However, if Ω_b has a value close to the upper limit, a significant fraction can be present in dark form, as the luminous part is significantly less ($\Omega_b^L \lesssim 0.02$) than that given by Equation (13.121). Some of it could be 'Massive Compact Halo Objects' (MACHOs) searched for by gravitational microlensing in the Milky Way halo. It is at least possible to use baryonic matter to explain the rotation curves of galaxies.

The predicted abundances (especially ^{4}He) depend on a number of parameters (Figure 13.14). These are, in principle, three: the lifetime of the neutron τ_n, the fraction of baryons to photons $\eta = n_B/n_\gamma$ and the number of relativistic degrees of freedom g_{eff}, where neutrinos contribute. In the following we only investigate the latter, for more details see [Ili15].

13.8.2 The relativistic degrees of freedom g_{eff} and the number of neutrino flavours

The expansion rate H is proportional to the number of relativistic degrees of freedom of the available particles (13.56). According to the Standard Model, at about 1 MeV, these are photons, electrons and three neutrino flavours. The dependence of the

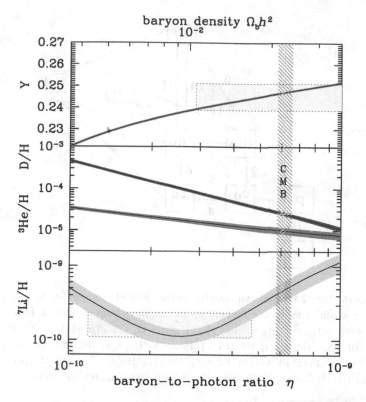

Figure 13.15. The primordial abundances of the light elements, as predicted by the Standard Model of Cosmology, as a function of the baryon density $\Omega_b h^2$ or of the baryon-to-photon ratio $\eta = n_B/n_\gamma$. The narrow and the wide vertical band indicate the CMB and the BBN constraint, respectively. The horizontal boxes show the experimentally observed element abundances. Despite the discrepancy in the lithium abundance, a consistent prediction is possible over 10 orders of magnitude (from [PDG18]). With kind permission from M. Tanabashi et al. (Particle Data Group).

freeze-out temperature on the number of degrees of freedom then results using Equation (13.110) in

$$H \sim g_{\text{eff}}^{1/2} T^2 \Rightarrow T_F \sim g_{\text{eff}}^{1/6}. \qquad (13.122)$$

Each additional relativistic degree of freedom (further neutrino flavours, axions, majorons, right-handed neutrinos, etc.) therefore resulting in an increase in the expansion rate and, therefore, a freezing-out of these reactions at higher temperatures. This again is reflected in a higher ^{4}He abundance. The number of neutrino flavours which can be determined from observations is $3.2 \pm 0.4(1\sigma)$ (see Figure 13.17) [Ioc09].

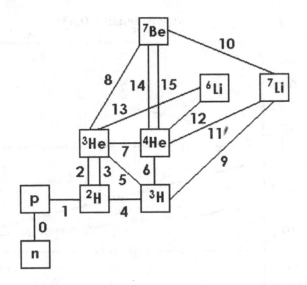

Figure 13.16. The 12 fundamental reactions in the chain of synthesis of the light elements, illustrating which elements can be built up in this way (from [Smi93b]). Labels indicate the following reactions: (0) $n \leftrightarrow p$, (1) $p(n,\gamma)d$, (2) $d(p,\gamma)\,^3\mathrm{He}$, (3) $d(d,n)\,^3\mathrm{He}$, (4) $d(d,p)t$, (5) $^3\mathrm{He}(n,p)t$, (6) $t(d,n)\,^4\mathrm{He}$, (7) $^3\mathrm{He}(d,p)^4\mathrm{He}$, (8) $^3\mathrm{He}(\alpha,\gamma)^7\mathrm{Be}$, (9) $t(\alpha,\gamma)\,^7\mathrm{Li}$, (10) $^7\mathrm{Be}(n,p)\,^7\mathrm{Li}$ (11) $^7\mathrm{Li}(p,\alpha)\,^4\mathrm{He}$. (12) $^4\mathrm{He}(d,\gamma)^6\mathrm{Li}$, (13) $^6\mathrm{Li}(p,\alpha)^3\mathrm{He}$, (14) $^7\mathrm{Be}(n,\alpha)^4\mathrm{He}$(15) $^7\mathrm{Be}(d,p)2^4\mathrm{He}$ (from [Ioc09]). © 2009 with permission from Elsevier.

13.9   Baryogenesis via leptogenesis

Under the assumption of equal amounts of matter and antimatter at the time of the Big Bang we observe today an enormous preponderance of matter compared with antimatter. If we assume that antimatter is not concentrated in regions that are beyond the reach of current observation, this asymmetry has to originate from the earliest phases of the universe. Here matter and antimatter destroy themselves almost totally except for a small excess of matter, leading to the current baryon asymmetry in the universe (BAU) of

$$Y_B = \frac{n_B - n_{\bar{B}}}{n_\gamma} \approx 10^{-10}. \tag{13.123}$$

In order to accomplish this imbalance, three conditions have to be fulfilled [Sac67b]:

(i)   both a C and a CP violation of one of the fundamental interactions,
(ii) non-conservation of baryon number and
(iii) thermodynamic non-equilibrium.

The production of the baryon asymmetry is usually associated with the GUT transition. The violation of the baryon number is not unusual in GUT theories, since, in these theories, leptons and quarks are placed in the same multiplet, as discussed in Chapter 5. That a CP violation is necessary can be seen in the following illustrative

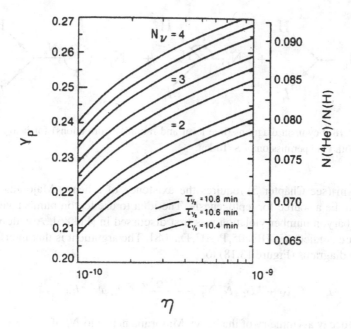

Figure 13.17. A more detailed illustration of the ^{4}He abundance as a function of the baryon/photon ratio $\eta = n_B/n_\gamma$. The influence of the number of neutrino flavours and the neutron lifetime on the predicted ^{4}He abundance can clearly be seen (from [Yan84]). With kind permission of J. Yang. © IOP Publishing. Reproduced with permission. All rights reserved.

set of reactions:

$$X \xrightarrow{r} u + u \qquad X \xrightarrow{1-r} \bar{d} + e^+ \tag{13.124}$$

$$\bar{X} \xrightarrow{\bar{r}} \bar{u} + \bar{u} \qquad \bar{X} \xrightarrow{1-\bar{r}} d + e^-. \tag{13.125}$$

In the case of CP violation $r \neq \bar{r}$. A surplus of u, d, e over $\bar{u}$, $\bar{d}$ and e^+ would follow therefore for $r \geq \bar{r}$. This is, however, possible only in a situation of thermodynamic non-equilibrium, as a higher production rate of baryons will otherwise also lead to a higher production rate of antibaryons. In equilibrium, the particle number is independent from the reaction dynamics. Theoretical estimates show that the CP-violating phase δ in the CKM matrix (see Chapter 3) is not sufficient to generate the observed baryon asymmetry and other mechanisms have to be at work.

13.9.1 Leptogenesis

The leptonic sector offers a chance for baryogenesis. In the case of massive neutrinos we have the PMNS matrix (see Chapter 5) in analogy to the CKM matrix. Moreover, for Majorana neutrinos with three flavours two additional CP-violating phases exist. Associated with Majorana neutrinos is lepton number violation. How this can be transformed into a baryon number violation is described later. Moreover, the seesaw

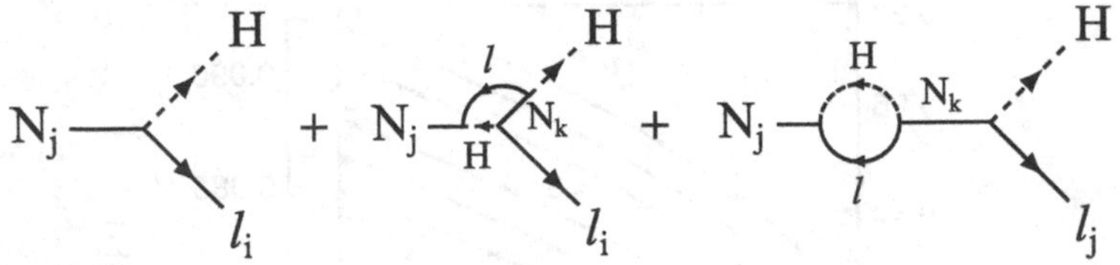

Figure 13.18. Feynman diagrams (tree level and radiative corrections) for heavy Majorana decays. With kind permission of S. Turkat.

mechanism (see Chapter 5) requires the existence of a heavy Majorana neutrino, which can be a source for leptogenesis. The idea to use lepton number violation to produce baryon number violation was first discussed in [Fuk86]. A wide variety of models are studied; see [Buc05, Pas07, Dav08]. The argument is that interference of one-loop diagrams (Figure 13.18) to

$$\mathcal{L} = \mathcal{L}_{\mathcal{EW}} + M_{Rij}\bar{N}_i^c N_j + \frac{(m_D)_{i\alpha}}{v}\bar{N}_i l_\alpha \phi^\dagger + h.c. \tag{13.126}$$

lead to a decay asymmetry of the heavy Majorana neutrino N_i of

$$\epsilon_i = \frac{\Gamma(N_i \to \phi l^c) - \Gamma(N_i \to \phi^\dagger l)}{\Gamma(N_i \to \phi l^c) + \Gamma(N_i \to \phi^\dagger l)}$$

$$= \frac{1}{8\pi v^2}\frac{1}{(m_D^\dagger m_D)_{ii}} \times \sum_{j\neq i}\text{Im}((m_D^\dagger m_D)_{ij}^2)f(M_J^2/M_i^2). \tag{13.127}$$

To fulfil the observations $\epsilon \approx Y_L \propto Y_B \approx 10^{-10}$. Unfortunately, there is no chance to explore the heavy Majorana neutrino sector directly; its only connection to experiment is via the seesaw mechanism to light neutrinos. Numerous models for neutrino masses and heavy Majorana neutrinos have been presented to reproduce the low-energy neutrino observations together with Y_B [Fla96, Rou96, Buc98]. As it turned out, there is no direct connection between low- and high-energy (meaning the heavy Majorana neutrino scale, normally related to the GUT scale) CP violation, unless there is a symmetry relating the light and heavy sectors [Bra01]. As a general tendency of most models, a strong dependence on the Majorana phases is observed [Rod02], while a possible Dirac CP-phase in the PMNS matrix seems to play a minor role. This fact makes the investigation of neutrinoless double β-decay (see Chapter 7) very important because this is the only known process where these phases can be explored. For more details see [Chu18].

The moment in the evolution of the universe at which lepton number violation is converted into baryon number violation is the electroweak phase transition. Its scale is characterized by the vacuum expectation value of the Higgs boson in the electroweak Standard Model and, therefore, lies around 250 GeV (see Chapter 3). It has been shown that non-Abelian gauge theories have non-trivial vacuum structures and with a different number of left- and right-handed fermions it can produce baryon and

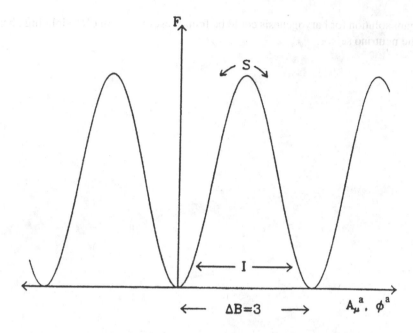

Figure 13.19. Schematic illustration of the potential with different vacua, which appear as different possible vacuum configurations of the fields A_μ and Φ in non-Abelian gauge theories. The possibility of the instanton tunnelling (I) *through* the barrier of height T_C, as well as the sphaleron (S) jumping *over* the barrier are indicated. In the transition $B + L$ changes by $2N_F$, and B by N_F, where N_F represents the number of families (in this case 3) (from [Kol90]). © Westview Press.

lepton number violation [t'Ho76, Kli84]. Figure 13.19 shows such vacuum configurations, which are characterized by different topological winding numbers and are separated by energy barriers of height T_C. In the case of $T = 0$ a transition through such a barrier can take place only by means of quantum mechanical tunnelling (*instantons*) and is, therefore, suppressed by a factor $\exp(2\pi/\alpha_w) \approx 10^{-86}$ ($\alpha_w =$ weak coupling constant). However, this changes at higher temperatures [Kuz85]. Now thermal transitions are possible and for $T \gg T_C$ the transition is characterized by a Boltzmann factor $\exp(-E_{\text{sph}}(T)/T)$. Here E_{sph} represents the sphaleron energy. The *sphaleron* is a saddle point in configuration space which is classically unstable. This means that the transition takes place mainly via this configuration. The sphaleron energy E_{sph} is equivalent to the height of the barrier T_C and, therefore, is also temperature dependent. If one proceeds from one vacuum to the next via the sphaleron configuration, the combination of $B + L$ changes by $2 \times N_F$, where N_F is the number of families, of which three are currently known. As, in addition, $B - L$ is free of gauge anomalies, which means that $B - L$ is conserved, the vacuum transition leads to $\Delta B = 3$. Calculations show that in the transition roughly half of the lepton number violation Y_L is converted into baryon number violation Y_B. In this way, an elegant solution for baryogenesis could be found resulting from CP-violating phases in the neutrino sector.

Chapter 14

Summary and outlook

DOI: 10.1201/9781315195612-14

Neutrino physics has experienced quite a massive boost in the last decade with the establishment of a non-vanishing rest mass of the neutrino. This allows for a variety of new features of neutrinos and has a huge impact on various physics processes. As discussed in more detail in the corresponding chapters of this book, all evidence stems from neutrino oscillation searches. There are two convincing pieces of evidence:

- A deficit in upward-going muons produced by atmospheric neutrinos. This has been confirmed by long-baseline accelerator experiments. This can be explained by ν_μ oscillations with $\Delta m^2 \approx 2.3 \times 10^{-3}$ eV^2 and $\sin^2 2\theta \approx 1$ into ν_τ.
- The observation of active neutrinos from the Sun besides ν_e. These results from observations performed by SNO and also Borexino together with the other solar neutrino data result. This has been confirmed with a completely different method by the KamLAND experiment using nuclear power plants resulting in combined best-fit values of $\Delta m^2 = (7.59 \pm 0.20) \times 10^{-5}$ eV^2 and $\theta \approx 34°$. This shows clearly that matter effects are at work in the Sun and the so-called LMA solution is correct.

Both discoveries were awarded the Nobel Prize in physics in 2015 to A. B. McDonald and T. Kajita. In the meantime a large number of experiments have provided more information. A graphical representation of all current results is shown in Figure 14.1. An incredible number of papers deducing neutrino mass models from the observed oscillation experiments exists. Within the context of mass models or underlying theories, a greater determination of the elements of the PMNS mixing matrix could be obtained due to more new experiments. The PMNS-matrix can be written in a suggestive way (assuming neutrinos are Dirac particles)

$$
\begin{pmatrix} \cos\theta_{12} & \sin\theta_{12} & 0 \\ -\sin\theta_{12} & \cos\theta_{12} & 0 \\ 0 & 0 & 1 \end{pmatrix} \begin{pmatrix} \cos\theta_{13} & 0 & \sin\theta_{13}e^{i\delta} \\ 0 & 1 & 0 \\ -\sin\theta_{13}e^{i\delta} & 0 & \cos\theta_{13} \end{pmatrix}
$$
$$
\times \begin{pmatrix} 1 & 0 & 0 \\ 0 & \cos\theta_{23} & \sin\theta_{23} \\ 0 & -\sin\theta_{23} & \cos\theta_{23} \end{pmatrix}. \tag{14.1}
$$

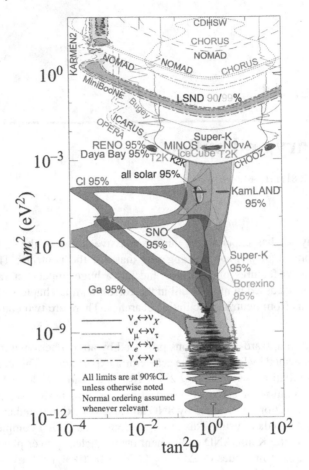

Figure 14.1. Compilation of most neutrino oscillation results: Shown is a plot in form of Δm^2 *versus* $\tan^2 \theta$ which is more reliable to present matter effects (from [PDG18]). With kind permission from M. Tanabashi et al. (Particle Data Group).

Current experimental global fits of all data (including Superkamiokande atmospheric data) are suggesting values for the mixing matrix elements [Est19] (see also discussion in [Est19]):

$$
\begin{pmatrix}
0.797 \to 0.842 & 0.518 \to 0.585 & 0.143 \to 0.156 \\
0.235 \to 0.484 & 0.458 \to 0.671 & 0.647 \to 0.781 \\
0.304 \to 0.531 & 0.497 \to 0.699 & 0.607 \to 0.747
\end{pmatrix}
\tag{14.2}
$$

This is quite different from the values of the CKM-mixing matrix in the quark sector showing a rather hierarchical ordering, diagonal terms are very large and the off-diagonal elements get a decrease in size. Clearly, this has to be understood and might guide theory toward physics beyond the Standard Model. With all the given oscillation evidences, now data must be analyzed in a full 3-flavour oscillation framework, also including matter effects. The OPERA experiment has successfully proven the ν_μ- ν_τ oscillation hypothesis for atmospheric neutrinos (Chapter 9). A large accelerator-based long-baseline program has been launched to investigate the neutrino mixing matrix elements in more detail and experiments like K2K, MINOS, T2K, MINOS+, NOvA and future DUNE are also getting closer to pin down the value of the CP-phase in the PMNS-matrix.

Potential Majorana phases (Chapter 7) are unobservable in oscillation experiments. In case of three Majorana neutrinos, this would add two more CP-violating phases to the PMNS matrix. They could play a significant role in leptogenesis (Chapter 13), the explanation of the baryon asymmetry in the universe with the help of lepton number violation, which will partly be transformed to a baryon number violation via the electroweak phase transition. Neutrinoless double β-decay is the most preferred process having sensitivity to prove the existence of a Majorana particle (Chapter 7). So, it is not only the main purpose to probe the neutrino mass but also the fundamental character of the neutrino. If we take the current oscillation results, a measurement in the region down to 50 meV would have discovery potential for this process. This would also support the inverted mass hierarchy with respect to the normal one (see Figure 14.2). To reach this mass region, a scaling up of existing experiments is still ongoing and half-life limits in the region of about 10^{26} years can be probed now. Experiments like GERDA, EXO, KamLAND-Zen and CUORE are or will be able to reach this half-life. As a remark, it should be mentioned that there is now also an analogous decay the other way around the isobar parabola, namely 2 neutrino double electron capture has been observed by Xenon1t (Chapter 7). Independent of the character of the neutrino, endpoint measurements either for the mass of the ν_e and $\bar{\nu}_e$ in form of beta decays or in form of internal bremsstrahlung (ν_e) must be performed. The KATRIN experiment has released first data and plans to improve the direct measurement of a neutrino mass in tritium β-decay by an order of magnitude (Chapter 6). The ECHO and HOLMES experiments is preparing for a measurement of ν_e. In addition, cosmology is constraining neutrino masses further, as from the observations the sum of all neutrino masses can be obtained (Chapter 13).

A rapidly expanding field is that of neutrino astrophysics (Chapter 12). The solution of the solar neutrino problem by SNO (Chapter 10) due to neutrino oscillations, which has been independently confirmed by KamLAND (Chapter 8) using nuclear reactors, is one of the major milestones in recent history. Borexino has also managed to see not only in real time the pp-neutrinos but could combine and fit all the major four neutrino production reactions. Finally, due to a degeneracy of the opacities and metallicities, an outstanding measurement would be the detection of the CNO in the Sun. This would for the first time experimentally prove that the CNO cycle is realised in nature and due to the neutrino energy up to almost 2 MeV will shed light on particle physics theories by covering a larger region of the survival probability curves. In addition, KamLAND and Borexino made the first observations of geoneutrinos. With greater statistics of geoneutrinos predictive models of the internal heat system of the Earth could be tested.

With new larger underground experiments operating in the MeV energy range,

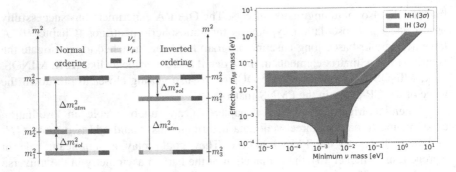

Figure 14.2. Left: The two hierarchical neutrino mass schemes discussed to explain the observation results. The absolute offset with respect to zero is unknown and must be explored by neutrino mass experiments. The choice of one of the schemes is described by a sign on Δm^2. Figure with kind permission of B. Zatschler. Right: The expected effective Majorana neutrino mass to be observed in neutrinoless double beta decay as a function of the lightest neutrino mass eigenstate m_1. The horizontal band at about 50 meV corresponds to the inverted hierarchy and the one splitting to the normal hierarchy. If the neutrino mass is larger than about 0.2 eV the hierachy vanishes and they are almost degenerate. Figure with kind permission of N. Barros.

we are fairly well equipped to observe nearby supernova explosions (Chapter 11). These are combined in the SNEWS network as an early warning system for other observers like astrophysical satellites and telescopes, which are joined by gravitational waves experiments. An extension to much higher neutrino energies arising from various astrophysical sources such as AGNs and GRBs is on its way (Chapter 12). The expected flux is much smaller which has to be compensated by a larger detector volume. Here, the obvious way is to use natural water resources such as lakes, oceans or Antarctic ice. Three so-called neutrino telescopes (Baikal, ANTARES, ICECUBE) are taking data for several years and have already produced spectacular results. All of them aim for upgrades of even larger arrays like GVD, KM3NET and IceCubeGen2. This very high end of the astrophysical neutrino spectrum is accompanied by giant cosmic air shower arrays like the Auger experiment in Argentina to probe the ultra-highest cosmic rays. Additionally, these projects are connected to various optical telescopes and satellites as well as gravitational wave detectors to make it a multi-messenger approach to optimise the gained knowledge. In this way, a first event has been found in the form of a blazar, which is a first good candidate source for the origin of ultra-high energy cosmic rays. On the low energy side, the detection of the cosmic relic neutrino background has not been observed. This 1.95 K relic neutrino background (Chapter 13) has not been detected yet and might be the ultimate challenge for experimental neutrino physics. But as the field has proven in the past, the vital excitement of neutrino physics stems from the fact that you always have to expect the unexpected.

References

[2dfgrs]	Webpage of the 2dFGRS collaboration, http://www.2dfgrs.net/ (Accessed: 18.04.2019)
[Aab18]	Aab A et al. 2018 *Science* **361** 147
[Aab19]	Aab A et al. 2019 *arXiv:1906.07422*
[Aab19a]	Aaboud M et al. 2019 *J. High Energy Phys.* **1901** 016
[Aad12]	Aad G et al., 2012 *Phys. Lett.* B **716** 1
[Aai18]	Aaij R et al., 2018 *Phys. Rev.* D **97** 031101
[Aal99]	Aalseth C E et al. 1999 *Phys. Rev.* C **59** 2108
	Aalseth C E et al. 2002 *Mod. Phys.* A **17** 1475
[Aal13]	Aaltonen T et al. 2013 *Phys. Rev.* D **88** 052018
[Aao19]	Webpage of the Australian Astronomical Observatory, http://aao.gov.au/news-media/media-releases/Supernova1987A-30 (Accessed: 07.09.2019)
[Aar15]	Aartsen M G et al. 2015 *Astrophys. J.* **809** 98
[Aar15a]	Aartsen M G et al. 2015 *Phys. Rev.* D **91** 122004
[Aar17]	Aartsen M G et al. 2017 *The European Physical Journal C* **77** 146 errat. 2019 *The European Physical Journal C* **79** 214
[Aar18]	Aartsen M G et al. 2018 *Science* **361** 6398
[Aar18a]	Aartsen M G et al. 2018 *Science* **361** 147
[Aar18b]	Aartsen M G et al. *Advances in Space Research*62102018
[Aar19]	Aartsen M G et al. 2019 *The European Physical Journal C* **79** 234
[Aba04]	Abazov V M et al. 2004 *Phys. Rev.* D **70** 092008
[Aba10]	Abazov V M et al. 2010 *Phys. Rev.* D **82** 032001
[Aba10a]	Abazov V M et al. 2010 *Phys. Rev. Lett.* **104** 064
[Abb08]	Abbasi R et al. 2008 *Phys. Rev. Lett.* **100** 101101
[Abb09]	Abbasi R et al. 2009 *Astrophys. J.* **701** L47
[Abb09a]	Abbasi R et al. 2009 *Phys. Rev.* D **79** 062001
[Abb09b]	Abbasi R et al. 2009 *Phys. Rev. Lett.* **102** 201302
[Abb11]	Abbasi R et al. 2011 *Astron. Astrophys.* **535** A19
[Abb16]	Abbott B P et al. 2016 *Phys. Rev. Lett.* **116** 061102
[Abb17]	Abbott B P et al. 2017 *Astrophys. J.* **849** L13
[Abb19]	Abbott T M C et al. 2019 *Astrophys. J. Lett.* **872** L30
[Abd96]	Abdurashitov J N et al. 1996 *Phys. Rev. Lett.* **77** 4708
[Abd99]	Abdurashitov J N et al. 1999 *Phys. Rev.* C **59** 2246
[Abd06]	Abdurashitov J N et al. 2006 *Phys. Rev.* C **73** 045805
[Abd09]	Abdurashitov J N et al. 2009 *Phys. Rev.* C **80** 015807
[Abe81]	Abela R et al. 1981 *Phys. Lett.* B **105** 263
[Abe05]	Abe K et al. 2005 *Preprint* hep-ex/0507045

[Abe06]   Abe K et al. 2006 *Phys. Rev. Lett.* **97** 171801
[Abe08]   Abe K et al. 2008 *Phys. Rev. Lett.* **100** 221803
[Abe08a]   Abe Y et al. 2008 *Phys. Rev* D **77** 052001
[Abe08a]   Abe K et al. 2011 *Nucl. Instrum. Methods* A **659** 106
[Abe08a]   Abe K et al. 2012 *Phys. Rev.* D **90** 072005
[Abe12]   Abe Y et al. 2012 *Phys. Rev.* D **86** 052008
[Abe14]   Abe K et al. 2014 *Phys. Rev.* D **90** 072005
[Abe14a]   Abe K et al. 2014 *Phys. Rev. Lett.* **112** 061802
[Abe16]   Abe K et al. 2016 *Phys. Rev.* D **94** 052010
[Abe16a]   Abe Y et al. 2016 *J. High Energy Phys.* **1601** 163
[Abe17]   Abe K et al. 2017 *Phys. Rev.* D **95** 012004
[Abe18]   Abe K et al. 2018 *Phys. Rev. Lett.* **121** 171802
[Abe18a]   Abeysekara A U et al. 2018 *Astrophys. J.* **861** L20
[Abe18b]   Abe K et al. 2018 *Phys. Rev.* D **97** 072001
[Abe19]   Abe K et al. 2019 *Phys. Rev.* D **100** 052006
[Abg14]   Abgrall N et al. 2014 *Adv. High. Energy Phys.* 365432
[Abg17]   Abgrall N et al. 2017 *arXiv:1709.01980*
[Abr82]   Abramowicz H et al. 1982 *Phys. Lett.* B **109** 115
[Abr99]   Abramowicz H and Caldwell A 1999 *Rev. Mod. Phys.* **71** 1275
[Abr07]   Abraham J et al. 2007 *Science* **318** 938
[Abr08]   Abraham J et al. 2008 *Phys. Rev. Lett.* **101** 061101
[Abt16]   Abt I 2016 *Annu. Rev. Nucl. Part. Phys.* **66** 377
[Acc15]   Acciarri R et al. 2015 *Preprint* arXiv:1512.06148
[Ace18]   Acero M A et al. 2018 *Phys. Rev.* D **98** 032012
[Ace19]   Acero M A et al. 2019 *Astropart. Phys.* **98** 032012
[Ach04]   Achert P et al. 2004 *Phys. Lett.* B **598** 15
[Ach07]   Achterberg A et al. 2007 *Phys. Rev.* D **76** 042008, err. ibid 77:089904 (2008)
[Ack98]   Ackerstaff K et al. 1998 *Eur. Phys. J.* C **5** 229
[Ack98a]   Ackerstaff K et al. 1998 *Phys. Rev. Lett.* **81** 5519
[Ack13]   Ackermann K H et al. 2013 *Eur. Phys. J.* C **73** 2330
[Acq09]   Acquafredda R et al. 2009 Journal*JINST*4P040182009
[Ada00]   Adams T et al. 2000 *Phys. Rev.* D **61** 092001
[Ada07]   Adamson P et al. 2007 *Phys. Rev.* D **76** 052003
[Ada08]   Adamson P et al. 2008 *Phys. Rev. Lett.* **101** 131802
[Ada08a]   Adamson P et al. 2008 *Phys. Rev.* D **77** 072002
[Ada08b]   Adamson P et al. 2008 *Phys. Rev. Lett.* **77** 221804
[Ada10a]   Adamson P et al. 2010 *Phys. Rev.* D **82** 051102
[Ada11]   Adamson P et al. 2011 *Phys. Rev.* D **83** 032001
[Ada13]   Adams S M et al. 2013 *Preprint* arXiv:1306.0559
[Ada13a]   Adamson P et al. 2013 *Phys. Rev. Lett.* **110** 171801
[Ada13b]   Adamson P et al. 2013 *Phys. Rev. Lett.* **110** 251801
[Ada16]   Adamson P et al. 2016 *Nucl. Instrum. Methods* A **806** 279
[Ada16a]   Adam R et al. 2016 *Astron. Astrophys.* **594** A1
[Ada16b]   Adamson P et al. 2016 *Phys. Rev.* D **94** 072006
[Ade86]   Aderholz M et al. 1986 *Phys. Lett.* B **173** 211
[Ade14]   Ade P A R et al. 2014 *Astron. Astrophys.* **571** A15
[Ade15]   Adey D et al. 2015 *Annu. Rev. Nucl. Part. Phys.* **65** 145
[Ade16]   Ade P A R et al. 2016 *Astron. Astrophys.* **594** A14
[Ade18]   Adey D et al. 2018 *Phys. Rev. Lett.* **173** 211

[Adl66]   Adler S L 1966 *Phys. Rev.* **143** 1144
[Adl11]   Adelberger E et al. 2011 *Rev. Mod. Phys.* **83** 195
[Adl00]   Adler S et al. 2000 *Phys. Rev. Lett.* **84** 3768
[Adu17]   Aduszkiewicz A et al., 2017 *The European Physical Journal C* **77** 671
[Aga18]   Agafonova N et al. 2018 *Phys. Rev. Lett.* **120** 211801
[Age07]   Ageron M et al. 2007 *Nucl. Instrum. Methods* A **581** 695
[Agl87]   Aglietta M et al. 1987 *Europhys. Lett.* **3** 1315
[Agl89]   Aglietta M et al. 1989 *Europhys. Lett.* **8** 611
[Aga14]   Agafonava N et al. 2014 *Eur. Phys. J. C* **74** 2933
[Agh18]   Aghanim N et al. 2018 *arXiv:1807.06209*
[Agh19]   Aghanim N et al. 2019 *arXiv:1907.12875*
[Ago15]   Agostini M et al. 2015 *Eur. Phys. J. C* **75** 416
[Ago15a]   Agostini M et al. 2015 *Phys. Rev. D* **92** 031101
[Ago17]   Agostini M et al. 2017 *Phys. Rev. D* **96** 091103
[Ago17a]   Agostini M et al. 2017 *arXiv:1707.09279*
[Ago17b]   Agostini M et al. 2017 *Astrophys. J.* **92** 21
[Ago18]   Agostini M et al. 2018 *Nature* **562** 505
[Ago18a]   Agostini M et al. 2018 *Phys. Rev. Lett.* **120** 132503
[Ago19]   Agostini M et al. 2019 *Science* **365** 1445
[Ago20]   Agostini M et al. 2020 *Phys. Rev. D* **101** 012009
[Agr96]   Agraval V et al. 1996 *Phys. Rev. D* **53** 1314
[Agu01]   Aguilar A A et al. 2001 *Phys. Rev. D* **64** 112007
[Agu07]   Aguilar-Arevalo A A et al. 2007 *Phys. Rev. Lett.* **98** 231801
[Agu08]   Aguilar-Arevalo A A et al. 2008 *Phys. Rev. Lett.* **100** 032301
[Agu08a]   Aguilar-Arevalo A A et al. 2008 *Phys. Lett.* B **664** 41
[Agu09]   Aguilar-Arevalo A A et al. 2009 *Nucl. Instrum. Methods* A **599** 28
[Agu09a]   Aguilar-Arevalo A A et al. 2009 *Phys. Rev. Lett.* **103** 061802
[Agu09b]   Aguilar-Arevalo A A et al. 2009 *Phys. Rev. Lett.* **103** 111801
[Agu09c]   Aguilar-Arevalo A A et al. 2009 *Phys. Rev. Lett.* **102** 101802
[Agu19]   Aguilar-Arevalo A A et al. 2019 2019 *Phys. Lett.* B **798** 134980
[Agu10]   Aguilar J A et al. 2010 *Astropart. Phys.* **33** 86 err. Journal*Astropart. Phys.*341852010
[Aha03]   Aharonian F A 2003 *Very High Energy Cosmic Gamma Radiation* (World Scientific)
[Aha04]   Aharmim B et al. 2004 *Phys. Rev. D* **70** 093014
[Aha05]   Aharmim B et al. 2005 *Phys. Rev. C* **72** 055502
[Aha06]   Aharmim B et al. 2006 *Astrophys. J.* **653** 1545
[Aha06a]   Aharonian F A et al. 2006 *Phys. Rev. Lett.* **97** 221102 err. 97:249901 (2006)
[Aha07]   Aharmim B et al. 2007 *Phys. Rev. C* **75** 045502
[Aha08]   Aharmim B et al. 2008 *Phys. Rev. Lett.* **101** 111301
[Aha09]   Aharmim B et al. 2009 *Phys. Rev. D* **80** 012001
[Aha10]   Aharmim B et al. 2010 *Phys. Rev. C* **81** 055504
[Ahl10]   Ahlen S et al. 2010 *Int. J. Mod. Phys.* **25** 1
[Ahl10a]   Ahlers M 2010 *FERMILAB-FN-0847-A*
[Ahm94]   Ahmed T 1994 *Phys. Lett.* B 324 241
[Ahm01]   Ahmad Q R et al. 2001 *Phys. Rev. Lett.* **87** 071301
[Ahm02]   Ahmad Q R et al. 2002 *Phys. Rev. Lett.* **89** 011301
[Ahm02a]   Ahmad Q R et al. 2002 *Phys. Rev. Lett.* **89** 011302
[Ahm04]   Ahmed S N et al. 2004 *Phys. Rev. Lett.* **92** 181301

[Ahm16]   Ahmadi M et al. 2016 2016 *Nature* **542** 506
[Ahm17]   Ahmadi M et al. 2017 2017 *Nature* **548** 66
[Ahm17a]   Ahmed S et al. 2017 *Pramana* **88** 79
[Ahm18]   Ahmadi M et al. 2018 *Nature* **561** 21
[Ahn03]   Ahn M H et al. 2003 *Phys. Rev. Lett.* **90** 041801
[Ahn06]   Ahn M H et al. 2006 *Phys. Rev.* D **74** 072003
[Ahn12]   Ahn J K et al. 2012 *Phys. Rev. Lett.* **108** 1918022
[Ahr87]   Ahrens L A et al. 1987 *Phys. Rev.* D **35** 785
[Ahr90]   Ahrens L A et al. 1990 *Phys. Rev.* D **41** 3297
[Ait89]   Aitchison I J R and Hey A J G 1989 *Gauge Theories in Particle Physics* (Bristol: Adam Hilger)
[Ait07]   Aitchison I J R 2007 *Supersymmetry in Particle Physics: An Elementary Introduction* (Cambridge:Cambridge University Press)
[Ake19]   Aker M et al. 2019 *arXiv:1909.06048*
[Akh88]   Akhmedov E Kh and Khlopov M Yu 1988 *Mod. Phys. Lett.* A **3** 451
[Akh97]   Akhmedov E Kh 1997 *Preprint* hep-ph/9705451, Proc. 4th Int. Solar Neutrino Conference, Heidelberg
[Akh03]   Akhmedov E Kh and Pulido J 2003 *Phys. Lett.* B **553** 7
[Akh09]   Akhmedov E Kh and Smirnov A Yu 2009 PAN **72** 1363
[Aki17]   Akimov D et al. 2017 *Science* **357** 1123
[Akr18]   Akrami Y et al. 2018 *arXiv:1807.06205*
[Alb82]   Albrecht A and Steinhardt P J 1982 *Phys. Rev. Lett.* **48** 1220
[Alb92]   Albrecht H et al. 1992 *Phys. Lett.* B **292** 221
[Alb00]   Albright C et al. 2000 *Preprint* hep-ex/0008064
[Alb06]   Albrecht A et al. 2006 *Preprint* astro-ph/0609591
[Alb15]   Albacete J L et al. 2015 *Phys. Rev.* D **92** 014027
[Alb18]   Albert J B et al. 2018 *Phys. Rev. Lett.* **120** 072701
[Alc00]   Alcaraz J et al. 2000 *Phys. Lett.* B **472** 215
[Alc07]   Alcaraz J et al. 2007 *Preprint* arXiv:0712:0929
[Ald18]   Alduino C et al. 2018 *Phys. Rev. Lett.* **120** 132501
[Ale87]   Alekseev E N, Alexeyeva L N, Krivosheina I V and Volchenko V I 1987 *JETP Lett.* **45** 589
[Ale87a]   Alekseev E N, Alexeyeva L N, Krivosheina I V and Volchenko V I 1987 *Proc. ESO Workshop SN1987A, (Garching near Munich, 6–8 July 1987)* p 237
[Ale88]   Alekseev E N Alexeyeva L N, Krivosheina I V and Volchenko V I 1988 *Phys. Lett.* B **205** 209
[Ale88a]   Alekseev E N, Alexeyeva I N, Krivosheina I V and Volchenko V I 1988 *Neutrino Physics* ed H V Klapdor and B Povh (Heidelberg: Springer) p 288
[Ale94]   Alexander D R and Ferguson J W 1994 *Astrophys. J.* **437** 879
[Ale96]   Alexander G et al. 1996 *Z. Phys.* C **72** 231
[Ale98]   Alessandrello A et al. 1998 *Phys. Lett.* B **420** 109
[Ale99]   Alessandrello A et al. 1999 *Phys. Lett.* B **457** 253
[Ale12]   Aleksic J et al. 2012 *Astropart. Phys.* **35** 435
[Ale13]   Alekhin S et al. 2013 *Phys. Lett.* B **720** 172
[Ale17]   Alekhin S et al. 2017 *Phys. Rev.* D **96** 014011
[Ale19]   Webpage of the ALEPH collaboration, http://aleph.web.cern.ch/aleph/aleph/newpub/physics/hadronline_aleph92.gif (Accessed: 20.06.2019)
[All84]   Allasia D et al. 1984 *Nucl. Phys.* B **239** 301

[All85a]   Allasia et al. 1985 *Z. f. Phys.* C **28** 321
[All87]   Allaby J V et al. 1987 *Z. Phys.* C **36** 611
[All88]   Allaby J V et al. 1988 *Z. Phys.* C **38** 403
[All93]   Allen R C et al. 1993 *Phys. Rev.* D **47** 11
[All97]   Allen C et al. 1997 *Preprint* astro-ph/9709223
[All08]   Allekotte I et al. 2008 *Nucl. Instrum. Methods* A **586** 409
[All18]   Allanach B 2018 European School of High Energy Physics
[Alp48]   Alpher R A and Herman R C 1948 *Nature* **162** 774
[Alp15]   Alpert B et al. 2015 *Eur. Phys. J.* C **75** 112
[Als03]   Alsharoa M M et al. 2003 *Phys. Rev. ST Acc. beams* 6,081001
[Alt77]   Altarelli G and Parisi G 1977 *Nucl. Phys.* B **126** 298
[Alt98]   Altegoer J et al. 1998 *Nucl. Instrum. Methods* A **404** 96
[Alt03]   Altarelli G and Feruglio F 2003 *Springer Tracts Mod. Phys.* 190,169
[Alt05]   Altmann M et al. 2005 *Phys. Lett.* B **616** 174
[Alt07]   Alt C et al. 2007 *Eur. Phys. J.* C **49** 897
[Alv19]   Alvis S I 2019 *arXiv*:1902.02299
[Ama87]   Amaldi U et al. 1987 *Phys. Rev.* D **36** 1385
[Ama91]   Amaldi U, deBoer W and Färstenau H 1991 *Phys. Lett.* B **260** 447
[Amb99]   Ambrosini G et al. 1999 *Eur. Phys. J.* C **10** 605
[Amb01]   Ambrosio M et al. 2001 *Astrophys. J.* **546** 1038
[Amb02]   Ambrosio M et al. 2002 *Nucl. Instrum. Methods* A **486** 663
[Ame04]   Amerio S et al. 2004 *Nucl. Instrum. Methods* A **527** 329
[Ams08]   Amsler C et al. 2008 Review of particle properties *Phys. Lett.* B 667 1.
[Ams97]   Amsler C et al. 1997 *Nucl. Instrum. Methods* A **396** 115
[An12]   An F P et al. 2012 *Phys. Rev. Lett.* **108** 171803
[An16]   An F P et al. 2016 *Nucl. Instrum. Methods* A **811** 16
[An16a]   An F P et al. 2016 *J. Phys.* G **43** 030401
[Anc10]   Anchordoqui L A, Montaruli T 2010 *Annu. Rev. Nucl. Part. Phys.* **60** 129
[And00]   Andres E et al. 2000 *Astropart. Phys.* **13** 1
[And01]   Andres E et al. 2001 *Nucl. Phys.* B *(Proc. Suppl.)* **91** 423(AMANDA-coll.)
[And03]   Ando S and Sato K 2003 *Phys. Rev.* D **67** 023004
[And05]   Ando S, Beacom J F and Yüksel H 2005 *Phys. Rev. Lett.* **95** 171101
[And16]   Andringa S et al. 2016 *Adv. High Energy Phys.* 6194250
[And18]   Andreev V et al. 2018 *Nature* **562** 355
[Ang98]   Angelopoulos A et al. 1998 *Phys. Lett.* B **444** 43
[Ang04]   Angrik J et al. 2004 KATRIN Design Report vol 7090 (Forschungszentrum
             Karlsruhe: Wissenschaftliche Berichte)
[Ani04]   Anisimovsky V V et al. 2004 *Phys. Rev. Lett.* **93** 031801
[Ann18]   Annala E et al. 2018 *Phys. Rev. Lett.* **120** 172703
[Ans92]   Anselmann P et al. 1992 *Phys. Lett.* B **285** 376
[Ans92a]   Anselmann P et al. 1992 *Phys. Lett.* B **285** 390
[Ans95a]   Anselmann P et al. 1995 *Phys. Lett.* B **342** 440
[Ans95b]   Anselmann P et al. 1995 *Nucl. Phys.* B *(Proc. Suppl.)* **38** 68
             Anselmann P et al. 1994 *Proc. 16th Int. Conf. on Neutrino Physics
             and Astrophysics, NEUTRINO '94* ed A Dar, G Eilam and M Gronau
             (Amsterdam: North-Holland)
[Ant97]   Antares Proposal 1997 *Preprint* astro-ph/9707136
[Ant99]   Antonioli P et al. 1999 *Nucl. Instrum. Methods* A **433** 104
[Ant04]   Antonioli P et al. 2004 *New J. Phys.* **6** 114

[Ant05]   Antoni T et al. 2005 *Astropart. Phys.* **24** 1

[Aok03]   Aoki M et al. 2003 *Phys. Rev.* D **67** 093004

[Apo98]   Apollonio M et al. 1998 *Phys. Lett.* B **420** 397

[Apo99]   Apollonio M et al. 1999 *Phys. Lett.* B **466** 415

[Apo02]   Apollonio M et al. 2002 *Preprint* hep-ph/0210192, CERN Yellow Report

[Apo03]   Apollonio M et al. 2003 *Eur. Phys. J.* C **27** 331

[App00]   Appel R et al. 2000 *Phys. Rev. Lett.* **85** 2877

[Apr19]   Aprile E et al. 2019 *Nature* **568** 532

[Ara97]   Arafune J and Sato J 1997 *Phys. Rev.* D **55** 1653

[Ara05]   Araki T et al. 2005 *Phys. Rev. Lett.* **94** 081801

[Ara05a]   Araki T et al. 2005 *Nature* **439** 499 Figure 3

[Ard06]   Ardellier F et al. 2006 *Preprint* hep-ex/0606025

[Arc15]   Arceo Diaz S et al. 2015 *Astropart. Phys.* **70** 1

[Ard06]   Ardouin D et al. 2006 *Astropart. Phys.* **26** 341

[Arg10]   Argyriades J et al. 2010 *Nucl. Phys.* A **847** 168

[Arm98]   Armbruster B et al. 1998 *Phys. Rev. Lett.* **81** 520

[Arm02]   Armbruster B et al. 2002 *Phys. Rev.* D **65** 112001

[Arn77]   Arnett W D 1977 *Astrophys. J.* **218** 815

            Arnett W D 1977 *Astrophys. J. Suppl.* **35** 145

[Arn78]   Arnett W D 1978 *The Physics and Astrophysics of Neutron Stars and Black
            Holes* ed R Giacconi and R Ruffins (Amsterdam: North-Holland) p 356

[Arn80]   Arnett W D 1980 *Ann. NY Acad. Sci.* **336** 366

[Arn83]   Arnison G et al. 1983 UA1-coll. *Phys. Lett.* B **122** 103

            Arnison G et al. 1983 *Phys. Lett.* B **126** 398

[Arn89]   Arnett W D et al. 1989 *Ann. Rev. Astron. Astrophys.* **27** 629

[Arn91]   Arnett W D 1991 *Astrophys. J.* **383** 295

[Arn94]   Arneodo M et al. 1994 *Phys. Rev.* D **50** R1

[Arn96]   Arnett W D 1996 *Supernovae and nucleosynthesis* (Princeton: Princeton
            University Press)

[Arn98]   Arnold R et al. 1998 *Nucl. Phys.* A **636** 209

[Arn99]   Arnold R et al. 1999 *Nucl. Phys.* A **658** 299

[Arn05]   Arnold R et al. 2005 *Phys. Rev. Lett.* **95** 18

[Arn06]   Arneodo F et al. 2006 *Phys. Rev.* D **74** 112001

[Arn14]   Arnold R et al. 2014 *Nucl. Phys.* A **925** 25

[Arn15]   Arnold R et al. 2015 *Phys. Rev.* D **92** 072011

[Arp96]   Arpesella C et al. 1996 *Nucl. Phys.* B *(Proc. Suppl.)* **48** 375

            Arpesella C et al. 1995 *Proc. 4th Int. Workshop on Theoretical and
            Phenomenological Aspects of Underground Physics (Toledo, Spain)* ed A
            Morales, J Morales and J A Villar (Amsterdam: North-Holland)

[Arp98]   Arpesella C et al. 1998 *Phys. Rev.* C **57** 2700

[Arp08]   Arpesella C et al. 2008 *Phys. Lett.* B **658** 101

[Arp08a]   Arpesella C et al. 2008 *Phys. Rev. Lett.* **101** 091302

[Arr94]   Arroyo C G et al. 1994 *Phys. Rev. Lett.* **72** 3452

[Art08]   Artamonov A V et al. 2008 *Phys. Rev. Lett.* **101** 191802

[Art15]   Artamonov A V et al. 2015 *Phys. Rev.* D **98** 052001 err. 2015 *Phys. Rev.* D **91**
            059903

[Ash62]   Asharyan G A 1962 *Sov. J. JETP* **14** 441

[Ash75]   Ashley G K et al. 1975 *Phys. Rev.* D **12** 20

            Asharyan G A 1965 *Sov. J. JETP* **21** 658

[Ash05]	Ashie Y et al. 2005 *Phys. Rev.* D **71** 112005
[Ask19]	Askins M et al. 2019 *arXiv:1911.03501*
[Asp09]	Asplund M et al. 2009 *Annu. Rev. Astron. Astrophys.* **47** 481
[Ass96]	Assamagan K et al. 1996 *Phys. Rev.* D **53** 6065
[Ast00]	Astier P et al. 2000 *Nucl. Phys.* B **588** 3
[Ast00a]	Astier P et al. 2000 *Phys. Lett.* B **486** 35
[Ast01]	Astier P et al. 2001 *Nucl. Phys.* B **605** 3
[Ast01a]	Astier P et al. 2001 *Nucl. Phys.* B **611** 3
[Ast02]	Astier P et al. 2002 *Phys. Lett.* B **526** 278
[Ast03]	Astier P et al. 2003 *Phys. Lett.* B **570** 19
[Asz06]	Asztalos S J et al. 2006 *Annu. Rev. Nucl. Part. Phys.* **56** 293
[Ath80]	Atherton H W et al. 1980 *Preprint* CERN 80-07
[Ath97]	Athanassopoulos C et al. 1997 *Nucl. Instrum. Methods* A **388** 149
[Ath98]	Athanassopoulos C et al. 1998 *Phys. Rev.* C **58** 2489
[Ath17]	Athron P et al. 2017 *The European Physical Journal C* **77** 824
[Aub04]	Aubert B et al. 2004 *Phys. Rev. Lett.* **93** 131801
[Aud03]	Audi G, Wapstra A H and Thibault C 2003 *Nucl. Phys.* A **729** 337
[Aue01]	Auerbach L B et al. 2001 *Phys. Rev.* D **63** 112001
[Auf94]	Aufderheide M et al. 1994 *Phys. Rev.* C **49** 678
[Aug96]	The Pierre Auger Design Report 1996, Fermilab-PUB-96-024
[Aut99]	Autin B, Blondel A and Ellis J (ed) 1999 *Preprint* CERN 99-02
[Avr09]	Avrorin A et al. 2009 *Preprint* arXiv:0909.5562
[Avr17]	Avrorin A et al. 2017 *PoS (ICRC17)*1034
[Atw09]	Atwood W B et al. 2009 *Astrophys. J.* **697** 1071
[Ayn09]	Aynutdinov V et al. 2009 *Nucl. Instrum. Methods* A **602** 14
[Ays01]	Aysto J et al. 2001 *Preprint* hep-ph/0109217
[Bab87]	Babu K S and Mathur V S 1987 *Phys. Lett.* B **196** 218
[Bab88]	Babu K S 1988 *Phys. Lett.* B **203** 132
[Bab95]	BaBar experiment 1995 *Tech. Design Report* SLAC-R-95-437
[Bab03]	Baby L T et al. 2003 *Phys. Rev. Lett.* **90** 022501
	Baby L T et al. 2003 *Phys. Rev.* C **67** 065805
[Bab15]	Babu K S and Khan S 2015 *Phys. Rev.* D **92** 075018
[Bac04]	Back H 2004 *Noon 2004 Conference Proceedings*
[Bag83]	Bagnaia P et al. 1983 UA2-coll. *Phys. Lett.* B **129** 130
[Bah88]	Bahcall J N and Ulrich R K 1988 *Rev. Mod. Phys.* **60** 297
[Bah89]	Bahcall J N 1989 *Neutrino Astrophysics* (Cambridge: Cambridge University Press)
[Bah92]	Bahcall J N and Pinsonneault M H 1992 *Rev. Mod. Phys.* **64** 885
[Bah95]	Bahcall J N and Pinsonneault M H 1995 *Rev. Mod. Phys.* **67** 781
[Bah97]	Bahcall J N and Ostriker J P (ed) 1997 *Unsolved Problems in Astrophysics* (Princeton, NJ: Princeton University Press)
[Bah01]	Bahcall J N, Pinseaunault M H and Basu S 2001 *Astrophys. J.* **550** 990
[Bah01a]	Bahcall J N and Waxman E 2001 *Phys. Rev.* D **64** 023002
[Bah02]	Bahcall J N 2002 *Phys. Rev.* C **65** 025801
[Bah03]	Bahcall J N 2003 *Nucl. Phys. B (Proc. Suppl.)* **118** 77
[Bah03a]	Bahcall J N, Gonzalez-Garcia M C and Pena-Garay C 2003 *J. High Energy Phys.* **0302** 009
[Bah03b]	Bahcall J N and Pena-Garay C 2003 *J. High Energy Phys.* **0311** 004
[Bah06]	Bahcall J N, Serenelli A M, Basu S 2006 *Astrophys. J. Suppl.* **165** 400

[Bak18]   Bak G et al. 2018 *Phys. Rev. Lett.* **121** 201801
[Bal94]   Baldo-Ceolin M et al. 1994 *Z. Phys.* C **63** 409
[Bal00]   Balkanov V et al. 2000 *Astropart. Phys.* **14** 61
[Bal01]   Balkanov V A et al. 2001 (Baikal-coll.) *Nucl. Phys. B (Proc. Suppl.)* **91** 438
[Bal06]   Balata M et al. 2006 *Eur. Phys. J.* C **47** 21
[Bal16]   Baldini A M et al. 2016 *Eur. Phys. J.* C **76** 434
[Bal18]   Baldini A M et al. 2018 *Eur. Phys. J.* C **78** 380
[Bam15]   Bambi C and Dolgov A D 2015 *Introduction to particle cosmology* Springer
[Ban83]   Banner M et al. (UA2-coll.) 1983 *Phys. Lett.* B **122** 476
[Ban09]   Bandyopadhyay A et al. 2009 *Rep. Prog. Phys.* **72** 1
[Bar76]   Barnett R M 1976 *Phys. Rev. Lett.* **36** 1163
[Bar83]   Baruzzi V et al. 1983 *Nucl. Instrum. Methods* A **207** 339
[Bar85]   Baron E, Cooperstein J and Kahana S 1985 *Phys. Rev. Lett.* **55** 126
[Bar87]   Barbiellini G and Cocconi G 1987 *Nature* **329** 21
[Bar88]   Barbieri R and Mohapatra R N 1988 *Phys. Rev. Lett.* **61** 27
[Bar95]   Barabash A S 1995 *Proc. ECT Workshop on Double Beta Decay and Related Topics (Trento)* ed H V Klapdor-Kleingrothaus and S Stoica (Singapore: World Scientific) p 502
[Bar97]   Barabash A S 1997 *Proc. Neutrino'96* (Singapore: World Scientific) p 374
[Bar98]   Barate R et al. (ALEPH-coll.)1998 *Eur. Phys. J.* C **2** 395
[Bar99]   Barger V et al. 1999 *Phys. Lett.* B **462** 109
[Bar01]   Barish B C 2001 *Nucl. Phys. B (Proc. Suppl.)* **91** 141
[Bar02]   Barger V, Marfatia D and Whisnant K 2002 *Phys. Rev.* D **65** 073023
[Bar05]   Barbier R et al. 2005 *Preprint* arXiv:0406039
[Bar06]   Barr G et al. 2006 *Phys. Rev.* D **74** 094009
[Bar06a]   Barwick S W et al. 2006 *Phys. Rev. Lett.* **96** 171101
[Bar07]   Barr G et al. 2007 *Eur. Phys. J.* C **49** 919
[Bar08]   Barabash A S et al. 2008 *Nucl. Phys.* A **807** 269
[Bar09]   Barabash A S et al. 2009 *Phys. Rev.* C **79** 045501
[Bar09a]   Barea J and Iachello F 2009 *Phys. Rev.* C **79** 044301
[Bar14]   Baron J et al. 2014 *Science* **343** 269
[Bar14a]   Barros N, Thurn J and Zuber K 2014 *J. Phys.* G **41** 115105
[Bar18]   Barabash A S et al. 2018 *Phys. Rev.* D **98** 044301
[Bar18a]   Baracchini E et al. 2018 *Preprint* arXiv:1808.01892
[Bas09]   Basu S et al. 2009 *Astrophys. J.* **699** 1403
[Bat99]   Battistoni G and Lipari P 1999 *Italian Phys.Soc.Proc.* **65** 547
[Bat03]   Battistoni G et al. 2003 *Astropart. Phys.* **19** 269
              Battistoni G et al. 2003 *Astropart. Phys.* **19** 291 (erratum)
[Bau97]   Baudis L et al. 1997 *Phys. Lett.* B **407** 219
[Bau09]   Baumann D 2009 *Preprint* arXiv:0907:5424
[Bay12]   Bays K et al. 2012 *Phys. Rev.* D **85** 052007
[Baz95]   Bazarko A O et al. 1995 *Z. Phys.* C **65** 189
[Bea01]   Beacom J F, Boyd R and Mezzacappa A 2001 *Phys. Rev.* D **63** 073011
[Bea02]   Beacom J F, Farr W M and Vogel P 2002 *Phys. Rev.* D **66** 033001
[Bea04]   Beacom J F and Vagins M R 2004 *Phys. Rev. Lett.* **93** 171101
[Bea10]   Beacom J F 2010 *Annu. Rev. Nucl. Part. Phys.* **60** 439
[Bea17]   Beacom J F 2017 *Chin.Phys.* G **41** 023002
[Bec92]   Becker-Szendy R et al. 1992 *Phys. Rev. Lett.* **69** 1010
[Bec08]   Becker J 2008 *Phys. Rev.* **458** 173

[Bed99]	Bednyakov V, Faessler A and Kovalenko S 1999 *Preprint* hep-ph/9904414
[Bed13]	Beda A G et al. 2013 *Phys. Part. and Nucl. Lett.* **10**, 139
[Beh69]	Behrens H and Janecke J 1969 in Landolt-Börnstein Bd.I/IV, Springer Publ.
[Beh09]	Behere A et al. 2009 *Nucl. Instrum. Methods* A **604** 784
[Bel95]	Belle experiment 1995 *Tech. Design Report* KEK-Rep 95-1
[Bel01]	Bellotti E 2001 *Nucl. Phys. B (Proc. Suppl.)* **91** 44
[Bel09]	Bellini G Talk presented at 31st Neutrino School, Erice 2009
[Bel10]	Bellini G et al. 2010 *Phys. Lett.* B **687** 299
[Bel11]	Bellini G et al. 2011 *Phys. Lett.* B **696** 191
[Bel14]	Bellini G et al. 2014 *Nature* **512** 383
[Bel14a]	Bellini G et al. 2014 *Phys. Rev.* D **89** 112007
[Bel15]	Bellini G et al. 2015 2015 *Phys. Rev.* D **92** 031101
[Bem02]	Bemporad C, Gratta G and Vogel P 2002 *Rev. Mod. Phys.* **74** 297
[Bem18]	Bemmerer D et al. 2018 *Preprint* arXiv:1810.08201
[Ben96]	Bennett C L et al. 1996 *Astrophys. J.* **464** L1
[Ben03]	Bennett C L et al. 2003 *Astrophys. J. Suppl.* **148** 1 WMAP Science Team
[Ben03a]	Benoit A et al. 2003 *Astron. Astrophys.* **398** L19
[Ben09]	Bennett G W et al. 2009 *Phys. Rev.* D **80** 052008
[Ber86]	Bernadi G et al. 1986 *Phys. Lett.* B **166** 479
[Ber87]	Berge J P et al. 1987 *Z. Phys.* C **35** 443
[Ber88]	Bernadi G et al. 1988 *Phys. Lett.* B **203** 332
[Ber90]	Berger C et al. 1990 *Phys. Lett.* B **245** 305
	Berger C et al. 1989 *Phys. Lett.* B **227** 489
[Ber90a]	Berezinskii V S et al. 1990 *Astrophysics of Cosmic Rays* (Amsterdam: North-Holland)
[Ber91]	Berezinksy V S 1991 *Nucl. Phys. B (Proc. Suppl.)* **19** 187
[Ber91a]	van den Bergh S and Tammann G A 1991 *Ann. Rev. Astron. Astrophys.* **29** 363
[Ber93]	Bernatowicz T et al. 1993 *Phys. Rev.* C **47** 806
[Ber93a]	Berthomieu G et al. 1993 *Astron. Astrophys.* **268** 775
[Ber98]	Bergstrom L, Edsjo J and Gondolo P 1998 *Phys. Rev.* D **58** 103519
[Ber99]	Bergström L and Goobar A 1999 *Cosmology and Particle Astrophysics* (New York: Wiley)
[Ber02]	Bertou X et al. 2002 *Astropart. Phys.* **17** 183
[Ber02a]	Bernard V, Elouadrhiri L and Meissner U G 2002 *J. Phys.* G **28** R1
[Ber02b]	Bersanelli M, Maino D and Mennella 2002 *Riv. Nuovo Cim.* **25** 1
[Ber05]	Bertone G, Hooper D and Silk J 2005 *Phys. Rep.* **405** 279
[Ber08]	Bertaina M et al. 2008 *Nucl. Instrum. Methods* A **588** 162
[Ber15]	Bernabeu J and Martinez-Vidal F 2015 *Rev. Mod. Phys.* **87** 165
[Ber16]	Bergström J et al. 2016 *J. High Energy Phys.* **1603** 132
[Bet38]	Bethe H A and Critchfield C L 1938 *Phys. Rev.* **54** 248, 862
[Bet39]	Bethe H A 1939 *Phys. Rev.* **55** 434
[Bet79]	Bethe H A, Brown G E, Applegate J and Lattimer J M 1979 *Nucl. Phys.* A **324** 487
[Bet82]	Bethe H A 1982 *Essays in Nuclear Astrophysics* ed C A Barnes, D D Clayton and D N Schramm (Cambridge: Cambridge University Press)
[Bet85]	Bethe H A and Wilson J F 1985 *Astrophys. J.* **295** 14
[Bet86]	Bethe H A 1986 *Phys. Rev. Lett.* **56** 1305
[Bet86a]	Bethe H A 1986 *Proc. Int. School of Physics 'Enrico Fermi', Course XCI (1984)* ed A Molinari and R A Ricci (Amsterdam: North-Holland) p 181

[Bet88]   Bethe H A 1988 *Ann. Rev. Nucl. Part. Sci.* **38** 1

[Bet90]   Bethe H A 1990 *Rev. Mod. Phys.* **62** 801

[Bet13]   Betts S et al. 2013 *arXiv:1307.473*

[Bha00]   Bhattachargjee P and Sigl G 2000 *Phys. Rep.* **327** 109

[Bie64]   Bienlein J K et al. 1964 *Phys. Lett.* **13** 80

[Bil78]   Bilenky S M and Pontecorvo B 1978 *Phys. Rev.* **41** 225

[Bil80]   Bilenky S M, Hosek J and Petcov S T 1980 *Phys. Lett.* B **94** 495

[Bil87]   Bilenky S and Petcov S T 1987 *Rev. Mod. Phys.* **59** 671

[Bil94]   Bilenky S M 1994 *Basics of Introduction to Feynman Diagrams and Electroweak Interactions Physics* (Gif sur Yvette: Éditions Frontières)

[Bil99]   Bilenky S M, Giunti C and Grimus W 1999 *Prog. Nucl. Part. Phys.* **43** 1

[Bil03]   Bilenky S M et al. 2003 *Phys. Rep.* **379** 69

[Bin07]   Binetruy P M R 2007 *Supersymmetry* (Oxford:Oxford University Press)

[Bio87]   Bionta R M et al.1987 *Phys. Rev. Lett.* **58** 1494

[Bir97]   Birnbaum C et al. 1997 *Phys. Lett.* B **397** 143

[Bjo64]   Bjorken J D and Drell S D 1964 *Relativistic Quantum Mechanics* (New York: McGraw-Hill)

[Bjo67]   Bjorken J D 1967 *Phys. Rev.* **163** 1767

               Bjorken J D 1969 *Phys. Rev.* **179** 1547

[Bla97]   Blanc F et al. 1997 ANTARES Proposal *Preprint* astro-ph/9707136

[Blo90]   Blondel A et al. 1990 *Z. Phys.* C **45** 361

[Blu92]   Bludman S et al. 1992 *Phys. Rev.* D **45** 1810

[Blu92a]   Bludman S A 1992 *Phys. Rev.* D **45** 4720 f

[Blü09]   Blümer J, Engel R, Hörandel J 2009 *Prog. Nucl. Part. Phys.* **63** 293

[Blu95]   Blundell S A, Johnson W R, Sapirstein J 1995 *Precision Test of the Standard Model* ed P Langacker (Singapore: World Scientific)

[Bod94]   Bodmann B E et al. 1994 *Phys. Lett.* B **339** 215

[Boe92]   Boehm F and Vogel P 1992 *Physics of Massive Neutrinos* (Cambridge: Cambridge University Press)

[Boe99]   Boezio M et al. 1999 *Astrophys. J.* **518** 457

[Boe00]   Boehm F 2000 *Current Aspects of Neutrino Physics* ed D O Caldwell (Springer)

[Boe01]   Boehm F et al. 2001 *Phys. Rev.* D **64** 112001

[Boe03]   Boerner G 2003 *The Early Universe, 2nd ed.* (Berlin: Springer)

[Bog00]   Boger J et al. 2000 *Nucl. Instrum. Methods* A **449** 172

[Bog17]   Bogomilov M et al. 2017 *Phys. Rev. Accel. Beams* **20** 063501

[Boh75]   Bohr A N and Mottelson B R 1975 *Nuclear Structure* 1st edn (Reading, MA: Benjamin)

               Bohr A N and Mottelson B R 1998 *Nuclear Structure* 2nd edn (Singapore: World Scientific)

[Bol90]   Bolton T et al. 1990 *Fermilab Proposal* P-815

[Bol12]   Bolozdynya A et al. 2012 *Preprint* arXiv:1211.5199

[Bon80]   Bond J R, Efstathiou G and Silk J 1980 *Phys. Rev. Lett.* **45** 1980

[Bon00]   Bonn J et al. 2000 *Physics of Atomic Nuclei* **63** 969 6

[Bon84]   Bond J R and Efstathiou G 1984 *Ap. J.* **285** L45

[Bor91]   BOREXINO Collaboration 1991 *Proposal for a Solar Neutrino Detector at Gran Sasso*

[Bos95]   Bosted P E 1995 *Phys. Rev.* C **51** 409

[Boe98]   Boettcher M and Dermer C D 1998 *Preprint* astro-ph/9801027

[Bot98a]   Bothun G 1998 *Modern Cosmological Observations and Problems* (London: Taylor and Francis)
[Bou74]   Bouichiat M and Bouichiat C J 1974 *J. Phys.* **35** 899
[Bou15]   Bouchet F R 2015 *Comptes Rendus Physique* **16** 10 891
[Boy06]   Boyanosvky D, deVega H J and Schwarz D J 2006 *Annu. Rev. Nucl. Part. Phys.* **56** 441
[Bra88]   Bratton C B et al. 1988 *Phys. Rev.* D **37** 3361
[Bra01]   Branco G et al. 2001 *Nucl. Phys.* B **617** 475
[Bra02]   Brash E J et al. 2002 *Phys. Rev.* C **65** 051001
[Bre69]   Breidenbach M et al. 1969 *Phys. Rev. Lett.* **23** 935
[Bri92]   Britton D I et al. 1992 *Phys. Rev.* D **46** R885
              Britton D I et al. 1992 *Phys. Rev. Lett.* **68** 3000
[Bro18]   Broggini C 2018 *EPJ Web Conf. 184, 01003*
[Bru85]   Bruenn S W 1985 *Astrophys. J. Suppl.* **58** 771
[Bru87]   Bruenn S W 1987 *Phys. Rev. Lett.* **59** 938
[Bry96]   Bryman D A and Numao T 1996 *Phys. Rev.* D **53** 558
[Buc98]   Buchmüller W and Plümacher M 1998 *Phys. Lett.* B **431** 354
[Buc05]   Buchmüller W, Peccei R D and Yanagida T 2005 *Annu. Rev. Nucl. Part. Phys.* **55** 311
[Bud03]   Budd M, Bodek A and Arrington J 2003 *Preprint* hep-ex/0308005
[Bur57]   Burbidge E M et al. 1957 *Rev. Mod. Phys.* **29** 547
[Bur86]   Burrows A and Lattimer J M 1986 *Astrophys. J.* **307** 178
[Bur92]   Burrows A, Klein D and Gandhi R 1992 *Phys. Rev.* D **45** 3361
[Bur95]   Burrows A and Hayes J 1995 *Ann. NY Acad. Sci.* **759** 375
[Bur95a]   Burrows A, Hayes J and Fryxell B A 1995 *Astrophys. J.* **450** 830
[Bur00]   Burrows A et al. 2000 *Astrophys. J.* **539** 865
[Bur01]   Burguet-Castell J et al. 2001 *Nucl. Phys.* B **608** 301
[Bur02]   Burrows A and Thompson T A 2002 *Preprint* astro-ph/0210212
[Bur05]   Burguet-Castell J et al. 2005 *Nucl. Phys.* B **725** 306
[Bur05a]   Burgess C P, Matias J and Quevedo F 2005 *J. High Energy Phys.* **0509** 052
[Bur06]   Burgess C P and Moore G 2006 *The Standard Model: A Primer* (Cambridge: Cambridge University Press)
[Bur13]   Burrows A 2013 2013 *Rev. Mod. Phys.* **85** 245
[But13]   Buttazzo D et al. 2013 *J. High Energy Phys.* **1312** 089
[Cad18]   Cadeddu M et al. 2018 *Phys. Rev.* D **98** 113010
[Cal69]   Callan C G and Gross D J 1969 *Phys. Rev. Lett.* **22** 156
[Cal98]   Caldwell R R, Dave R and Steinhardt P J 1998 *Phys. Rev. Lett.* **80** 1582
[Cal09]   Caldwell R R and Kamionkowski M 2009 *Annu. Rev. Nucl. Part. Phys.* **59** 397
[Cap98]   Capelle K S et al. 1998 *Astropart. Phys.* **8** 321
[Cap18]   2018 Cappuzzello F et al. *arXiv:18.11.08693*
[Car93]   Carone C D 1993 *Phys. Lett.* B **308** 85
[Car06]   Carr J, Lamanna G and Lavalle J 2006 *Rep. Prog. Phys.* **69** 2475
[Car08]   Carey R M et al. 2008 *Proposal* Fermilab-0973
[Car14]   Carroll B W and Ostlie D A 2014 *An introduction to modern astrophysics, 2nd edition* (Cambridge University Press)
[Cas48]   Casimir H G B 1948 *Proc. Kon. Akad. Wet.* **51** 793
[Cas03]   Cassisi, S, Salaris M and Irwin A 2003 *Astrophys. J.* **588** 862
[Cat05]   Cattadori C M et al. 2005 *Nucl. Phys.* A **748** 333
[Cat08]   Catanesi G et al. 2008 *Astropart. Phys.* **30** 124

[Cau96]   Caurier E et al. 1996 *Phys. Rev. Lett.* **77** 1954
[Cau08]   Caurier E et al. 2008 *Phys. Rev. Lett.* **100** 052503
[Cec11]   Cecchini S et al. 2011 *Astropart. Phys.* **34** 486
[Cha14]   Chadwick J 1914 *Verh. der Deutschen Physikalischen Ges.* **16** 383
[Cha32]   Chadwick J 1932 *Nature* **129** 312
[Cha39, 67] Chandrasekhar S 1939 and 1967 *An Introduction to the Study of Stellar Structure* (New York: Dover)
[Cha12]   Chatryan S et al., 2012 *Phys. Lett.* B **716** 30
[Cha18]   Chang L J et al. 2018 *Phys. Rev.* D **98** 123004
[Cha19]   Chabanier S, Millea M and Palanque-Delabrouille N 2019 *Mon. Not. R. Astron. Soc.* **489** 2
[Che92]   Chevalier R A 1992 *Nature* **355** 691
[Chi80]   Chikashige Y, Mohapatra R H and Peccei R D 1980 *Phys. Rev. Lett.* **45** 1926
[Chr64]   Christenson J H et al. 1964 *Phys. Rev. Lett.* **13** 138
[Cho16]   Choi J H et al. 2016 *Nucl. Instrum. Methods* A **810** 100
[Chu89]   Chupp E L, Vestrand W T and Reppin C 1989 *Phys. Rev. Lett.* **62** 505
[Chu02]   Church E D et al. 2002 *Phys. Rev.* D **66** 013001
[Chu18]   Chun E J 2018 *Int. J. Mod. Phys.* A **33** 1842005
[Cin98]   Ammar R et al. 1998 *Phys. Lett.* B **431** 209
[Cla68]   Clayton D D 1968 *Principles of Stellar Evolution* (New York: McGraw-Hill)
[Cle98]   Cleveland B T et al. 1998 *Ap. J* **496** 505
[Cli92]   Cline D 1992 *Proc. 4th Int. Workshop on Neutrino Telescopes (Venezia, 10–13 March 1992)* ed M Baldo-Ceolin (Padova University) p 399
[Cli94]   Cline D B et al. 1994 *Phys. Rev.* D **50** 720
[Cli18]   Cline J M 2018 *TASI lectures, arXiv:1807.08749*
[Clo79]   Close F E 1979 *An Introduction to Quarks and Partons* (New York: Academic)
[Cno78]   Cnops A M et al. 1978 *Phys. Rev. Lett.* **41** 357
[Col90]   Colgate S A 1990 *Supernovae, Jerusalem Winter School* vol 6, ed J C Wheeler, T Piran and S Weinberg (Singapore: World Scientific) p 249
[Col90a]   Colgate S A 1990 *Supernovae* ed S E Woosley (Berlin: Springer) p 352
[Col05]   Cole S et al. 2005 *Mon. Not. R. Astron. Soc.* **362** 505
[Com83]   Commins E D and Bucksham P H 1983 *Weak Interaction of Leptons and Quarks* (Cambridge: Cambridge University Press)
[Con98]   Conrad J M, Shaevitz M H and Bolton T 1998 *Rev. Mod. Phys.* **70** 1341
[Con08]   Constantini H et al. 2008 *Nucl. Phys.* A **814** 144
[Coo84]   Cooperstein J, Bethe H A and Brown G E 1984 *Nucl. Phys.* A **429** 527
[Cos88]   Costa G et al. 1988 *Nucl. Phys.* B **297** 244
[Cos95]   Cosulich E et al. 1995 *Nucl. Phys.* A **592** 59
[Cou02]   Couvidat S, Turck-Chieze S and Kosovichev A G 2002 *Preprint* astro-ph/0203107
[Cow72]   Cowsik R and McClelland J 1972 *Phys. Rev. Lett.* **29** 669
[Cox68]   Cox J P and Guili R T 1968 *Stellar Structure and Evolution* (New York: Gordon and Breach)
[Cyb08]   Cyburt R H, Fields B D and Olive K A 2008 *JCAP* **0811** 012
[Cyb16]   Cyburt R H et al. 2016 *Rev. Mod. Phys.* **88** 015004
[Cre11]   Creighton J D E and Anderson W G 2011 *Gravitational-Wave Physics and Astronomy* (Weinheim: Wiley-VCH)
[Cr016]   Croce M P et al. 2016 *J. Low Temp. Phys.* 184 938
[Dan62]   Danby G et al. 1962 *Phys. Rev. Lett.* **9** 36

[Dan03]	Danevich F A et al. 2003 *Phys. Rev.* C **68** 035501
[Dar05]	Daraktchieva Z et al. 2005 *Phys. Lett.* B **615** 153
[Das09]	Dasgupta B et al. 2009 *Phys. Rev. Lett.* **103** 051105
[Dau87]	Daum M et al. 1987 *Phys. Rev.* D **36** 2624
[Dau95]	Daum K et al. 1995 *Z. Phys.* C **66** 417
[Dav55]	Davis R 1955 *Phys. Rev.* **97** 766
[Dav64]	Davis R 1964 *Phys. Rev. Lett.* **12** 303
[Dav68]	Davis R, Harmer D S and Hoffman K C 1968 *Phys. Rev. Lett.* **20** 1205
[Dav83]	Davis M and Peebles P J E 1983 *Astrophys. J.* **267** 465
[Dav94]	Davis R 1994 *Prog. Part. Nucl. Phys.* **32** 13
[Dav94a]	Davis R 1994 *Proc. 6th Workshop on Neutrino Telescopes (Venezia, Italy)* ed M Baldo-Ceolin (Padova University)
[Dav96]	Davis R 1996 *Nucl. Phys. B (Proc. Suppl.)* **48** 284
[Dav08]	Davidson S, Nardi E and Nir Y 2008 *Phys. Rep.* **466** 105
[Ded97]	Dedenko L G et al. 1997 *Preprint* astro-ph/9705189
[deG00]	de Gouvea A, Friedland A and Murayama H 2000 *Phys. Lett.* B **490** 125
[deG06]	de Gouvea A, Jenkins J 2006 *Phys. Rev.* D **74** 033004
[deG16]	de Gouvea A 2016 *Annu. Rev. Nucl. Part. Phys.* **66** 197
[deH03]	deHolanda P C and Smirnov Y A 2003 *JCAP* **0302** 001
[deH04]	deHolanda P C and Smirnov Y A 2004 *Astropart. Phys.* **21** 287
[deK19]	deKerret H et al. 2019 *arXiv:1901.09445*
[Del03]	Della Valle M et al. 2003 *Astron. Astroph.* **406** L33
[Den90]	Denegri D, Sadoulet B and Spiro M 1990 *Rev. Mod. Phys.* **62** 1
[deR81]	de Rujula A 1981 *Nucl. Phys.* B **188** 414
[Der93]	Derbin A I et al. 1993 *JETP* **57** 768
[Dev04]	Devenish R, Cooper-Sarkar A 2004 *Deep Inelastic Scattering* (Oxford:Oxford University Press)
[Die91]	Diemoz M 1991 *Neutrino Physics* ed K Winter (Cambridge: Cambridge University Press)
[Die01]	Dienes K R 2001 *Nucl. Phys. B (Proc. Suppl.)* **91** 321
[Dig00]	Dighe A S and Smirnov A Y 2000 *Phys. Rev.* D **62** 033007
[Dig10]	Dighe A S 2010 *J. Phys. Conf. Ser.* **203**:012015
[Din07]	Dine M 2007 *Supersymmetry and String Theory: Beyond the Standard Model* (Cambridge: Cambridge University Press)
[Dis09]	Distefano C 2009 *Nucl. Phys. B (Proc. Suppl.)* **190** 115
[Dit13]	Dittmaier S, Schuhmacher M 2013 PNPP **70** 1
[Diw16]	Diwan M V et al. 2016 ARNPS **66** 47
[Dod03]	Dodelson S 2003 *Modern Cosmology* (Academic Presss)
[Doi83]	Doi M, Kotani T and Takasugi E 1983 *Prog. Theor. Phys.* **69** 602
[Doi85]	Doi M, Kotani T and Takasugi E 1985 *Prog. Theo. Phys. Suppl.* **83** 1
[Doi88]	Doi M, Kotani T and Takasugi E 1988 *Phys. Rev.* D **37** 2575
[Doi92]	Doi M and Kotani T 1992 *Prog. Theor. Phys.* **87** 1207
[Doi93]	Doi M and Kotani T 1993 *Prog. Theor. Phys.* **89** 139
[Dok77]	Dokshitser Y L 1977 *Sov. J. JETP* **46** 641
[Dol97]	Dolgov A D 1997 *Proc. 4th Paris Cosmology Meeting* (Singapore: World Scientific)
[Dol02]	Dolgov A D 2002 *Phys. Rep.* **370** 333
[Dol10]	Dolgov A D 2010 *Phys. Atom. Nucl.* **73** 815

416   *References*

[Dol19]   Dolinski M J, Poon A W P, Rodejohann W 2019 *Annu. Rev. Nucl. Part. Phys.* **69**
[Dom71]   Domogatskii G V and Nadezhin D K 1971 *Sov. J. Nucl. Phys.* **12** 678
[Dom78]   Domogatskii G V and Nadezhin D K 1978 *Sov. Ast.* **22** 297
[Dom93]   Domokos G, Elliott B and Kovesi-Domokos S J 1993 *J. Phys. G* **19** 899
[Dom02]   Domogatskii G V 2002 *Nucl. Phys. B (Proc. Suppl.)* **110** 504
[Don92]   Donoghue J F, Golowich F and Holstein B R 1992 *Dynamics of the Standard Model* (Cambridge: Cambridge University Press)
[Dor89]   Dorenbosch J et al. 1989 *Z. Phys. C* **41** 567
             Dorenbosch J et al. 1991 *Z. Phys. C* **51** 142(E)
[Dor04]   Doroshkevich A et al., 2004 *Astron. Astrophys.* **418** 7
[Dor08]   Dore U and Orestano D 2008 *Rep. Prog. Phys.* **71** 106201
[Dor13]   Doro M et al., 2013 *Astropart. Phys.* **43** 189
[Dor18]   Dore U, Loverre P and Ludovici L 2018, *Preprint* acc-ph/1805.01373
[Dra87]   Dragon N, Ellwanger U and Schmidt M G 1987 *Prog. Nucl. Part. Phys.* **18** 1
[Dre94]   Drexlin G et al. 1994 *Prog. Nucl. Part. Phys.* **32** 351
[Dre17]   Drewes M and Garbrecht B 2017 *Nucl. Phys. B* **921** 250
[Dua06]   Duan H et al. 2006 *Phys. Rev. D* **74** 105014
[Dua06a]   Duan H, Fuller G M and Qian Y Z 2006 *Phys. Rev. D* **74** 123004
[Dua09]   Duan H and Kneller J P 2009 *J. Phys. G* **36** 113201
[Dua10]   Duan H, Fuller G M and Qian Y Z 2010 *Annu. Rev. Nucl. Part. Phys.* **60** 69
[Dub11]   Dubbers D and Schmidt M G 2011 *Rev. Mod. Phys.* **83** 1111
[Due11]   Duerr M, Lindner M and Merle M 2011 *J. High Energy Phys.* **1106** 091
[Due11a]   Dueck A, Rodejohann W and Zuber K 2011 *Phys. Rev. D* **83** 113010
[Dur08]   Durrer R 2008 *The Cosmic Microwave Background* (Cambridge: Cambridge University Press)
[Ebe16]   Ebert J et al. 2016 *Phys. Rev. C* **94** 024603
[Efs83]   Efstathiou G and Silk J 1983 *Fund. Cosmic Phys.* **9** 1
[Egu03]   Eguchi K et al. 2003 *Phys. Rev. Lett.* **90** 021802
[Egu04]   Eguchi K et al. 2004 *Phys. Rev. Lett.* **92** 071301
[Egu04a]   Eguchi K et al. 2004 *Phys. Rev. Lett.* **92** 071301
[Ein17]   Einstein A 1917 *Sitzungsberichte Preuß. Akad. Wiss.* **142**
[Eis86]   Eisele F 1986 *Rep. Prog. Phys.* **49** 233
[Eis05]   Eisenstein D J et al. 2005 *Astrophys. J.* **633** 560
[Eit01]   Eitel K 2001 *Nucl. Phys. B (Proc. Suppl.)* **91** 191
[Eit08]   Eitel K 2008 *J. Phys. G* **35** 014055
[Eji97]   Ejiri H 1997 *Proc. Neutrino'96* (Singapore: World Scientific) p 342
[Eji00]   Ejiri H et al. 2000 *Phys. Rev. Lett.* **85** 2917
[Eji01]   Ejiri H et al. 2001 *Phys. Rev. C* **63** 065501
[Eji19]   Ejiri H, Suhonen J and Zuber K 2019 *Phys. Rep.* **x** x
[Elg02]   Elgaroy O et al. 2002 *Phys. Rev. Lett.* **89** 061301
[Elg03]   Elgaroy O and Lahav O 2003 *JCAP* 0304:004
[Eli11a]   Eliseev S et al. 2011 *Phys. Rev. Lett.* **106** 052504
[Eli11b]   Eliseev S et al. 2011 *Phys. Rev. C* **84** 012501
[Eli15]   Eliseev S et al. 2015 *Phys. Rev. Lett.* **115** 062501
[Ell27]   Ellis C D and Wooster W A 1927 *Proc. R. Soc. A* **117** 109
[Ell87]   Elliott S R, Hahn A A and Moe M 1987 *Phys. Rev. Lett.* **59** 2020
[Ell90]   Elliott S R, Hahn A A and Moe M K 1988 *Nucl. Instrum. Methods A* **273** 226

[Ell98]   Ellis J 1998 *Preprint* hep-ph/9812235, Lecture presented at 1998 European School for High Energy Physics
[Els98]   Elsener K (ed) 1998 *NGS Conceptual Technical Design Report* CERN 98-02 INFN/AE-98/05
[Enb08]   Enberg R, Hall M H and Sarcevic I 2008 *Phys. Rev.* D **43** 043005
[Eng64]   Englert F and Brout R 1964 *Phys. Rev. Lett.* **13** 321
[Eng88]   Engel J, Vogel P and Zirnbauer M R 1988 *Phys. Rev.* C **37** 731
[Eng03]   Engel J, McLaughlin G C and Volpe C 2003 *Phys. Rev.* D **67** 013005
[Eno05]   Enomoto S et al. 2005 *Preprint* hep-ph/0508049
[Eri94]   Ericson M, Ericson T and Vogel P 1994 *Phys. Lett.* B **328** 259
[Esf17]   Esfahani A A *et al* 2017 *J. Phys.* G **44** 054004
[Esk97]   Eskut E et al. 1997 *Nucl. Instrum. Methods* A **401** 7
[Esk01]   Eskut E et al. 2001 *Phys. Lett.* B **503** 1
[Esk08]   Eskut E et al. 2008 *Nucl. Phys.* B **793** 326
[Esp07]   Espinal X and Sanches F 2007 *AIP Conf. Proc.* **967** 117
[Est19]   Esteban I et al. 2019 *J. High Energy Phys.* **1901** 106
[Fae95]   Faessler A 1995 *Proc. ECT Workshop on Double Beta Decay and Related Topics (Trento)* ed H V Klapdor-Kleingrothaus and S Stoica (Singapore: World Scientific) p 339
[Fae15]   Faessler A et al. 2015 *J. Phys.* G **42** 015108
[Fai78]   Faissner H et al. 1978 *Phys. Rev. Lett.* **41** 213
[Fal03]   Falcke H and Gorham P W 2003 *Astropart. Phys.* **19** 477
[Fal05]   Falcke H et al. 2005 *Nature* **435** 313
[Far97]   Farine J 1997 *Proc. Neutrino'96* (Singapore: World Scientific) p 347
[Fed18]   Fedynitsch A et al. 2018 *arXiv:1806.04140*
[Fei88]   von Feilitzsch F 1988 *Neutrinos* ed H V Klapdor (Berlin: Springer)
[Fec09]   Fechner M et al. 2009 *Phys. Rev.* D **79** 112010
[Fen10]   Fenaz J F and Nachtman J 2010 *Rev. Mod. Phys.* **82** 699
[Fen17]   Fenu F et al. 2017 *Proc. Sci.* 486 (ICRC17)
[Fer34]   Fermi E 1934 *Z. Phys.* **88** 161, *Nuovo Cim.* **11** 1
[Fer19]   Webpage of Fermilab, https://news.fnal.gov/wp-content/uploads/2015/10/14-0024-13D.hr_.jpg (Accessed: 06.05.2019)
[Fey58]   Feynman R P and Gell-Mann M 1958 *Phys. Rev.* **109** 193
            Feynman R P and Gell-Mann M 1958 *Phys. Rev.* **111** 362
[Fey69]   Feynman R P 1969 *Phys. Rev. Lett.* **23** 1415
[Fie97]   Fields B D, Kainulainen K and Olive K A 1997 *Astropart. Phys.* **6** 169
[Fil83]   Filippone B W et al. 1983 *Phys. Rev. Lett.* **50** 412
            Filippone B W et al. 1983 *Phys. Rev.* C **28** 2222
[Fio67]   Fiorini E et al. 1967 *Phys. Lett.* B **45** 602
[Fio95]   Fiorentini G, Kavanagh R W and Rolfs C 1995 *Z. Phys.* A **350** 289
[Fio98]   Fiorini E 1998 *Phys. Rep.* **307** 309
[Fio01]   Fiorini E 2001 Int. Workshop on neutrino masses in the sub-eV range, Bad Liebenzell 2001, http://www-ik1.fzk.de/tritium/liebenzell
[Fio07]   Fiorentini G, Lissia M and Mantovani F 2007 *Phys. Rep.* **453** 117
[Fis10]   Fischer T et al. 2010 *Astron. Astrophys.* **517A** 80F
[Fis18]   Fischer V 2018 *arXiv:1809.05987*
[Fix96]   Fixsen D J et al. 1996 *Ap. J.* **473** 576
[Fla96]   Flanz M et al. 1996 *Phys. Lett.* B **389** 693
[Fla00]   Flanz M, Rodejohann W and Zuber K 2000 *Eur. Phys. J.* C **16** 453

418   *References*

[Fla00a]   Flanz M, Rodejohann W and Zuber K 2000 *Phys. Lett.* B **473** 324
[Fle01]   Fleming B T et al. 2001 *Phys. Rev. Lett.* **86** 5430
[Fog99]   Fogli G L et al. 1999 hep-ph/9912237
[Fog03]   Fogli G L et al. 2003 *Phys. Rev.* D **67** 073002
[Fog03a]   Fogli G L et al. 2003 *Phys. Rev.* D **67** 093006
[Fog05]   Fogli G L et al. 2005 *Phys. Lett.* B **623** 80
[Fow75]   Fowler W A, Caughlan G R and Zimmerman B A 1975 *Ann. Rev. Astron.*
            Astrophys. **13** 69
[Fra95]   Franklin A 1995 *Rev. Mod. Phys.* **67** 457
[Fra09]   Franco D et al. 2009 *Nucl. Phys. B (Proc. Suppl.)* **188** 127
[Fre74]   Freedman D Z 1974 *Phys. Rev.* D **9** 1389
[Fre02]   Freedman W 2002 *Int. J. Mod. Phys.* A **1751** 58
[Fre05]   Frekers D 2005 Talk presented at IPPP meeting on Matrix elements for double
            beta decay, Durham
[Fre15]   Frekers D et al. 2015 *Phys. Lett.* B **722** 233
[Fri75]   Fritzsch H and Minkowski P 1975 *Ann. Phys.* **93** 193
[Fri08]   Frieman J A, Turner M S and Huterer D 2008 *Annu. Rev. Astron. Astrophys.*
            46 385
[Frö06]   Fröhlich C et al. 2006 *Phys. Rev. Lett.* **96** 142502
[Fry04]   Fryer C L 2004 *Stellar collapse* (Berlin: Springer)
[Fuk86]   Fukugita M and Yanagida T 1986 *Phys. Lett.* B **174** 45
[Fuk87]   Fukugida M and Yanagida T 1987 *Phys. Rev. Lett.* **58** 1807
[Fuk88]   Fukugita M et al. 1988 *Phys. Lett.* B **212** 139
[Fuk94]   Fukuda Y et al. 1994 *Phys. Lett.* B **235** 337
[Fuk98]   Fukuda Y et al. 1998 *Phys. Rev. Lett.* **82** 2644
[Fuk03]   Fukuda S et al. 2003 *Nucl. Instrum. Meth.* A **501** 418
[Fuk03a]   Fukugita M and Yanagida T 2003 *Physics of Neutrinos* (Berlin: Springer)
[Ful92]   Fuller G M et al. 1992 *Astrophys. J.* **389** 517
[Ful98]   Fuller G M, Haxton W C and McLaughlin G C 1998 *Phys. Rev.* D **59** 085055
[Ful99]   Fulgione W 1999 *Nucl. Phys. B (Proc. Suppl.)* **70** 469
[Ful01]   Fuller G M 2001 *Current Aspects of Neutrino Physics* ed D Caldwell (Berlin:
            Springer)
[Fur39]   Furry W H 1939 *Phys. Rev.* **56** 1184
[Fut99]   Futagami T et al. 1999 *Phys. Rev. Lett.* **82** 5192
[Gai95]   Gaisser T K, Halzen F and Stanev T 1995 *Phys. Rep.* **258** 173
[Gai98]   Gaisser T K and Stanev T 1998 *Phys. Rev.* D **57** 1977
[Gai02]   Gaisser T K and Honda M 2002 *Annu. Rev. Nucl. Part. Phys.* **52** 153
[Gai04]   Gaitskell R J 2004 *Annu. Rev. Nucl. Part. Phys.* **54** 315
[Gai09]   Gaisser T K 2009 *Preprint* arXiv:0901.2386, *Earth Planets Space* **62** 195
[Gai14]   Gaisser T K and Halzen F 2014 *Annu. Rev. Nucl. Part. Phys.* **64** 101
[Gai16]   Gaisser T K, Engel R and Risconi E 2016 *Cosmic Rays and Particle Physics*
            2nd. ed. (Cambridge: Cambridge University Press)
[Gal98]   Galeazzi M 1998 *Nucl. Phys. B (Proc. Suppl.)* **66** 203
[Gal11]   Gallagher H, Garvey G and Zeller G P 2011 *Annu. Rev. Nucl. Part. Phys.* **61**
            355
[Gam36]   Gamow G and Teller E 1936 *Phys. Rev.* **49** 895
[Gam38]   Gamow G 1938 *Phys. Rev.* **53** 595
[Gam46]   Gamow G 1946 *Phys. Rev.* **70** 572
[Gan96]   Gandhi R et al. 1996 *Astropart. Phys.* **5** 81

[Gan98]   Gandhi R 1998 *Phys. Rev.* D **58** 093009
[Gan01]   Gandhi R 2001 *Nucl. Phys. B (Proc. Suppl.)* **91** 453
[Gan03]   Gando Y et al. 2003 *Phys. Rev. Lett.* **90** 171302
[Gan12]   Gando A et al. 2012 *Astrophys. J.* **745** 193
[Gan15]   Gando A et al. 2015 *Phys. Rev.* D **88** 033001
[Gar57]   Garwin R L, Ledermann L M, Weinrich M 1957 *Phys. Rev.* **105** 1415
[Gar19]   Garrappa S et al. 2019 *Astrophys. J.* **880** 103
[Gas17]   Gastaldo L et al. 2017 *Eur.Phys.J.ST* 2261623
[Gat97]   Gatti F et al. 1997 *Phys. Lett.* B **398** 415
[Gat99]   Gatti F et al. 1999 *Nature* **397** 137
[Gat01]   Gatti F 2001 *Nucl. Phys. B (Proc. Suppl.)* **91** 293
[Gav03]   Gavrin V N 2003 *Nucl. Phys. B (Proc. Suppl.)* **118**
[Gav03a]   Gavrin V 2003 *TAUP 2003 Conference, Seattle*
[Gee98]   Geer S 1998 *Phys. Rev.* D **57** 6989
             Geer S 1999 *Phys. Rev.* D **59** 039903 (erratum)
[Gee07]   Geer S and Zisman M S 2007 *Prog. Nucl. Part. Phys.* **59** 631
[Gee09]   Geer S 2009 *Annu. Rev. Nucl. Part. Phys.* **59** 347
[Geh09]   Gehrels N, Ramirez-Ruiz E and Fox D B 2009 *Annu. Rev. Astron. Astrophys.* **47** 567
[Gei90]   Geiregat G et al. 1990 *Phys. Lett.* B **245** 271
[Gei93]   Geiregat D et al. 1993 *Nucl. Instrum. Methods* A **325** 91
[Gel64]   Gell-Mann M 1964 *Phys. Lett.* **8** 214
[Gel78]   Gell-Mann M, Ramond P and Slansky R 1978 *Supergravity* ed F van Nieuwenhuizen and D Freedman (Amsterdam: North-Holland) p 315
[Gel81]   Gelmini G B and Roncadelli M 1981 *Phys. Lett.* B **99** 411
[Geo74]   Georgi H and Glashow S L 1974 *Phys. Rev. Lett.* **32** 438
[Geo75]   Georgi H 1975 *Particles and Fields* ed C E Carloso (New York: AIP)
[Geo76]   Georgi H and Politzer H D 1976 *Phys. Rev.* D **14** 1829
[Ger56]   Gershtein S S and Zeldovich Ya B 1956 *Sov. J. JETP* **2** 596
[Giu98]   Giunti C, Kim C W and Monteno M 1998 *Nucl. Phys.* B **521** 3
[Giu07]   Giunti C, Kim C W 2007 *Fundamentals of Neutrino Physics and Astrophysics* (Oxford:Oxford University Press)
[Giu09]   Giunti C, Studenikin A 2009 *Phys. Rev.* D **80** 013005
[Giu15]   Giunti C, Studenikin A 2015 *Rev. Mod. Phys.* **87** 531
[Giz17]   Gizhko A et al. 2017 *Phys. Lett.* B **775** 233
[Gla60]   Glashow S L 1960 *Phys. Rev.* **118** 316
[Gla61]   Glashow S L 1961 *Nucl. Phys.* **22** 579
[Gle00]   Glenzinski D A and Heintz U 2000 *Annu. Rev. Nucl. Part. Phys.* **50** 207
[Glu95]   Glück M, Reya E and Vogt A 1995 *Z. Phys.* C **67** 433
[Gni00]   Gninenko S N and Krasnikov N V 2000 *Phys. Lett.* B **490** 9
[Goe35]   Goeppert-Mayer M 1935 *Phys. Rev.* **48** 512
[Gol58]   Goldhaber M, Grodzin L and Sunyar A W 1958 *Phys. Rev.* **109** 1015
[Gol80]   Goldstein H 1980 *Classical Mechanics* (New York: Addison-Wesley)
[Gol17]   Goldstein A et al. 2017 *Astrophys. J.* **848** L14
[Gon99]   Gonzalez-Garcia M C et al. 1999 *Phys. Rev. Lett.* **82** 3202
[Gon00]   Gondolo P et al. 2000 *Preprint* astro-ph/0012234
[Gon00a]   Gondolo P and Silk J 2000 *Nucl. Phys. B (Proc. Suppl.)* **87** 87
[Gon08]   Gonzalez-Garcia M C and Maltoni M 2008 *Phys. Rep.* **460** 1

[Gon19]   Gonzalez-Alonso M, Naviliat-Cuncic O and Severijns N 2019 *Prog. Nucl. Part. Phys.* **104** 165
[Gor07]   Gorham P W et al. 2007 *Phys. Rev. Lett.* **99** 171101
[Got67]   Gottfried K 1967 *Phys. Rev. Lett.* **18** 1174
[Gou92]   Gould A 1992 *Astrophys. J.* **388** 338
[Gou98]   Gould A and Uza O 1998 *Astrophys. J.* **494** 118
[Gra03]   Grainge K et al. 2003 *Mon. Not. R. Astron. Soc.* **341** L23
[Gra06]   Grahn R et al. 2006 *Phys. Rev. D* **74** 052002
[Gre66]   Greisen K 1966 *Phys. Rev.* **118** 316
[Gre86]   Greiner W and Müller B 1986 *Theoretische Physik Bd 8 Eichtheorie der Schwachen Wechselwirkung* (Frankfurt: Harri Deutsch)
[Gre93]   Grevesse N 1993 *Origin and Evolution of Elements* ed N Prantzos, E Vangioni Flam and M Casse (Cambridge: Cambridge University Press) pp 15–25
[Gre93a]   Grevesse N and Noels A 1993 *Phys. Scr.* T **47** 133
[Gre94]   Greife U et al. 1994 *Nucl. Instrum. Methods* A **350** 326
[Gre12]   Grebenev S A et al. 2012 *Nature* **490** 373
[Gri72]   Gribov V N and Lipatov L N 1972 *Sov. J. Nucl. Phys.* **15** 438, 675
[Gri96]   Grimus W and Stockinger P 1996 *Phys. Rev. D* **54** 3414
[Gri01]   Grieder P K F 2001 *Cosmic Rays on Earth* (Amsterdam: Elsevier Science)
[Gri08]   Griffiths D J 2008 *Introduction to Elementary Particles* (Weinheim: Wiley-VCH)
[Gro69]   Gross D J and Llewellyn-Smith C H 1969 *Nucl. Phys.* B **14** 337
[Gro73]   Gross D J and Wilczek F 1973 *Phys. Rev. Lett.* **30** 1343
[Gro79]   deGroot J G H et al. 1979 *Z. Phys.* C **1** 143
[Gro90]   Grotz K and Klapdor H V 1990 *The Weak Interaction in Nuclear, Particle and Astrophysics* (Bristol: Hilger)
[Gro97]   Grossman Y and Lipkin H J 1997 *Phys. Rev. D* **55** 2760
[Gro00]   Groom D et al. 2000 Review of particle properties Eur. Phys. J. C 3 1
[Gro04]   Grossheim A and Zuber K 2004 *Nucl. Instrum. Methods* A **533** 532
[Gul00]   Guler M et al. 2000 *OPERA Proposal* LNGS P25/2000 CERN SPSC 2000-028
[Gun90]   Gunion G F, Haber H E, Kane G and Dawson S 1990 *The Higgs–Hunter Guide, Frontiers in Physics* vol 80 (London: Addison-Wesley)
[Guo07]   Guo X et al. 2007 *Preprint* hep-ex/0701029
[Gut81]   Guth A H 1981 *Phys. Rev. D* **23** 347
[Gut89]   Guth A H 1989 *Bubbles, Voids and Bumps in Time: The New Cosmology* ed J Cornell (Cambridge: Cambridge University Press)
[Gyu95]   Gyuk G and Turner M S 1995 *Nucl. Phys. B (Proc. Suppl.)* **38** 13
[Hab85]   Haber H E and Kane G L 1985 *Phys. Rep.* **117** 75
[Hab01]   Habig A 2001 *Preprint* hep-ex/0106026
[Hag02]   Hagiwara K et al. 2002 Review of particle properties *Phys. Rev. D* 66 010001
[Hai88]   Haidt D and Pietschmann H 1988 *Landolt-Boernstein I/10* (Berlin: Springer)
[Hal83]   Halprin A, Petcov S T and Rosen S P 1983 *Phys. Lett.* B **125** 335
[Hal84]   Halzen F and Martin A D 1984 *Quarks and Leptons* (London: Wiley)
[Hal94]   Halzen F, Jacobsen J E and Zas E 1994 *Phys. Rev. D* **49** 1758
[Hal96]   Halprin A, Leung C N and Pantaleone J 1996 *Phys. Rev. D* **53** 5356
[Hal96a]   Halzen F, Jacobsen J E and Zas E 1996 *Phys. Rev. D* **53** 7359
[Hal97]   Halzen F and Zas E 1997 *Astrophys. J.* **488** 669

[Hal98]	Halzen F 1998 *Preprint* astro-ph/9810368, Lecture presented at TASI School, 1998
[Hal98a]	Halzen F and Saltzberg D 1998 *Phys. Rev. Lett.* **81** 4305
[Hal01]	Halzen F 2001 *AIP Conf. Proc.* **558** 43
[Hal02]	Halzen F and Hooper D 2002 *Rep. Prog. Phys.* **65** 1025
[Hal02a]	Halverson N W et al. 2002 *Ap. J.* **568** 38
[Ham93]	Hampel W 1993 *J. Phys. G: Nucl. Part. Phys.* **19** 209
[Ham96]	Hampel W et al. 1996 *Phys. Lett.* B **388** 384
[Ham98]	Hammache F et al. 1998 *Phys. Rev. Lett.* **80** 928
[Ham99]	Hampel W et al. 1999 *Phys. Lett.* B **447** 127
[Han49]	Hanna G C and Pontecorvo B 1949 *Phys. Rev.* **75** 983
[Han98]	Hannestad S and Raffelt G G 1998 *Astrophys. J.* **507** 339
[Han02]	Hannestad S 2002 *Preprint* astro-ph/0208567
	Hannestad S 2003 *Nucl. Phys. B (Proc. Suppl.)* **118** 315
[Han03a]	Hannestad S 2003 *JCAP* **0305** 004
[Han03a]	Hannestad S 2003 *Phys. Rev.* D **67** 085017
[Han06]	Hannestad S 2006 *Annu. Rev. Nucl. Part. Phys.* **56** 137
[Han10]	Hannestad S 2010 *Prog. Nucl. Part. Phys.* **65** 185
[Han16]	Hannestad S, Schwetz T 2016 *JCAP* **1611** 035
[Har96]	Hargrove C K et al. 1996 Journal*Astropart. Phys.*51831996
[Har99]	HARP - Proposal, Catanesi M G et al. 1999 *Preprint* CERN-SPSC-99-35
[Har15]	Hardy J C and Towner I S 2015 *Phys. Rev.* C **91** 022501
[Har16]	Hardy J C and Towner I S 2016 *Proceedings of Science* (CKM2016) 028
[Har18]	Hardy J C and Towner I S 2018 *Preprint* arXiv:1807.01146
[Has73]	Hasert F J et al. 1973 *Phys. Lett.* B **46** 121
[Has73a]	Hasert F J et al. 1973 *Phys. Lett.* B **46** 138
[Has74]	Hasert F J et al. 1974 *Nucl. Phys.* B **73** 1
[Has05]	Hasegawa M et al. 2005 *Phys. Rev. Lett.* **95** 252301
[Has18]	Hashim I H et al. 2018 *Phys. Rev.* C **97** 014617
[Hat94]	Hata N and Langacker P 1994 *Phys. Rev.* D **50** 632
[Hax84]	Haxton W C and Stephenson G J 1984 *Prog. Nucl. Part. Phys.* **12** 409
[Hax08]	Haxton W C and Serenelli A M 2008 *Astrophys. J.* **687** 678
[Hay99]	Hayato Y et al. 1999 *Phys. Rev. Lett.* **83** 1529
[Hay16]	Hayes A C and Vogel P 2016 *Annu. Rev. Nucl. Part. Phys.* **66** 219
[Hay19]	Hayen L et al. 2019 *Phys. Rev.* C **99** 031301
[Heg05]	Heger A et al. 2005 *Phys. Lett.* B **606** 258
[Hei80]	Heisterberg R H et al. 1980 *Phys. Rev. Lett.* **44** 635
[Hel56]	Helm R 1956 *Phys. Rev.* **104** 1466
[Her94]	Herant M et al. 1994 *Astrophys. J.* **435** 339
[Her94a]	HERA-B, DESY/PRC 94-02 1994 *Proposal*
[Her95]	Herczeg P 1995 *Precision Tests of the Standard Model* ed P Langacker (Singapore: World Scientific)
[Heu95]	Heusser G 1995 *Annu. Rev. Nucl. Part. Phys.* **45** 543
[Hig64]	Higgs P W 1964 *Phys. Lett.* **12** 252
[Hig64a]	Higgs P W 1964 *Phys. Rev. Lett.* **13** 508
[Hik92]	Hikasa K et al. 1992 *Phys. Rev.* D **45** S1
[Hik92a]	Hikasa K et al. 1992 *Phys. Rev.* D **45** S1
[Hil72]	Hillas A M 1972 *Cosmic Rays* (Oxford: Pergamon)
[Hil88]	Hillebrandt W 1988 *NEUTRINOS* ed H V Klapdor (Berlin: Springer) p 285

[Hin09]	Hinton J A, Hofmann W 2009 *Annu. Rev. Astron. Astrophys.* **47** 523
[Hin09a]	Hinshaw G et al. 2009 *Astrophys. J. Suppl.* **180** 225
[Hir87]	Hirata K S et al. 1987 *Phys. Rev. Lett.* **58** 1490
[Hir88]	Hirata K S et al. 1988 *Phys. Rev. D* **38** 448
[Hir94]	Hirsch M et al. 1994 *Z. Phys. A* **347** 151
[Hir95]	Hirsch M, Klapdor-Kleingrothaus H V and Kovalenko S 1995 *Phys. Rev. Lett.* **75** 17
[Hir96]	Hirsch M et al. 1996 *Phys. Lett. B* **374** 7
[Hir96a]	Hirsch M, Klapdor-Kleingrothaus H V and Kovalenko S 1996 *Phys. Lett. B* **352** 1
[Hir96b]	Hirsch M, Klapdor-Kleingrothaus H V and Kovalenko S 1996 *Phys. Lett. B* **378** 17
	Hirsch M, Klapdor-Kleingrothaus H V and Kovalenko S 1996 *Phys. Rev. D* **54** 4207
[Hir96c]	Hirsch M et al. 1996 *Phys. Lett. B* **372** 8
[Hir96d]	Hirsch M, Klapdor-Kleingrothaus H V and Kovalenko S G 1996 *Phys. Rev. D* **53** 1329
[Hir02]	Hirsch M et al. 2002 *Preprint* hep-ph/0202149
[Hir06]	Hirsch M, Kovalenko S G and Schmidt I 2006 *Phys. Lett. B* **642** 106
[Hjo03]	Hjorth J et al. 2003 *Nature* **423** 847
[Hoe03]	Hörandel J 2003 *Astropart. Phys.* **19** 193
[Hoe04]	Hörandel J 2004 *Astropart. Phys.* **21** 241
[Hol78]	Holder M et al. 1978 *Nucl. Instrum. Methods* **148** 235
	Holder M et al. 1978 *Nucl. Instrum. Methods* **151** 69
[Hol92]	Holzschuh E 1992 *Rep. Prog. Phys.* **55** 851
[Hol99]	Holzschuh E et al. 1999 *Phys. Lett. B* **451** 247
[Hon95]	Honda M et al. 1995 *Phys. Rev. D* **52** 4985
[Hon04]	Honda M et al. 2004 *Phys. Rev. D* **70** 043008
[Hor82]	Horstkotte J et al. 1982 *Phys. Rev. D* **25** 2743
[Hor09]	Horiuchi S, Beacom J F and Dwek E 2009 *Phys. Rev. D* **79** 083013
[Hor18]	Horiuchi S et al. 2018 *Mon. Not. R. Astron. Soc.* **475** 1363
[Hos06]	Hosaka J et al. 2006 *Phys. Rev. D* **73** 112001
[Hu95]	Hu W, Sugiyama N and Silk J 1995 *Preprint* astro-ph/9504057
[Hu98]	Hu W, Eisenstein D J and Tegmark M 1998 *Phys. Rev. Lett.* **80** 5255
[Hu02]	Hu W and Dodelson S 2002 *Ann. Rev. Astron. Astroph.* **40** 171
[Hu03]	Hu W 2003 *Annals. Phys.* **303** 203
[Hub29]	Hubble E 1929 *Proc. Nat. Acad.* **15** 168
[Hub02]	Huber P, Lindner M and Winter W 2002 *Nucl. Phys. B* **645** 3
[Hug09]	Hughes S A 2009 *Annu. Rev. Astron. Astrophys.* **47** 107
[IAU17]	IAU Sypmposium 2017 331 *SN1987,30 years later*
[Ibe13]	Iben I, 2013 *Stellar Evolution Physics Vol. 1 and 2* (Cambridge: Cambridge University Press)
[Ice17]	IceCube Collaboration 2017 *Proceedings of Science* (ICRC17) 974
[Ice19]	Webpage of the IceCube collaboration, https://icecube.wisc.edu/ (Accessed: 11.12.2019)
[Igl96]	Iglesias C A and Rogers F J 1996 *Astrophys. J.* **464** 943
[Ili15]	Iliadis C 2015 *Nuclear Physics of Stars* 2nd ed. (Cambridge: Cambridge University Press)

[Imm07]   Immler S (ed.), Weiler K (ed.) and McCray R (ed) 2007 *Supernova1987A:20 years after: Supernovae and Gamma Ray Bursters*, AIP Conf. Proc. 937
[Ing96]   Ingelman G and Thunman T 1996 *Phys. Rev.* D **54** 4385
[Ing96a]   Ingelman G and Thunman T 1996 *Preprint* hep-ph/9604286
[Ioc09]   Iocco F et al. 2009 *Phys. Rep.* **472** 1
[Iva15]   Ivanov D et al. 2015 *Proc. Sci 349* ICRC 2015
[Jac57]   Jackson J D, Treiman S B and Wyld H W 1957 *Phys. Rev.* **106** 517
[Jac57a]   Jackson J D, Treiman S B and Wyld H W 1957 *Nucl. Phys.* **4** 206
[Jae10]   Jaeger T R, Mutel R L and Gayley K G 2010 *Astroparticle Phys.*, vol. 34, issue 5 December 2010 pp. 293-303
[Jam09]   James C W et al. 2009 *Preprint* arXiv:0906.3766
[Jan89a]   Janka H Th and Hillebrandt W 1989 *Astron. Astrophys. Suppl. Series* **78** 375
[Jan89b]   Janka H Th and Hillebrandt W 1989 *Astron. Astrophys.* **224** 49
[Jan95]   Janka H Th and Müller E 1995 *Astrophys. J. Lett.* **448** L109
[Jan95a]   Janka H Th and Müller E 1995 *Ann. NY Acad. Sci.* **759** 269
[Jan96]   Janka H Th and Müller E 1996 *Astron. Astrophys.* **306** 167
[Jan07]   Janka H Th et al. 2007 *Phys. Rep.* **442** 38
[Jan12]   Janka H Th 2012 *Annu. Rev. Nucl. Part. Phys.* **62** 407
[Jan16]   Janka H Th 2016 *Annu. Rev. Nucl. Part. Phys.* **66** 341
[Jec95]   Jeckelmann B et al. 1995 *Phys. Lett.* B **335** 326
[Jon82]   Jonker M et al. 1982 *Nucl. Instrum. Methods* A **200** 183
[Jon06]   Jones W C et al. 2006 *Astrophys. J.* **647** 823
[Jun96]   Jungmann G, Kamionkowski M and Griest K 1996 *Phys. Rep.* **267** 195
[Jun01]   Jung C K et al. 2001 *Annu. Rev. Nucl. Part. Phys.* **51** 451
[Jun02]   Junghans A R et al. 2002 *Phys. Rev. Lett.* **88** 041101
[Jun03]   Junghans A R et al. 2003 *Phys. Rev.* C **68** 065803
[Kac01]   Kachelriess M, Tohas R and Valle J W F 2001 *J. High Energy Phys.* **0101** 030
[Kac19]   Kachelriess M, Semikoz D V 2019 *Preprint* arXiv:1904.08160
[Kae10]   Kaether F et al. 2010 Journal*Phys. Lett.* B685472010
[Kaf94]   Kafka T et al. 1994 *Nucl. Phys. B (Proc. Suppl.)* **35** 427
[Kah86]   Kahana S 1986 *Proc. Int. Symp. on Weak and Electromagnetic Interaction in Nuclei (WEIN'86)* ed H V Klapdor (Heidelberg: Springer) p 939
[Kaj01]   Kajita T and Totsuka Y 2001 *Rev. Mod. Phys.* **73** 85
[Kaj14]   Kajita T 2014 *Annu. Rev. Nucl. Part. Phys.* **64** 343
[Kam01]   Kampert K H 2001 *Preprint* astro-ph/0102266
[Kan87]   Kane G 1987 *Modern Elementary Particle Physics* (New York: Addison-Wesley)
[Kap00]   Kaplinghat M, Steigman G and Walker T P 2000 *Phys. Rev.* D **62** 043001
[Kar02]   Karle A 2002 *Nucl. Phys. B (Proc. Suppl.)* **118**
[Kas96]   Kasuga S et al. 1996 *Phys. Lett.* B **374** 238
[Kat01]   KATRIN collaboration 2001 *arXiv:hep-ex/0109033*
[Kat05]   KATRIN collaboration, 2005 KATRIN Design Report, FZKA scientific report 7090
[Kau98]   Kaulard J et al. 1998 *Phys. Lett.* B **422** 334
[Kav69]   Kavanagh R W et al. 1969 *Bull. Am. Phys. Soc.* **14** 1209
[Kaw93]   Kawashima A, Takahashi K and Masuda A 1993 *Phys. Rev.* C **47** 2452
[Kay81]   Kayser B 1981 *Phys. Rev.* D **24** 110
[Kay89]   Kayser B, Gibrat-Debu F and Perrier F 1989 *Physics of Massive Neutrinos* (Singapore: World Scientific)

[Kay02]   Kayis-Topasku A et al. 2002 *Phys. Lett.* B **527** 173
[Kay08]   Kayis-Topasku A et al. 2008 *Nucl. Phys.* B **798** 1
[Kea02]   Kearns E T 2002 *Preprint* hep-ex/0210019
[Kei03]   Keil M T, Raffelt G G and Janka H T 2003 *Astrophys. J.* **590** 971
[Kha10]   Khachatryan V et al. 2010 *Phys. Lett.* B **692** 83
[Kha15]   Khachatryan V et al. 2015 *The European Physical Journal C* **75** 186
[Kho99]   Khohklov et al. 1999 *Astrophys. J.* **524** L107
[Kib67]   Kibble T W B 1967 *Phys. Rev.* **155** 1554
[Kid14]   Kidd M et al. 2014 *Phys. Rev.* C **90** 055501
[Kim93]   Kim C W and Pevsner A 1993 *Neutrinos in Physics and Astrophysics* (Harwood Academic)
[Kin03]   King B 2003 *Nucl. Phys. B (Proc. Suppl.)* **118** 267
[Kin13]   King S F and Luhn C, 2013 *Rep. Prog. Phys.* **76** 056201
[Kip90]   Kippenhahn R and Weigert A 1990 *Stellar Structure and Evolution* (Berlin: Springer)
[Kir67]   Kirsten T, Gentner W and Schaeffer O A 1967 *Z. Phys.* **202** 273
[Kir68]   Kirsten T, Gentner W and Schaeffer O A 1968 *Phys. Rev. Lett.* **20** 1300
[Kir86]   Kirsten T 1986 *Proc. Int. Symp. on Nuclear Beta Decays and Neutrinos (Osaka)* ed T Kotani, H Ejiri and E Takasugi (Singapore: World Scientific) p 81
[Kir99]   Kirsten T A 1999 *Rev. Mod. Phys.* **71** 1213
[Kit83]   Kitagaki T et al. 1983 *Phys. Rev.* D **28** 436
[Kla95]   Klapdor-Kleingrothaus H V and Staudt A 1995 *Non Accelerator Particle Physics* (Bristol: IOP)
[Kla97]   Klapdor-Kleingrothaus H V and Zuber K 1997 *Particle Astrophysics* (Bristol: IOP) revised paperback version 1999
[Kla97a]   Klapdor-Kleingrothaus H V and Zuber K 1997 *Teilchenastrophysik* (Berlin: Springer)
[Kla98]   Klapdor-Kleingrothaus H V, Hellmig J and Hirsch M 1998 *J. Phys. G* **24** 483
[Kla99]   Klapdor-Kleingrothaus H V (ed) 1999 *Baryon and Lepton asymmetry* (Bristol: IOP)
[Kla01]   Klapdor-Kleingrothaus H V (ed) 2001 *Seventy Years of Double Beta Decay* (Singapore: World Scientific)
[Kle08]   Klein M and Yoshida R 2008 *Prog. Nucl. Part. Phys.* **61** 343
[Kle19]   Kleesiek M et al. 2019 2019 *The European Physical Journal C* **79** 204
[Kli84]   Klinkhammer D and Manton N 1984 *Phys. Rev.* D **30** 2212
[Kli17]   Klijnsma T et al. 2017 *The European Physical Journal C* **77** 778
[Kna17]   Knapen S, Lin T and Zurek K M 2017 *Phys. Rev.* D **96** 115021
[Kni02]   Kniehl B A and Zwirner L 2002 *Nucl. Phys.* B **637** 311
[Kob73]   Kobayashi M and Maskawa T 1973 *Prog. Theor. Phys.* **49** 652
[Kob05]   Kobayashi K et al. 2005 *Phys. Rev.* D **72** 052007
[Kod01]   Kodama K et al. 2001 *Phys. Lett.* B **504** 218
[Kod08]   Kodama K et al. 2008 *Phys. Rev.* C **72** 052002
[Kog96]   Kogut A et al. 1996 *Astrophys. J.* **464** L5
[Kol87]   Kolb E W, Stebbins A J and Turner M S 1987 *Phys. Rev.* D **35** 3598
[Kol90]   Kolb E W and Turner M S 1990 *The Early Universe (Frontiers in Physics 69)* (Reading, MA: Addison-Wesley)
[Kol93]   Kolb E W and Turner M S 1993 *The Early Universe* (Reading, MA: Addison-Wesley)

[Kol01]   Kolbe E and Langanke K 2001 *Phys. Rev.* C **63** 025802
[Kom09]   Komatsu E et al. 2009 *Astrophys. J. Suppl.* **180** 330
[Kon66]   Konopinski E J 1966 *Beta Decay* (Oxford: Clarendon)
[Kop04]   Kopp S E 2004 *Fermilab Conf-04-300*
[Kos92]   Koshiba M 1992 *Phys. Rep.* **220** 229
[Kos11]   Kostelecky V A, Russell N 2011 *Rev. Mod. Phys.* **83** 11
[Kot06]   Kotake K, Sato K and Takahashi K, 2006 *Rep. Prog. Phys.* **69** 971
[Kot11]   Kotake K, 2011 *Preprint* arXiv:1110.5107
[Kot11a]   Kotera K and Olinto A V 2011 *Annu. Rev. Astron. Astrophys.* **49** 119
[Kow08]   Kowalkski M et al. 2008 *Astrophys. J.* **686** 749
[Kra88]   Krane K S 1988 *Introduction to Nuclear Physics* (John Wiley & Sons, New
             York)
[Kra90]   Krakauer D A et al. 1990 *Phys. Lett.* B **252** 177
[Kra91]   Krakauer D A et al. 1991 *Phys. Rev.* D **44** R6
[Kra05]   Kraus C et al. 2005 *Eur. Phys. J.* C **40** 447
[Kra06]   Kravchenko I et al. 2006 *Phys. Rev.* D **73** 082002
[Kri16]   Krief M, Feigel A and Gazit D 2016 *Astrophys. J.* **821** 45
[Kum13]   Kumar K S et al. 2013 *Annu. Rev. Nucl. Part. Phys.* **63** 237
[Kun01]   Kuno Y and Okada Y 2001 *Rev. Mod. Phys.* **73** 151
[Kun13]   Kuno Y et al. 2013 *PTEP* **2013** 022C01
[Kuo04]   Kuo C et al. 2004 *Astrophys. J.* **600** 32
[Kuo89]   Kuo T K and Pantaleone J 1989 *Rev. Mod. Phys.* **61** 937
[Kuz66]   Kuzmin V A 1966 *Sov. Phys.–JETP* **22** 1050
[Kuz85]   Kuzmin V A, Rubakov V A and Shaposhnikov M E 1985 *Phys. Lett.* B **155** 36
[Lam97]   Lamoreaux S K 1997 *Phys. Rev. Lett.* **78** 5
[Lam00]   Lampe B and Reya E 2000 *Phys. Rep.* **332** 1
[Lan52]   Langer L M and Moffat R D 1952 *Phys. Rev.* **88** 689
[Lan56]   Landau L D 1956 *JETP* **32** 405
             Landau L D 1956 *JETP* **32** 407
[Lan75]   Landau L D and Lifschitz E M 1975 *Lehrbuch der Theoretischen Physik* vol 4a
             (Berlin: Academischer)
[Lan81]   Langacker P 1981 *Phys. Rep.* **72** 185
[Lan86]   Langacker P 1986 *Proc. Int. Symp. Weak and Electromagnetic Interactions in
             Nuclei, WEIN '86* ed H V Klapdor (Heidelberg: Springer) p 879
[Lan88]   Langacker P 1988 *Neutrinos* ed H V Klapdor (Berlin: Springer)
[Lan91]   Langer N 1991 *Astron. Astrophys.* **243** 155
[Lan93b]   Langacker P and Polonsky N 1993 *Phys. Rev.* D **47** 4028
[Lan94]   Langanke K 1994 *Proc. Solar Modeling Workshop (Seattle, WA)*
[Lan09]   Langacker P 2009 *The Standard Model and Beyond* (Taylor & Francis Inc.)
[Lan12]   Langacker P 2012 *Annu. Rev. Nucl. Part. Phys.* **62** 215
[Lan18]   Langacker P 2018 arXiv:1811.07396 (ICHEP 2018)
[Lat88]   Lattimer J M and Cooperstein J 1988 *Phys. Rev. Lett.* **61** 24
[Lat12]   Lattimer J M 2012 *Annu. Rev. Nucl. Part. Phys.* **62** 485
[Lat16]   Lattimer J M and Prakash M 2016 *Phys. Rev.* **621** 127
[Lea95]   Learned J G and Pakvasa S 1995 *Astropart. Phys.* **3** 267
[Lea96]   Leader E and Predazzi E 1996 *An Introduction to Gauge Theories and Modern
             Particle Physics* (Cambridge: Cambridge University Press)
[Lea00]   Learned J G and Mannheim K 2000 *Annu. Rev. Nucl. Part. Sci.* **50** 679

[Lea01]   Learned J 2001 *Current Aspects of Neutrino Physics* ed D Caldwell (Berlin: Springer)

[Lea03]   Learned J G 2003 *Nucl. Phys. B (Proc. Suppl.)* **118** 405

[Lee56]   Lee T D and Yang C N 1956 *Phys. Rev.* **104** 254

[Lee57]   Lee T D and Yang C N 1957 *Phys. Rev.* **105** 1671

[Lee72]   Lee B W and Zinn-Justin J 1972 *Phys. Rev.* D **5** 3121

[Lee77]   Lee B W and Shrock R E 1977 *Phys. Rev.* D **16** 1444

[Lee77a]   Lee B W and Weinberg S 1977 *Phys. Rev. Lett.* **39** 165

[Lee95]   Lee D G et al. 1995 *Phys. Rev.* D **51** 229

[Lee01]   Lee A T et al. 2001 *Astrophys. J.* **561** L1

[Lee12]   Lee B W and Weinberg S 2012 *Phys. Rev. Lett.* **109** 211801

[Leh02]   Lehtinen N G 2002 *Astropart. Phys.* **17** 279

[Lei01]   Leibundgut B 2001 *Astronom. Astrophys. Review* **39** 67

[Lei03]   Leibundgut B and Suntzeff N B 2003 *Preprint* astro-ph/0304112, to be published in *Supernovae and Gamma Ray Bursters (Lecture Notes in Physics)* Springer

[Lei06]   Leitner T, Alvarez-Ruso L and Mosel U 2006 *Phys. Rev.* C **73** 065502

[Lem06]   Lemut A et al. 2006 *Phys. Lett.* B **634** 483

[Len98]   Lenz S et al. 1998 *Phys. Lett.* B **416** 50

[LEP06]   The LEP collaborations 2006 *Preprint* hep-ex/0612034

[Les06]   Lesgourgues J and Pastor S 2006 *Phys. Rep.* **429** 307

[Les13]   Lesgourgues J et al. 2013 *Neutrino cosmology* (Cambridge Univ. Press)

[Let03]   Letessier-Selvon A 2003 *Nucl. Phys. B (Proc. Suppl.)* **118**

[Li18]   Li Z et al. 2018 *Phys. Rev.* D **98** 052006

[Lid15]   Liddle A 2015 *An introduction to modern cosmology* (New York: Wiley)

[Lig17]   LIGO coll. et al. 2017 *Astrophys. J.* **848** L12

[Lim88]   Lim C S and Marciano W J 1988 *Phys. Rev.* D **37** 1368

[Lin82]   Linde A D 1982 *Phys. Lett.* B **108** 389

[Lin84]   Linde A D 1984 *Rep. Prog. Phys.* **47** 925

[Lin88]   Lin W J et al. 1988 *Nucl. Phys.* A **481** 477 and 484

[Lin89]   Lindner M et al. 1989 *Geochim. Cosmochim. Acta* **53** 1597

[Lin02]   Linde A D 2002 *Preprint* hep-th/0211048

[Lip91]   Lipari P and Stanev T 1991 *Phys. Rev.* D **44** 3543

[Lip93]   Lipari P 1993 *Astropart. Phys.* **1** 193

[Lip95]   Lipari P, Lusignoli M and Sartogo F 1995 *Phys. Rev. Lett.* **74** 4384

[Lip99]   Lipkin H J 1999 *Preprint* hep-ph/9901399

[Lip00a]   Lipari P 2000 *Astropart. Phys.* **14** 153

[Lip00b]   Lipari P 2000 *Nucl. Phys. B (Proc. Suppl.)* **14** 171

[Lip01]   Lipari P 2001 *Nucl. Phys. B (Proc. Suppl.)* **91** 159

[Lis97]   Lisi E, Montanino D 1997 *Phys. Rev.* D **56** 1792

[Liu04]   Liu D W et al. 2004 *Phys. Rev. Lett.* **93** 021802

[Lle72]   Llewellyn Smith C H 1972 *Phys. Rep.* **3** 261

[Lob85]   Lobashev V M et al. 1985 *Nucl. Instrum. Methods* A **240** 305

[Lob99]   Lobashev V M et al. 1999 *Phys. Lett.* B **460** 227

[Lob03]   Lobashev V M 2003 *Nucl. Phys.* A **719** 153

[Lon92, 94]   Longair M S 1992, 1994 *High Energy Astrophysics* (Cambridge: Cambridge University Press)

[Lop96]   Lopez J L 1996 *Rep. Prog. Phys.* **59** 819

[Lor02]   Loredo T J and Lamb D Q 2002 *Phys. Rev.* D **65** 063002

[Los97] Celebrating the neutrino 1997 *Los Alamos Science* vol 25 (Los Alamos: Los Alamos Science)

[Lud13] Ludhova L and Zavatareeli S 2013 *Adv. High Energy Phys.* 2013, 425693

[Lue98] Luescher R et al. 1998 *Phys. Lett.* B **434** 407

[Lun01] Lunardini C and Smirnov A Y 2001 *Nucl. Phys.* B **616** 307

[Lun03] Lunardini C and Smirnov A Y 2003 *Preprint* hep-ph/0302033, *J. CAP* **0306** 009

[Lun03a] Lundberg B, Niwa K and Paolone V 2003 *Annu. Rev. Nucl. Part. Phys.* **53** 199

[Lun04] Lunardini C and Smirnov A Y 2004 *Astropart. Phys.* **21** 703

[Lun09] Lunardini C 2009 *Phys. Rev. Lett.* **102** 231101

[Lun16] Lunardini C 2016 2016 *Astropart. Phys.* **79** 49

[Mac84] McFarland D et al. 1984 *Z. Phys.* C **26** 1

[Mag08] Maggiore M 2008 *Gravitational Waves. Vol1: Theory and experiments* (Oxford: Oxford University Press)

[Maj37] Majorana E 1937 *Nuovo Cimento* **14** 171

[Mak62] Maki Z, Nakagawa M and Sakata S 1962 *Prog. Theor. Phys.* **28** 870

[Mal93] Malaney R A and Mathews G J 1993 *Phys. Rep.* **229** 145, 147

[Mal97] Malaney R A 1997 *Astropart. Phys.* **7** 125

[Mal03] Malek M et al. 2003 *Phys. Rev. Lett.* **90** 061101

[Man95] Mann A K 1995 *Precision Test of the Standard Model* ed P Langacker (Singapore: World Scientific)

[Man01] Mann W A 2001 *Nucl. Phys. B (Proc. Suppl.)* **91** 134

[Man01a] Mannheim K, Protheroe R J and Rachen J P 2001 *Phys. Rev.* D **63** 023003

[Man04] Mantovani F L et al. 2004 *Phys. Rev.* D **69** 013001

[Man07] Manchester R N 2007 *AIP Conf. Procs.* **937** 134

[Mar59] Marklund I and Page L P 1959 *Nucl. Phys.* **9** 88

[Mar69] Marshak R E et al. 1969 *Theory of Weak Interactions in Particle Physics* (New York: Wiley–Interscience)

[Mar77] Marciano W J and Sanda A I 1977 *Phys. Lett.* B **67** 303

[Mar92] Marshak R E 1992 *Conceptional Foundations of Modern Particle Physics* (Cambridge: Cambridge University Press)

[Mar98] Martel H, Shapiro P R and Weinberg S 1998 *Astrophys. J.* **492** 29

[Mar99] Marciano W J 1999 *Phys. Rev.* D **60** 093006

[Mar03] Marciano W J and Parsa Z 2003 *J. Phys. G* **29** 2629

[Mar08] Martinez V J 2008 *Preprint* arXiv:0804.1536

[Mar09] Marek A and Janka H Th 2009 *Astrophys. J.* **694** 664

[Mar10] Martin S P 2010 *Adv. Ser. Direct. High Energy Physics* **21**,*1*

[Mar16] Martin J 2016 *Astrophys. Space Sci. Proc.* 45,41

[Mas95] Masterson B P and Wiemann C E 1995 *Precision Test of the Standard Model* ed P Langacker (Singapore: World Scientific)

[Mat88] Matz S M et al. 1988 *Nature* **331** 416

[Mat01] Matsuno S 2001 *27th Int. Cosmic Ray Conference (ICRC01)* (Hamburg)

[Mau00] Mauskopf P D et al. 2000 *Astrophys. J.* **536** L59

[May87] Mayle R, Wilson J R and Schramm D N 1987 *Astrophys. J.* **318** 288

[May02] Mayer-Kuckuk T 2002 *Kernphysik* (Berlin: Springer)

[McC16] McCray R and Fransson C 2016 *Annu. Rev. Astron. Astrophys.* **54** 19

[McD99] McDonough W F 1999 in *Encyclopedia of Geochemistry* ed. C. P. Marshall and R. F. Fairbridge (Dordrecht:Kluwer Academics Publ.)

[McD01] McDonald K T 2001 *Preprint* hep-ex/0111033

[McD03]   McDonough W F 2003 in *The Mantle and Core* ed. R. W. Carlson, Vol. 2 of
               Treatise in geochemistry (Oxford:Elsevier-Pergamon)
[McG93]   McGray R 1993 *Annu. Rev. Astron. Astrophys.* **31** 175
[McG99]   McGrew C et al. 1999 *Phys. Rev.* D **59** 051004
[McN87]   McNaught R M 1987 *IAU Circ. No* **4316**
[Mei30]   Meitner L and Orthmann W 1930 *Z. Phys.* **60** 143
[Mel00]   Melchiorri A et al. 2000 *Astrophys. J.* **536** L63
[Mes02]   Meszaros P 2002 *Annu. Rev. Astron. Astrophys.* **40** 137
[Mes08]   Meshik A P et al. 2008 *Nucl. Phys.* A **809** 275
[Meu98]   Meunier P 1998 *Nucl. Phys. B (Proc. Suppl.)* **66** 207
[Mey67]   Meyerhof E 1967 *Elements of Nuclear Physics* (McGraw Hill)
[Mey19]   Meyer M and Zuber K 2019 *Proc. 5th Int. Solar Neutrino Conference
               (Dresden, Germany)* (Singapore: World Scientific)
[Mez98]   Mezzacappa A et al. 1998 *Astrophys. J.* **495** 911
[Mez01]   Mezzacappa A et al. 2001 *Phys. Rev. Lett.* **86** 1935
[Mez05]   Mezzacappa A et al. 2005 *Annu. Rev. Nucl. Part. Phys.* **55** 467
[Mic03]   Michael D 2003 *Nucl. Phys. B (Proc. Suppl.)* **118** 1
[Mic08]   Michael D G 2008 *Nucl. Instrum. Methods* A **596** 190
[Mik86]   Mikheyev S P and Smirnov A Y 1986 *Nuovo Cimento* C **9** 17
[Mil01]   Mills G F 2001 *Nucl. Phys. B (Proc. Suppl.)* **91** 198
[Min98]   Minakata H and Nunokawa H 1998 *Phys. Rev.* D **57** 4403
[Min00]   Minakata H and Nunokawa H 2000 *Phys. Lett.* B **495** 369
[Min01]   Minakata H and Nunokawa H 2001 *Phys. Lett.* B **504** 301
[Min02]   Minakata H, Nunokawa H and Parke S J 2002 *Phys. Rev.* D **66** 093012
[Mis73]   Misner C, Thorne K and Wheeler J 1973 *Gravitation* (London: Freeman)
[Mir15]   Mirea M, Pahomi T and Stoica S 2015 *Rom. Rep. Phys.* 67, 872 (2015)
[Mis91]   Mishra S R et al. 1991 *Phys. Rev. Lett.* **66** 3117
[Mis94]   Missimer J H, Mohapatra R N and Mukhopadhyay N C 1994 *Phys. Rev.* D **50**
               2067
[Moe91]   Moe M 1991 *Phys. Rev.* C **44** R931
[Moe91a]   Moe M 1991 *Nucl. Phys. B (Proc. Suppl.)* **19** 158
[Moh80]   Mohapatra R N and Senjanovic G 1980 *Phys. Rev. Lett.* **44** 912
[Moh86, 92]   Mohapatra R N 1986 and 1992 *Unification and Supersymmetry* (Heidelberg:
               Springer)
[Moh86]   Mohapatra R N 1986 *Phys. Rev.* D **34** 3457
[Moh88]   Mohapatra R N and Takasugi E 1988 *Phys. Lett.* B **211** 192
[Moh91]   Mohapatra R N and Pal P B 1991 *Massive Neutrinos in Physics and
               Astrophysics* (Singapore: World Scientific)
[Moh96a]   Mohapatra R N 1996 *Proc. Int. Workshop on Future Prospects of Baryon
               Instability Search in p-Decay and n–n̄ Oscillation Experiments (28–30
               March, 1996, Oak Ridge)* ed S J Ball and Y A Kamyshkov (US Dept of
               Energy Publications) p 73
[Moh01]   Mohapatra R N 2001 *Current Aspects of Neutrino Physics* (Berlin: Springer)
[Moh06]   Mohapatra R N, Smirnov A Y 2006 *Annu. Rev. Nucl. Part. Phys.* **56** 569
[Moh07]   Mohapatra R N et al. 2007 *Rep. Prog. Phys.* **70** 1757
[Moh18]   Mohayai T A 2018 *arXiv:1806.01807*
[Mon00]   Agafonova N Y et al. 2000 *MONOLITH Proposal* LNGS P26/2000,
               CERN/SPSC 2000-031
[Mon02]   Montaruli T 2002 *Nucl. Phys. B (Proc. Suppl.)* **110** 513

[Mon09]	Monreal B and Formaggio J A 2009 *Phys. Rev.* D **80** 051301
[Mor73]	Morita M 1973 *Beta Decay and Muon Capture* (Reading, MA: Benjamin)
[Mon09]	Monreal B and Formaggio J A 2009 *Phys. Rev.* D **80** 051301
[Mos16]	Mosel U 2016 *Annu. Rev. Nucl. Part. Phys.* **66** 171
[Mou09]	Mount J B, Redshaw M and Myers E G 2009 *Phys. Rev. Lett.* **103** 122502
[Muf15]	Mufson S et al. 2015 *Nucl. Instrum. Methods* A **799** 1
[Mül04]	Müller E et al. 2004 *Astrophys. J.* **603** 221
[Mut88]	Muto K and Klapdor H V 1988 *Neutrinos* ed H V Klapdor (Berlin: Springer)
[Mut91]	Muto K, Bender E and Klapdor H V 1991 *Z. Phys.* A **39** 435
[Mut92]	Muto K 1992 *Phys. Lett.* B **277** 13
[Mye15]	Myers G et al. 2015 *Phys. Rev. Lett.* **114** 013003
[Nac90]	Nachtmann O 1990 *Elementary Particle Physics* (Berlin: Springer)
[Nag00]	Nagano M and Watson A A 2000 *Rev. Mod. Phys.* **72** 689
[Nak02]	Nakamura S et al. 2002 *Nucl. Phys.* A **707** 561
[Nak05]	Nakayama S et al. 2005 *Phys. Lett.* B **619** 255
[Nar93]	Narlikar J V (ed) 1993 *Introduction to Cosmology* 2nd edn (Cambridge: Cambridge University Press)
[Nar03]	Narison S 2003 *QCD as a Theory of Hadrons* (Cambridge: Cambridge University Press)
[Nes13]	Nesti F and Salucci P 2013 *JCAP* **1307** 016
[Nes14]	Nesterenko D A et al. 2014 *Phys. Rev.* C **90** 042501
[Net02]	Netterfield C B et al. 2002 *Ap. J.* **571** 604
[Nie03]	Niessen P 2003 *Preprint* astro-ph/0306209
[Nil84]	Nilles H P 1984 *Phys. Rep.* **110** 1
[Nil95]	Nilles H P 1995 *Preprint* hep-th/9511313
	Nilles H P 1995 *Conf. on Gauge Theories, Applied Supersymmetry and Quantum Gravity (July 1995, Leuven, Belgium)* ed B de Wit et al. (Belgium: University Press)
[Noe18]	Noether E 1918 *Kgl. Ges. Wiss. Nachrichten. Math. Phys. Klasse* (Göttingen) p 235
[Noe87]	Nötzold D 1987 *Phys. Lett.* B **196** 315
[Nol09]	Nolta M R et al. 2009 *Astrophys. J. Suppl.* **180** 296
[Nom84]	Nomoto K 1984 *Astrophys. J.* **277** 791
[Nor84]	Norman E B and DeFaccio M A 1984 *Phys. Lett.* B **148** 31
[Nov07]	NOvA Technical Design Report 2007 No. FERMILAB-DESIGN-2007-1
[Nuf01]	Proc. 3rd International Workshop on neutrino factories based on muon storage rings, 2003 *Nucl. Inst. Meth.* A **503** 1
[Nuf02]	Proc. 4th International Workshop on neutrino factories 2003 *J. Phys.* G **29** 1463
[Nuf08]	Proc. 9th International Workshop on neutrino factories, superbeams and beta beams 2008 AIP Conf. Proc. **981** (2008) 1
	McNutt J et al. 1996 *Proc. 4th Int. Workshop on Theoretical and Phenomenological Aspects of Underground Physics (TAUP'95) (Toledo, Spain)* ed A Morales, J Morales and J A Villar (Amsterdam: North-Holland)
[Obe87]	Oberauer L et al. 1987 *Phys. Lett.* B **198** 113
[Obe93]	Oberauer L et al. 1993 *Astropart. Phys.* **1** 377
[Ödm03]	Ödman C J et al. 2003 *Phys. Rev.* D **67** 083511
[Oh09]	Oh Y 2009 *Nucl. Phys. B (Proc. Suppl.)* **188** 109
[Oli99]	Olive K 1999 *Preprint* hep-ph/9911307

[One05]   Önengüt G et al. 2005 *Phys. Lett.* B **614** 155
[One06]   Önengüt G et al. 2006 *Phys. Lett.* B **632** 65
[Orl02]   Orloff J, Rozanov A and Santoni C 2002 *Phys. Lett.* B **550** 8
[Ort00]   Ortiz C E et al. 2000 *Phys. Rev. Lett.* **85** 2909
[Ott95]   Otten E W 1995 *Nucl. Phys. B (Proc. Suppl.)* **38** 26
[Ott06]   Ott C D et al. 2006 *Phys. Rev. Lett.* **96** 201102
[Ott08]   Otten E W and Weinheimer C 2008 *Rep. Prog. Phys.* **71** 086201
[Oya98]   Oyama Y 1998 *Preprint* hep-ex/9803014
[Oya15]   Oyama Y 2015 *ArXiv:1510.07200*
[Pad93]   Padmanabhan T 1993 *Structure Formation in the Universe* (Cambridge: Cambridge University Press)
[Pae99]   Päs H et al. 1999 *Phys. Lett.* B **453** 194
             Päs H et al. 2001 *Phys. Lett.* B **498** 35
[Pag97]   Pagel B E J 1997 *Nucleosynthesis and Chemical Evolution of Galaxies* (Cambridge: Cambridge University Press)
[Pag09]   Pagliaroli G et al. 2009 *Astropart. Phys.* **31** 163
[Pak03]   Pakvasa S and Zuber K 2003 *Phys. Lett.* B **566** 207
[Pak03a]   Pakvasa S and Valle J W F 2003 *Preprint* hep-ph/0301061
[Pal92]   Pal P B 1992 *Int. J. Mod. Phys.* **A7** 5387
[Pan91]   Panagia N et al. 1991 *Astrophys. J.* **380** L23
[Pan95]   Panman J 1995 *Precision Tests of the Standard Model* ed P Langacker (Singapore: World Scientific)
[Par94]   Parker P D 1994 *Proc. Solar Modeling Workshop (Seattle, WA)* ed A B Balantekin and J N Bahcall (Singapore: World Scientific)
[Par95]   Partridge R B 1995 *3K: The Cosmic Microwave Background Radiation* (Cambridge: Cambridge University Press)
[Par01]   Parke S and Weiler T J 2001 *Phys. Lett.* B **501** 106
[Pas97]   Passalacqua L 1997 *Nucl. Phys. B (Proc. Suppl.)* **55C** 435
[Pas00]   Pascos E A, Pasquali L and Yu J Y 2000 *Nucl. Phys.* B **588** 263
[Pas02]   Pascoli S, Petcov S T and Rodejohann W 2002 *Phys. Lett.* B **549** 177
[Pas07]   Pascoli S, Petcov S T and Riotto A 2007 *Nucl. Phys.* B **744** 1
[Pat74]   Pati J C and Salam A 1974 *Phys. Rev.* D **10** 275
[Pat17]   Patton K et al. 2017 *Astrophys. J.* **851** 6
[Pau61]   Pauli W 1961 *Aufsätze und Vorträge über Physik und Erkenntnistheorie* (Wiesbaden: Springer)
[Pau91]   Pauli W 1991 On the earlier and more recent history of the neutrino (1957) *Neutrino Physics* ed K Winter (Cambridge: Cambridge University Press)
[Pay15]   Payez A et al. 2015 *JCAP* **1502** 006
[PDG00]   Groom D et al. 2000 Review of particle properties *Eur. Phys. J.* C **3** 1
[PDG02]   Hagiwara K et al. 2002 Review of particle properties *Phys. Rev.* D **66** 010001
[PDG06]   Yao W-M et al. 2006 Review of particle properties *J. Phys.* G **33** 1
[PDG08]   Amsler C et al. 2008 Review of particle properties *Phys. Lett.* B **667** 1
[PDG16]   Patrignani C et al. 2016 *Chin.Phys.* G **40** 100001Review of particle properties
[PDG18]   Tanabashi M et al. 2018 *Phys. Rev.* D **98** 030001Review of particle properties
[Pea98]   Peacock J A 1998 *Cosmological Physics* (Cambridge: Cambridge University Press)
[Pea02]   Peacock J A et al. 2002 *Preprint* astro-ph/0204239
[Pea03]   Pearson T J et al. 2003 *Ap. J.* **591** 556

[Pee80]   Peebles P J E 1980 *The Large-Scale Structure of the Universe* (Princeton, NY: Princeton University Press)

[Pee93]   Peebles P J E 1993 *Principles of Physical Cosmology, Princeton Series in Physics* (Princeton, NJ: Princeton University Press)

[Pen65]   Penzias A A and Wilson R W 1965 *Astrophys. J.* **142** 419

[Pen15]   Pendlebury M *et al* 2015 *Phys. Rev.* D **92** 092003

[Per88]   Perkins D H 1988 *Proc. IX Workshop on Grand Unification (Aix-les-Bains)* (Singapore: World Scientific) p 170

[Per95]   Percival J W et al. 1995 *Astrophys. J.* **446** 832

[Per97]   Perlmutter S et al. 1997 *Astrophys. J.* **483** 565

[Per99]   Perlmutter S et al. 1999 *Astrophys. J.* **517** 565

[Per00]   Perkins D H 2000 *Introduction to High Energy Physics* (New York: Addison-Wesley)

[Per04]   Peres O L G and Smirnov A Y 2004 *Nucl. Phys.* B **680** 479

[Pet90]   Petschek A G (ed) 1990 *Supernovae* (Heidelberg: Springer)

[Pet99]   Petrov A A and Torma T 1999 *Phys. Rev.* D **60** 093009

[Pet06]   Petrovic J 2006 *J. Phys. Conf. Ser.* **39** 471

[Phi16]   Philips II D G et al. 2016 PRP **612** 1

[Pic92]   Picard A et al. 1992 *Nucl. Instrum. Methods* B **63** 345

[Pik02]   Pike R 2002 *Scattering* (Amsterdam: Elsevier)

[Pil99]   Pilaftsis A 1999 *Int. J. Mod. Phys.* **A14** 1811

[Pit18]   Pitrou C et al. 2018 PR **754** 1

[Pol73]   Politzer H D 1973 *Phys. Rev. Lett.* **30** 1346

[Pon60]   Pontecorvo B 1960 *Sov. J. Phys.* **10** 1256

[Pre78]   Prescott C et al. 1978 PLB **77** 347

[Pre85]   Press W H and Spergel D N 1985 *Astrophys. J.* **296** 79

[Pre09]   Presani E 2009 *Nucl. Phys. B (Proc. Suppl.)* **188** 270

[Pri68]   Primakoff H and Rosen S P 1968 α, β, γ *Spectroscopy* ed K Siegbahn (Amsterdam: North-Holland)

[Pro17]   Profumo S 2017 *Introduction to particle dark matter* World Scientific

[Pry88]   Pryor C, Roos C and Webster, M 1988 *Astrophys. J.* **329** 335

[Qui83]   Quigg C 1983 *Gauge Theories of the Strong, Weak and Electromagnetic Interactions (Frontiers in Physics 56)* (New York: Addison-Wesley)

[Qui93]   Quirrenbach A 1993 *Sterne und Weltraum* **32** 98

[Rab17]   Raby S 2017 *Lect. Notes Phys. 939,1*

[Rac37]   Racah G 1937 *Nuovo Cimento* **14** 322

[Raf85]   Raffelt G G 1985 *Phys. Rev.* D **31** 3002

[Raf90]   Raffelt G G 1990 *Astrophys. J.* **365** 559

[Raf96]   Raffelt G 1996 *Stars as Laboratories for Fundamental Physics* (Chicago, IL: University of Chicago Press)

[Raf99]   Raffelt G G 1999 *Annu. Rev. Nucl. Part. Phys.* **49** 163

[Raf01]   Raffelt G G 2001 *Astrophys. J.* **561** 890

[Raf02]   Raffelt G G 2002 *Nucl. Phys. B (Proc. Suppl.)* **110** 254

[Raf03]   Raffelt G G, Keil M T and Janka H T 2003 in Suzuki, Y et al. Neutrino Oscillators and Their Origin: *Proceedings of the 4th International Workshop*, Kanazawa, Japan 10-14 February 2003, World Scientific Publishing Company (June 2004) ISBN 9789812384294

[Rag76]   Raghavan R S 1976 *Phys. Rev. Lett.* **37** 259

[Rag94]   Raghavan R S 1994 *Phys. Rev. Lett.* **72** 1411

[Rag97]	Raghavan R S 1997 *Phys. Rev. Lett.* **78** 3618
[Raj05]	Raja R 2005 *Nucl. Instrum. Methods* A **553** 225
[Raj05a]	Rajguru N et al. 2005 *Mon. Not. R. Astron. Soc.* **363** 1125
[Ram00]	Rampp M and Janka H T 2000 *Astrophys. J. Lett.* **539** L33
[Rea92]	Readhead A C S and Lawrence C R 1992 *Ann. Rev. Astron. Astrophys.* **30** 653
[Ree94]	Reeves H 1994 *Rev. Mod. Phys.* **66** 193
[Rei53]	Reines F and Cowan C L Jr 1953 *Phys. Rev.* **92** 330
[Rei56]	Reines F and Cowan C L Jr 1956 *Nature* **178** 446, 523 (erratum)
	Reines F and Cowan C L Jr 1956 *Science* **124** 103
[Rei76]	Reines F et al. 1976 *Phys. Rev. Lett.* **37** 315
[Rei81]	Rein D and Sehgal L M 1981 *Ann. Phys.* **133** 79
[Rei83]	Rein D and Seghal L M 1983 *Nucl. Phys.* B **223** 29
[Rei09]	Reichardt C L et al. 2009 *Astrophys. J.* **694** 1200
[Rei10]	Reid B A et al. 2010 *Mon. Not. R. Astron. Soc.* **404** 60
[Rei10a]	Reid B A et al. 2010 *JCAP* **1001** 003
[Rei18]	Reimann R 2018 *Proceedings of Science* (ICRC2017) 715
[Rho94]	Rhode W et al. 1994 *Astropart. Phys.* **4** 217
[Ric01]	Rich J 2001 *Fundamentals of Cosmology* (Berlin: Springer)
[Ric16]	Richard E et al. 2016 *Phys. Rev.* D **94** 052001
[Rie98]	Riess A G et al. 1998 *Astronom. J.* **116** 1009
[Ril13]	Riles K 2013 *Prog. Nucl. Part. Phys.* **68** 1
[Rit00]	van Ritbergen T and Stuart R G 2000 *Nucl. Phys.* B **564** 343
[Rob88]	Robertson R G H and Knapp D A 1988 *Annu. Rev. Nucl. Part. Phys.* **38** 185
[Rod52]	Rodeback G W and Allen J S 1952 *Phys. Rev.* **86** 446
[Rod00]	Rodejohann W and Zuber K 2000 *Phys. Rev.* D **62** 094017
[Rod01]	Rodejohann W and Zuber K 2001 *Phys. Rev.* D **63** 054031
[Rod01a]	Rodejohann W 2001 *Nucl. Phys.* B **597** 110
[Rod02]	Rodejohann W 2002 *Phys. Lett.* B **542** 100
[Rod08]	Rodriguez A et al. 2008 *Phys. Rev.* D **78** 032003
[Rod11]	Rodejohann W 2011 *Int. J. of Modern Phys.* E **20** 1833
[Rod11]	Rodriguez T R, Martinez-Pinedo G 2011 *Phys. Rev. Lett.* **105** 252503
[Rog02]	Rogers F and Nayfonov A 2002 *Astrophys. J.* **576** 1064
[Rol88]	Rolfs C E and Rodney W S 1988 *Cauldrons in the Cosmos* (Chicago, IL: University of Chicago Press)
[Ros84]	Ross G G 1984 *Grand Unified Theories (Frontiers in Physics 60)* (New York: Addison-Wesley)
[Rou96]	Roulet E, Covi L and Vissani F 1996 *Phys. Lett.* B **384** 169
[Ruh03]	Ruhl J E et al. 2003 *Astrophys. J.* **599** 786
[Saa13]	Saakyan R 2013 *Annu. Rev. Nucl. Part. Phys.* **63** 503
[Sac67b]	Sacharov A D 1967 *JETP Lett* **6** 24
[Sah20]	Saha M 1920 *Phil. Mag. Ser. 6* 40
[Sah00]	Sahni V and Starobinsky A 2000 *Int. J. Mod. Phys.* **D9** 373
[Sak90]	Sakamoto W et al. 1990 *Nucl. Instrum. Methods* A **294** 179
[Sal57]	Salam A 1957 *Nuovo Cimento* **5** 299
[Sal64]	Salam A and Ward J C 1964 *Phys. Lett.* **13** 168
[Sal68]	Salam A 1968 *Elementary Particle Theory* ed N Swarthohn, Almquist and Wiksell (Stockholm) p 367
[Sal01]	Saltzberg D et al. 2001 *Phys. Rev. Lett.* **86** 2802
[Sal12]	Salathe M, Ribordy, M and Demirors L 2012 *Astropart. Phys.* **35** 485

[Sal18]   Salamanna G et al. 2018 *arXiv:1809.03821*
[Sam07]   Samtleben D, Staggs S and Winstein B 2007 *Annu. Rev. Nucl. Part. Phys.* **57** 245
[Sam13]   Samoylov O et al. 2013 *Nucl. Phys.* B **876** 339
[San88]   Santamaria R and Valle J W F 1988 *Phys. Rev. Lett.* **60** 397
[San00]   Sanuki T et al. 2000 *Astrophys. J.* **545** 1135
[San03]   Sanchez M et al. 2003 *Phys. Rev.* D **68** 113004
[Sat09]   Sathyaprakash B S and Schutz B F 2009 *Preprint* arXiv:0903:0338, *Living Rev. Rel.***122**
[Sau94]   Saulson P R 1994 *Fundamentals of Interferometric Gravitational Wave Detectors* (Singapore:World Scientific)
[Sch60]   Schwartz M 1960 *Phys. Rev. Lett.* **4** 306
[Sch66]   Schopper H 1966 *Weak Interactions and Nuclear Beta Decay* (Amsterdam: North-Holland)
[Sch82]   Schechter J and Valle J W F 1982 *Phys. Rev.* D **25** 2591
[Sch97]   Schmitz N 1997 *Neutrinophysik* (Stuttgart: Teubner)
[Sch98]   Schmidt B P et al. 1998 *Astrophys. J.* **507** 46
[Sch98a]   Schramm D N and Turner M S 1998 *Rev. Mod. Phys.* **70** 303
[Sch00]   Schlickeiser R 2000 *Cosmic Ray Astrophysics* (Heidelberg: Springer)
[Sch01]   Scholberg K 2001 *Nucl. Phys. B (Proc. Suppl.)* **91** 331
[Sch03]   Schoenert S 2003 *Nucl. Phys. B (Proc. Suppl.)* **118** 62
[Sch06]   Schael S et al 2006 *Phys. Rep.* **427** 257
[Sch06a]   Schneider P 2006 *Extragalactic Astronomy and Cosmology: An Introduction* (Heidelberg: Springer)
[Sch97]   Schmitz N 1997 *Neutrinophysik* Stuttgat: Teubner Figure 5.15
[Sch08]   Schiffer J P et al. 2008 *Phys. Rev. Lett.* **100** 112501
[Sch12]   Scholberg K 2012 *Annu. Rev. Nucl. Part. Phys.* **62** 81
[Sch18]   Scholberg K 2018 *Proceedings of Science* (NuFact2017) 020
[Sco03]   Scott P F et al. 2003 *Mon. Not. R. Astron. Soc.* **341** 1076
[Sco18]   Scolnic D et al. 2018 *Astrophys. J.* **852** 1,L3
[Sec91]   Seckel D, Stanev T and Gaisser T K 1991 *Astrophys. J.* **382** 652
[Sem00]   Semertzidis Y K et al. 2000 *Preprint* hep-ph/0012087
[Ser09]   Serenelli A M 2009 *Preprint* arXiv:0910.3690, *Astroph. Space Sci.***328**13
[Ser16]   Serenelli A M 2016 *The European Physical Journal A* **52** 78
[Seu01]   Seunarine S et al. 2001 *Int. J. Mod. Phys.* A **16** 1016
[Sev06]   Severijns N, Beck M and Naviliat-Cuncic O 2006 *Rev. Mod. Phys.* **78** 991
[Sev11]   Severijns N, Naviliat-Cuncic O 2011 *Annu. Rev. Nucl. Part. Phys.* **61** 23
[Sev17]   Severijns N, Blank, B 2017 *J. Phys. G* **44** 074002
[Sex87]   Sexl R U and Urbantke H K 1987 *Gravitation und Kosmologie* (Mannheim: B I Wissenschaftsverlag) and 1995 (Heidelberg: Spectrum Academischer)
[Sha83]   Shapiro S L and Teukolsky S A 1983 *Black Holes, White Dwarfs and Neutron Stars* (London: Wiley)
[Sie68]   Siegbahn K (ed) 1968 α, β, γ *Ray Spectroscopy* vol 2
[Sie19]   Siegel D. M, Barnes J and Metzger B D 2019 *Nature* **569** 241
[Sil67]   Silk J 1967 *Nature* **215** 1155
[Sil68]   Silk J 1968 *Ap. J.* **151** 459
[Sil02]   Silk J 2002 *Preprint* astro-ph/0212305
[Sim09]   Simeone F 2009 *Preprint* arXiv:0908.0862
[Sim01]   Simkovic F et al. 2001 *Part. Nucl. Lett.* **104** 40

[Sim08]   Šimkovic F et al. 2008 *Phys. Rev.* C **77** 045503
[Sin98]   Singh B et al. 1998 *Nucl. Data Sheets* **84** 487
[Sis04]   Sisti M et al. 2004 *Nucl. Instrum. Methods* A **520** 125
[Sma09]   Smartt S J 2009 *Annu. Rev. Astron. Astrophys.* **47** 63
[Smi72]   Smith R A and Moniz E J 1972 *Nucl. Phys.* B **43** 605, err. 1975 *Nucl. Phys.* B
           101 547
[Smi90]   Smith P F and Lewin J D 1990 *Phys. Rep.* **187** 203
[Smi93b]   Smith M S, Kawano L H and Malaney R A 1993 *Astrophys. J. Suppl.* **85** 219
[Smi94]   Smirnov A, Spergel D and Bahcall J N 1994 *Phys. Rev.* D **49** 1389
[Smi96]   Smirnov A Y, Vissani F 1996 *Phys. Lett.* B **386** 317
[Smi97]   Smith P F 1997 *Astropart. Phys.* **8** 27
[Smi01]   Smith P F 2001 *Astropart. Phys.* **16** 75
[Smo90]   Smoot G F et al. 1990 *Astrophys. J.* **360** 685
[Smo95]   Smoot G F 1995 *Preprint* astro-ph/9505139
           Smoot G F 1994 *DPF Summer Study on High Energy Physics: Particle and
           Nuclear Astrophysics and Cosmology in the Next Millennium, Snowmass*
           p 547
[Smo17]   Smorra C et al. 2017 *Nature* **550** 371
[Sne55]   Snell A H and Pleasonton F 1955 *Phys. Rev.* **97** 246
           Snell A H and Pleasonton F 1955 *Phys. Rev.* **100** 1396
[Sob01]   Sobel H 2001 *Nucl. Phys. B (Proc. Suppl.)* **91** 127
[Sok89]   Sokolsky P 1989 *Introduction to Ultrahigh Energy Cosmic Ray Physics
           (Frontiers in Physics 76)* (London: Addison-Wesley)
[Sor07]   Sorel M 2007 *AIP Conf. Procs* **967** 17
[Spe03]   Spergel D N et al. 2003 *Astrophys. J. Suppl.* **148** 175
[Spi96]   Spiering C 1996 *Nucl. Phys. B (Proc. Suppl.)* **48** 463
           Spiering C 1996 *Proc. 4th Int. Workshop on Theoretical and Phenomeno-
           logical Aspects of Underground Physics (TAUP'95) (Toledo, Spain)* ed
           A Morales, J Morales and J A Villar (Amsterdam: North-Holland)
[Spi01]   Spiering C 2001 *Nucl. Phys. B (Proc. Suppl.)* **91** 445
[Spr87]   Springer P T, Bennett C L and Baisden R A 1987 *Phys. Rev.* A **35** 679
[Spr06]   Springel V, Frenk C S and White S D 2006 *Nature* **440** 1137
[Spu18]   Spurio M 2018 *Probes of Multimessenger Astrophysics* (Springer Int. Publ.)
[Sra12]   Sramek O, McDonough W F and Learned, J G 2012 *Adv. High. Energ.* 235686
[Sta90]   Staudt A, Muto K and Klapdor-Kleingrothaus H V 1990 *Europhys. Lett.* **13** 31
[Sta03]   Stanek K Z et al. 2003 *Astrophys. J.* **591** L17
[Sta04]   Stanev T 2004 *High Energy Cosmic Rays* (Berlin: Springer)
[Sta15]   Stahl A 2015 *Proceedings of Science* (LeptonPhoton2015) 014
[Ste07]   Steigman G 2007 *Annu. Rev. Nucl. Part. Phys.* **57** 463
[Sti02]   Stix M 2002 *The Sun* 2nd version (Heidelberg: Springer)
[Str03]   Strumia A and Vissani F 2003 *Phys. Lett.* B **564** 42
[Str06]   Strumia A and Vissani F 2006 *Preprint* hep-ph/0606054
[Suh93]   Suhonen J and Civitarese O 1993 *Phys. Lett.* B **312** 367
[Suh98]   Suhonen J and Civitarese O 1998 *Phys. Rep.* **300** 123
[Suh07]   Suhonen J 2007 *From Nucleons to Nucleus: Concepts of Microscopic Nuclear
           Theory* (Berlin: Springer)
[Suj04]   Sujkowski Z and Wycech S 2004 *Phys. Rev.* C **70** 052501
[Sun92]   Suntzeff N B et al. 1992 *Astrophys. J. Lett.* **384** 33

[Sup19]   Webpage of the Super Kamiokande collaboration, http://www-sk.icrr.u-
               tokyo.ac.jp/sk/gallery/wme/sk_01h-wm.jpg (Accessed: 11.12.2019)
[Sut92]   Sutton C 1992 *Spaceship neutrino* Cambridge University Press
[Swi97]   Swift A M et al. 1997 *Proc. Neutrino'96* (Singapore: World Scientific) p 278
[Swo02]   Swordy S P et al. 2002 *Astropart. Phys.* **18** 129
[Tak84]   Takasugi E 1984 *Phys. Lett.* B **149** 372
[Tak02]   Takahashi K and Sato K 2002 *Phys. Rev.* D **66** 033006
[Tak18]   Takács M 2018 *Nucl. Phys.* A **970** 78
[Tam94]   Tammann G, Löffler W and Schröder A 1994 *Astrophys. J. Suppl.* **92** 487
[Tan93]   Tanaka J and Ejiri H 1993 *Phys. Rev.* D **48** 5412
[Tat97]   Tata X 1997 *Preprint* hep-ph/9706307, Lectures given at IX Jorge A Swieca
               Summer School, Campos do Jordao, Brazil, February 1997
[Tay94]   Taylor J H 1994 *Rev. Mod. Phys.* **66** 711
[Taz03]   Tazzari S and Ferrario M, 2003 *Rep. Prog. Phys.* **66** 1045
[Teg95]   Tegmark M 1995 *Preprint* astro-ph/951114
[Ter04]   Terranova F et al. 2004 *Eur. Phys. J.* C **38** 69
[Tew18]   Tews I, Margueron J and Reddy S 2018 *Phys. Rev.* C **98** 045804
[The58]   Theis W R 1958 *Z. Phys.* **150** 590
[Thi12]   Thies J H et al. 2012 *Phys. Rev.* C **86** 044309
[Tho95]   Thorne K 1995 *Ann. NY Acad. Sci.* **759** 127
[Tho01]   Thompson L E 2001 (ANTARES-coll.) *Nucl. Phys. B (Proc. Suppl.)* **91** 431
[Tho08]   Thomas H V et al. 2008 *Phys. Rev.* C **78** 054608
[Tho09]   Thompson I J and Nunes F M 2009 *Nuclear Reactions for Astrophysics:*
               Principles, Calculations and Applications of Low Energy Reactions
               (Cambridge: Cambridge University Press)
[Tho13]   Thomson M 2013 *Modern Particle Physics* (Cambridge: Cambridge
               University Press)
[t'Ho72]   't Hooft G and Veltman M 1972 *Nucl. Phys.* B **50** 318
[t'Ho76]   't Hooft G 1976 *Phys. Rev.* D **14** 3432
[Thu96]   Thunman M, Ingelman G, and Gondolo P 1996 *Astropart. Phys.* **5** 309
[ToI98]   Firestone R B, Chu S Y F and Baglin C M (ed) 1998 *Table of Isotopes* (New
               York: Wiley)
[Tom88]   Tomoda T 1988 *Nucl. Phys.* A **484** 635
[Tom00]   Tomoda T 2000 *Phys. Lett.* B **474** 245
[Ton03]   Tonry J L et al. 2003 *Astrophys. J.* **594** 1
[Tor99]   Torbet E et al. 1999 *Astrophys. J.* **521** L79
[Tos01]   Toshito T 2001 *Preprint* hep-ex/0105023
[Tot95]   Totani T and Sato K 1995 *Astropart. Phys.* **3** 367
[Tot96]   Totani T, Sato K and Yoshii Y 1996 *Ap. J.* **460** 303
[Tot98]   Totani T et al. 1998 *Astrophys. J.* **496** 216
[Tre79]   Tremaine S and Gunn J E 1979 *Phys. Rev. Lett.* **42** 407
[Tre95]   Tretyak V I and Zdesenko Y G 1995 *At. Data Nucl. Data Tables* **61** 43
[Tre02]   Tretyak V I and Zdesenko Y G 2002 *At. Data Nucl. Data Tables* **80** 83
[Tri97]   Tripathy S C and Christensen-Dalsgaard J 1997 *Preprint* astro-ph/9709206
[Tsa08]   Tsagas C G, Challinor A and Maartens R 2008 *Phys. Rep.* **465** 61
[Tis13]   Tishchenko V et al. 2013 *Phys. Rev.* D **87** 052003
[Tur88]   Turck-Chieze S et al. 1988 *Astrophys. J.* **335** 415
[Tur92]   Turkevich, A I, Economou T E and Cowen G A 1992 *Phys. Rev. Lett.* **67** 3211

[Tur92a]   Turner M S 1992 *Trends In Astroparticle Physics* ed D Cline and R Peccei (Singapore: World Scientific) p 3

[Tur93a]   Turck-Chieze S et al. 1993 *Phys. Rep.* **230** 57
Turck-Chieze S 1995 *Proc. 4th Int. Workshop on Theoretical and Phenomenological Aspects of Underground Physics (TAUP'95) (Toledo, Spain)* ed A Morales, J Morales and J A Villar (Amsterdam: North-Holland)

[Tur93b]   Turck-Chieze S and Lopes I 1993 *Astrophys. J.* **408** 347

[Tur96]   Turck-Chieze S 1996 *Nucl. Phys. B (Proc. Suppl.)* **48** 350

[Tur02]   Turner M S 2002 *Annales Henri Poincare* **4** S333, astro-ph/0212281

[Tyt00]   Tytler D et al. 2000 *Phys. Scr.* **T85** 12

[Ulm15]   Ulmer S et al. 2015 *Nature* **524** 196 2015

[Uns92]   Unsöld A and Baschek B 1992 *The New Cosmos* (Heidelberg: Springer)

[Vaa11]   Vaananen D and Volpe C 2011 *JCAP* **1110** 019

[Val03]   Valle J W F 2003 *Nucl. Phys. B (Proc. Suppl.)* **118** 255

[Val15]   Valle J W F and Romão J C 2015 *Neutrinos in High Energy and Astroparticle Physics* (Weinheim: Wiley-VCH)

[Van06]   Vandenbrouke J, Gratta G, Lethinen N 2006 *Astrophys. J.* **621** 301

[Vin17]   Vinyoles N et al. 2017 *Astrophys. J.* **835** 202

[Vel18]   Velghe B et al. 2018 *arXiv:1810.06424*

[Vie98]   Vietri M 1998 *Phys. Rev. Lett.* **80** 3690

[Vil94]   Vilain P et al. 1994 *Phys. Lett. B* **335** 246

[Vil95]   Vilain P et al. 1995 *Phys. Lett. B* **345** 115

[Vin17]   Vinyoles et al. 2017 *Astrophys. J.* **835** 202

[Vir99]   Viren B 1999 *Preprint* hep-ex/9903029, Proc. DPF Conference 1999

[Vis15]   Vissani F 2015 *J. Phys. G* **42** 013001

[Vog86]   Vogel P and Zirnbauer M R 1986 *Phys. Rev. Lett.* **57** 3148

[Vog88]   Vogel P, Ericson M and Vergados J D 1988 *Phys. Lett. B* **212** 259

[Vog89]   Vogel P, Ericson M and Engler J 1989 *Phys. Rev. D* **39** 3378

[Vog95]   Vogel P 1995 *Proc. ECT Workshop on Double Beta Decay and Related Topics (Trento)* ed H V Klapdor-Kleingrothaus and S Stoica (Singapore: World Scientific) p 323

[Vog99]   Vogel P and Beacom J 1999 *Phys. Rev. D* **60** 053003

[Vol86]   Voloshin M, Vysotskii M and Okun L B 1986 *Sov. Phys.–JETP* **64** 446
Voloshin M, Vysotskii M and Okun L B 1986 *Sov. J. Nucl. Phys.* **44** 440

[Vol07]   Volpe C 2007 *J. Phys. G* **34** R1

[Vos15]   Vos K, Wilshut H W and Timmermans H G E 2015 *Rev. Mod. Phys.* **87** 1483

[Wan02]   Wang L et al. 2002 *Ap. J.* **579** 671

[Wan15]   Wang W et al. 2015 *Mon. Not. R. Astron. Soc.* **453** 377

[Wan17]   Wang M et al. 2017 *Chin.Phys. G* **41** 030003

[Wat08]   Watanabe H et al. 2008 *Preprint* arXiv:0811.0735

[Wax97]   Waxman E and Bahcall J N 1997 *Phys. Rev. Lett.* **78** 2292

[Wax99]   Waxman E and Bahcall J N 1999 *Phys. Rev. D* **59** 023002

[Wax03]   Waxman E 2003 *Nucl. Phys. B (Proc. Suppl.)* **118** 353

[Wea80]   Weaver T A and Woosley S E 1980 *Ann. NY Acad. Sci.* **336** 335

[Web11a]   Webber D M et al. 2011 *Phys. Rev. Lett.* **106** 041803

[Web11b]   Webber D M et al. 2011 *Phys. Rev. Lett.* **106** 079901

[Wei35]   von Weizsäcquire C F 1935 *Z. Phys.* **96** 431

[Wei37]   von Weizsäcker C F 1937 *Z. Phys.* **38** 176

[Wei67]   Weinberg S 1967 *Phys. Rev. Lett.* **19** 1264

	Weinberg S 1971 *Phys. Rev. Lett.* **27** 1688
[Wei72]	Weinberg S 1972 *Gravitation and Cosmology* (New York: Wiley)
[Wei79]	Weinberg S 1979 *Phys. Rev. Lett.* **43** 1566
[Wei82]	Weiler T J 1982 *Phys. Rev. Lett.* **49** 234
	Weiler T J 1982 *Astrophys. J.* **285** 495
[Wei89]	Weinberg S 1989 *Rev. Mod. Phys.* **61** 1
[Wei99]	Weinheimer C et al. 1999 *Phys. Lett.* B **460** 219
[Wei00]	Weinberg S 2000 *The Quantum Theory of Fields. Vol. 3: Supersymmetry* (Cambridge: Cambridge University Press)
[Wei02]	Weinheimer C 2002 *Preprint* hep-ex/0210050
[Wei03]	Weinheimer C 2003 *Nucl. Phys.* B *(Proc. Suppl.)* **118** 279
[Wei03a]	Weinheimer C 2003 *Preprint* hep-ph/030605
[Wei05]	Weinberg S 2005 *The Quantum Theory of Fields, Volume 3: Supersymmetry* (Cambridge: Cambridge University Press)
[Wei08]	Weinberg S 2008 *Cosmology* (Cambridge: Cambridge University Press)
[Wel17]	Welker A et al. 2017 *The European Physical Journal A* **53** 153
[Wes74]	Wess J and Zumino B 1974 *Nucl. Phys.* B **70** 39
[Wes86, 90]	West P 1986, 1990 *Introduction to Supersymmetry* 2nd edn (Singapore: World Scientific)
[Whe90]	Wheeler J C 1990 *Supernovae, Jerusalem Winter School for Theoretical Physics, 1989* vol 6, ed J C Wheeler, T Piran and S Weinberg (Singapore: World Scientific) p 1
[Whe03]	Wheeler J C 2003 *Am. J. Phys.* **71** 11
[Whi94]	White M, Scott D and Silk J 1994 *Ann. Rev. Astron. Astrophys.* **32** 319
[Whi16]	Whitehead L H et al. 2016 *Nucl. Phys.* B **908** 130
[Wie96]	Wietfeldt F E and Norman E B 1996 *Phys. Rep.* **273** 149
[Wie01]	Wieser M E and DeLaeter J R 2001 *Phys. Rev.* C **64** 024308
[Wie17]	D. Wiedner, 2017 *JINST* **12** C07011
[Wil86]	Wilson J R et al. 1986 *Ann. NY Acad. Sci.* **470** 267
[Wil93]	Wilson T 1993 *Sterne und Weltraum* **3** 164
[Wil01]	Wilkinson J F and Robertson H 2001 *Current Aspects of Neutrinophysics* ed D O Caldwell (Springer)
[Wil14]	Wildner E et al. 2014 *Phys.Rev.ST Accel.Beams* **17** 071002
[Win00]	Winter K 2000 *Neutrino Physics* ed K Winter (Cambridge: Cambridge University Press)
[Wis99]	Wischnewski R et al. 1999 *Nucl. Phys.* B *(Proc. Suppl.)* **75** 412
[Wis03]	Wischnewski R 2003 *Preprint* astro-ph/0305302, Proc. 28th ICRC Tsukuda, Japan
[Wis08]	Wischnewski R et al. 2008 *Preprint* arXiv:0811.1109
[Wol78]	Wolfenstein L 1978 *Phys. Rev.* D **17** 2369
[Wol81]	Wolfenstein L 1981 *Phys. Lett.* B **107** 77
[Wol83]	Wolfenstein L 1983 *Phys. Rev. Lett.* **51** 1945
[Won07]	Wong H T a et al. 2007 *Phys. Rev.* D **75** 012001
[Woo82]	Woosley S E and Weaver T A 1982 *Supernovae: A Survey of Current Research, Proc. NATO Advanced Study Institute, Cambridge* ed M J Rees and R J Sonteham (Dordrecht: Reidel) p 79
[Woo86]	Woosley S E and Weaver T A 1986 *Ann. Rev. Astron. Astrophys.* **24** 205
[Woo88]	Woosley S E and Phillips M M 1988 *Science* **240** 750
[Woo88a]	Woosley S E and Haxton W C 1988 *Nature* **334** 45

[Woo90]	Woosley S E et al. 1990 *Ast. J.* **356** 272
[Woo92]	Woosley S E and Hoffman R D 1992 *Astrophys. J.* **395** 202
[Woo95]	Woosley S E et al. 1995 *Ann. NY Acad. Sci.* **759** 352
[Woo97]	Woosley S E et al. 1997 *Preprint* astro-ph/9705146
[Woo05]	Woosley S E and Janka H Th 2005 *Nature* **1** 147
[Woo06]	Woosley S E and Bloom J S 2006 *Annu. Rev. Astron. Astrophys.* **44** 507
[Wri94]	Wright E L et al. 1994 *Astrophys. J.* **420** 450
[Wri03]	Wright E Z 2003 *Preprint* astro-ph/0305591
[Wu57]	Wu C S et al. 1957 *Phys. Rev.* **105** 1413
[Wu66]	Wu C S and Moszkowski S A 1966 *Beta Decay* (New York: Wiley)
[Wu08]	Wu Q et al. 2008 *Phys. Lett.* B **660** 19
[Wur17]	Wurm M 2017 *Phys. Rep.* **685** 1
[Yam06]	Yamamoto S et al. 2006 *Phys. Rev. Lett.* **96** 181801
[Yan54]	Yang C N and Mills R 1954 *Phys. Rev.* **96** 191
[Yan84]	Yang J et al. 1984 *Astrophys. J.* **281** 493
[Yan09]	Yang B S 2009 *Preprint* arXiv:0909.5469
[Yan17]	Yang T 2017 *35th ICRC conference, Bexco, Korea*
[Yao06]	Yao W-M et al. 2006 Review of particle properties *J. Phys.* G **33** 1
[Yas94]	Yasumi S et al. 1994 *Phys. Lett.* B **334** 229
[Yos06]	Yoshida Y et al. 2006 *Phys. Rev. Lett.* **96** 091101
[Zac02]	Zach J J et al. 2002 *Nucl. Instrum. Methods* A **484** 194
[Zal03]	Zaldarriaga M 2003 *Preprint* astro-ph/0305272
[Zas92]	Zas E, Halzen F and Stanev T 1992 *Phys. Rev.* D **45** 362
[Zat66]	Zatsepin G T and Kuzmin V A 1966 *JETP* **4** 53
[Zee80]	Zee A 1980 *Phys. Lett.* B **93** 389
[Zel02]	Zeller G P et al. 2002 *Phys. Rev. Lett.* **88** 091802
[Zha16]	Zhang B et al. 2016 *Space. Sci. Rev.* **202** 3
[Zin19]	Zinatulina D et al. 2019 *Phys. Rev. C*
[Zub92]	Zuber K 1992 *Ann. NY Acad. Sci* **688** 509 Talk presented at PASCOS'92
[Zub97]	Zuber K 1997 *Phys. Rev.* D **56** 1816
[Zub98]	Zuber K 1998 *Phys. Rep.* **305** 295
[Zub00]	Zuber K 2000 *Phys. Lett.* B **485** 23
[Zub00a]	Zuber K 2000 *Phys. Lett.* B **479** 33
[Zub01]	Zuber K 2001 *Phys. Lett.* B **519** 1
[Zub03]	Zuber K 2003 *Phys. Lett.* B **571** 148
[Zub05]	Zuber K (ed.) 2005 *Preprint* nucl-ex/0511009
[Zuc02]	Zucchelli P 2002 *Phys. Lett.* B **532** 166
[Zwe64]	Zweig G 1964 *CERN-Reports* Th-401 and Th-412

Index

Printed in the United States
by Baker & Taylor Publisher Services

Printed in the United States
by Baker & Taylor Publisher Services